The Textbook of Pistol Technology and Design

Production
Principles
Progress

COVER IMAGE

Barrel and slide components of a WALTHER *CREED* pistol
and ATF Form 4590 – Factoring Criteria for Weapons.
Cover Image: © Stefan Allerhand
Cover Design: Henryk Ochmann and Helen Kranhold

ABOUT THE AUTHOR

Peter Dallhammer studied mechanical engineering
in Germany, as well as in England and attained a
postgraduate degree in business. In his graduate
career, he contributed considerably to the
development of the P99/PPQ polymer pistol line of
Carl Walther GmbH. Today, in addition to analytical
tasks, the development and marketing of handguns
are his primary motivation. The author's particular
objective is to pass on knowledge and experience.
Parallel to his professional activities, Peter continued graduate studies
and obtained a Master of Engineering degree in technology-management
with an emphasis on production engineering. Subsequently, he earned
his Doctor of Mechanical Engineering degree from the Helmut-Schmidt-
University, University of German Federal Armed Forces, Hamburg.
Peter's doctoral thesis focuses on the improvement of handguns and the
implementation of new technologies to ease user benefits.

Peter Dallhammer

The Textbook of Pistol Technology and Design

Production – Principles – Progress

2nd Edition

Shaker Verlag
Düren 2020

Bibliographic information published by the Deutsche Nationalbibliothek
The Deutsche Nationalbibliothek lists this publication in the Deutsche
Nationalbibliografie; detailed bibliographic data are available in the Internet at
http://dnb.d-nb.de.

Printed in Germany.

ISBN 978-3-8440-7465-9
ISSN 1866-1742

Shaker Verlag GmbH • Am Langen Graben 15a • 52353 Düren
Phone: 0049/2421/99011-0 • Telefax: 0049/2421/99011-9
Internet: www.shaker.de • e-mail: info@shaker.de

The Textbook of Pistol Technology and Design

REMARKS

Insofar as the masculine form is used in the contents of this book, it is assumed that this refers to all genders on equal terms.

Throughout this book, terms referencing direction, such as left, right, upward, downward, front, back, and so on, are expressed to illustrate in relative position of a pistol being held in the normal firing position. In other words, the firearm is pointed away from the shooter, with the barrel horizontal to the ground with the grip of the pistol pointing downward.

Explanations are presented utilizing right hand operation, left-handed users should operate conversely.

To facilitate readability, proper names are presented in SMALL CAPS and product names are called out in *italics*. Occasionally, technical terms, names of individuals and last names referenced in the Bibliography are also presented in italics.
Product names comprised by a combination of abbreviations and or numbers are published affording no spaces between letters and numbers; such as *P38*.

Disclaimer: All information is provided without guarantee.

The utmost care was taken during the compilation of the content of this book. Nevertheless, errors must not be ruled out. The information referred to in this work corresponds to the state upon completion of the manuscript. Please refer to the latest editions of official guidelines, regulations, laws, patents, etc.

Brands and trademarks are used without guarantee of free use. Trademarks present in this book belong to their respective owners. The use of such marks is for descriptive purposes only to accurately describe the subject matter covered in this book. Any term used which may possibly be a protected brand or trademark is unrestrictedly subject to the provisions of the valid trademark law and the ownership rights of the respective registered owners.

PREFACE

The *Textbook of Pistol Technology and Design* is aimed at all those who deal with the technical aspects in this field. It addresses those individuals who are involved with manufacturing handguns, as well as readers interested in technology in general who would like to broaden their horizons.

The snapshot presented here offers both a retrospective view and the opportunity either to gain a comprehensive overview of the current status or to just answer individual questions and thus provide orientation. The concise presentation and the comprehensive sources have resulted in a reference work which is intended to complement the extensive monographic publications published in recent years. For this second edition, graphics have been made more attractive, statistics have been updated, and the content of individual sections has been refined. In addition, 3D printing technology has also been taken into account; a truly promising innovation.

When we think of progress, or to be more specific innovation[1], we see a product or a novelty that is readily available to markets and successful. According to this definition, a novel idea or invention may not represent an innovation; it is only through the success of an idea or product and its subsequent impact upon society that it can be considered to be an innovation. Popular gun-related examples are the self-contained metallic cartridge or injection molded polymer frames.

In contrast to the background of this definition, we could ask ourselves if innovative new products are feasible in the field of handguns[2], or whether a major advancement in weapons technology can take place only through a breakthrough in a core area. For example, ammunition is one of these core areas. According to George M. Chinn, *The Machine Gun*, substantial progress can only be achieved when a new breakthrough in ammunition takes place. As stated by Chinn in 1987, the burning of a propellant agent is the only way to transform energy in order to operate a self-loading firearm, and either the resulting gas pressure is applied to a mechanism or acceleration of moveable parts is caused by net force. Chinn counted more than 3,000 registered patents since 1884, which are based on these two traditional principles of operation.[3]

We are living in a world of change. At a time where life is strongly influenced by electronics, the considerations presented here strive to contribute to a comprehension of a complex and demanding – yet at the same time fascinating – world of mechanical technology by trying to grasp it from its historical evolution before looking ahead to develop visions.

The history of the engineered definition of emotion and fascination is marked by a number of trends that continued throughout the 20th century and into the present day. In addition to purely technical aspects, the non-functional and non-offensive approach to the subject is also noticeable, as is the dominant role of the American gun culture, which in a way also has a formative affect on the rest of the gun-world and occasionally acts as a growth driver.

From the very beginning, pistols were perceived as aesthetic objects – attention-grabbing, fascinating, and beautiful. – Although the way of perception may have changed, there is still a demand for aesthetics, but also for fundamental top performance in the categories of safety, reliability, operability, performance, accuracy and longevity.

We are dealing with a very special economic good, which is truly multifarious and still just a mechanical device, and in the following a panoramic view of it is provided. Taking as many perspectives as possible can help to better understand a sometimes very emotional relationship to this object of cultural value. The relationship of man – whether a professional user, including those in the military or law enforcement, passionate hunter, marksman, collector, artisan, student, journalist, or reader of relevant literature – to the gun. All the emotions that are connected to it, when handling a firearm, during training or even just the pure fascination, have to be taken just as seriously as the industrial genesis that has to be completed prior to this: a design to compliment manufacturing, the industrial large-scale production, and the successful sales of these high-precision and purely mechanical spring-mass oscillators.

The engineer mostly maintains a rational approach to the topic. Apart from more or less lifeblood that flows into designs, the end product itself remains for him emotionless mechanics, at best aesthetics. For the expert, this durable consumer good consists essentially of technical components, the exterior and the associated ergonomics.

What tinkerers sometimes ignore when working on the single mass oscillator is what counts on the bottom line: profitability and liquidity. Products must be well received by the consumer: the greater the acceptance, the better the prospects for success in the market.

Economic markets, the economy and sentiment in the market are cyclical. An outstanding example is the United States of America, which enjoys enormous attention amongst suppliers worldwide. Sales in this distinguished country are repeatedly driven by political developments and fluctuations. If the political climate is favorable, demand can go through the roof and even start-ups may have good opportunities in the market.

Even with the best efforts, in declining markets there is little companies can do to regain market share. This is also due to the fact that dealers offer consumers a multitude of similar solutions to satisfy what they think would be the consumer's needs. In addition to interchangeability of products, saturation phenomena, and fiercer competition on complex competitive markets with high volatility, there has been an acceleration in extending the range of products within the respective industry under consideration for years. Consequently, shorter product development lead times and faster product launches can be observed.

In addition, markets and consumers are strongly networked. This promotes the spreading of information among market participants. In some cases, there is a virtually unrestrained courting for attention on social media platforms. If hype arises, new products can take off literally overnight. This can happen surprisingly quickly for anyone involved; or on the contrary may just not happen at all. The acceptance of new brands can now be influenced in one way or another by virtual communication. Seriousness and the quality of information suffer as a result. The cult status of some influencers is not always comprehensible to the naked eye.

As indicated previously, a holistic overview is offered, and in particular it traces the individual technological achievements as well as fundamental ideas. The details become clear through the interrelationship between practice, illustration and description. Simultaneously, the portrait of a technical utility object is shown, which reflects man's unique individual relationship to a very controversial device. A product that sometimes exerts an enormous attraction on people. This very special affinity is what distinguishes this commodity from any other technical object.

About This Book

It begins with a looking at the manufacturing techniques of handguns. Pistol principles are then discussed to orientate the reader for the subsequent analysis of guns. In addition, an overview of the requirements for the importation of handguns into the United States is provided. The focus is on the American firearm market as this is regarded to be one of the world's largest commercial markets for semi-automatics. However, not everyone knows that weapons – including those that were previously exported – may only be imported when meeting certain factoring criteria. In this context, a consideration of legislation of some of the most important U.S. states is also provided. These states are of particular importance for gun manufacturers, since they may only offer their products in there if they comply with the local regulations.

Finally, an analysis of the U.S. market will show which gun companies are the market leaders. Examples of selected pistols from some of these firms form a sampling of popular products. Holding to the fact that established market leaders are the well-known big players and their products also tend to be conservative-looking pistols, some other products that are innovative and promising, but from less well-known firms, are examined additionally.

The "weapon system" in question consists of various components, for example the operator, holster, ammunition and pistol. In the following we concentrate on the firearm and neither special applications are discussed nor trends in the development of holsters, ammunition, suppressors, weapon lights or lasers, red dot sights or other equipment. Moreover, product liability and revolvers are not considered. The emphasis is rather on the analysis of technically interesting aspects that grasp one's attention, which could be from a production standpoint, or from a general perspective of a consumer or user.

In the middle of the 20th century *Lugs* already drew the following conclusion:

When studying the individual constructions, it can be clearly seen modern weapons were often created just through a conversion of older constructions. Every inventor must continue to work creatively on the basis of earlier designs, and only thorough evaluation of older ideas guarantees progress. An all too strongly emphasized originality leads to an originality at any price and often to unusual and practically unusable designs.[4]

Following the idea of Lugs concerning "originality at any price," it is intended to generate exploitable ideas. Ideas that are designed to provide benefits to consumers. The question is raised regarding which innovations in handgun technology are currently identifiable, with handgun manufacturers in mind who develop guns for the U.S. market, especially those who produce outside the United States, and intend to import their products to America.

Commonly, two aspects are linked to the United States: a free market economy and the Constitutional right to keep and bear arms. This perspective is common outside the United States and the extent to which the gun laws in the individual states differ from each other is less well known. From exceptionally pro-Second Amendment or gun-friendly states, including Kentucky, Tennessee and Texas; to those states, such as California and Massachusetts, that have strict gun-control policies and laws.

Outside the United States these conditions are often just known by those directly affected, in respect to this; there are no publications within the relevant industry in Europe.

It is no secret for European Union residents and those directly affected enterprises that increasing restrictions on the granting of export licenses to countries outside the European Union make the presently accessible U.S. market more attractive as an export destination. Non-compliance with the relevant U.S. laws causes serious consequences for exporters. An understanding of the relationship can be decisive for success and is a survival factor, which may prevent a failure in the U.S. market.

Furthermore, the explanations given want to show how a gun must be designed, such that it meets the legal requirements applicable in the United States, and also possibly meets the demands of a broad variety of potential buyers. The descriptions close a gap of what is rarely covered in today's research on pistols, which previously centered around the end-product and wasn't considered in legal requirements of U.S. legislation, or other commercial success factors, such as the demands of the consumers. The reader will see the opportunity to create a unique customer benefit with a new combination of existing product properties and to differentiate new products from those of competitors.

Apart from an introductory chapter to reflect some of the basics, this work is divided into four parts. With the chapters on manufacturing technology as Chapter 2 and pistol principles as Chapter 3, a review takes place, which also explains other issues on the current state of the art and technical terms. The research carried out in Chapter 3 is based on the relevant literature. In some places this chapter goes into quite a bit of detail, and when individual protagonists – including Brandeis (1881), Wille (1896), Fleck (1901), Kaisertreu[5] (1902), Cranz (1926), Chinn (1955) and Mootz (1989) – have their say, a panorama is unfolded. The result is a sharply delineated picture that is at once rich in contrast and particularly vivid. These reflections were considered necessary when questions of importance in professional practice could be answered at such points, or where the information seemed to be of general interest.

Chapter 4 is a summary of regulations and legal requirements in the United States, which contains provisions for the states of California, Massachusetts, Maryland and New York, as well as the testing program of the National Institute of Justice. Also included are design requirements and testing for SAAMI. It also covers what pistols must look like to meet IPSC Production Division rules. Mainly Internet sources of competent bodies were used for the regulations and

laws in the United States. This approach provided the required timeliness of the information.

The content of Chapter 4 is to enable gunmakers to develop products that can be offered more successfully in a wide variety of markets within the United States. As regulations have always been, this chapter is unfortunately somewhat dry for a practitioner. While Chapter 5 then provides an extensive reference to practice, a further selection of pistols is evaluated with respect to commercially useful properties.

Since specific legislation exists on the one hand and a wide variety of user requirements also exists for different types of firearms on the other hand, the research carried out here is limited to a single handgun type and cartridge combination: i.e. semi-automatic pistols chambered for the 9 mm Luger cartridge. This limitation was because pistols are in demand, the specific cartridge type is very common, and the popularity of semi-autos chambered for this cartridge is currently increasing. A good commercial usability of the results can thereby be assumed.

A notable number of consumers in the firearms market are not early adopters and the experienced reader may have guessed it already: trends and individual product enhancements are common, but there are rarely any true changes or groundbreaking technological advances.

The analysis presented here will show a potential that contributes to the emergence of technological advance. The reader is given the opportunity to get to the bottom of things. He grasps, how small but beneficial improvements to products can lead to competitive advantage in the market. Supplementary information in the endnotes is meant to give some food for thought, and illustrates what has been created so far in pistol technology. It provides a better understanding of where we currently stand.

Usually a new product is created through the unique assimilation of that which already existed and improving it. In our work-sharing organized world, success is rarely the work of an individual. It is needless to say that I have benefited from many insights and experiences of other authors and my fellow engineers. In accordance with my own interests, I have drawn on this knowledge and at the same time tried to be as neutral as possible when formulating my summary. Accordingly, a broad spectrum of questions is covered, whose explanation in this combination was not known so far in the literature.

To conclude this preface, I would like to take this opportunity to sincerely say thanks to you: the reader and all those who contribute to progress.[6]

Peter Dallhammer
Ulm, Germany

Don't try this at home:
A 9 mm case, which was fired out of a .40 cal. chamber.

CONTENTS

Preface   vii

Abbreviations and Acronyms   xvii

Characters, Symbols and Subscripts   xxi

1   Introduction   1

2   Production Technologies   17

2.1   Additive Manufacturing   19

2.2   Machining   27

2.3   Injection Molding   29

2.4   Investment Casting   32

2.5   Forging   34

2.6   Metal Injection Molding   35

2.7   Stamping   39

2.8   Surface Treatment   41

       2.8.1   Nitrocarburizing   41

       2.8.2   Diamond-Like Carbon   48

       2.8.3   High-Temperature Ceramic Coating   48

2.9   Summary "Production Technologies"   49

3   Pistol Principles   51

3.1   Methods of Operation   52

3.2   Barrel   60

3.3   Slide   72

       3.3.1   Breech Face   74

       3.3.2   Sights   78

3.4   Frame   84

3.5   Magazine Release   91

3.6   Fire-control Mechanism   92

3.7   Springs   112

3.8   Safety Devices   126

       3.8.1   Manual Safeties   130

       3.8.2   Automatic Safeties   131

       3.8.3   Passive Safeties   133

3.9		Microstamping	136
3.10		Electronics	138
	3.10.1	User Authorization Technologies	141
	3.10.2	Firearm Accessories and Current Applications	143
	3.10.3	Summary "Electronics"	150
3.11		Summary "Pistol Technology"	151

4   **U.S. Regulations and Legal Requirements**   **155**

4.1		Review of the Historical Development of the U.S. Firearms Act	156
4.2		Federal Firearms Regulations	157
	4.2.1	Factoring Criteria for Weapons	162
	4.2.2	Marking Requirements	167
	4.2.3	Semi-Automatic Assault Weapons	170
	4.2.4	Undetectable Firearms Act	171
4.3		State Legislations of Selected States	172
	4.3.1	California	175
	4.3.2	Commonwealth of Massachusetts	192
	4.3.3	Maryland	199
	4.3.4	New York	201
4.4		National Institute of Justice Performance Requirements	203
4.5		Evaluation of New Firearms, ANSI/SAAMI Z299.5	208
4.6		IPSC Production Division	213
4.7		Summary "Requirements"	214

5   **Survey of Various Pistols**   **217**

5.1	Analysis of the U.S. Market	217
5.2	Methodology	220
5.3	CARACAL *F* (9 mm Luger)	238
5.4	GLOCK *G19 Gen4* (9 mm Luger)	245
5.5	HECKLER & KOCH *VP9* (9 mm Luger)	254
5.6	KEL-TEC *PF-9* (9 mm Luger)	262
5.7	RUGER *LC9* (9 mm Luger)	267
5.8	SMITH & WESSON *M&P9* (9 mm Luger)	275
5.9	SPRINGFIELD *XD^{M}* (9 mm Luger)	282
5.10	TAURUS *PT 24/7 PRO* (9 mm Luger)	286

5.11   WALTHER *PPX* (9 mm Luger)   293

5.12   Brief Discussion of Random Pistol Models   300

5.13   Results and Discussions   311

Appendices   **327**

Appendix 1   Principle of Linear Momentum   329

Appendix 2   Checklist, Specs Sheet (Template)   333

Appendix 3   ATF Form 4590 (5330.5)   335

Appendix 4   Positive Manually Operated Safety Device, Definition   337

Appendix 5   HTS Alloy, Definition   339

Appendix 6   Certification of Microstamping Technology   341

Appendix 7   GLOCK Files Amicus Brief   345

Appendix 8   Maryland Roster   347

Appendix 9   Compilation of U.S. Regulations   349

Appendix 10 U.S. Handgun Production and Net Supplies   353

Appendix 11 Comparison: Flat Wire vs. Round Wire Spring   367

Appendix 12 Influence of the Frame Material on the Total Weight of Pistols   369

Appendix 13 Slide Mass, 9 mm Luger   373

Appendix 14 Mean Magazines Between Failures   379

Appendix 15 Barrel Heating   383

Appendix 16 Trigger Pull Analysis   387

Appendix 17 Slide Racking Force Analysis   399

Appendix 18 Muzzle Velocity and Muzzle Energy   409

Appendix 19 Slide Motion Analysis   413

Appendix 20 MSRP and Street Prices   461

Appendix 21 Technical Data of Sample Pistols   463

Appendix 22 Technical Data, Summary   473

Appendix 23 Heydenreich's Method   475

Appendix 24 Lock Time and Striker Energy, Striker-Fired Pistols   489

Bibliography   **495**

Endnotes   **517**

ABBREVIATIONS AND ACRONYMS

3D	Three dimensional
AFMER	Annual Firearms Manufacturing and Export Report. This report exhibits firearms, including separate frames or receivers, actions or barreled actions, manufactured and disposed of in commerce during the calendar year. ATF compiles the submitted data and releases it each January, with a one-year delay to comply with the Trade Secrets Act.
ANSI	The American National Standards Institute.
ATF	United States Department of Justice; Bureau of Alcohol, Tobacco, Firearms and Explosives; also known as BATF or BATFE.
BAAINBw	Bundesamt für Ausrüstung, Informationstechnik und Nutzung der Bundeswehr (Federal Office of Bundeswehr (German Armed Forces [translated by PD]) Equipment, Information Technology and In-Service Support)
BATF	see ATF
CA	California
CAD	Computer-aided design
cf	from Latin: *confer*, compare
CFR	Code of Federal Regulations
CIP	Commission Internationale Permanente pour l'Epreuve des Armes à Feu portatives (<English>: Permanent International Commission for the Proof of Small Arms).
CMR	Code of Massachusetts Regulations
CPSC	Consumer Product Safety Commission
DA	Double-Action; NIJ-definition: A mode of operation that permits a single pull of the trigger to cock and fire the pistol.
DAO	Double-Action-Only; DA for each shot.
DIN	Deutsches Institut für Normung (<English>: German Institute for Standardization)
DIY	Do it yourself
DOD	Department of Defense
DOJ	Department of Justice
EN	Europäische Norm (<English>: European Standard)

ER	Erprobungsrichtlinien zur Technischen Richtlinie Pistolen im Kaliber 9 mm × 19 (ER "Pistolen"); German law enforcement testing program, specifies test procedures pistols should meet to qualify for police use; supplement of the *Technische Richtlinie "Pistolen"* (Technical Standards for Autoloading Pistols for Police Officers, see TR).
FIN	Firearm Identification Number
fps	Foot per second (also ft/s), unit of velocity, 1 ft/s is equal to 0.3048 m/s
FPS	Frames per second
FTISB	Firearms Technology Industry Services Branch; this ATF Division is responsible for technical determinations concerning types of firearms approved for importation into the United States under GCA § 925(d)(3). Also known as Firearms and Ammunition Technology Division (FATD).
GCA	Gun Control Act of 1968, also known as GCA68
GmbH	Gesellschaft mit beschränkter Haftung (limited liability company [translated by PD]), in Austria also abbreviated as Ges.m.b.H.
gr	The grain is commonly used to measure the mass of projectiles and propellants. There are 7000 grains per pound.
HK	HECKLER & KOCH; German firearms manufacturer
HS	Headspace; the distance from the face of the closed breech of a firearm to the surface in the chamber on which the cartridge case seats.
HTS Alloy	High Tensile Strength Alloy, non-ferrous metals, e.g. aluminum alloys.
ibid	from Latin: *ibidem*, in the same place
in	Inch (also "), unit of length, 1 in is equal to 25.4 mm
ISO	International Organization for Standardization
KG	Kommanditgesellschaft (private limited partnership company [translated by PD])
lb	from Latin: *libra*, unit of mass, 1 lb is equal to approx. 453.59 grams
lbs	Plural form of lb
l.c.	from latin: *loco citato*, in the passage just quoted
LCI	Loaded Chamber Indicator
LE	Law Enforcement
MA	Massachusetts
MD	Maryland
MGL	Massachusetts General Laws

MIM	Metal Injection Molding
MMBF	Mean Magazines Between Failures
MRBF	Mean Rounds Between Failures
MRBS	Mean Rounds Between Stoppages; aka MRBF; see MRBF.
MSRP	Manufacturer's Suggested Retail Price
MTBF	Mean Time Between Failures
N/A	Not available
NFA	The National Firearms Act was originally enacted in 1934. Similar to the current NFA, the original Act imposed a tax on the making and transfer of firearms defined by the Act, as well as a special (occupational) tax on persons and entities engaged in the business of importing, manufacturing, and dealing in NFA firearms. Title II of the GCA amended federal firearms laws.
NIJ	The National Institute of Justice is a component of the Office of Justice Programs.
NRA	National Rifle Association
NSSF	National Shooting Sports Foundation; the NSSF is an American national trade association for the firearms industry.
NY	New York
oz	Avoirdupois ounce, unit of mass, 1 oz is equal to approx. 28.35 grams
PD	Peter Dallhammer
psi	Pounds to the square inch; 1 psi is equal to approx. 6,894.757 Pa
q.v.	from Latin: *quod vide*, which see; used to reference material used in text
RFID	Radio Frequency Identification
RHT	Right Hand Twist
SAO	Single-Action-Only; consistent SA trigger.
SAAMI	Sporting Arms and Ammunition Manufacturers' Institute
SAF	Second Amendment Foundation; American nonprofit organization that supports gun rights.
SHOT Show	Shooting, Hunting and Outdoor Trade Show. Largest firearms trade show in the world. Open to manufacturers, dealers and gun shops; not open to public.
S/N	Serial Number
S&W	SMITH & WESSON; U.S. firearms manufacturer

TR	Technische Richtlinie Pistolen im Kaliber 9 mm × 19 (TR "Pistolen") (Technical Standards for Autoloading Pistols for Police Officers; as used in Germany [translated by PD]); see also ER.
URL	Uniform Resource Locator
USC	United States Code
USD	The United States Dollar is the official currency of the USA.
WTS	Wehrtechnische Studiensammlung – BAAINBw Koblenz (Reference Collection of Military Engineering [translated by PD])
ZMSBw	Zentrum für Militärgeschichte und Sozialwissenschaften der Bundeswehr (German Armed Forces Center of Military History and Social Sciences [translated by PD])

CHARACTERS, SYMBOLS AND SUBSCRIPTS

Characters in alphabetic order

Symbol	Units	Interpretation
a	m/s²	Acceleration
A	m²	Area
A		Absolute term
C		Absolute term
d		Differential
d	m	Diameter, or distance
D	m	Diameter
e	1	Euler's number in calculus, $e = 2.71828\ldots$
E	J	Energy
f	s⁻¹	Frequency, or frame rate
F	N	Force
g	m/s²	Acceleration of gravity, $g = 9.80665$ m/s²
G	N/mm²	Shear Modulus
h	m	Large cross-section side
H	m	Height
j		Imaginary unit
k		Capacity
k	1	Coefficient of restitution
K	s⁻¹	Firing rate
L	m	Length
m	kg	Mass
n		Quantity, or counter
N		Total number
p	N s	Momentum (engineering mechanics)
p	Pa	Pressure (thermodynamics)
P	1	Malfunction rate
r	N/m	Spring rate
R	N/m	Spring rate
s	m	Displacement, e.g. projectile travel
t	s	Time, or duration
T	s	Periodic time
v	m/s	Velocity
w	m	Short cross-section side
W	m	Width
W	J	Work
x	m	Displacement, e.g. slide travel

xxi

Characters in alphabetic order, Greek letters

Symbol	Units	Interpretation
δ	m	Displacement, deflection (Delta)
ε	1	Coefficient (Epsilon)
ζ	1	Mass ratio (Zeta)
η	1	Pressure ratio (Eta)
ϑ	°C	Temperature (Theta)
Θ	1	Multiplier (Theta)
λ	1	Relative distance, or constant (Lambda)
μ	1	Coefficient of friction (Mu)
π	1	Archimedes' constant (Pi), $\pi = 3.14159\ldots$
Π	1	Multiplier (Pi)
P	1	Multiplier (Rho)
Σ	1	Multiplier (Sigma)
T	1	Multiplier (Tau)
φ	rad	Phase angle (Phi)
Φ	1	Uncoupling ratio, or multiplier (Phi)
Ψ	1	Multiplier (Psi)
ω	s^{-1}	Angular Frequency (Omega)
Ω	1	Multiplier (Omega)

Symbols and Operations

Symbol	Interpretation
$\equiv$	Is defined as
$:=$	Assignment operation. The value of the right-hand side of the expression is assigned to its left-hand side
$\approx$	Approximately equal
$\Rightarrow$	Consequential
Δx	Delta, difference between two values
€	Euro, the euro (EUR) is the official currency of the European Union
"	Inch, unit of length, one inch (1" or 1 in) is equal to 25.4 mm
Σ	Summation
$\hat{x}$	Amplitude
$\dot{x}, \ddot{x}$	First or second derivative respectively
$\bar{x}$	Mean value
x'	Value, after an incident
x^*	Value under consideration of cocking action
$\# x$	Number

Subscripts in alphanumeric order

Subscript	Interpretation
x_α	Sequence, e.g. α, β, γ, δ
x_a	Active
x_B	Barrel, or sear break
x_{BB}	Blowback, i.e. straight blowback
x_c	Solid length (of a spring)
x_C	Charge (i.e. propellant charge)
x_{CC}	Cartridge case
x_d	Displacement, total slide recoil travel
x_e	Empirical
x_f	Flat wire, or active coils
x_{fl}	Frictional losses
x_F	Frame, receiver, or grip
x_G	Rifling groove
x_h	Flat wire, long cross-cutting side of the material is perpendicular
x_i	Inertia
x_K	Striker spring (i.e. mainspring)
x_{kin}	Kinetic
x_ℓ	projectile's base leaves the muzzle
x_L	Land
x_m	Average
x_{mag}	Magazine
x_{max}	Maximum
x_n	Sequence number
x_p	Pre-cocking
x_{pot}	Potential
x_P	Projectile (bullet)
x_{Ps}	Projectile seating depth
x_r	Recoiling, reciprocating, or round wire
x_R	Spring, e.g. recoil spring
x_{res}	Resulting
x_S	Slide
x_{std}	Standard
x_t	Theoretical
x_{tot}	Total
x_T	Striker
x_u	Uncoupling
x_0	At the muzzle, or at the beginning
x_1	Sequence number, e.g. #1

Defensive handgun training, Nevada

Whether the little elves inside the gun that make it work have blue hats or green hats doesn't matter to me. What matters is that the gun works the way I want it to.

(Forum's Member "*ToddG*", 2010)

1   Introduction

Patents are regularly filed in the field of small arms technology, and almost every day new weapons are rolled out, but true innovations now seem rare since the large-scale introduction of polymer frames in the 1980s. *The Further Development of the Self-Loading Pistol* by Mötz & Schuy identifies three innovation advancements within the years 1891 to 2010. These three episodes of innovation built the basis for our modern guns. The authors refer to the years ranging from 1891 to 1900 as the early days. During this decade, the self-loading function, magazines and rimless bottle necked pistol cartridges were developed. Additionally, between the years 1929 to 1939, double-action triggers found their way from revolvers to pistols while the magazine capacity was significantly expanded. Ground-breaking was the FN *Hi-Power* pistol's 13-round magazine capacity, courtesy of a double-stack magazine.[7] In their presence, a further development of pistols took place from 1975 through 1984 with regard to operational safety, fire power, modern materials and economic production. Within the 12 decades starting from 1891 to 2010, only 30 years appeared formative, while 90 years were marked by stagnation, facelifts and attention to detailed improvements.[8]

Perhaps we are just expecting too much progress in the field of weapons. However, in the 21st century, at a time where the sky is the limit for new electronic products and a period that experienced increasingly shortened product life cycles, we should expect more.

Some products of the "arms" industry may seem like "dinosaurs" when compared to the life cycles of today's rapidly "innovated" electronic products. For example, today's most popular and widely used pistol cartridge[9], i.e. the 9 mm Luger, also known as 9 mm × 19, 9 mm Para(bellum) or 9MM, was developed in 1902[10] and has remained virtually unchanged ever since.[11] The introduction of this fairly powerful cartridge with a cylindrical straight case was technically remarkable at this moment in history because semi-automatic pistols with positive locking used bottle-necked cartridges to promote functional reliability.[12] There are other examples of even earlier innovations.

In the mid-19th century a noticeable acceleration of technological progress was caused by the industrialization, which had significant impact on the metalworking industry in general, and on small arms in particular.

Two innovations were developed independently but at the same time, and together these would change firearm development. The first was the Bessemer process, named after its inventor, the Englishman Sir Henry Bessemer. This was an industrial process increasing the scale and speed of production of steel utilizing air blown through the molten iron to remove impurities by oxidation. Other improvements in steel production followed and contributed to the making of the railroads in Europe and the United States, larger bridges and ships as well as they enabled quality tools to be made and allowed for easier mass production of artillery and small arms.

While steel could be mass-produced in increasingly better quality, the other innovation was the percussion cap, which led to major changes in weapons technology.

At the beginning of the American Civil War (1861–65), the opponents still faced each other equipped with smooth bore flintlock muzzle-loaders, but by the end of the conflict some units were issued with breech-loading rifles chambered for metallic cartridge ammunition[13], while weapons capable of firing multiple shots[14] began to appear on the battlefield.

At almost the same time in Europe Nicolaus von Dreyse's needle gun – the *M/41* – was adopted by Prussia as its infantry rifle.[15] This single shot, bolt-action breech loader was chambered for a self-contained paper cartridge with papier-mâché sabot (see Fig. 1.1). The rifle's use during the Austro-Prussian War of 1866 gave an advantage to the Prussian soldiers, who could reload even while laying on the ground. It thus offered a greater rate of fire than the Austrian muzzle-loading counter-

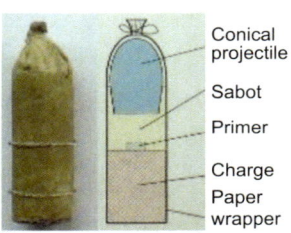

Fig. 1.1: Paper cartridge for a needle gun.
(© WTS)

Conical projectile
Sabot
Primer
Charge
Paper wrapper

parts, which had to be reloaded while the soldiers were standing. This, and the ballistic accuracy of their rifles, allowed the Prussians to sweep the field. The DREYSE remained in service for 30 years, but was still long outdated when its replacement, the Infantry Rifle *M/71*, was introduced. Prussia followed the example of Bavaria, which had introduced the WERDER *M/69* rifle, chambered for a metal cartridge.

The *M/71* was still a single shot service rifle, but was chambered for a cartridge with a brass case, black powder and paper wrapped lead bullet.[16] It may be hard to imagine, but its introduction made the arms industry face new challenges: foremost in the area of mass production of cartridges with metal casings. Various countries were experimenting with self-contained metallic cartridges at that time,[17] however the quality of 1860s production was not up to today's standards, as reliable manufacturing methods had yet to be invented.[18]

Apart from the invention of metal cartridges, the importance of the jacketed bullet was not realized at first. The German military rejected this idea due to the cost, arguing that it would not "make the bullets of pure gold."[19] Instead, the German military subsequently provided the Prussian single-shot *M/71* with a tubular magazine (*M71/84*) and thought the strength of the army was thus sufficiently increased. Only after a French deserter presented the Germans with the newly-introduced French "Lebel" Model 1886 rifle and its "small-caliber" ammunition, did the German military thinkers recognize the potential technical and ballistic virtues of a jacketed bullet, and began activities that culminated in the introduction of the *Model 1888* "Commission" Rifle.[20]

There were in fact three innovations that were combined in the *Model 88* for a significant increase in efficiency of the infantry rifle: a box magazine, smokeless powder and smaller caliber jacketed bullets.[21]

The *Model 88* combined the aforementioned innovations, but, owing to some design flaws, it does not present an example of true technical progress in the history of gun technology.[22] In contrast, there were other weapons that shaped generations and significantly advanced weapon technology as well as manufacturing.

An example is the Maxim-based *MG 08/15* machine gun. Now more than 100 years old it coined the term "08/15" in its days as a figure of speech in everyday language in Germany. The main features of this machine gun are a modular design and the interchangeability of components. The manufacturing process of components was carried out by a variety of factories which resulted in a reorganization of industrial weapon production.[23]

The goal of achieving interchangeability of parts is now a matter of course, which at that time represented a major challenge because technical standards as well as industry standards did not exist at the end of the 19th century. Just at the beginning of the 20th century, before World War I, a go-no-go gauging system (Grenzlehrsystem Schlesinger-Loewe) was introduced by private engineering companies in Germany, which was improved during the war and eventually became an industry standard (Deutsche Passungsnorm).[24]

Far beyond this industry standard, efforts to introduce a tolerance system started in 1917 as a standard for infantry and artillery equipment. This resulted in an industry-standard after the Great War. From the cooperation of authorities, the industry, the Royal Production Offices for Infantry and Artillery in Spandau (Königliches Fabrikationsbüro, Fabo-I and Fabo-A) and the Association of German Engineers (VDI)

emerged the Standardization Committee of German Industry, Deutscher Normenauss-
chuß, that has been known since 1975 as DIN, short for Deutsches Institut für
Normung.[25]

"The introduction of the metal cartridge which also seals the breech and especially
the development of the industrial manufacturing of guns with interchangeable parts are
important milestones in the history of firearms."[a,26] With this, Jaroslav Lugs summarized the above statement on that point in 1956. However, on the quest to produce
semi-automatic pistols as they are known today, three more challenges had to be met.
These included the introduction of rimless cases to improve flawless reloading, eliminating black powder to reduce fouling and an ammunition feeding device.[27] Integrating
an interchangeable magazine into the pistol grip was achieved by Hugo Borchardt during the development of his *C93* pistol. The removable magazine in the handgun's grip
is still a characteristic design for today's pistols.[28] Five years later another success
followed: a slide stop lever appeared on Andreas Wilhelm Schwarzlose's *M98* pistol,
which would hold the bolt in the open position, if the magazine was empty.[29]

At the time it might have seemed as if all the technical challenges were resolved. At
the beginning of the 21st century, hardly could anyone remember times when the interrelation of rapid growth in gun making, the technological progress of weapons, and
their tactical deployment decided victory or defeat of an entire nation.[30] Today manufacturers produce their products worldwide by similar means. These products often
differ only in size, ergonomics, and trigger characteristic, or even just in the perception
of the brand by the consumer; and depending on how reasonable a country's weapon
law is, the harder the competition among producers. Gone are the days when the decision to purchase a certain weapon was purely based on objective and technical criteria. The United States of America represents a good example of this change.

The American firearms industry contributes about $32 billion annually to the eco-
nomic performance of the country.[31] In spite of the recession that began in 2008,[32]
there were 5.5 million arms produced in year 2010 in America alone, and on top of this,
3.3 million more were imported. Estimates show approximately 300 million firearms in
U.S. households, which means circa 89 firearms per 100 citizens.[33]

---

[a] Reliable obturation, i.e. sealing the breech to keep the hot gases inside the barrel and to reduce fouling
inside the gun, was an important step on the way to self-loading firearms. Developers of firearms with
caseless ammunition still find a technological challenge in this very respect.

Firearms are a matter of course for many Americans, and the issue of private ownership has created a deep divide. Proponents of the "keep and bear arms law" justify their claim with the validity of the Second Amendment to the Constitution dating back to 1791. Furthermore, in June 2008, the Supreme Court upheld the validity of the Second Amendment, in the judgment of "District of Columbia v. Heller."[34]

The Second Amendment is a response to the battle between American militia forces and British General Thomas Gage in the then British colony of Massachusetts, where the militia forces won a victory against the British Crown in 1775 with the "first shot heard round the world." It would still take six years of hard fighting until the Battle of Yorktown insured American freedom from the British, and proponents of the Second Amendment argue that the right to keep and bear arms ensured that this freedom could be maintained.

Fig. 1.2: At the beginning of the great crisis in 2020, bad news drives sales.

(© NRA)

The Second Amendment is a part of the "Bill of Rights," which was written in 1791 by founding father and future President James Madison in response to calls from several states for greater constitutional protection for individual liberties. The Bill of Rights lists specific prohibitions on governmental power, and the Second Amendment states: "A well regulated Militia being necessary to the security of a free State, the right of the people to keep and bear Arms shall not be infringed."[35]

In the verdict, the Supreme Court confirmed that the Second Amendment protects the individual right to possess a firearm and to use the weapon within legal boundaries, e.g. for self-defense regardless from serving in a militia.[36] Previous court rulings never clearly identified the Second Amendment as an individual right to carry firearms independently of belonging to a militia. Now this has been clearly defined and it has been followed by a number of liberalizations for the carrying of guns in many states around the country.

Golden times for the gun industry began around the same time as the 2008 presidential election of Barrack Obama. Due in part to fears of new restrictions and regulations, retailers posted sales increases up to year 2013, accompanied by an increase in the number of first-time buyers, whose share in early 2014 amounted to about 25 percent of gun sales.[37]

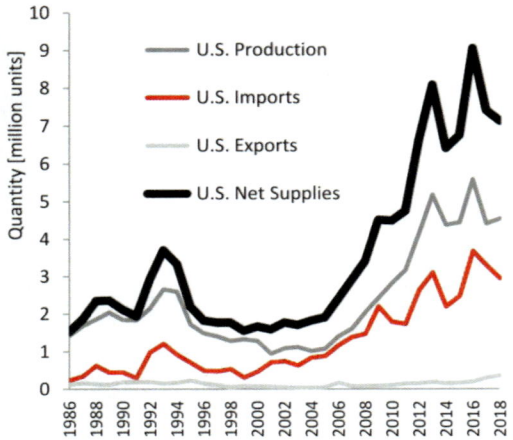

Fig. 1.3: U.S. Net Handgun Supplies, 1986–2018 U.S. Production, U.S. Import and U.S. Export of Handguns. (Compiled from ATF AFMER and ATF Commerce Report)

When the government moved from pro-gun Republicans to anti-gun Democrats under Obama in the year 2008, it contributed to an increase in firearms sales. This trend continued even during Obama's second term. After his re-election in 2012 more guns were sold than ever before (Fig. 1.3). Since the term of office as a President in the United States is limited to a maximum of two terms, some consumers feared the Democrat Obama would reinforce restrictions on the acquisition of firearms and introduce stricter gun laws after his re-election.

By the middle of the year 2013 the shelves in retail stores were literally emptied out.[38] In order to satisfy the increasing demand, gun manufacturers expanded production capacities accordingly. After the collapse of the market in 2014, these capacities were broken and companies fought for market share to be able to take advantage of fixed cost degression effects[39] with the shares they had once achieved.[40]

During the highly contested 2016 presidential election, the industry once again benefitted from a significant increase in sales, in part because of high expectations that former Secretary of State Hillary Clinton would be elected President.

After Republican candidate Donald Trump took office as President of the United States in January 2017, consumers calmed and demand slowed down. Once again the industry was in a market with weaker demand and as often occurs in situations of increasing competition, manufacturers attempted to maintain market share by sales promotion, for instance rebates (Fig. 1.4).

Market participants are connected through the common market, and each of their actions in the fight for benefits causes interactions. Michael E. Porter speaks in this context of the "Five Competitive Forces" that determine industry profitability. These five forces are:

- New Entrants – threat of new competitors,
- Substitutes – alternative products or services,
- Industry Competitors – rivalry among existing firms,
- Suppliers – bargaining power of suppliers, and
- Buyers – bargaining power of buyers.[41]

To protect themselves against risks posed by the Five Competitive Forces and to achieve an advantageous defensive position, companies must succeed in developing a competitive strategy. Porter believes that companies have the choice between three generic competitive strategies such as cost leadership, differentiation and focus (Fig. 1.5).[42]

If a company is not able to develop a corporate strategy in one or more of the three areas, it risks maneuvering itself into a critical situation. Porter refers to this position as "stuck in the middle," and warns that companies could run into trouble in the long term. Also, the previously mentioned approach of companies to secure market position and size-related economies by

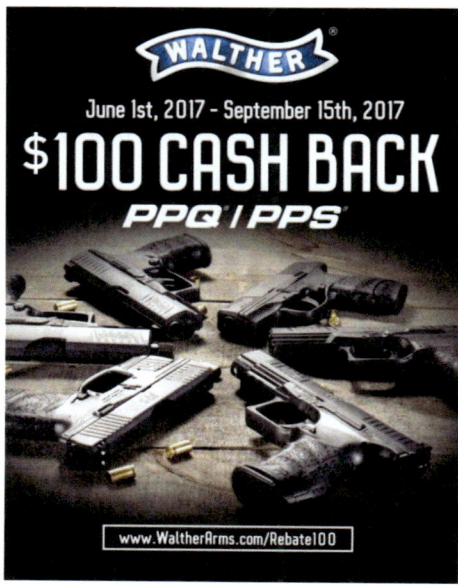

Fig. 1.4: Rebates in 2017.
(© REMINGTON, S&W, WALTHER ARMS)

means of price reductions and risking danger of entering a price war could only be an alternative for a company if it was a cost-leader and was able to generate profit during the race to the bottom, and in particular also after the price war has ended.

During a price war, the manufacturer risks eroding market price and long-term profitability, thus jeopardizing future initiatives by diminishing the financial resources required.[43] Additionally, the support of retailers could be lost as a consequence over de-valued or heavy inventory positions, forcing retailers to sell their inventory at a very low profit margin, or even at a loss.

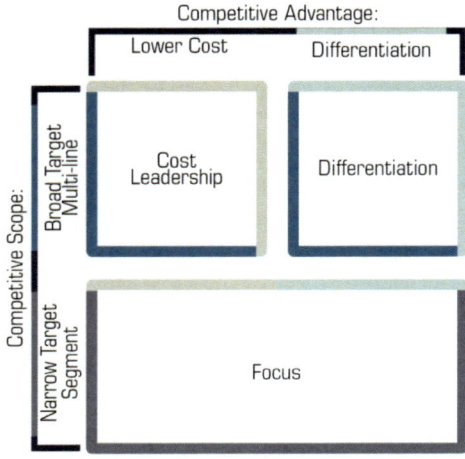

Fig. 1.5: *Porter*'s Generic Strategies. (Porter, 1999 p. 75) and (Porter, 2010 p. 38)

The competitive position on the market for firearms now decisively depends to what extent companies are able to differentiate their respective products from the competition, to open up markets with attractive new products and perform more effectively to meet existing requirements through rationalization and process optimization. It is a constant process of renewal and optimization of the existing market, to take into account the interrelation of customer benefit and production technology. Manufacturers of handguns are affected by market conditions in the same way as any other industry. There are young companies with new and promising products, but also renowned producers, that try to stay on top with more conservative offerings. A description of such can be found at Jurgen Brauer's survey on the U.S. firearms industry.

The survey considers supply and demand of firearms in the United States, it provides analysis of market entry and exit, as well as a rank order and inter-rank mobility analysis; and a discussion of recent mergers and acquisitions in the industry. According to the findings of the survey's author, the gun market is highly cyclical, particularly for the pistol segment of the market. Manufacturers faced severe challenges when the industry went through rigorous business cycles in the years from 1980 to 2010. This required gunmakers either to reduce or to raise output by 50 percent within a very short time.

Manufacturers were further challenged by non-U.S. companies, which became firmly established in the American market, particularly in the pistol market where they had a strong impact. In 2010, Brauer describes that three among the top five manufacturers were non-U.S. brands, namely BERETTA, SIG SAUER and TAURUS. Furthermore, the survey notes considerable market entry and exit, i.e. new brands enter the market and known players may have to cease their activities. In each year's top 20 ranking of gunmakers, Brauer observes an up and down among the companies listed.

Over time, according to his observations, the very top ranks of pistol manufacturers have remained fairly stable. In his conclusion, Brauer points out that companies may not take their customers or market position for granted and firms would be competing with each other in a situation where the brand loyalty of customers is not self-evident. Smart management would be necessary in order to compete in this heavily cyclical, competitive market with domestic and foreign challengers entering with relative ease.[44]

Companies and Their Business Environment

Every company is established and exists to achieve a purpose. Companies are defined within the framework of the economic cycle as manufacturing units. They offer services or combine various goods and services by means of technological and organizational procedures, thereby creating a new commodity that has a value greater than the total value of individual services that were incorporated into the production of such goods. This *added value* could be a result of the direct application of a technical production process, but also by other corporate actions that increase the usefulness and the value of a good.

Production can thus be understood as any transformation of goods in factual, spatial and temporal terms. Resources used in the production process are called *factors of production*.[45] In contrast to this, the term *technology* covers anything from techniques, engineering, processing, machining, manufacturing, tools and software, to specific technical know-how, methods and skills required for the development, production or use of goods. These terms will come into play any time when a company needs the approval by the government for the export of technology or know-how.

Entrepreneurs are part of a dynamic environment and the decisions they take are often based on incomplete information, for instance issues like procurement costs and sales prices may not be fully known or which intermediate inputs should be produced in-house. Also, which preliminary products can be bought and what kind of product configuration should be offered in which markets. The objective is clear, which is the company's long-term survival on the market and profits.

Within their field of interest companies require comprehensive knowledge of available technologies and costs involved, as well as information on obtainable revenues of their products or services. Based on a company's given production technology, the production costs mainly depend on the production volume, which ultimately depends on the performance of the market and thus the production costs may be influenced only within narrow limits. Therefore, the market price will decide whether a product can be offered profitably and noting that the production costs is an important basis for the success on the market.[46]

Factors of production in the economic sense are known to be land, raw materials and preliminary products; but above all is labor and capital. The economy is growing, either because existing factors of production are utilized more efficiently or by making a better use of labor and capital in the production process, which goes hand-in-hand with improving the quality of the factors of production. Generally, the latter is known as technological progress in the broadest sense.[47] Technological progress and growth are interdependent. A new production technology allows a company in a competitive market to produce at a lower cost, e.g. demonstrating the interchangeability of parts by investing in the total uniformity provided by a combination of CNC machining and injection molding in the 1980s.[48]

Very often new technologies are tied to an investment in machinery and equipment, and not all market participants will invest in new technology at the same time. Some companies are innovation-friendly, while others are more resistant and will wait to replace their older production technology. Since the dissemination of technological progress takes time, different economic conditions apply for manufacturers within an industry branch. Users of modern technologies can produce at lower costs and as a courtesy of their lower cost structure, they may generate profits even when competitive market prices are low. The margins serve as a signal for other manufacturers to follow with the introduction of new technology. This will increase the supply of goods into the market at lower prices. Manufacturers with outdated production technology might come under pressure if they make less money once the competitive price drops as a result of technological progress and if the price no longer covers their variable cost per unit. With continuously falling prices, companies are forced to streamline production or face being pushed out of the market.

When a technological and structural change is completed, the industry reaches a new equilibrium in which all companies produce at a similar minimum of costs. If perfect transparency of a market exists, all efficiency gains of the companies will be to the consumer's benefits in terms of lower prices and greater supply in the long term.[49]

Another objective is therefore clearly known: the need to minimize costs. To satisfy the demand in the market at the lowest possible costs, companies must be aware of the production methods that are economically advantageous. However, the demand is subject to fluctuations and the ups and downs have an impact on the cost of *factors of production* and even the size of the company. Sometimes existing workforce and machines can be enough to satisfy an increasing demand. In this case, a higher output can be reached with the existing factors of production.

If the factors of production can be expanded in such a way where costs rise slower than the output then a company has *economies of scale* in production. In the best case, unit costs decrease as production output increases. But normally, from a certain production quantity onward, unit cost will not continue to decrease but remain steady. In the automotive industry this effect is expected when the size of enterprise is around 500,000 units per year; gun manufacturers have not published any information on this. If the quantities are raised further, production cost per unit may increase again (Fig. 1.6). Problems in procurement, qualified staff, loans, rising costs for planning, management and

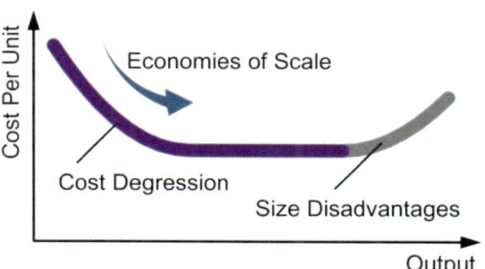

Fig. 1.6: Long-term unit costs diagram (exaggerated for clarity).

control of operations may be reasons for this effect. So, if a specific size is exceeded, size disadvantages will be noticeable (*diseconomies of scale*).[50]

Minimizing production costs is not limited to production technology but may extend equally over all areas of a company. The approach of thinking "lean" across all entities of a company started at the beginning of the 1960s at Japanese automaker TOYOTA and triggered a revolution in the automotive industry, an industry which had been a prime example of mass production.

John Krafcik coined the term "lean" in a *MIT Sloan Management Review* article *Triumph of the Lean Production System*, that was published in 1988. Womack, Jones, and Roos followed in 1990 with their renowned book *The Machine That Changed the World*, which explained how TOYOTA became "lean" under the objective to reduce the timeline to a minimum from the moment the customer gave an order to the point when it collected the cash.[51] This later work highlights the evolution of industrial production by the example of the automotive industry. That industry, along with many others, had its origin where products were practically custom assembled in job-shop fashion by master craftsmen, so termed craft-based production.

Craft-based production describes the beginnings of manufacturing as one that required talented workers who would use relatively simple tools and devices, just as we know it from old-fashioned gunsmithing, early industrial gun manufacturing or the beginning of the automotive industry. Craftsmen did not have the ability to create innovations, costs per unit were high, and above all the components made were not interchangeable.[52] During the transition phase to affordable mass-produced cars, Henry Ford's achievement wasn't the introduction of the moving assembly line – an idea that Ford had derived from the "disassembly lines" of meatpackers in Chicago and Cincinnati – but he pushed toward the introduction of easy to install interchangeable parts, which in turn was known from gun arsenals.[53]

From then on, the *mass production* of automobiles was possible. Semi-skilled workers would perform only a few specific tasks repetitively and these workers were replaceable at any time. This structure however also involved leagues of other employees who were not adding value to the product, including industrial engineers, foremen, supervisors and other indirect workers.[54] The focus of mass producers was the production of a small variety of products in large quantities. At the same time, they focused on vertical integration and in-house production, trying to avoid dependence on suppliers, because their failure would have caused an interruption of production. To keep the metal continuously moving was paramount and this often required the establishment of buffers for the deployment of additional components, which should allow continuous-flow production.

The provision of material buffers became a decisive factor. However, stocks hinder the flow of materials, which increases the throughput time of individual components. The work pace was hard and despite all efforts, only the quality inspection at the end of the assembly line would show if a car was without assembly defects. Even in 1986 an average of 130 assembly defects could be found per 100 cars manufactured.[55] Unacceptable even for today's gun manufacturers. At that time, a car consisted of round about 10,000 discrete parts and rework could be time consuming. Apart from this, the high output generated could cause an oversupply of products. As a result of this, these products had to be pushed into the market (push strategy), especially if demand slowed down. The effects can still be seen in the United States over a wide range of different industries, in which large numbers of goods on the dealer's sites are standard.

Today customers are conditioned to expect an oversupply. They want to choose from a wide variety of products, demand a choice in phones, TVs and most electronic products and want to instantly take home whatever they buy; not just guns or cars, but almost anything. They do not have any patience for delivery times unless it means significant savings.

Another cause for excess inventories on dealers is the sales strategy of mass producers. A manufacturer may use new, high-demand products as leverage to sell older, less popular models. For example, manufacturers may put together package deals featuring new guns as well as some older models, with a discount or free goods as the incentive for the dealer. Another strategy is that discounts may be tied to a minimum purchase. This method of "buy 10, get one free" and the like is also used in the U.S. for the sales of guns between manufacturers and dealers.

Lean combines the best of craft production and mass production and thereby reduces cost and improves the product quality with a greater flexibility in production at the same time. The idea behind lean is to improve the flow of products through the operations; or equivalently lead time. This approach is to eliminate the major disruptions to flow. It promotes to tackle problems, i.e. the detection of issues and elimination of waste and thus avoids waste and defects in the long term. The continuous improvement across all parts of a company's infrastructure improves efficiency and delegates responsibility clearly to those who add value to the product. Teamwork is a major component of lean.

Generally, any action is based on an awareness of processes. If a disruption is observed, it is tracked down to its ultimate cause. In doing so, failures, defective parts and associated rework (or costs) can be eliminated and will not recur. Costs are also reduced by producing goods with minimum lead time, in such a quantity that will meet the demand of the market. Buffers of inventory ties up capital and are kept as small as possible, in which short machine set-up times and just-in-time are common instead. Group work is also an integral part, where everyone from management to production

and also suppliers are part of the whole setup. Suppliers in particular are not just ven-
dors, and they do not have to fear being replaced by another vendor. Instead, they
develop sophisticated components and are connected to their subcontractors in a similar way.[56]

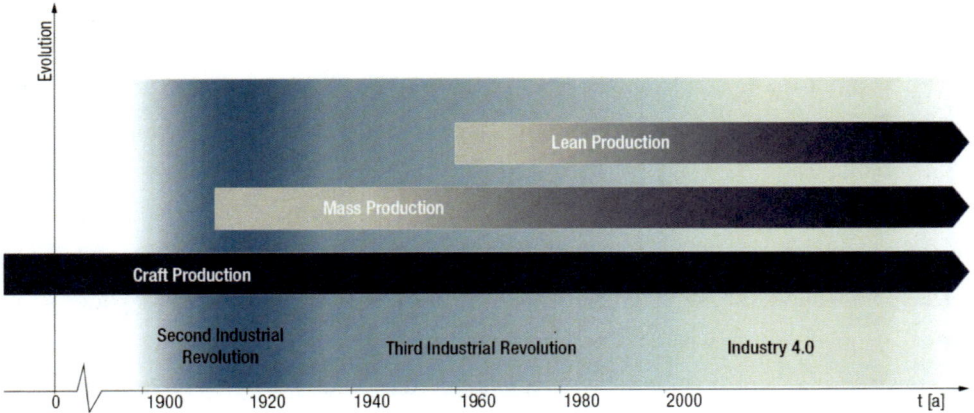

Fig. 1.7: Evolution of industrial manufacturing.

Fig. 1.7 outlines the time course of the evolution from craft production to lean production. In addition, the transition to the third and fourth industrial revolution is shown. Both
transitions extended across all areas of industrial production. The fourth industrial revolution is also referred to as Industry 4.0.

In *The Fourth Industrial Revolution*, economist Klaus Schwab explains the first industrial revolution spanned from about 1760 to sometime around 1840 as railroads
and the steam engine were introduced. It was followed by the introduction of electricity
and the assembly line, which are considered one of the traits of the second industrial
revolution, which in turn lasted from the end of the 19th century to the early 20th century and made mass production possible.

The digital revolution began in the 1960s, starting with semiconductors and main-
frame computing, subsequently personal computing came in, and in the 1990s the Internet started the digital revolution.

The Founder and Executive Chairman of the World Economic Forum classifies the
progress currently taking place as the fourth industrial revolution. This development

started at the turn from the 20th to the 21st century and is a continuation of the digital revolution. It is characterized by new technologies merging the physical, digital and biological worlds, more ubiquitous and mobile Internet, aiding the spread of artificial intelligence (AI), with smaller and more powerful sensors and machine learning. In total, the changes would be much wider than just linking smart machines and systems, explains Schwab.

The economist points out that the characteristics of the fourth revolution would be the fusion and interaction across the physical, digital and biological domains. It is comprised of any field of technology, such as from gene sequencing to nanotechnology, also from renewables up to quantum computing. What is known in Germany since 2011 as Industry 4.0 is the organization of global value chains, which through an adaptive link of virtual and physical systems of manufacturing would cooperate globally with each other in a flexible way, allowing a variety of new business approaches for the customization of products and how customers are served. In Schwab's opinion, it will impact all disciplines, economies and industries. Any manufacturer will have to face fundamental changes, and "the question is no longer 'am I going to be disrupted?' but 'when is disruption coming, what form will it take and how will it affect me and my organization?'"[57]

We are experiencing a transformation with partly tectonic changes in the global economy, a transition from closed to open, from static to dynamic, from stable to changing, from simple to complex, from stationary retail to ecosystems with platform providers that operate worldwide and are permanently accessible online. Cultural Change 4.0 involves a paradigm shift to digitization and decentralization with less capital-intensive knowledge economies, from mechanical devices to digital products. – Firearms will certainly continue to be an interesting issue of concern in the future.

Fig. 1.8: STANDARD MFG. *S333 Volleyfire* – two shots one pull.

2   Production Technologies

The manufacturing of firearms is more demanding than some users might think. Henning Hoffmann looks at the technology through the eyes of the user:

> *In theory, any handgun may be considered to be a technically relatively simple consumer good. The requirements for a self-defense weapon can be summarized easily: It must a) be at hand, b) function reliably and c) reasonably fulfil its purpose. With an emphasis on reasonably.*[58]

From a manufacturer's point of view, the reality looks different. Consumers are more cost-conscious than ever and competition among the market is fierce. It isn't enough to try to differentiate a product based on performance alone, so manufacturers utilize special promotions, persuasion and sale price.[59]

Generally, it is easier to promote and sell new products: the media prefers news and will cover these products on their front pages or even on the magazine covers. The pressure for new products increases and leads to an acceleration of the dynamics of product lifecycles and an increasing variety of models.[60] In comparison to this, a description of the production from the 1950s sounds almost fussy; however, it still holds true:

> *The tasks of production often seem to be very easy to the layman, however, even today the question of tolerances, i.e. the permissible deviations, belong to the most difficult problems of modern weapon technology. A flawless and consistently guaranteed function of a weapon requires the determination of minimum play, which may not be reduced thereafter in any system of manufactured gun parts. The plus-minus tolerances of the individual components must be chosen in a way so that play is increased.*
>
> *Every single component intended for serial production demands a specific manufacturing process dictated already by its dimensions, and the process can be changed hardly ever and if so, then only with the knowledge of the product engineer. The product engineer, who has reviewed the manufacturing process, must check whether the necessary tolerances as specified by the blueprint are met after a modification. Without such a consultation, a whole series of workpieces may have to be scrapped, if the problem is not identified soon enough.*
>
> *Once a component is released for serial production it cannot be changed easily. Scarcely anywhere are the consequences of a modification more irritating and harmful as in serial production. Consequently, the manufacturing process, the fixtures, the gauges, the tools or machines need to be revised, even the material and its processing. This generally causes great loss of time and money.*[61]

Continuous manufacturing processes and downstream compatibility of components are more important than ever in gun production, and it is necessary to employ state of the art manufacturing processes. The selection of the manufacturing process in companies takes into consideration a streamlined production, production costs, and amortization, as well as technical feasibility; such as production techniques, which allow undercuts or internal corners. In addition to the technical feasibility, the production costs and thus the planned production quantity and anticipated sales numbers have a significant influence on the decision for or against a particular manufacturing process. If an investment in machinery is to be taken, that investment only pays for itself in the case of a high degree of capacity utilization.

Manufacturing processes that use a liquid, a powder, or other amorphous material and fuse it to manufacture a solid workpiece, are classified as primary shaping. Additive manufacturing technologies are included in this group of techniques (Fig. 2.1). A separate section covering this relatively new manufacturing technology is given below.

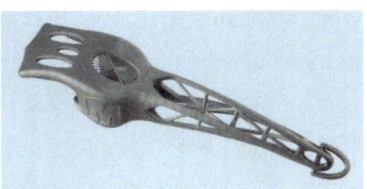

Fig. 2.1: Primary shaping: laser engineered metal part of intricate shape.

Gun production often utilizes parts made of metal or plastic, and even today sometimes wood. Common manufacturing processes include: injection molding, die casting, investment casting, forging, stamping, and machining. Each one leaves salient marks on the parts produced. Based on these characteristic marks one can tell how the part was produced.

Three primary shaping techniques are commonly used in the manufacturing of weapon workpieces: injection molding (polymer parts), die casting of (casting of non-ferrous metals) and investment casting (lost-wax casting). The siblings injection molding and die casting differ mainly in the processed material – injection molding is characterized by forcing polymer material under pressure into a mold cavity, die casting is characterized by injecting molten metal into a mold.

Process temperature for injection molding would be below 390 °F (200 °C), whereas die casting is typical for a liquid melt of metal materials with low melting point between 760 and 1400 °F (400 and 760 °C), normally non-ferrous metals, such as zinc, copper, aluminum, magnesium, or Zamak (a family of alloys with zinc as a base metal and alloying elements of aluminum, magnesium, and copper). Special know-how is required to keep the required shape and dimensional accuracy of the components, because after molding while the components cool down to room temperature, a volume shrinkage occurs that needs to be taken into account during the lay out of the molds.[62]

2.1   Additive Manufacturing

In subtractive manufacturing and in conventional manufacturing in general, the defined removal of material or using molds to form shapes is standard, and the process often is composed of a sequence of different steps. Basic raw materials or semi-finished parts in the correct dimensions must be provided to sustain production, followed by operations such as turning, milling, honing and, if necessary, further machining and various steps of post-processing to optimize surfaces. Process steps can be cost drivers when several machines are required, inaccuracies and downtimes are apparent or when stored inventories of material are necessary to ensure smooth production. Not to speak of the trained workers and skilled engineers needed to transform a design into reality efficiently. The success of traditional mass production usually lies in the fact that the costs decrease with the number of units. It is therefore attractive for a company to make as many products as possible in the same configuration.

Generative manufacturing, three-dimensional (3D) printing, additive manufacturing, rapid prototyping, rapid technology, direct manufacturing, digital manufacturing or similar descriptions are different kinds of labels of the same flavor. They describe the opposite of subtractive manufacturing, namely the joining of volume elements until the desired shape of a workpiece is achieved.[63] A body is fabricated generatively by arranging a layer of raw material and converting it into a defined contoured solid layer, i.e. depositing or binding raw material into layers to form a solid, three-dimensional object. This layered production process and the characteristic activation of the raw material is repeated until the desired object is finished.[64]

Before we get too far ahead, let us quickly focus on terminology. In technical jargon, the use of "rapid" in combination with a more descriptive second technical term has been overused, which might cause confusion. Before the reader is subject to the fallacy that the described technology delivers shorter production times, this perception must be corrected. The speed of rapid manufacturing processes becomes apparent when one considers the entire process chain, i.e. from 3D drafting an idea on the computer to product in hand. A time advantage is noticeable with customized products and is particularly evident when the design and manufacture of tools and fixtures is not required anymore.[65]

Initially, 3D printing was regarded as a niche production method in the field of prototype development, i.e. for custom made workpieces in small quantities. The fabrication of prototypes and individual parts is an essential, often time-critical, aspect during the development phase of new products. Realizing and testing mechanically resilient work-pieces within a short time represents an immense competitive advantage. The acceleration in the creation of series-identical prototypes enables, among other things, shorter project lead times and more mature designs, which contributes to avoid refinements and technical fixes to the end product once production is in full swing.

The early users of additive manufacturing were prototype builders as well as tool and mold makers. In the meantime, this revolutionary manufacturing technology has made great progress, opens up completely new horizons, is in the process of being transferred to daily series production on an industrial scale and is used in key industries. For example, particularly weight-saving components are produced in automotive engineering or in the field of aerospace technology. The medical and dental industries are demanding custom-made body parts that are made to precisely fit an individual's body.[66] Printers are also used in the production of miniature components, commonly referred to as microstructures, or miniature elements for mechatronic or precision mechanical assemblies or micromechanical components. Bore diameters of up to 0.1 mm are achievable.

The layered manufacturing process does away with previously seen restrictions of traditional machining and enjoys the reputation of being particularly resource-saving. In particular, the batch size-independent production of customer-specific series with single piece customization offers possibilities that were previously unthinkable or could only be carried out at great expense. Particularly advantageous processes allow the simultaneous production of various work pieces which can be positioned close to and also on top of each other. The greatest possible freedom of complex geometries, such as integrated channels for conformal cooling, meandering or helical channels, and internal cavities in general, is created as a by-product, so to speak. The advantages of generative manufacturing extend far beyond the actual machining process. Components that are supposedly easy to manufacture but have been conventionally machined so far can be given new properties through creative approaches in design. Topology optimization using computer-based algorithms leads to advantageous design of component bodies with regard to geometry and material distribution. New freedoms in design and construction enable bionic shaping that reduces the basic shape to the essentials while taking into account the expected mechanical stress, functions and connection points. The result is lightweight structures reminiscent of natural structures.

3D printing is becoming more and more popular. Meanwhile, printing speed[a] has gone up, precision and surface finish of the printing result have been improved. A variety of specialized printing technologies allow a wide choice of materials, including wood, stone, cement, food, living cells but also especially metals, such as metals that are difficult to machine, or they combine various materials, such as metal and ceramics. In reality, raw material manufacturers secure their market power by providing proprietary material that must match their hardware. Production is still dominated by standard materials such as plastics, aluminum, titanium or steel alloys. The mechanical properties of printed parts composed of standard material are close to those of the base material.

In the early 1980s various inventors were developing similar systems, but in 1983 American Charles Hull filed a patent for a stereolithography fabrication system, in which layers are added by curing photopolymers with ultraviolet light lasers. The American was looking for a solution to the problem of transforming a computer-generated design directly to physical objects using a time-saving process. Although this project did not fall within his daily work, Hull centered his research work on the application of ultraviolet light. UV light was a common technology in the company he worked for. The idea of his pet project sounds simple: raw material, which is in liquid condition under ambient conditions, solidifies when exposed to UV light (photopolymerization), resulting in cured plastic. The cured area is then covered with new raw material and the upper layer is exposed to UV light again. The process keeps repeating until the workpiece is finished. This method became known as stereolithography (SL). It uses viscous non-crosslinked or low-crosslinked monomers based on acrylate or epoxy resins, which are interspersed with photo inhibitors.[67] In addition to the chemical curing described above, depending on the actual material physical curing and melting processes are currently used for plastics, synthetic resins, ceramics and metals.

When processing metal raw material, some sort of powdered metal or metal granulate is used and arranged in a powder bed while the raw material is stacked into layers by a selective binding process to fuse or bind raw material into consecutive layers, by re-layering and melting, so to speak. In general, all materials that behave thermoplastically can be used, i.e. materials which melt by the application of heat and solidify

---

[a] For the comparison purposes of the processing speed, the build-up rate is either measured in cubic centimeters per hour [cm³/h] or in mm/h. The speed depends on the machine type as well as on geometry, material and the method of activation. Professional equipment achieves 2–100 cm³/h with SLM. The speed of SL and LS fabricators is in the range of 10–30 mm/h.

again when cooled down. Accordingly, a wide variety of plastics or multi-component metal-polymer powders can be processed as well. The thermal activation of the material is done by high-energy laser or electron beams, or possibly as a chemical reaction induced by a heat lamp. A generic term is *powder bed fusion*. The term *selective laser melting* (SLM) refers to the processing of metals, whereas "sintering" refers to plastics as well, when the material is melted in the same way as noted. One speaks of laser sintering (LS), selective laser sintering (SLS) or generally of beam melting or simply melting.[68]

Increasing productivity and reliable quality are the dominant challenges. Besides, the costs for printed parts are, in most cases, still more expensive than conventionally produced alternatives designed for mass production. Basically, the technology is divided into applications for the production of concept models and functional models (rapid prototyping) and for the production of end-use parts (rapid manufacturing). Rapid tooling, i.e. tool and mold making and the manufacture of tool inserts, combines the features of the first two fabrication types.[69] – To ex-

Fig. 2.2: KODAK *Portrait*
(© KODAK)

haustively demonstrate the complete range of additive manufacturing processes would go beyond the scope of our consideration. Meanwhile, changes are to be expected that will overtake today's methods. For this reason, we will only deal here with the basics and roughly outline the probable effects of this technology.

The starting point for every generative manufacturing process is a three-dimensional volume model designed on a computer. For the sake of compatibility when exporting digital design files, the Computer Aided Design (CAD) model is reduced to a description of the outer shell. Today's industry standard for data transfer is the STL file format.[70] In such a file, each surface is defined by a mesh of smallest triangles. Before the file can be used for 3D printing, the triangle meshes might have to be repaired and optimized. Finally, the conversion of the file into a layer model is done by a process called *slicing*.[71] If necessary, a final conversion into the respective system-specific code of the printer takes place before printing.[72] The latter is carried out automatically.

The goal of 3D printing is to enable consistency across all steps of the value chain and to deliver high-quality functional components with the required properties. Some companies still have a wait-and-see attitude when it comes to 3D printing. Usually, firms in the field of mechanical engineering have a great deal of process competence with standard materials and conventional processing. Some companies try to gain a first experience by using less expensive printers in the beginning, for example in the field of rapid prototyping. Additive manufacturing, and each of the various 3D manufacturing methods in itself, behaves differently than is known from traditional machining and influences the process with regard to unwanted deformation due to lack of heat dissipation, commonly named warping, which is why the necessary know-how must be built up.

The application of additive manufacturing processes has an immense potential in the field of industrial applications and can be considered to be a huge contribution to the technological landscape. It may be assumed that there will be a vast variety of growth areas, which in the future may shape large areas of our everyday life. It seems as if the technology, which is still in its infancy, enables a broad spectrum of changes.

Generative manufacturing processes enable manufacturers to produce custom-made products in industrial quantities or even creations that would not be possible using conventional machining processes. This opens up new possibilities in the design and fabrication of components as well as it promotes lean supply chains. It is conceivable to reduce the effort required for transportation and storage if in the future production can be carried out on site and just in time, i.e. at almost any location in the world. As a result, this could have an impact on the supply chain, including the freight-forwarding business and logistics industry, as well as the original manufacturers of products.[73]

The essential finding for original manufacturers is that if they handle 3D data too lightly, they give up control of the artwork and the intellectual property it contains to non-licensed third parties. This involves data, data exchange, copyrights and protection against counterfeiting. Once the original file has been handed over, first-class copies can be made. Black marketeers are no longer forced to 3D scan an object or to re-measure a product and to transfer the dimensions into a CAD system manually. As the quality of the possibilities increases, the appeal of imitating objects will become more attractive for third parties and may increase. Secure and traceable ways of exchanging design files are urgently needed. Keywords here: Digital Rights Management (DRM) and block chain technology.[74]

In addition to the jeopardy of profit loss and misuse by putting illegal copies on the market, manufacturers are confronted with product liability issues. What is the situation, for example, if a file is handed over to consumers? Or who is liable for the manufacture and the sales of spare parts? Are they original spare parts, spare parts in original equipment manufacturer's quality or simply an imitation, if parts are manufactured and made available by third parties on site by printing? Weapon manufacturing is currently not at the point where risks of this kind or the loss of control over the actual manufacturing process call for clarification.[75] But this particular product seems to have a very special appeal and printed firearms are an issue.

The direct transformation of CAD data into a physical part is the unifying trait of all generative manufacturing processes.[76] The relevance for printing hobbyists is understandable. There are different options of home-scale printers: Either the purchase of a ready to use personal fabricator or the do-it-yourself option. Similar to the concept of open source software, it is possible to build a printer using open source hardware.[77] If one prefers a little more comfort, he can opt for a ready-made DIY kit. Even simpler kits are available for about 200 Euro. Personal printers will neither use an electron beam nor are metal powder particles fused with a laser beam; instead these consumer-grade machines use a

Fig. 2.3: ANYCUBIC delta printer.

heated nozzle to melt a plastic thread, the printing filament, which is applied layer by layer.[78] Generic technical names for the technology and printers that deposit raw material through some kind of syringe or nozzle print head is *fused deposition modeling*, abbreviated FDM, and *selective deposition printers* or FDM printers.

Not yet available at attractive prices for home use: the filament is a mixture of plastic, wax and a metal powder filler. After printing, the resulting green part must be chemically freed from the binder. This brown part is now treated in a sintering furnace; non-metallic components are burned off in the process. After sintering, the material density is at least 97 percent. Honeycomb structures inside a closed workpiece enable savings in component weight without losing the required stiffness.

Lot sizes starting from one unit and objects manufactured within a short time with a high degree of individualization are significant strengths of 3D printing. One side of the coin is mass customization, shortened lead times for the production of tools and fixtures during set-up of industrial series production, faster market entry in general, the just-in-time based production of spare parts, groundbreaking problem solutions and other advantages. Logically, however, there is a downside: product piracy and other legal criteria. In this context one may remember the discussion on printed firearms. There is no doubt that hobbyists have always found ways and means to build a device for firing live ammunition. Using printing technology, illegal replicas have recently reached a new level. Hardly any gun-related know-how is required. A computer is commonly found in almost every household, personal printers are portable and inconspicuous, and printing costs are reasonable. The special feature for sufficiently experienced hobbyists: firearms that are not recognizable as such because they simulate another object, e.g. guns that are disguised as a mobile phone.[79]

Characteristics

The application possibilities are versatile, hence manufacturing specific signatures may vary. Organic designs or "grown" patterns are possible. Due to a layered approach there is almost no limitation in shape. Honeycomb structures, facet geometries in general, as well as parts optimized to weigh less occur. Novel designs, complex geometries, cavities as well as internal structures or geometries which were previously too complex to make or would be extremely complex to do with conventional manufacturing methods, may be found. Almost all kinds of materials can be used.

Depending on the material used, the printing process, and the contouring, sometimes stepped edges and stepped or grainy surface finishes or post-processing marks might be found. The latter may originate from support structures that are fabrication-related and whose contact points were removed and smoothed by grinding, blasting or vibratory grinding during post-processing.

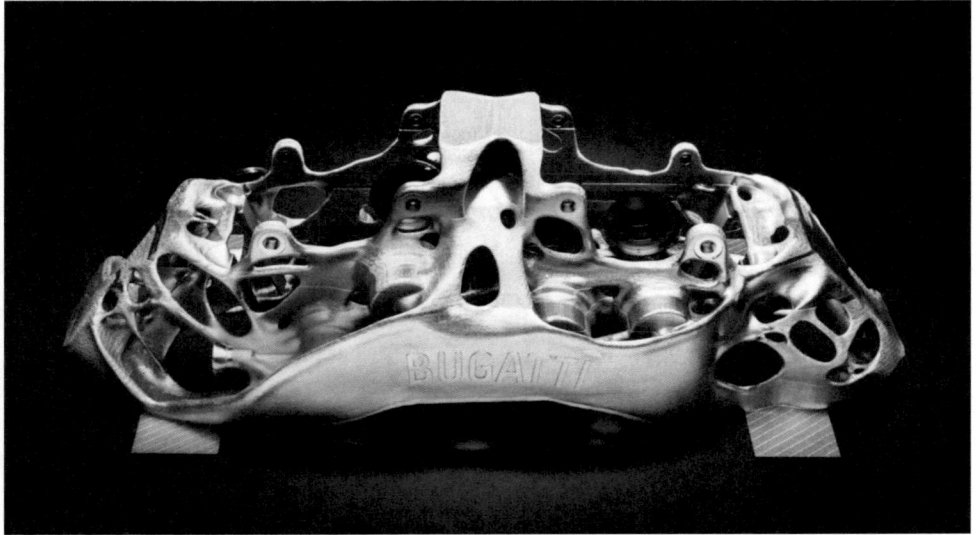

Fig. 2.4: Largest titanium functional component produced by additive manufacturing as of 2018: BUGATTI eight-piston monobloc brake caliper.

(© BUGATTI AUTOMOBILES S.A.S.)

2.2 Machining

Milling, turning, drilling, broaching, slotting, reaming and grinding are processes in which a piece of raw material is transferred into a desired final shape and size by a controlled material-removal process.[80] Slides are often machined by using CNC milling machines (Fig. 2.5). These machines are expensive to purchase and operate; usually manufacturing costs are in a range from $50 to $70 U.S. per hour based on a two-shift operation. The costs for cutting tools have to be considered on top of this. Depending on the individual CNC machine, its tool magazine can store 120 or 240 cutters. This sums up to a total capital of about 15,000, or 30,000 Dollar for a fully loaded tool magazine, given an estimate of about $125 U.S. for each tool and its toolholder (as of 2014). A cost-oriented manufacturer reduces this potential by limiting production to a minimum of different cutters during the development phase of new products.

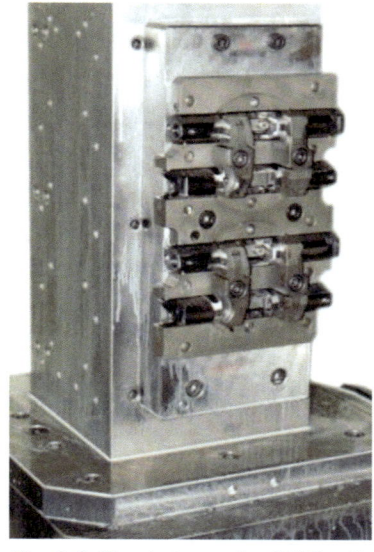

Fig. 2.5: Tombstone of a CNC milling center with slide blanks waiting to be machined.

Broaching and slotting are machining techniques applied to cut square internal corners. Broaching is done with a multiple-tooth cutter (a broach), where each blade is a little larger than its predecessor (Fig. 2.6). Magazine wells in metal frames or breech faces with square guide blocks used to be machined by broaching. In some cases, this would have involved a series of expensive broaches that were used one after the other.

Since broaching is a machining process where the removal of chips takes place at a very slow pace – and must also be performed on a separate broaching machine – it requires the workpieces to be put

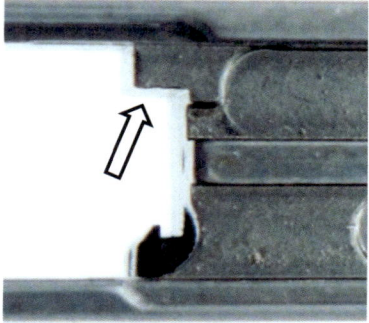

Fig. 2.6: Detail view of a pull broach (above) and guide block (arrow) next to breech face.

one by one into another fixture. In modern production the breech face of a slide is machined on CNC
machines by slotting, a cutting operation with repeated – mostly straight cutter movement – which
puts a demanding load of shocks onto the milling
machine. However, manufacturing tolerances are
reduced considerably by avoiding the need to repeatedly zero the workpieces in a variety of different
fixtures and enable a more efficient production process and parts' interchangeability (Fig. 2.7).[81]

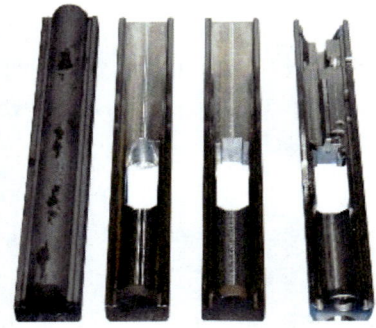

Fig. 2.7: Production steps of a pistol slide: raw material, rough-
ened, broached, finished (from
left).

Characteristics

Machined surfaces carry toolmarks that are caused by the teeth of the cutters and are
in an orientation in the cutting direction of the teeth. These toolmarks occur either in
straight or curved orientation, depending on whether the milling was done with the side
or the front face of the cutter. The orientation of the marks further depends on how
progressively the removal of material was carried out, and as a result the marks can
be anything from rough to a smooth surface. However, the latter can also be the result
of a secondary operation, such as grinding.

Grinding (Fig. 2.8), polishing or beading can be used to remove tool marks and might
make it difficult to clearly tell what machining process was used to produce the workpiece when the toolmarks are no longer visible.

Fig. 2.8: Pistol slide with ground surface finishing (shiny).

2.3 Injection Molding

The use of polymer parts is widespread in gun manufacturing. Plastic parts are ideal for both integral designs and functional integration, and are beneficial in the construction of pistols. There were efforts in the early 20th century to replace machined parts by those made of Bakelite® [a],[82], and later by the use of polymers. The former, Bakelite, is a thermosetting phenol formaldehyde resin that is formed from a condensation reaction of phenol with formaldehyde, and which was commonly reinforced with fabric fibers or sawdust.

In 1907 Leo Baekeland filed for the first patent for the production of the synthetic material after he had developed the manufacture of phenolic resins. Three years later the first industrial production was carried out by the BAKELITE GMBH in Erkner near Berlin, Germany. Items such as bayonet grip handles, and pistol grip side plates, as well as parts of telephones, radios, and light switches, etc. were made until the middle of the 20th century of Bakelite. Despite its limited use today, the term Bakelite is still a registered trademark.

The high benefits of modern polymer injection molding technology lay within the dimensional accuracy. It offers virtually unlimited possibilities in the shaping of the parts as compared to the machining of workpieces, as well as in its repeatable reproducibility; once a required injection mold was made.

Engineers can use a wide variety of polymers with different properties and advantages and choose between various fillers, i.e. basically between unreinforced, and carbon- or glass-fiber reinforced polymers. Pistol frames are often made of glass fiber reinforced polyamide (PA 6 and PA 12) since these – in contrast to carbon fiber reinforced plastics – are non-hygroscopic, meaning that this polymer does not absorb moisture from the environment. This is true especially with changing climatic conditions, as might occur in tropical countries. It is a huge advantage for good reliability of the gun if the properties of the material do not change and functional parts stay within the acceptable tolerances.

Fig. 2.9: Injection molding machine
(© KAUT GMBH)

---

[a] Bakelite is a registered trademark of HEXION.

Injection molding machines are used for the production of plastic parts (Fig. 2.9). An injection molding machine heats up polymer material until it is in thermoplastic condition and then presses it into a mold. These molds are made of solid steel and carry one or several cavities inside. The polymer raw material is in pellet form of neutral color, and in most cases, it will be mixed with color dye before it is poured into the molding machine. The material is heated up to a temperature of approximately 390 °F (200 °C) where it turns into a pasty substance, after which it can be hydraulically pressed into the mold.

A cooling process must then take place before the mold opens and the molded part can be removed. After the cooling the two halves of the mold are spread apart, and ejectors push out the new part from one half of the cavity. Alternatively, the removal can be carried out by a handling device (Fig. 2.10).[83]

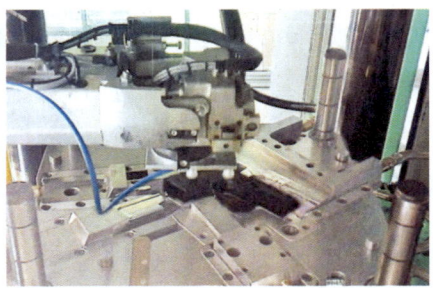

Fig. 2.10: A handling device removes a pistol frame from the mold.
(© KAUT GMBH)

After the ejection, there is a danger of warpage that may occur while the part cools down to room temperature. If harmful warpage cannot be avoided, the molder may put the polymer part in a fixture while the part is still warm to put a defined preload on the part until it cools down to room temperature.

However, for an optimum flow of production, preloading should be avoided. This can be achieved when the product design engineer takes care to

Fig. 2.11: Parting line on a SIG SAUER *Mosquito* pistol frame.

maintain the same wall thickness during the design of the part; where favorable geometries are utilized to avoid material accumulation, and an injection point that supports easy filling of the cavity and the flow of the polymer while it fills it.[84] Moreover, we know that even over several weeks a conditioning of the material takes place, which changes the properties of the component and has an effect on its dimensions.

Characteristics

Typical traits of injection-molded parts are ejector marks and fine parting lines or seams (Fig. 2.11) caused by parting lines of the mold halves or tooling slides. At the gating, i.e. the gate where the polymer is injected from the sprue or runner into the cavity, a "gate mark," that is a small nub or projection may be evident on the molded piece, which is leftover after the cast was removed. The gate might be recessed, so the molder does not have to manually deburr this area (see pos. 1 in Fig. 2.12).

If parts came off a mold with several cavities, they usually have a marking, such as numbers or letters, which is the mapping of components to the respective cavity. This identification is useful if a part does not meet design dimensions and the matching cavity needs rework. In this case, the toolmaker must be able to identify the cavity in question.

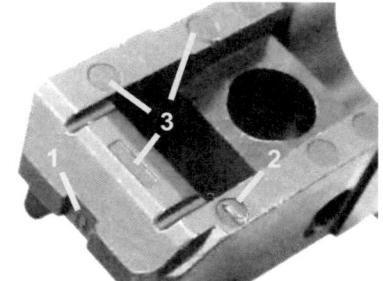

Fig. 2.12: Workpiece with recessed gating (1), cavity number (2), and ejector pin markings (3).

Besides, a date stamp may be found on injection-molded parts, mostly on the inside. It enables the molder, the manufacturer and the end-user to identify when the part was actually molded, now that traceability and accountability have become increasingly important for all concerned. Date stamps or inserts are round mold inserts with the capability of indicating different pieces of information depending on what the manufacturer needs.

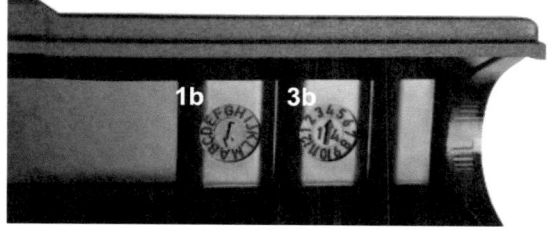

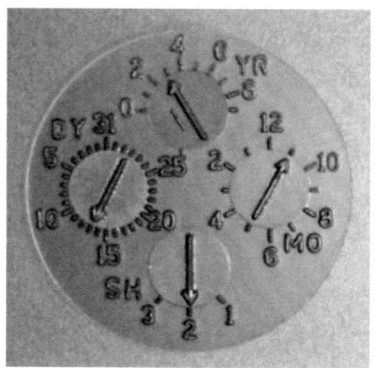

Fig. 2.13: Date stamps:
Date stamp found on a WALTHER *PPS* and *PPQ* (shown below), and on a Pelican Case indicating year (YR), month (MO), shift (SH), and day (DY).

The most common information required is year, month, day, and shift or drawing revision. Typically, a date stamp will have an outer ring of information and an inner insert with an arrow pointing to the desired year, month, day, etc. Tool makers have the opportunity to choose from a variety of different inserts, consequently, different marks can be found on polymer parts.

In Fig. 2.13 date stamps marked as pos. 1a and 1b indicate the revision level of the drawing, on which the polymer part (here, revision G and A) was molded. Based on the markings on pos. 3a and 3b month and year of manufacturing can be identified, here September '13 and April '14. Position #2 shows the calendar day.

2.4 Investment Casting

In contrast to sand castings, which are molded by using sand as the mold material, investment casting or lost-wax casting is valued for its ability to produce components. Investment casting offers accuracy, repeatability, versatility, and integrity in a variety of metals and high-performance alloys. It derives its name from the pattern being invested with a refractory material to form a one-piece ceramic shell.

During the 1930s investment casting was considered to be a useful but special production process, which was still not widespread. This changed during World War II, when machining capacities were no longer sufficient to meet the demands of the industry. As well as cost-savings, investment casting fulfilled the needs of repeatability, accuracy and applicability for metals with a high melting point and high metallurgical grade.[85] The high surface quality of castings allows for the production of ready to use parts or parts that only need secondary operations in certain areas with critical dimensional tolerances.

Fig. 2.14: Wax injection mold with two wax patterns (square, lower end of left image), and empty mold after wax patterns were ejected (right image).

(© ECRIMESA)

The more complicated the shape, the more attractive investment casting will be because machining costs can be saved. Using investment casting, parts made of steel, as well as iron, titanium, or aluminum-based alloys can be produced with a weight ranging from a few grams up to 10 kg. Heavier parts of up to several hundred kilograms may even be possible.[86]

Specific to investment casting is the creation of wax patterns, which are attached to wax sprues to create a pattern cluster or tree. The wax patterns are molded just in the same way as a polymer part would be injection molded, however, a suitable wax is used instead of a polymer. Depending on the number of cavities in the injection mold tool, multiple wax patterns can be molded at the same time (Fig. 2.14). Complex geometries, e.g. undercuts or tubes, can be done using pre-formed positive cores made of ceramic or soluble wax (Fig. 2.15).

Fig. 2.15: Wax patterns (dark) with soluble cores inside (bright).

(© ECRIMESA)

The ceramic mold, known as the *investment*, is made of a number of layers produced by repeating a series of steps where the tree (Fig. 2.16) is dipped into a slurry of fine refractory material and then drained to create a uniform surface coating (Fig. 2.17). This surface coating will also be covered in sand and then cured. Coating, stuccoing, and hardening are repeated 6 to 12 times until the desired mold thickness is achieved.

Fig. 2.16: Creating a tree of a number of wax patterns.

(© ECRIMESA)

Once the ceramic mold has fully cured, it is put in a furnace at a temperature of about 300 °F (150 °C) to melt out the wax. The mold is then subjected to a burnout to sinter the mold at temperatures from 1740 to 2010 °F (950 to 1100 °C).[87] Molten metal is then poured into the hot mold. After the metal has cooled down the shell is removed by destroying it. The casting may then be cleaned up by sand blasting. Finally, the castings are separated from the sprue and the gate marks ground off.[88]

Fig. 2.17: Tree dipped into refractory material.

(© ECRIMESA)

In general, extremely intricate parts are castable. However, their mechanical properties do not quite reach the quality of forgings, since pores may exist in the microstructure of castings, which reduce their load-carrying capacity. In addition, the surfaces of castings have a characteristic texture (see Fig. 2.18).[89] RUGER used to have the reputation of using lots of investment castings, which were easy to recognize by their unique surface finish.[90]

Characteristics

The characteristic surface tension of liquid metals is the reason why investment castings originally will not have sharp edges. However, sharp edges can be machined in a secondary operation.

Cavity number marking

Ejector mark

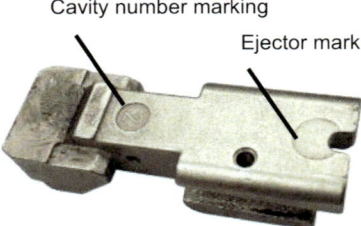

Depending on how much effort a manufacturer puts into the finishing of castings, parts may be recognized by their rough surface, which in some instances might be found on internal surfaces only (Fig. 2.18).

Fig. 2.18: Investment cast REMING-TON *R51* breechblock with markings and typical investment cast surface finish.

The gate mark may appear as a roughly ground off surface. Also, a mark for the number of the cavity may be visible or parting lines or ejector markings, which come from the cavities of the wax pattern mold and are typically about 0.004 in (0.1 mm) sunken or proud. In addition, cast logos, item numbers or other markings may be visible.

2.5 Forging

Forged blanks, made of steel or aluminum alloys, are often used as raw material for further processing to make slides and barrels. The advantage lies in the forming process of the material, which produces a homogenous grain structure of the metal following the outline of the component and the associated densification of the material whereby pores in the raw material are eliminated. Forged components are especially suitable for power and movement transition; especially when high static and dynamic loads are likely to be encountered.[91]

Furthermore, the use of forged blanks reduces the amount of raw material to be removed in manufacturing since the outline is already very similar to that of the finished part. As a consequence of manufacturing tolerances, the final outline of forged blanks can't be exactly produced to blueprint dimensions, so before machining of forged blanks, datum surfaces have to be made first, which are used in the downstream manufacture for reference.

The processing time of forged blanks in production can be similar to solid raw material. It is left to cost accounting, whether forged blanks offer a price advantage. For large series with high volume output, forgings will have cost advantages. Smaller manufacturers may choose not to use forgings, if their manufacturing of price sensitive low volume products depends on a streamlined inventory of raw material, and if they want to maintain certain flexibility in production for future product updates.

Fig. 2.19: Surface finish and parting line (see arrow) of a die-forged part.

Characteristics

Circumferential flash is a typical trait of impression-die forgings, where the metal is placed in a die that resembles a mold. During forging, the metal will flow and fill the die cavities. Excess metal is squeezed out of the cavities, forming flash. This flash has to be removed after forging, leaving a parting line at its center (Fig. 2.19).

2.6 Metal Injection Molding

Metal injection molding (MIM) or powder injection molding, is a combination of molding and sintering of metal powders. The origin of the MIM goes back to the 1940s, when ceramic components were manufactured by means of injection molding technology.

In the 1970s the procedure was finally transferred to the processing of metallic materials.[92] Today the metal injection molding is very popular in the firearms industry. It is usually used for the efficient production of components with a mass from 0.2 to 100 grams, where machining would be complicated and correspondingly high costs would be involved. Using MIM, parts with complex geometries are produced in large quantities and tighter dimensional tolerances can be achieved than with investment castings. However, their geometric complexity as well as their size and raw material costs, which is dependent on the mass of the component, limit the applicability of this technology. In practice, three points appear as particularly critical: cost of raw material and thus of the finished component, cracks, and the dimensional stability during the manufacturing process.

A blend (feedstock) of vitrifiable metal alloy powder and thermoplastic materials, and waxes as a binder and free-flow agents is the raw material for the MIM production processes. Because of the metal powder's grain size of less than 20 μm, the feedstock can be handled initially in a molding process that resembles polymer injection: the feedstock is pressed – in a doughy state – into the cavity of an injection mold and removed after a cool-down process. Because the feedstock is a composite of metal powder and polymer, the component is initially larger than the later finished part. At this stage, the component is known as a green part (Fig. 2.20).

The next step is debinding, where the existing plastic portion of the green part is removed. Depending on the manufacturing process, this is done thermally, catalytically or chemically. The debinding starts from the outside going inwards at a speed of one to three millimeters per hour. What is left at the end of this process is a porous and thus fragile component, which consists of metal particles only, referred to as brown state. After a complete debinding the metal particles are only held together by van der Waals forces, i.e. intermolecular forces.[93]

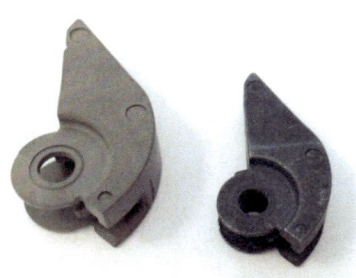

Fig. 2.20: Green part and finished part (right).

Sintering under a protective atmosphere or in a vacuum follows at a temperature of 2280 to 2640 °F (1250 to 1450 °C) (Fig. 2.21), usually in a computer-controlled furnace. A stabilization and homogenization of the proportion of carbon content as well as the elimination of organic residues in combination with a reduction of the amount of metal oxides can be achieved through uniform production parameters. During sintering, the metal molecules bind to-

Fig. 2.21: Sintering done in a continuous furnace.

(© MIMECRISA)

gether and will form a solid, homogeneous body, while the porosities of the brown state disappear. This condenses and shrinks the component which reaches the required dimensions. The shrinkage during sintering depends on the alloy powder, ranges from 18 to 25 percent and must be taken into consideration during the construction of the injection mold for the green part.[94]

The density of MIM'ed parts is lower than that of solid material, but a density of at least 96 percent can be realized. Depending on the geometry of the parts and the optimization of the manufacturing process, MIM'ed components can tend to crack despite their still relatively high material density. If, for example, the green part is misaligned while it is ejected from the mold, a micro crack might be caused, which will remain in the component and cause the part to break at the affected area. Special care taken in the manufacturing process reduces the risk of demolding cracks. For economic reasons, a check for cracks on the green part is only feasible in production in special cases.

If a MIM'ed part breaks during the later use of the finished product, examination of the fracture surface should show if it was caused by a defect introduced during the MIM process: If a micro crack already existed during the MIM process, then the surface in this pre-damaged area will be of the same color as on the outside of the component. In other words, the

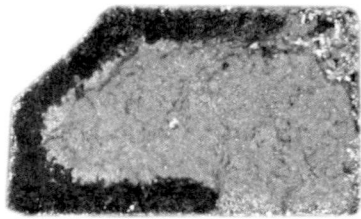

Fig. 2.22: Overload breakage with pre-damaged area (dark).

surface of the ruptured cross-section is not entirely metallic (Fig. 2.22).

The danger of deformation of a part in the brown state is another risk factor during sintering. In the furnace, the parts are heated up high enough to nearly melt the metal and thus are doughy soft. If the components have a delicate shape – maybe there are thin flanges – these can deform. The challenge is to support such places and at the same time to take into account the shrinking. In addition, a part in the brown state will move relatively to the surface it is located on during sintering. This movement must happen with as limited friction as possible, otherwise, cracks may be caused in this step of production.

Advantages of the MIM process are the short production time of fewer than 24 hours, a reproducible quality and complex shape possibilities in connection with material properties approaching those of components made of solid material. The achievable surface finish is better than that of investment cast parts. Close dimensional tolerances and good surface finish (roughness of R_a 1.6) help to reduce subsequent processing and polishing costs.

Adversely high tooling costs have a negative effect. A high precision of the injection molding tool plays an important role for the precision and surface quality of MIM'ed parts. Until the production process is finalized, usually considerable costs for tool changes and dimensional corrections have to be covered. This rework can hardly be avoided, as dimensional deviations or non-uniform shrinkage during sintering can only be identified after first production samples are made. Repeated need for adjustments is often the result until the full production of high-quality parts gets underway.[95]

The objective of modern manufacturing techniques lies in the reduction of manufacturing steps by manufacturing near net shape parts and even trying to produce barrels this way.[96] However, in the latter case, there are still limits to overcome. In addition to the tooling and material costs per component, it has to be considered that with increasing component weight the reject rate goes up and with it the overall costs per production lot. The larger and more complex MIM'ed components are, the more challenging their production process is in cost sensitive production where the failure rate must be kept at a minimum. The special expertise of the product engineer is to avoid material accumulation and to design parts with consistent wall thickness to prevent warpage, which is a typical effect found on injection molded parts, such as the green part.

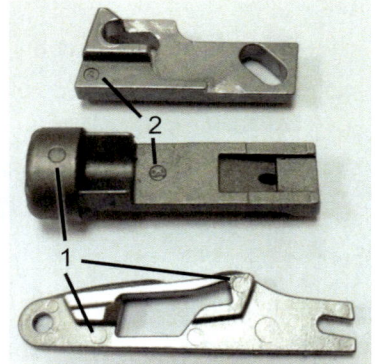

Fig. 2.23: Ejector mark (1) and number of cavity (2).

Characteristics

Because MIM is related to injection molding, manufacturing characteristics similar to injection molded polymer parts can be found on MIM'ed components including: ejector marks, fine parting lines or seams caused by parting lines of the mold halves or tooling slides, and also gate marks. If parts are from tools with multiple cavities, an appropriate identification mark may be found, usually located somewhere on the inside, i.e. not easily visible from the outside (Fig. 2.23).

Another trait is a surface with geometry that could not be made by machining, such as inner wells with corners instead of radii (Fig. 2.24).

Common examples of MIM'ed parts include: front and rear sights, trigger parts, as well as safety, take-down and other levers.

Fig. 2.24: MIM features shown on a WALTHER *PK380* slide: Ejector mark (1), parting line (2), gate mark (3), and square breech face (right), which could not be machined this way.

2.7 Stamping

The heyday of drawing, stamping and forming of metals in the fabrication of sheet metal weapons began during the 1940s, and found its way into all kinds of weapons by the end of the Second World War including machine guns, submachine guns, assault rifles and tentatively even last ditch semi-auto pistols, which were produced by MAUSER and WALTHER almost entirely in sheet metal stamping.[97]

Stamping is now in competition with other low-cost manufacturing methods, such as MIM or polymer injection molding. Injection molding enables the production of attractive-looking parts, whose geometry does not have to exactly follow the manufacturing process and whose wall thickness may be partly different. Nevertheless, stampings are still common in weapon manufacturing, because just where polymer parts are approaching their limit, sheet metal stampings are used instead, and may also be used as reinforcement of larger polymer assemblies.

In stamping, flat sheet metal parts represent the simplest type of workpieces. They are produced by shear cutting, colloquially known as die-cutting. More complex shaped stampings are produced by progressive stamping.[98] Here a strip of sheet metal is fed from a large coil, and the process of stamping is combined with cold forming processes.

In the tool, integrated forming operations can be bending, punching, coining and other shaping steps, combined with an automatic feeding system. As a result, the workpiece is not just flat, but can be of a three-dimensional (3D) shape. Cutting and bending is done in a series of steps, i.e. in different stations of a progressive die in a continuous process. Each station performs one operation or more until the part is complete. If certain areas need to be particularly smooth and burr free or a high dimensional accuracy is required, it is possible to make clean, 100 percent sheared edges by means of fineblanking. Fineblanking avoids rough edges and should be used in the production of trigger bars, slide rails, slide stop levers, and ejectors.[99]

The strip shown in Fig. 2.25 details the production steps of an ejector insert made by progressive stamping. The illustrated strip was produced progressively from right to left using a reciprocating stamping press. The individual operations are carried out one after the other, the feeding system advances the strip forward very precisely as it moves from station to station. In the first step index holes are punched, together with an outer contour, before the strip is moved to the next station, where bending, embossing, and additional cuts follow. The final station is a cutoff operation, which separates the finished part from the carrying web. With each stroke of the press, a completed part drops into a container.[100]

As a result of high manufacturing and maintenance costs of the progressive die, this production technology is only suitable for high volume output of components to a fixed pattern. Product engineers should avoid any tool changes as reworking the die is costly and might affect its service life.

Fig. 2.25: Workpiece of a progressive stamping strip.

Characteristics

Stampings used in today's guns are one to three mil-limeters thick, 3D shaped metal parts which can be recognized by the circumferential cutting surface caused by shearing (Fig. 2.26): proprietary for stamping is the die roll zone, i.e. a plastic defor-mation of the sheet edge, which causes permanent edge rounding, followed by a sheared surface and a surface fracture zone with burrs where the material

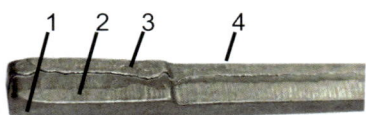

Fig. 2.26: Stamping surface: Die roll zone (1), sheared surface (2), fracture zone (3), and burr (4).

is irregularly stripped away from the matching surface, at the end of the shearing op-eration. The fracture zone may be left with burrs, unless they are not removed in a secondary operation during production. Apart from the predominantly smooth cut and fracture zone seen on the edges, embossing marks or index holes might also be visi-ble.

Slide stop levers, trigger bars, ejectors, and slide rails are often to be found as sheet metal stampings.

2.8 Surface Treatment

Dark coloring in general and specifically the classic bluing of metal gun parts has a long tradition in the field of gun making. Unsatisfactory corrosion resistance of blued ferrous metals required regular application of a thin oil film on the surfaces during weapon maintenance. Today's weapons require less attention from the users, and new surface coatings provide many kinds of colors. Three common methods are explained below: nitrocarburizing, diamond-like carbon, and high-temperature ceramic coatings.

2.8.1 Nitrocarburizing

The corrosion protection of weapon parts improved significantly around 1975, when German gunmakers began to replace bluing with salt bath nitrocarburizing with subsequent oxidation. Nitrocarburizing creates a thermochemical case hardening. It is applicable for all types of ferrous materials and was first used in 1929 for engine components, that is crankshafts, which, despite the special treatment, still did not last long enough and therefore gave the impetus for research into the theoretical foundations of the nitrocarburizing process.[101]

In addition, attempts by MAUSER in 1944 are known, where they tried to increase the service life of ball bearings by nitriding.[102] Today nitrocarburizing is used on an industrial scale for components with high demands on wear resistance, fatigue strength and mechanical properties; e.g. engine valves, gearboxes and differential parts, or when it comes to pistols: slides.

The nitrocarburizing process has the advantage of little or no warpage of the parts during treatment. The risk of uncontrolled warpage cannot be avoided during classic heat treatment, i.e. hardening and tempering, and warpage might exceed the required manufacturing tolerances and require subsequent straightening operations, or lead to scrapping of the parts in the worst case. Both would lead to additional costs in manufacturing which calls into question traditional treatment.[103] In addition, induction hardening and grinding as secondary operations can be omitted when nitrocarburizing is used.[104]

Nitrocarburizing can be accomplished using three different media: gas, plasma or liquid salt bath. In the latter, nitrogen-bearing salts produce a controlled and highly uniform release of nitrogen at the interface of the workpiece, providing an efficient manufacturing process and a higher nitrogen content in the surface of the workpiece, which achieves a better wear and corrosion protection than during gas- or plasma nitrocarburizing.[105] Plasma nitrocarburizing, however, makes it possible to exclude certain areas of the workpiece from the hardening by covering these spots before the treatment. This advantage is however detrimental. If the parts are handled in bulk or attached to a fixture for the process (Fig. 2.27) these areas will remain without corrosion protection at the end.[106]

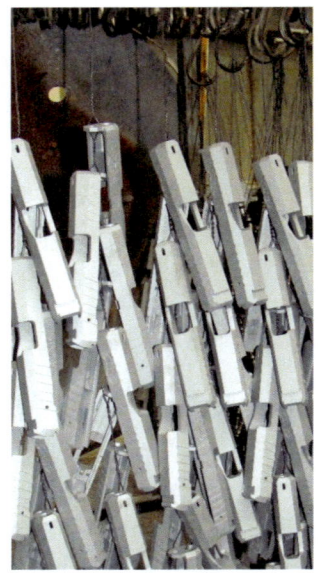

Fig. 2.27: Workpieces carefully arranged for salt bath nitrocarburizing: slides attached to wires.

Salt bath nitrocarburizing (Fig. 2.28) is usually the first choice in the gun industry today. It provides attractive process times from 60 to 120 minutes, large penetration depths with deep black coloring and high corrosion resistance, in combination with a tough substrate beneath the hard surface. One drawback is that salt residues might remain on the workpiece, which must be removed with considerable effort. Gas-nitrided parts, however, are immediately ready for installation, but their surface appears somewhat grey instead of black. Liquid nitriding and gas nitriding systems can be integrated into the manufacturing process on site and, in turn, enable a more efficient production process. In addition, an integrated process will only be commercially attractive when large production volumes are to be produced. A gas nitriding oven takes, for example, about 300 workpieces. Only manufacturers with large production numbers can invest in correspondingly larger or multiple gas nitriding furnaces.[107]

Fig. 2.28: Salt bath application: oxidizing bath, rinsing, and salt furnace (from left).
(© Donau-Härterei)

The working principle of nitrocarburizing is based on the formation of a compound layer. The achievable results and thickness of this compound layer are subject to the process chosen. For example, 24 hours of plasma nitrocarburizing of 42CrMo4 steel will result in a circa 4 µm of compound layer, 12 hours of gas nitrocarburizing will give roughly 12 µm, and just a few hours of salt bath nitrocarburizing will provide something in the neighborhood of 20 µm. Note: corrosion resistance depends on the thickness of the compound layer, and best results are achieved with a layer of 12 µm or more.

The salt bath nitrocarburizing treatment cycle consists of six basic steps:[108]

Step 1:   Preheating of the workpieces in an air preheat furnace.

Step 2:   Nitrocarburizing: immersion in a molten nitrogen-rich salt solution.

Step 3:   Oxidizing; as a result, the workpieces go black.

Step 4:   Cooling and cleaning.

Step 5:   Mechanical finishing, e.g. polishing or blast finishing.

Step 6:   Post oxidizing; same oxidizing bath as in step 3.

This thermochemical treatment leads to a transformation of the chemical composition in the boundary layer of the material: During the process, a nitrocarburized layer is formed consisting of a thin outer compound layer and a diffusion layer thereunder. In Fig. 2.29, for example, a 9 µm thick compound layer can be seen as well as a large diffusion layer (dark). The chemically transformed nitrocarburized area shown here reaches about 120 µm deep into the material.

Fig. 2.29: Micrograph of the boundary layer.

(© DURFERRIT)

Compound layer with pore zone

The compound layer is created during step 2, when nitrogen diffuses into, and chemically combines with, nitride-forming elements in the metal, producing, through a catalytic reaction, a tough, hard compound layer at the interface of the workpiece. In the case of nitrocarburizing, the compound layer catalytic reaction combines iron with nitrogen and carbon. Thus, there is limited extent for oxygen to react with the iron left within this layer (corrosion), as the iron is already bound to nitrogen or carbon. The layer, typically just about 20 µm thick, no longer has metallic properties but ceramic characteristics. This non-metallic structure reduces friction and thus the tendency to wear significantly.

When the nitrogen diffuses into the metal, nitrides are formed in the metal which have hardness and wear properties that can be in the range from 800 to 1500 HV, if the material offers a sufficient content of nitride forming alloy elements.[109]

What is known as the porous layer is located on the outside of the compound layer. This pore zone is located directly on the surface of the workpiece (Fig. 2.29, for example, a 3 µm thick pore zone can be seen). The oxidizing in step 3 fills the pores with magnetite, which braces the protective black oxide layer. The porosity allows good adhesion of lubricants and consequently better running characteristics. The porous layer can also serve as a lubricant reservoir and thus supports the dry-running operation properties of treated components. Prerequisite for this is that an adequately thick compound layer exists in addition to a sufficiently thick porous layer.[110]

A good corrosion resistance can be obtained if the compound layer consists of almost mono phase ε-iron carbon nitride. In addition, the compound layer should be at least 12 µm thick and the black iron oxide, which is a result of the oxidizing treatment in step 3, should exist on the outermost layer and in the pores.[111] This is ensured by the post oxidizing in step 6 where the oxide layer is re-applied and a uniform dark color of the surface is achieved at the same time.

Apart from the treatment conditions – time, temperature, composition of the salt bath – the content of carbon and alloying elements of the workpiece material significantly affect the achievable thickness of the compound layer. With the usual treatment time from an hour or two in the salt bath, the layer will have a thickness from 0.0004 to 0.0008 in (10 to 20 µm). In comparison to classic case hardening, the nitrocarburized layer has a significantly higher heat resistance; the compound layer can withstand temperatures close to its formation temperature.[112]

Diffusion layer

The area below the compound layer, in which the nitrogen content does not suffice for the formation of iron nitride, is called diffusion layer. The obtainable thickness and hardness of the diffusion layer is material-dependent on the same extent as at the compound layer.

Advantages of the treatment and corrosion resistance

Nitrocarburized components are generally characterized by a particularly good resistance to corrosion, wear and galling. In principle, with decreasing surface roughness, the corrosion resistance of any steel part increases. This principle applies also for this treatment. Therefore, the workpieces must be provided with sufficient surface finish before the nitrocarburizing can begin; this is also because rework of already nitrocarburized components is not easy. Above that, chances are that any mechanical rework, such as for dimensional corrections, would compromise the compound layer and the black color and corrosion protection would no longer exist.

Fig. 2.30 shows the six process steps of the salt bath nitrocarburizing treatment. The oxidizing treatments (steps 3 and 6) are of particular importance in addition to the actual nitrocarburizing (step 2). Here a black iron oxide layer is created that causes a significant increase of corrosion resistance on the surface of the treated parts.[113]

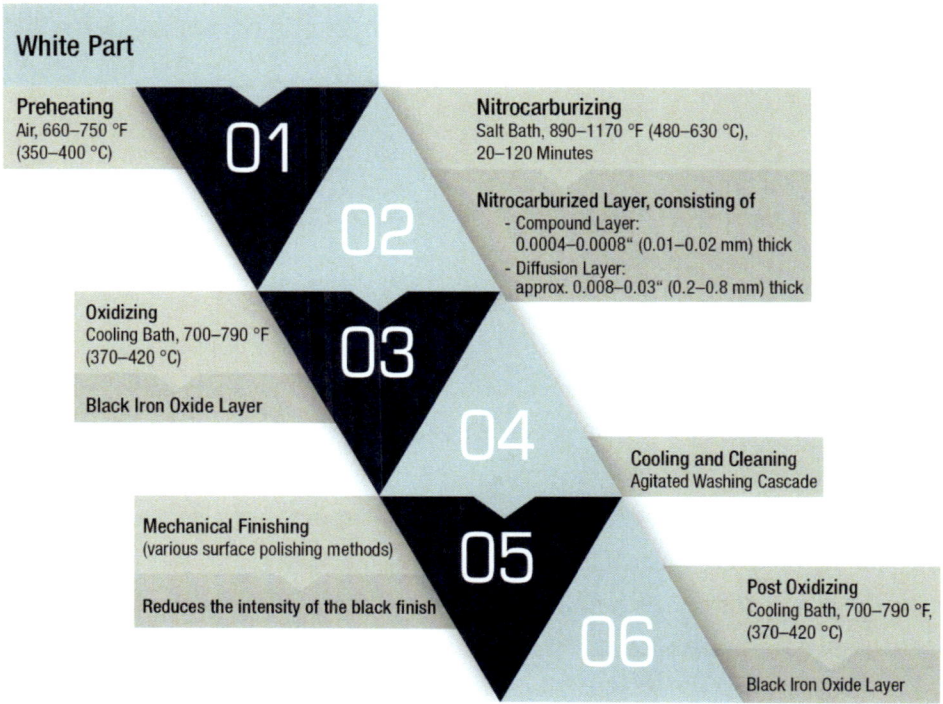

White Part

Preheating
Air, 660–750 °F
(350–400 °C)

01

Nitrocarburizing
Salt Bath, 890–1170 °F (480–630 °C),
20–120 Minutes

02

Nitrocarburized Layer, consisting of
- Compound Layer:
  0.0004–0.0008" (0.01–0.02 mm) thick
- Diffusion Layer:
  approx. 0.008–0.03" (0.2–0.8 mm) thick

Oxidizing
Cooling Bath, 700–790 °F
(370–420 °C)

03

Black Iron Oxide Layer

04

Cooling and Cleaning
Agitated Washing Cascade

Mechanical Finishing
(various surface polishing methods)

05

Reduces the intensity of the black finish

Post Oxidizing
Cooling Bath, 700–790 °F,
(370–420 °C)

06

Black Iron Oxide Layer

Fig. 2.30: Nitrocarburizing treatment; schematic overview.

The mechanical finishing (step 5 in Fig. 2.30, also Fig. 2.31) is to reduce the roughness of the surface and to obtain a uniformly black finish following the second oxidizing treatment (step 6).

Depending on the size and shape of the workpiece, as well as the required surface quality, different methods can be used for mechanical finishing. Common procedures are polishing, mechanical grinding, fine grinding and blast finishing with metal shot or glass beads (Fig. 2.32). At this point of the process, reduction of corrosion protection is normal and correctable, if this finishing does not destroy the compound layer and post oxidizing follows (step 6).

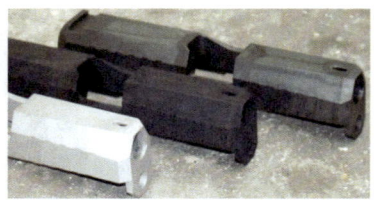

Fig. 2.31: Slides: white part, nitrocarburized, and after mechanical finishing (from left).

Rounding off of edges represents a particular danger during step 5, especially if treated with blast pressures of more than 4 bar.[114] A spot test with copper ammonium chloride can show if the compound layer still exists. If a drop of copper ammonium chloride on the surface of a metal sample goes red, free iron and therefore no compound layer is present on this spot.[115]

Fig. 2.32: Setting up workpieces on beading equipment.

Generally, any mechanical removal of, or damage to, the compound layer will affect the corrosion resistance and protection from corrosion is no longer given where the compound layer is perforated or where it is removed completely. This also applies to secondary operations, such as laser etching the serial numbers on weapon parts. To protect the metal against corrosion here, a thin layer of gun lubricant should be applied.

Also, nitrocarburized gun parts can show wear on the typical spots when carried in a holster, for example on the front edges of the slide where the black iron oxide layer will wear off. If required, it can be determined whether the compound layer on the metal surfaces and thus the corrosion protection is still available by applying the abovementioned copper ammonium chloride test.[116] If shiny areas are found to be visually disturbing, affected parts should not be subjected to another nitrocarburizing (step 2) to get the black color back, because this would increase the thickness of the compound layer and the material might tend to get too brittle. In this case post oxidizing (step 6) or classic bluing can be done to freshen up the black coloring. Note that the visibility of existing serial numbers or other markings could be affected if they are only on the surface of the metal and were not engraved or stamped before the nitrocarburizing.

Different synonymous trade-names are used for the same nitrocarburizing treatment, and they are commonly used instead of a more neutral term. The company DURFERRIT GMBH offers its salt under the name Tenifer in Europe, Tuftride on the Asian market and under the trade-name Melonite in the USA. Salt bath nitrocarburizing in Europe is often referred to as Tenifer treatment, Tenifer-QPQ, or QPQ (Quench, Polish, Quench) – apparently with reference to DURFERRIT. QPQ comprises nitrocarburizing and oxidizing (Q), mechanical finishing (P), and post oxidizing (Q). Tenifer, Tuftride, Melonite, and QPQ are registered trademarks of DURFERRIT GMBH.

Characteristics of Nitrocarburized Metal Parts

Typically, nitrocarburizing is used for slides, barrels, sights and metal stampings, such as trigger bars. Nitrocarburized gun parts are characterized by an extremely hard and dark surface. The color of the surface is part of the metal, which means the color is not a separate external layer, as is the case of coated surfaces. However, nitrocarburized parts might have been coated in a secondary operation, which might make it difficult to recognize the compound layer.

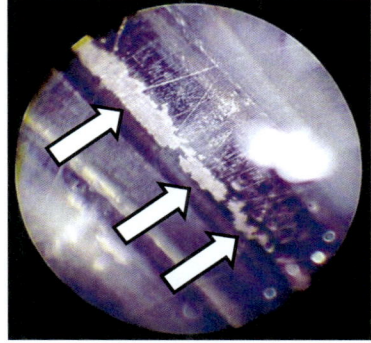

Fig. 2.33: Flaking inside a barrel with hexagonal rifling.

Salt bath nitrocarburized parts come with a deep black color, whereas gas nitride parts appear more greyish after degreasing.

Also flaking might be encountered on the lands of nitrocarburized barrels (Fig. 2.33). It is a phenomenon in nitrocarburized barrels where flaking on the compound layer of just a few microns deep occurs. This might cover larger areas in barrels with higher round counts (Fig. 2.34), which however does not affect the accuracy of the barrel.

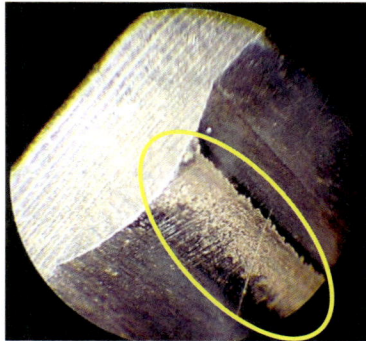

Fig. 2.34: Continuous line of flaking seen on a "land" at the muzzle end of a barrel with hexagonal rifling.

2.8.2 Diamond-Like Carbon

Various physical vapor deposition coating technologies are available for the application of thin surface layers of highly resistant coatings. CERTESS®, a trademark name registered of H.E.F. is a blanket term regarding a range of thin film applications including *Diamond-like Carbon*, also known as DLC. This 1 to 3 µm thin layer of black carbon is tribological in nature resulting in an amorphous coating that is dense and chemically inert which exhibits enhanced surface hardness consequently improving resistance to wear and abrasion.

Typically, weapon components will be subjected to a heat treatment process to achieve the desired surface hardness. In a second step the compound layer will have to be removed, if a nitrocarburizing process was used. The following is a multi-stage vacuum process that finally separates the actual carbon layer with a diamond-like structure. The layer is stain-resistant and contributes to a high-quality finish of the workpiece. Additional features of this coating are pleasant surface feel and attractive appearance (Fig. 2.35). It improves haptics and the velvet black finish may at first glance appear like polymer.[117]

Fig. 2.35: Wristwatch with DLC-coated bezel.

2.8.3 High-Temperature Ceramic Coating

So far only enamels such as ILAFLON *Resist K2*[118] or hard coatings, like physical vapor deposition (PVD), have been ways to alter the appearance of weapons, however high-temperature ceramic coatings are becoming a popular way to replace traditional black finish with attractive-looking surfaces. The brands of DURACOAT,[119] and more recently CERAKOTE[a] represent two such coatings (Fig. 2.36).

CERAKOTE is a ceramic-based finish that can be applied to metals, polymers and wood. According to

Fig. 2.36: Customized GLOCK *G21* for Marcus Luttrell, Lone Survivor.

---

[a] CERACOTE is a registered trademark of NIC INDUSTRIES, INC. White City, Oregon.

the manufacturer, it provides protection against wear and corrosion and gives impact strength. Hardness and chemical resistance are supposed to be better than with other high-temperature ceramic coatings.

The application is done in four steps: After soaking the part in degreaser and grit blasting, a 0.001 in (25 μm) thick layer of paint is applied by spraying, then followed by curing. The company CERAKOTE FIREARMS COATINGS recommends to cure metallic materials for an hour at an object temperature of 300 °F (150 °C), polymer parts are to be cured two hours at 195 °F (90 °C), and wood at 150 °F (65 °C). The company also offers air-drying coatings, but the drawback is lowered resistance.[120]

Characteristics of Coated Surfaces

Because coatings are first sprayed on, sometimes areas will be missed. Dry sprayed finish, rough surfaces, or other kinds of weakly sprayed surfaces (Fig. 2.37) may be found, especially on the insides of the workpieces. Depending on the quality of the spray job, runs or rounded-off edges with too much paint might be seen, or the opposite: too little coating on edges.

In general, the surface does not give the typical metallic feel anymore, but the Cerakote has almost a soft feeling compared to untreated metal.

Fig. 2.37: Weakly sprayed surface on the inside of a cerakoted slide.

2.9 Summary "Production Technologies"

Sales are market driven, and hence production costs are of great importance to the success of a product when it is put on the market. Design for manufacturability is key: The objective is to design for lower cost.

The cost is driven by time, so the design must minimize the time required to manufacture a product. Production processes are required which deliver ready-to-install components or workpieces that require only small secondary operations; accordingly, polymer injection molding and MIM technology are extremely popular technologies in today's production. Apart from molding, the use of sheet metal stampings and surface finishing such as nitrocarburizing is widespread.

3   Pistol Principles

The last 40 years has seen much innovation in the world of consumer electronics prod-ucts – from the VCR and SONY *Walkman* in the 1970s to the CD in the 1980s to the great advances in the past decade with mobile phones and HDTV. At the same time there also have been advances in the field of weapons technology. New trends bring about a certain evolution, where some innovations disappear from the market and oth-ers prevail in the long run. For example, at the end of the 1980s, the revolutionary concept of the HECKLER & KOCH *G11* with caseless ammunition failed in Germany be-cause of political reasons,[121] and caseless ammunition in general never caught on. Obturation of the casing to seal off the breech, wear of the firing pin nose, and heat dissipation in the chamber area had been challenges for the engineers,[122] but tradi-tional metal cased cartridges have been the solution to these problems for ages.

Introduced in the 1980s, but still going strong are pistols with polymer frames. This innovation is closely linked with the name of Gaston Glock[123]. More recently there has been a trend with new calibers of ammunition, including 10 mm Auto and .40 S&W, and currently more deep conceal carry guns have been introduced in the American market.

With a combination of historical perspective and an examination of the current state of the art, we will be looking at the big picture in handgun technology. This is an ad-vantage for the survey in Chapter 5 of this book, when we inspect modern pistols with regard to commercially useful features. We will examine the characteristics of barrels and slides with respect to weapon technology and how these are made, as well as the specifics of polymer frames. Technical features such as magazine releases, firing mechanisms and individual components such as springs and safeties as well as elec-tronics on guns and a schematic view of the shot cycle completes this section.

One additional comment bears note as we examine the auto-loading mechanisms in the chapter ahead. Some say it is an academic question; whether to include the blow-back system in the general class of recoil-operated systems or in the gas-operated systems. Others suggest it should be considered as a class on its own.[124] However, as we shall see in the next section, this book strongly disagrees and classifies it as recoil-actuated. We are aware that Richard Wille started this discussion in 1896[125] and *Kaisertreu* took it to the next level in 1900.[126] Since then, much has been written about operating systems and thus the wide range of breech locking systems connected with it requires a focus on established solutions.

3.1   Methods of Operation

Firearms are heat engines that use the energy provided by a burning propellant charge. Compared to other internal combustion engines, the energy transfer in semi-automatic weapons is not continuous, but rather temporary within fractions of a second and is associated with high temperatures, enormous forces and accelerations. Fifty percent of the total energy of the propellant charge is transformed into heat and internal energy of the gases, while 10 percent is wasted as heat transfer to the gun, leaving about 40 percent in the form of the projectile's kinetic energy and for the self-loading function.[127]

This auto-loading function replaces the empty cartridge casing with a new round and resets the fire-control mechanism. Each time the trigger is pulled, another shot is fired until the last cartridge is spent. In describing the self-loading function, terms such as gas pressure, momentum, and recoil are often used and that is why we need to examine these terms before moving ahead in this chapter.

Force, Inertia, Momentum, Impulse, Recoil, Kinetic Energy and Pressure
Every object either remains at rest or in motion at a constant velocity unless acted upon by a *force* (Newton's first law). The term *inertia* describes this natural tendency of objects to resist changes in their state of motion, and is related to their mass.[128] The greater the mass of an object, indicates the more inertia it possesses and as a result relatively more force is required to get it into motion or to change its state of motion.[129]

Once in motion, the product of an object's mass and velocity is termed *momentum*. A change of momentum will only occur if a force is applied to an object, and that force is directly proportional to its rate of change of momentum.[130] Assuming the object's mass remains constant, this gives us the relation $F = m\,\Delta v/\Delta t$ for the force. The force F is proportional to the mass m of the object and its change in velocity Δv during a period of time Δt. Change in velocity over time is acceleration a, so we can re-write the formula as $F = m\,a$ (Newton's second law). All of these imply that if a body is accelerating, then there must be a force applied to it.

The application of a force for a period of time (instantaneous) – like a hammer blow, – is called *impulsive force* (see Appendix 1) and as we can see from the above formula, it is equal to the change in an object's momentum: $F\,\Delta t = m\,\Delta v$.[131] This formula serves as a basis for the calculation of the velocity of certain weapon parts, which get accelerated when a cartridge is fired. In our case, where a projectile of known mass is accelerated from an initial velocity of zero until it leaves the barrel at muzzle velocity.

All forces between two objects exist in equal magnitude and in opposite direction, i.e. forces appear in pairs (Newton's third law: action = reaction). Applying the phenomenon to pistols, a force acting upon a projectile, causes an equal and opposite force on the gun which we call *recoil*, i.e. ejecta momentum = recoiling momentum.[132] The recoiling momentum can be employed to accelerate the gun's breech-block; hence the term recoil-operation.[133] In our case the impulsive force causing the projectile to move down the barrel, also appears on the breech face instantly with the same amount, and in the opposite direction.

Fig. 3.1: Circular breech face (arrow) of a turning bolt head of a *Desert Eagle* pistol.

In the context of the continued movement of bodies after a force had acted upon them, we must look at what is referred to as *kinetic energy*. Kinetic energy is the energy possessed by an object due to its motion. Having gained this energy during its acceleration, the object maintains this kinetic energy unless its velocity or weight changes. The same amount of work is done by the object in decelerating from its current speed back to its state of rest.[134]

Finally, we need to look at the concept of *pressure* to describe the action of the propellant gases, which consist of molecules in constant random motion. Gases will expand to fill their surroundings; and their condition is defined in the field of thermodynamics by three state variables: temperature, volume and pressure. The magnitude of the state variable pressure p is a product of particle density and kinetic energy.[135]

In other words: pressure describes the energy density and its magnitude is a measure of energy per volume. Pressure is responsible for the physical effect of gases on a body. This effect is caused by the transfer of the particles' momentum onto their surroundings. As stated earlier, change in momentum involves the application of a force F, which in this case can be calculated as the product of the pressure p and the area A over which it is applied ($F = pA$). The intensity of this impulse force transmission increases with the kinetic energy of the gas particles, i.e. the higher the pressure and the hotter the gas, the more force it exerts.[136]

Self-Loading Principles

Small arms in general, can either be powered internally with the source of energy provided by the propellant, or make use of an external power. For example, an electric motor or hand crank. For semi-automatic pistols there is only one source of energy that imparts velocity to the projectile and at the same time actuates the self-loading mechanism; pistols employ the energy derived from burning a propellant charge contained in the cartridge. During this exothermic reaction, the propellant charge is converted to gaseous products, i.e. propellant gases, and the kinetic energy of the gas particles is transferred onto its surroundings. This primary source of energy can be used to actuate the cycle of operation by either one of the following two options:[137]

- employing the propellant gases' particles momentum inside of the barrel; or
- some energy from the propellant gases is used externally.

In the first case, the kinetic energy of the propellant gases is transferred to the projectile, and in the opposite direction (action = reaction) via the head of the cartridge case onto the breech face.[138] – Remember, if a body is accelerating, then there is a force on it and its acceleration is directly proportional to the *net force*. Also remember that forces always come in pairs, equal and in opposite directions of action-reaction, and employing this phenomenon to actuate a pistol's cycle of operation is referred to as recoil-operation. – In the second case, propellant gases are tapped off the barrel and applied externally to power the auto-loading mechanism. This is described as gas-operation.[139]

Classification

When discussing the actuation systems of pistols, we will make use of the following terms:[140]

- *Recoil-Actuated* – this method of actuation employs the recoiling momentum acting against the breech face to cause a backward movement of the breech face and the components attached to it. Recoil-operation depends on the inertia of moveably mounted components.
- *Gas-Actuated* – energy from the propellant gases is harnessed from a fixed barrel to power a self-loading mechanism outside of the barrel.[141]

In this publication, breech locking systems will be restricted to the following:[142]

- Force locked slide,[143]
- Form locked breech: positive locking with coupled members, and
- Form locked breech, positive locking with stationary components.[144]

Diagram Fig. 3.2 illustrates schematically the essential elements.[145]

Pistol Operating Principles

	Recoil-Actuated (employs recoiling momentum acting on breech face)		Gas-Actuated (gas is tapped from the barrel)
	moveable barrel	fixed barrel	fixed barrel
	form locked breech, coupled members	force locked slide	form locked breech
	Recoil-Operated	**Blowback**	**Gas-Operated**
System	Popular with pistols chambered in 9 mm Luger or larger calibers	Mainly used in pistols up to .380 cal.; rarely for 9 mm Luger	Not widely used in pistols
Operating Energy	Exothermic reaction; resultant forces accelerate the projectile and the coupled members at the same time	Exothermic reaction; resultant forces accelerate the projectile, cartridge case and slide at the same time	Exothermic reaction; burning propellant charge produces gases. These high-pressure gases are tapped off and used as a source of energy
Operation	Inertia: Coupled barrel and slide recoil together over a certain distance until they are mechanically uncoupled and only the slide continues to reciprocate	Inertia: As the casing gets accelerated its head is backed up by the inertia of a moveably mounted body	Gas-actuation: Bolt is locked to a non-moveable part of the gun. Unlocking is initiated through a gas-actuated mechanism

Schematic diagram of common systems (springs not shown):

Recoil-Operated   Blowback   Gas-Operated

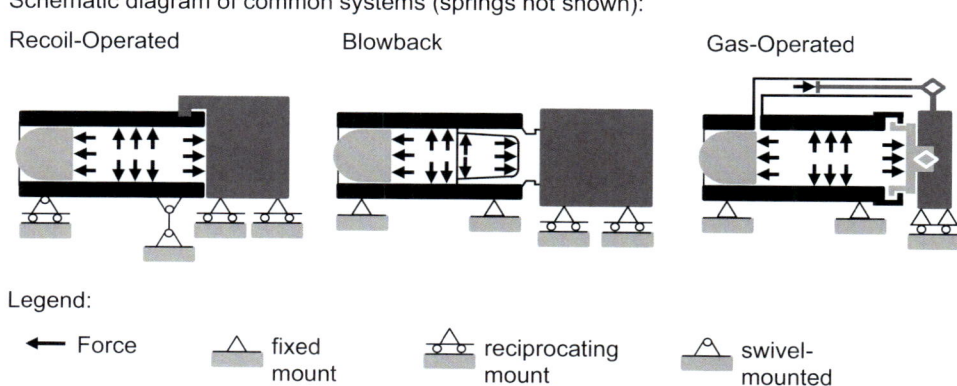

Legend:

← Force   △ fixed mount   ⏚ reciprocating mount   △ swivel-mounted

Fig. 3.2: Operating principles of auto-loading pistols and their characteristics.

Semi-automatic operation

Modern gun designs must ensure the safety of the
user as well as offering reliability and durability. The
stresses resulting from the rearward thrust of recoiling parts and high-pressure propellant gases
caused by an explosive combustion have a strong
impact on the durability of pistols.

Fig. 3.3: Components of a recoil-
operated pistol: slide, barrel,
and recoil spring assembly.

One of the challenges encountered when designing semi-auto pistols is to keep the stresses within
manageable limits. The gun recoils violently and the
parts which have to take the load need to be designed in a proper way to reliably carry
the stresses applied to them. Choosing the correct weight of recoiling masses allows
the engineer to optimize the velocities they will achieve while they recoil (see Appendix 19, Slide Motion Analysis).[146]

Semi-automatic operation, recoil-actuated pistols

Any gun will be subjected to some recoil action when a round is fired, but unless the
recoil is put to use in actuating the cycle of operation, the gun does not employ recoil-
actuation.[147] Most of today's duty pistols are recoil-operated semi-automatics chambered for all kinds of ammunition types.[148] In the preferred method of operation, barrel
and slide are coupled and the members are secured together against uncontrolled
separation during the first millimeters of their rearward motion at the moment of firing.

When the projectile is accelerated down the bore, a backward movement of the coupled barrel and slide occurs. The coupling remains
safeguarded until the projectile has cleared the muzzle. Once the projectile has cleared the muzzle, the
barrel is being uncoupled from the slide. Shortly after the uncoupling took place, the motion of the barrel is terminated abruptly while the slide continues
rearward under its momentum. Recoil-operated pistols with a barrel that moves only over a short distance, that is, shorter than the length of the casing,
are referred to as short-recoil designs.

Fig. 3.4: The exception proves the
rule: GLOCK *G42* as an example of a pistol chambered in
.380 ACP with coupled barrel
and slide.

The coupling of recoiling components provides a
technically interesting benefit with regard to the size

of the coupled members: by coupling the components, the masses are also combined, i.e. the total inertia of the coupled members results in a moderate velocity even though each recoiling member has a moderate mass. This principle of recoil-operated pistols with coupled barrel and slide is the dominant type on the market when it comes to pistols using higher powered cartridges which provide more recoil, e.g. 9 mm Luger.

For lower powered cartridge types, i.e. .380 ACP or smaller calibers, a method is used where the breech is force-locked at the moment of firing. One of the characteristics of these "blowback" pistols is their fixed barrel and their simplicity in general, due to the lack of a locking mechanism. In most cases, the barrel is fixed to the frame and the slide rests against the rear of the barrel. The instant a shot is fired the recoil force begins to accelerate the casing, which is only resisted by the slide's inertia. This causes the case and slide to simultaneously move backward. Straight blowback, simple blowback or plain blowback are terms being also used for this kind of operation.[149]

Fig. 3.5: An example of a blow-back-operated pistol chambered in 9 mm Luger: HK *VP70*.

Blowback-operation of pistols chambered for low-powered ammunition is mainly based on the inertia of the slide's mass and to a negligible extend on the opposing force exerted by the recoil spring and its mass. It does not require any mechanically-challenging design to securely lock or unlock the breech, which is usually associated with locking mechanisms that involve a control of barrel and slide engagement. The impulse force acting in one direction and causing the projectile to clear the bore also initiates a recoiling momentum at the same time and with the same amount in the opposite direction onto the cartridge case. The instant backward movement of the cartridge case relative to the chamber of the barrel is the distinguishing characteristic of blowback-operation.

Blowback-operated pistols are mainly chambered in .380 ACP or smaller calibers, but in some cases higher-powered cartridge types might be also used. These tend to use a delaying mechanism to retard the blowback-operation, e.g. roller delay in the HK *P9*, or gas delay in the HK *P7*; both chambered in 9 mm Luger. Only in rare cases – for instance, STRASSELL'S MACHINE, INC. HI-POINT *C9* – straight blowback is used for high-powered cartridges like the 9 mm Luger.[150]

Semi-automatic operation, gas-actuated pistols

Gas-actuated pistols are not very popular in the market. Other designs are simpler and tapping off hot propellant gases may cause considerable heating of some of the components in gas-actuated pistols.[151] A modern example of a gas-operated pistol is the Desert Eagle, introduced by MAGNUM RESEARCH, INC. during the 1980s.[152]

Besides harvesting the energy from propellant gases to power the operation of components outside of the fixed barrel (Fig. 3.6), gas-operated pistols are characterized by one more distinct feature: a positive locked breech, i.e. a breechblock which is form-joined with either the barrel or the receiver. This locked breech gets unlocked and a breechblock or bolt moves relative to the gun once activated by said components from outside of the barrel.

The breech of gas-actuated guns is in most cases a group of parts consisting of at least a moveable bolt carrier, which carries and controls the bolt. The bolt has a breech face and locking lugs, which prevent axial movement and transmit the force exerted on the breech face to the component in which the locking lugs are engaged. The bolt remains locked until the bolt carrier – itself driven by the tapped off gas – introduces the unlocking movement to it, which disengages the positive locking between the bolt lugs and the locking surfaces on their counterpart.[153]

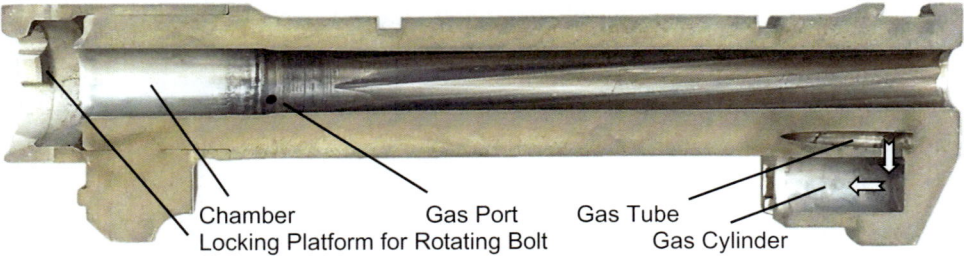

Chamber   Gas Port   Gas Tube
Locking Platform for Rotating Bolt   Gas Cylinder

Fig. 3.6: Gas-actuation as found in a *Desert Eagle* pistol: As opposed to other gas-operated firearms, this pistol uses a gas port near the chamber. A portion of high-pressure propellant gases is routed through the gas port into a gas tube, which is linked back to the action via a gas cylinder. Inside this gas cylinder the gases impinge on a piston head (not shown) to provide motion for the unlocking of the action.

Residual Bore Pressure

In the design of any firearm, great care is exercised to achieve maximum muzzle velocity of the projectile while keeping the peak propellant pressure in the bore as low as possible. Under ideal conditions, the pressure remains at a moderate and constant level while the projectile is accelerated in the bore.[154] As the projectile moves down the bore, a continuous force is generated as long as the progressive burning propellant produces gas faster than the volume behind the projectile increases. This requires a careful tuning of many parameters.[155] As the projectile is accelerated down the bore its velocity might get so high that the volume behind it grows faster than the propellant gases can fill the void, and so the acceleration drops. Ideally, the projectile leaves the muzzle right at the instant of attaining its maximum velocity.[156] Two extreme cases may be seen: muzzle flash and barrel obstruction.[157] The former is a phenomenon where unburnt particles of propellant burn off after exiting the muzzle, whereas the latter occurs when a projectile gets stuck in the barrel[a] as a result of the gas pressure being insufficient to drive the bullet down the bore.

While the projectile is accelerated down the bore, tremendous gas pressure is built up. This changes as the projectile leaves the muzzle. The gas pressure quickly drops to the ambient level as the propellant gas leaves the muzzle. The pressure that exists from the time the projectile leaves the muzzle until it reaches the ambient pressure level is called *residual pressure*.[158]

In this context, the bore pressure that exists at the instant when the projectile leaves the barrel, which is referred to as *muzzle pressure*, is of interest. The magnitude of the muzzle pressure is a result of the ammunition type, barrel length and obturation of the projectile and cartridge case. Cartridge case obturation is a phenomenon caused by the propellant gas pressure inside the bore, which acts uniformly in all directions including the inside of the cartridge case. The radial components of the pressure act on the wall of the case to expand it against the chamber, thus creating a seal which prevents the escape of the propellant gases to the rear.[159]

As the pressure of the propellant gas decreases, the case contracts enough to free itself from its grip on the chamber wall. This contraction will reduce friction while the residual pressure continues to drive the case to the rear and out of the chamber. Ideally, the forces on the case at this point in time are below the material limit of the brass case in order to avoid any danger of it failing by separation or rupture while exiting the chamber.[160]

The major problem encountered specifically in blowback-operation is to control the cartridge case motion until the pressure has dropped to a safe level with regard to the

---

[a] A pistol might complete the cycle of operation even though the projectile has not left the bore. Firing the next round can cause a bulging of the barrel or other type of catastrophic failure which may cause injury or even death to the shooter.

elastic limit of the case material. The same conditions apply to recoil-actuated pistols with coupled slide and barrel as well as gas-actuated pistols in the instant where the case exits the chamber. From this moment on, the conditions are the same as encountered in blowback-operation. When the projectile of a 9 mm Luger cartridge leaves the muzzle, the magnitude of the muzzle pressure is above 20 MPa (2900 psi)[161], more likely in the range from 30 to 40 MPa (4350 to 5800 psi, cf. Fig. A23.9, p. 488), hence the residual bore pressure while the case is exiting the chamber may be equally high or just below this. The residual pressure is likely to push the case out of the chamber, and depending on the specific design of the individual gun, the case may even exert a rearward thrust on the breech face at the same time.

3.2   Barrel

Put in simple terms, a pistol barrel (Fig. 3.7) is a steel cylinder, through which a projectile is given linear momentum. The rear end has a chamber (Fig. 3.11, p. 63) wherein the cartridge is placed. The cartridge is loaded from the magazine via the feed ramp and through the chamber opening (chamber mouth) into the chamber itself.

Fig. 3.7: Pistol barrel with cam and feed ramp (WALTHER *P99*).

The chamber is concentric to the bore, and its diameter must be within strict tolerances. In the design of any chamber and mouth, great care must be exercised to ensure that the thin and soft material of the cartridge case is adequately supported as the propellant pressure acts upon the wall of the case to expand it against the chamber. Ideally, the forces on the case should be below the ultimate strength limit of the brass to avoid any danger of the case failing by rupture.

All current ammunition uses cartridge cases that expand easily in radial direction to give efficient obturation at the

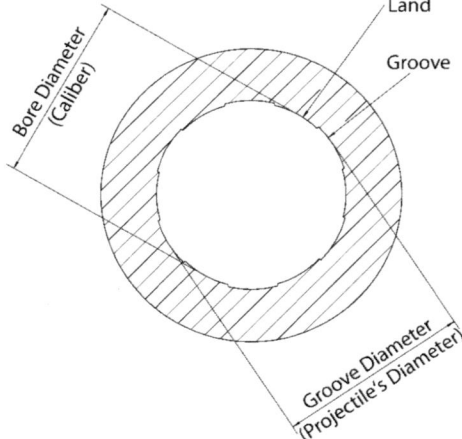

Fig. 3.8: Barrel profile: Cross-section of a barrel with lands and grooves rifling profile.

breech end of the barrel. That is, the cartridge case seals the rear end of the barrel and prevents the propellant gases from leaking when the weapon is fired.[162] Defects in the chamber, such as helical tool marks or corrosion may cause failures to extract because of increased friction between the cartridge case and the chamber wall.[163]

The purpose of the barrel is to impart velocity and direction to the projectile. Helical grooves inside the bore (rifling) are used to give the projectile a spin around an axis coinciding with its initial line of flight to keep it stable in flight.[164] The type of rifling of the bore is called the rifling profile or pattern. The main dimension of a barrel is its caliber, which is the dimension of the bore before the rifling profile is machined into it (Fig. 3.8).

In most barrels, the cross-section of the rifling appears as a number of lands and grooves. The angle of twist of the rifling is often described by the distance for one complete revolution in millimeters or inches, for instance 254 mm, i.e. 1:10 in. Pistol barrels normally come with right hand twist and six grooves. The raised portions between the grooves are called lands and are referred to as bore diameter or caliber.

Fig. 3.9: WALTHER *P5* barrel shown from raw material to white part (from bottom to top).

To achieve forward obturation, the projectile is made to the same diameter as the grooves of the rifling, which means that the projectile's diameter is larger than the caliber. In most cases, caliber designations of guns and cartridges are for identification purposes only[165] and specific standard dimensions as shown in SAAMI (Sporting Arms and Ammunition Manufacturers' Institute) or CIP (Commission Internationale Permanente pour l'Epreuve des Armes à Feu portatives) apply.

In order to guarantee that a barrel will accept any projectile it is specified for, the barrel dimensions shown in the standards are minimum dimensions, as opposed to cartridge and projectile dimensions, which are maximum dimensions. CIP barrel dimensions for the 9 mm Luger cartridge are for instance, 8.82 mm (lands) and 9.02 mm (grooves) respectively. If a barrel was manufactured exactly to this dimension, the depth of the rifling grooves would be 0.1 mm or 0.004 in.

The diameter of the area immediately ahead of the chamber, i.e. forward of the case mouth, equals groove diameter or even slightly larger, and extends forward in the form of a tapering cone to the commencement of the rifling. The distance with straight walls and no taper, from the chamber's front end up to the start of the rifling is called free-bore.[166]

The tapered section is known as the leade, throat, chamber throat or forcing cone. As the projectile is pushed into the leade, the rifling lands deform the projectile, a process called engraving. The throat is designed to provide smooth engraving of the projectile and has two important measurements; length and angle. The gentle transitioning of the projectile is critical to keep the rifling engraving forces and the propellant gas pressure (engraving pressure) low at this point. Generally, the projectile is first blown out of the case (starting pressure or short shot) by overcoming the holding force of the case neck, and thereafter hits the leade during the firing process.[167]

If the throat depth is too short or too steep, the projectile might get jammed up into it during chambering. This might make it difficult to remove a chambered cartridge from the pistol during manual unloading, and can cause excessive starting pressure because the propellant gas pressure will have to increase greatly in order to blow the projectile out of the case and force it into the rifling all at once.[168]

To avoid bullet jamming, factory freebore is typically made long. Once the cartridge is fired, the projectile travels freely in the bore until it contacts the leade. This is called bullet jump[169] and the distance is called bullet jump clearance. In some parts of the firearms world, the distance from the beginning of the bullet's motion to the point of throat contact is called non-spin bullet travel.

Free travel, where a projectile leaves the case completely before encountering the leade, would be an occasion where the projectile is neither guided by the case mouth nor forcing cone. It is also part of the projectile's non-spin travel.[170] However, these small technical details do not matter when it comes to the over-all barrel length. Barrel length is the dis-

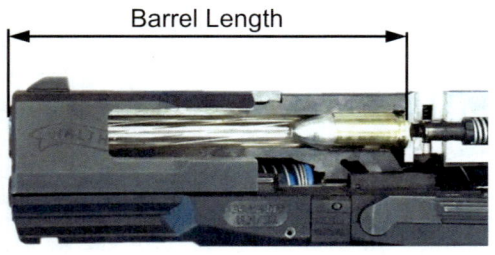

Fig. 3.10: Barrel Length dimension.

tance between the slide's breech face and barrel's muzzle end (Fig. 3.10).

Some barrels designed for straight wall cases have a step in the chamber, about 4 mm (^{5}/32 in) from the seating point (Fig. 3.11). Initially, this stepped chamber was the subject of Georg Luger's patent DRP 237,192, dated February 16, 1910. The use of a step in the chamber was intended to improve obturation of straight or slightly conical cartridge cases, in order to prevent the rearward escape of propellant gases. Since obturation of high-powered cylindrical straight cases has never been an issue in locked breech pistols for design engineers, today only a few models use the stepped chamber to reduce fouling inside the pistol.[171]

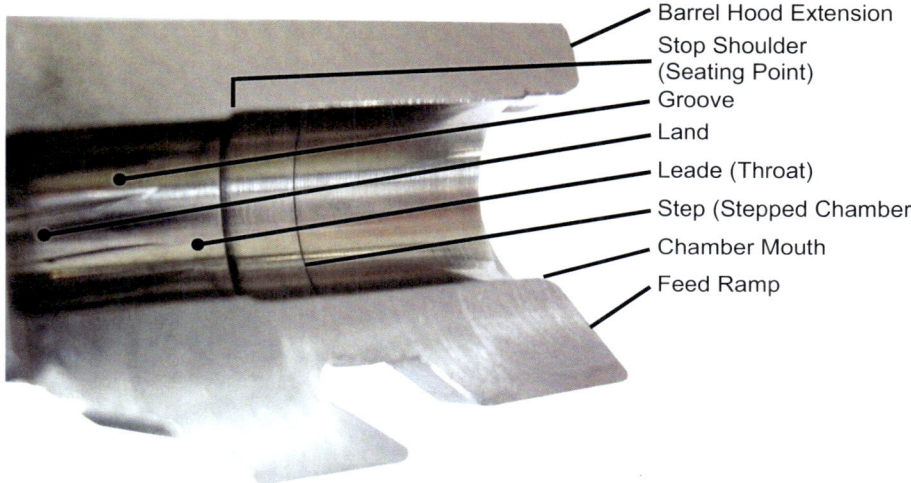

Barrel Hood Extension
Stop Shoulder (Seating Point)
Groove
Land
Leade (Throat)
Step (Stepped Chamber)
Chamber Mouth
Feed Ramp

Fig. 3.11: Chamber for a straight case (9 mm Luger).

Fig. 3.12: Stepped chamber ring-marks of HECKLER & KOCH *P30*, *Pistole 08*, and WALTHER *PPQ* (from left).

Nickel, chromium, and tungsten are used in alloys to improve the quality of barrel steel. Nickel increases the strength and tensile ductility of the steel, chrome increases its hardness, while tungsten increases both strength and hardness. Nickel also makes the steel more resistant to corrosion, as corrosion might be caused by propellant residues.[172]

CIP states that it is the manufacturer's responsibility to choose an appropriate barrel material[173] and lists ISO 683 and ÖNORM M-170 – 1981 for reference, where the latter refers to handgun barrels. The content of ISO 683 is more general and does not specifically apply to gun barrel steel. For appropriate tensile ductility, CIP recommends 0.025 to 0.036 percent for the maximum content of phosphor and sulfur. According to CIP, chromium-molybdenum-vanadium steel alloys are popular because of their yield strength and toughness up to temperatures of 550 °C (1000 °F).[174]

Fig. 3.13: Cracked barrel cam; Pistol barrels rarely get worn out.

The most common forms of rifling are created by either cut rifling, button rifling, hammer forging, or electrochemical rifling. The original way was to cut the rifling with more or less rectangular lands and grooves with a special tool. Starting in the first half of the 20th century button rifling, or that done by hammer forging, became popular while electrochemical rifling then came into the picture, but neither it nor cut rifling are widely used today.

Button Rifling

Single point cut rifling and broach rifling are classic manufacturing techniques dating back centuries, where steel is actually removed from the inside of the bore using a cutting tool. The button rifling method does not cut the rifling – it is formed by cold working the bore – and was introduced much later. One of the first patents was filed in 1920.[175]

Fig. 3.14: 9 mm button

The carbide tool used (Fig. 3.14) is essentially a short button with the mirror image of the barrel, i.e. the grooves to be cut are carved in relief on the button surface. The button can be pushed into or pulled through the barrel. The method used depends upon the length of the rod to which the button is attached and the rod's tendency for buckling. For pistol barrels, the button is more likely to be pushed through the bore.

After deep bore drilling, the bore must be honed to enable the button to be pushed through it with high precision. Honing is an abrasive machining process, which eliminates inaccuracies inside the bore and produces a precision surface with exact diameter and form.

In the next step, the rifling button is pushed through the bore by a hydraulic ram applying a load in the range of 25 kN (2.8 tons). As it moves, the hydraulic head also turns at the desired rifling pitch. This requires plenty of lubrication of both button and bore to avoid jamming. In addition, the passing of the button through the bore must be done while the blank is still a uniform cylinder. If the outside of the barrel was already contoured, the rifling would be uneven because the rifling button displaces metal and puts a lot of stress on the steel.

Fig. 3.15: Displaced material at muzzle end after button rifling.

The button engraves a reverse of itself inside the bore, hardening and smoothing the inside of the barrel as it progresses. The deflection of the steel (Fig. 3.15) is done from the inside of the bore to the outside, which is the opposite of hammer forging.

Hammer Forging

The hammer forging method for barrel rifling profiles was developed during World War II when MAUSER developed a low budget *Volkspistole*. However, the production of the barrel proved to be a bottleneck. One of the methods tested in this context was hammer forging of the rifling profile, which was carried out by inserting a mandrel into the bore and cold hammer forging the barrel blank around it from the outside.[176]

A more general term for this manufacturing technique is "rotary swaging"[177], which was developed by GUSTAV APPEL MASCHINENFABRIK, Berlin-Spandau, Germany. APPEL used this method in the production of cleaning rods made of solid material, and it was able to modify this technique to make use of tubes instead of solid rods and to swage an internal pattern into these tubes.[178] It was Mr. Gustav Appel who designed a machine to do the rotary swaging of internal shapes.

Until the end of World War II, Appel's company manufactured and sold these machines. One of the military supervisors who had been responsible for GUSTAV APPEL

MASCHINENFABRIK, Mr. Bruno Kralowetz[179] from Austria, took the idea back home with him and established his own company GESELLSCHAFT FÜR FERTIGUNGSTECHNIK UND MASCHINENBAU (GFM) in Steyr, Austria.[180]

Hammer forging of barrel blanks not only forms rifling inside the barrel, but it also takes care of contouring the outside of the barrel and in some cases, the chamber is created in the same operation. It uses a mandrel for the swaging of the rifling. The mandrel is shorter than the barrel. The rifling is swaged by moving the mandrel down the bore and hammer forging the appropriate spot with radial forging hammers at the same time (Fig. 3.16).[181]

Fig. 3.16: Characteristic surface texture of hammer forged barrel blanks.

An advantage of hammering is the process-related material compression. The compression is caused by high radial forces, which are applied on the outside of the blank and must be high enough to swage the internal shape by pressing the barrel steel hard enough onto the mandrel so that the material copies the shape of the mandrel during the forming process.[182] As the barrel is being hammered, a 15 to 20 percent reduction in diameter results while its length increases. The hammering process also work hardens the barrel,[183] including the rifling, achieving 10 to 25 percent increase in material strength.[184]

Electrochemical Rifling

Controlled etching of the bore to produce rifling was mentioned for the first time in the patent application US 2,848,401 by James C. Hartley filed May 7, 1953. Hartley mentions that one of the objectives of his invention is the efficient mass production of accurate electrolytically formed rifling, regardless of whether the bores are tapered or non-tapered and regardless of the rifling pitch and the type of twist (constant, gain, or loss) required of the rifling. The grooves would also be uniform in depth, the rifling can be done even with hardened steel, and an electrolytical polishing can be included in the process. Furthermore, Hartley sees the benefits of the method in maximum efficiency and its suitability for mass production at low cost.

More patents by other inventors followed, e.g. Beck/Birnby, US 3,429,798 (1967), Haggerty, US 3,769,194 (1969), Vishnitzky, US 4,690,737 (1986), among which the latter uses the term "Electrochemical Rifling of Gun Barrels" for the first time for this technology in his patent. Vishnitzky discloses a method, where the barrel is the anode, and the metal rifling rod mounted in the barrel is the cathode. The working gap between the barrel and rod is filled with an electrolyte fluid flowing under high velocity and pressure.

Electrochemical Rifling (ECR) is not widely used by manufacturers. It is used, for example, by SMITH & WESSON.

Concepts to Extend Barrel Life

Barrel life is the expected number of rounds through a barrel beyond which the performance of the weapon falls below the required level, for instance, reduced accuracy, keyholing, or variations and reductions in muzzle velocity. Barrel wear is caused by two effects: (1) the abrasive effect of the projectiles, and (2) barrel erosion caused by direct impingement of rapidly flowing propellant gases, solid particles included in these gases traveling over the bore surface, and chemical processes that occur at the bore surface under the extreme heat of the propellant gases.[185]

Hard chrome plating[186] of the bore and polygonally rifled[a] barrels are commonly used on assault rifle or machine gun barrels to improve longevity, but are less common for pistol barrels. Both techniques are employed to improve the durability of barrels used in fully automatic guns. Such guns will shoot occasionally large amounts of ammunition in one sitting, causing their barrels to heat up which may induce wear more quickly.

During the Second World War MAUSER did extensive research on what caused barrel wear when it needed to improve the service life of barrels on large caliber machine guns. Research showed that wear is not distributed uniformly along the barrel. The most critical region where wear and erosion are most visible is the throat, and from the beginning of the rifling[187] up to the near point of pressure peak, mainly in the first one-third part of the barrel, where the acceleration of the projectile and the heating[b] of the bore are at their maximum.[188] It was found that in barrels with constant twist rate, the rotation forces on the lands are proportional to the gas pressure.[189]

---

[a] In mathematics, the term polygon describes a plain figure that is bounded by a number of straight line segments closing in a loop to form a closed chain, and the bore of a polygonal barrel has a polygonal profile with tangentially transferring areas. NB: The term polygon describes neither the number of existing areas nor any shape.

[b] The combustion temperature of nitro powder is between 2500 °C to 3000 °C [4530 to 5400 °F; PD]. (Kneubuehl, 2013 p. 52) Hence, barrel and adjacent components will heat up considerably during shooting and it may be dangerous to touch the hot parts. The shooter might get burned or a chambered round might cook off in a hot barrel.

During the development of the *MK213C*, MAUSER found an average increase in temperature by 1 °C [34 °F; PD] per shot. (Schmid, 2003 p. 11f). Pistol barrels are likely to heat up at a similar rate.

As the temperature gradient of the barrel's inside surface is also proportional to the gas pressure, the main rotation forces created by the projectile on the lands are found in the rear third of the barrel. While the bullet is receiving a spin around its axis it creates high stresses on the driving wall of the rifling. These forces generate friction on the side of the walls of the lands but decrease towards the other side. The friction forces wear down the rifling.

MAUSER concluded that the remedy was either to conduct away the high temperature created in the first third of the barrel as quickly as possible for barrels with constant rifling pitch, or to decrease forces on the driving wall of the lands in the region of the highest temperature by changing the twist rate of the rifling. The second solution was found to be the easier one and led to the introduction of gain twist.[190] The idea was not new as some expensive barrels in the 1800s had a gain twist.[191]

Barrels with gain twist were designed to induce the spin to the projectile after the maximum propellant pressure was reached. Here, the rifling starts to spin the projectile gently (initial twist rate). The rifling pitch increases as the projectile runs down the bore reaching a constant twist rate shortly before the muzzle (final twist rate). This final twist will result in increased accuracy.[192]

The introduction of a progressive twist decreased forces on the driving wall of the lands on the most critical region of the bore, but it made polishing them with a lead lap, emery and oil no longer possible. The lead lap would have to be very short and thus the effect of polishing was very small as well.[193] Experiments were made to finish the bore by sand blasting. Sand blasting reduced the time needed for polishing to one third, and showed good results. Lead and oil were saved, it deburred the corners of the lands and the surface finish was found to be ideal for chromium plating.[194] With this combination of techniques, barrel life increased from 2,000 rounds to 30,000 rounds.[195]

The application of polygon profiles initially appeared in general mechanical engineering applications. In the 1930s, studies were carried out to find ways to reduce stress spikes on corners which had been typical on spline shaft-hub connections. These studies led to the introduction and continuous improvement of polygon profiles for this kind of connections. However, the application of polygon profiles for rifling had been around in the gun world long before. Various polygonal patterns (Fig. 3.17) were occasionally used from the middle of the 17th century, but were not widespread because they brought no real advantage to these early firearms. This situation changed after the appearance of machine guns. In the 1960s, polygonal rifling in barrels on *MG3* machine guns was introduced by HECKLER & KOCH. Since then polygonal rifling has found its way into some hunting rifles, and even pistols. Hammer forging is often used to economically produce barrels with polygonal rifling. The number and shape of the surfaces of the polygonal rifling profile can vary depending on the manufacturer. Common polygonal rifling profiles have six sides. A three-sided polygon was used in the STEYR *GB80* barrel; and, depending on the caliber, GLOCK uses six or eight lands which look like circular arches rising between curved grooves akin to the sides of a polygon.[196]

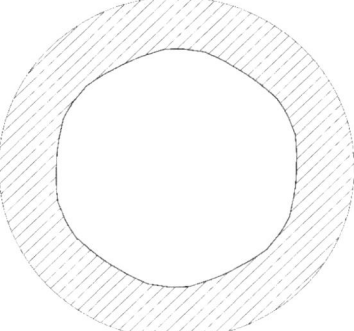

Fig. 3.17: Polygonal rifling profile.

In addition to reduced stresses on the driving wall of the rifling, polygonal rifling is said to provide good forward obturation of the projectile preventing propellant gas blow-by. The gas seal around the projectile may reduce erosion caused by the combination of direct impingement of gas and solid particulate flow on the bore surface. Also, good obturation between projectile and polygonal rifling is thought to deliver more efficient use of the combustion gases trapped behind the projectile, in theory resulting in higher muzzle velocity of the projectile compared to barrels with conventional rifling. In reality, the type of ammunition used, the length of the barrel and the specific bore dimensions will have a considerable effect on the bullet's muzzle velocity.

An important difference between polygonal and conventional rifling profiles lies in the definition of the CIP bore dimensions. Since the usual lands and grooves caliber gauges cannot be used in bores with polygonal rifling and no standard for a polygonal rifling pattern exists, the bore dimension is checked during CIP proofing in a different way. For barrels with polygonal rifling CIP mandates the following:

> *Weapons with a rifling profile which departs from conventional grooves and lands (polygonal barrel) may be accepted if the cross sectional area of the barrel is not more than 0.7% less than the value Q given in CIP Tables.*
>
> *The use of cartridges loaded with solid projectiles having a core hardness figure greater than that of lead [HV1 ≤ 40; PD] is not allowed in weapons with barrels having polygonal rifling. [...]*
>
> *This ban need not be applied if an applicant provides demonstrable data to the proof house that the pressure of the cartridges in the weapons with polygonal rifling having a*

cross sectional area of the barrel bore 0.7% less than the value Q given in the CIP Tables, remains with the limits given in the Tables.[197]

Table IV of CIP shows 62.61 mm² for the inner cross-section Q of a conventional barrel chambered in 9 mm Luger. Hence the inner cross-section of the same barrel with polygonal rifling must be 62.17 mm² (62.61 mm² · 0.993) or more. SAAMI specifies for a *Bore & Groove Area* ≥ 62.387 mm² on barrels chambered in 9 mm Luger.

Trends in Barrel Manufacturing

Manufacturing costs of a barrel depend on the amount of raw materials needed, machine time and how demanding the manufacturing of the barrel is in relation to its design. The cost of the barrel ultimately affects the cost of the entire product. It is an advantage if the production of a barrel starts from a small blank and only a short period of time is required for machining.

To produce a barrel faster with less effort and in good quality, its design must be simple and allow dimensional variations in the manufacturing processes involved in order to meet the final assembly quality and cost targets. A not overengineered design of a gun does traditionally stand for good reliability. With increasing complexity, the risk of manufacturing errors is going up and will lead to expensively produced waste. Hammer forging, for instance, where the hammering process takes care of the rifling and also the chamber at the same time can be done at a high speed and in turn reduces waste.

Another way to simplify the manufacturing of barrels is the use of a barrel tube, i.e. a rifled "tube" which is inserted into an already prefabricated barrel block. This technique was used on WALTHER *P38* pistols, some revolvers, e.g. DAN WESSON, and later on WALTHER *P22* pistols.

Fig. 3.18: Production stages of a modified Browning type barrel (WALTHER *PPS*).

In its *PPX* pistol, WALTHER takes the concept to the next level and may achieve cost savings by using a barrel that is partially made out of MIM'ed parts. The barrel tube is screwed into the MIM'ed locking block which also holds a MIM'ed feed ramp. The idea of this arrangement comes from rifles, where it is common to screw the barrel into the action. In comparison, the pistol barrel with this kind of

Fig. 3.19: *PPX* barrel consists of three separate parts: barrel tube, barrel block and feed ramp.

arrangement is made of three easy-to-produce parts, and utilizes technologically advanced manufacturing which improves efficiency and quality (Fig. 3.19).

In addition to the cost and manufacturing aspects of the barrel production, the safety of the shooter is of even more importance. How safe a barrel is under extreme conditions can be tested with bore obstruction tests. Two types of obstructions are considered the most likely to be encountered on a pistol when a shot is fired while another projectile is stuck in the bore: (i) a projectile just forward of the chamber and (ii) a projectile right behind the muzzle. It is recommended to use heavy weight projectiles which are above the typical weights for the standard cartridge for these tests – usually the projectiles used are pulled out of live cartridges.

For the first test, a projectile is placed into the weapon chamber and then forced forward into the rifling just far enough to permit a live cartridge to be inserted and the breech to be closed. For the second test, a projectile is forced forward until its tip is in line with the muzzle.[198] Normally, the latter is the more dangerous test (see Fig. 3.20), which may cause a catastrophic weapon failure.

Fig. 3.20: Burst pistol barrel, obstructing projectile at muzzle end, .380 ACP.

When pistols are tested in Germany to determine if they meet the performance requirements for service sidearms, barrel obstruction tests are carried out among other tests. The pistols have successfully completed testing, if there is no detectable chipping on the slide, barrel or frame. Ruptures, bulging or other permanent impairments are acceptable.[199]

With modern guns, a vast amount of the barrel is usually covered by the slide. Thus, the slide can serve as a shield[200] of the enclosed barrel and, in the case of a burst barrel, the slide can protect against hot gases and barrel fragments. However, bore obstructions often cause a bulged barrel and the slide will jam. Pistols with an exposed barrel have an advantage in this respect because they offer certain emergency operating properties in the event of a bulged barrel.

3.3 Slide

The operating principle of recoil-actuated guns requires a certain mass, which is why slides are usually made of steel. Since the machining time of steel and therefore the manufacturing costs are high, efforts were made early to reduce the production cost of slides. The *Jäger-Pistol*[201] uses an economically producible slide with a separate breechblock, and other components on this gun resemble sheet metal technology. MAUSER and WALTHER followed during WWII when developing their Volkspistols. HECKLER & KOCH used a deep drawn component on the *P9* pistol's slide.[202]

The slide of early SIG SAUER *P226* pistols is a cold deep-drawn sheet metal part with a butt welded front end, where the front end is originally produced by milling.[203] WALTHER uses zinc die casting technology for the slides of pistols for low-powered ammunition types (*P22* pistol) and uses MIM'ed slides for the *PK380* chambered in .380 ACP. Currently, the latter MIM method is also used for the *CCP* pistol chambered in 9 mm Luger.

Fig. 3.21: *Jäger-Pistol*, 1914–1917, slide and breechblock (© *B.J. Martens*)

Fig. 3.22: Polymer covered metal stamping as found on a FN *Five-seveN* slide.

Another way to reduce costs in the production of pistol slides is to use an insert, similar to the metal breechblock insert of a SIG Sauer *P226* (Fig. 3.23). Requirements for such an insert are low-cost and a reliable manufacturing process, particularly dimensional stability. The WALTHER *PPX* uses an insert made of polymer, however its slide and breech face are milled out of one piece of solid steel. The polymer insert carries components needed for the safety features and the ignition, as well as the extractor on one side and a guide block on the opposing side where the base of the cartridge rests upon while the extractor claw keeps pressing sideways on the rim of the case. A side effect of inserts that carry all functional components is a self-contained appearance of the slide, without any cut-outs for the extractor and other functional parts.

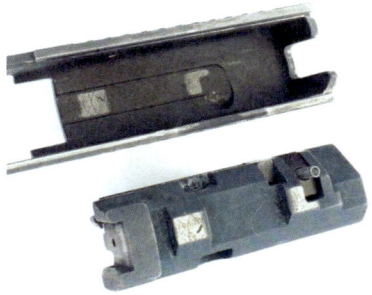

Fig. 3.23: SIG SAUER *P226* slide (shown above) and breech-block insert with functional parts.

For low-powered ammunition types, for example SIMUNITION training guns, GLOCK uses a slide construction whose body forms a steel skeleton which is subsequently fitted with polymer inserts. Patent application EP 2,660,551 A2, *Small Caliber Gun*, discloses the manner of installing these plastic inserts by means of molding or groove-spring and snap connections (Fig. 3.24).

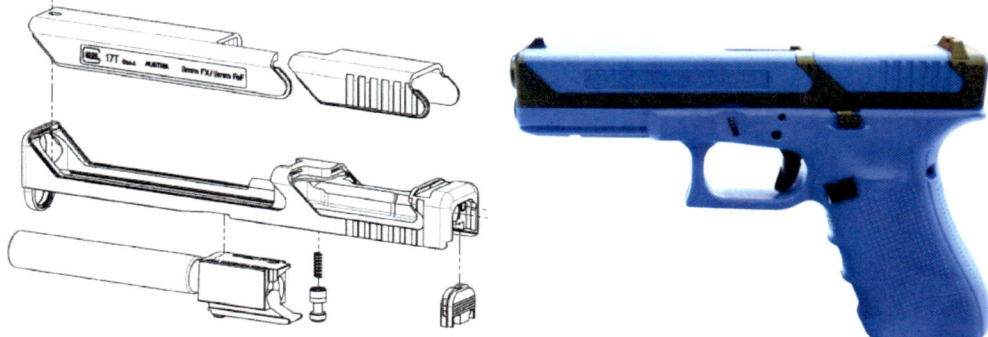

Fig. 3.24: Slide and polymer inserts, as well as GLOCK training gun (right).
(EP 2,660,551 A2, GLOCK, Fig. 11)

3.3.1 Breech Face

When a cartridge is discharged, the base of the case sets back until its head gets to rest against the breech face. Being the part that backs up the casing in the firing chamber, the breech face must be designed to withstand considerable forces when a cartridge is fired (Fig. 3.25), which poses a challenge for pistol engineers from a production standpoint and that of adequate longevity of the slide. If the breech face were to "settle" during the lifetime of the weapon, or if its geometry should change in any other way, it would have an immediate impact on the headspace[204] dimension and hence on the safety of the weapon.

Fig. 3.25: Circular breech face on a WALTHER *PPK/S* slide.

In firearms, headspace (HS) is the distance between the seating point of the cartridge and the breech face. In practice, HS is not actually measured but checked by gauging and tells how much space is left between the case head and the breech face when the slide is closed. If a pistol has a form locked breech, headspace is not just a result of the actual seating point and its distance to the breech face alone, but the total number of components in-

Fig. 3.26: WALTHER *Mod. 3* made in 1910. NB: Left side ejection port and extractor.

volved in the form locking of barrel and breechblock is included. Every single component involved has its dimensional tolerances and headspace must stay within given dimensions (e.g. CIP standards) despite the tolerance stack up and wear and tear over a pistol's service years. For instance, CIP headspace dimensions for a 9 mm Luger are 19.15 mm minimum ("Go" gauge) and 19.45 mm maximum ("No-Go" gauge).

A casing might separate if the tolerances of HS dimensions are exceeded. When the primer cup is struck, the cartridge is pushed into the chamber until it hits the seating point. During discharge, pressure expands the thin brass near the mouth of the cartridge case against the chamber wall, thus forming the seal which prevents the powder gases from escaping to the rear. The thin brass is expanded heavily against the chamber wall, producing a high-pressure metal-to-metal contact which results in a very high friction between the case's "neck" and the chamber. In fact, the forward portion of the case may be gripped so tightly during the period of high chamber pressure that it sticks to the chamber wall. Since the radial pressure is not extremely high at the relatively thick-walled base of the casing during this phase, the axial component of the pressure causes the entire case to stretch in the chamber so that it takes up any excess head-space which may exist. In this event, the forward portion of the case still sticks to the chamber wall but the rear portion continues to move, causing the case to stretch plas-tically. If the headspace is far too excessive, the breech face can not put a stop to the stretching of the case. If the tensile yield strength of the case wall is exceeded, i.e. when the allowable elongation of the case is beyond its limit, the case will separate.[205]

The breech face has a form that offers an ideal support to the case head as the case sets back against it. Known forms include circular recesses (Fig. 3.25) or rectangular recessed breech faces, the latter are often groove shaped, with vertical ori-entation (Fig. 3.27, also Fig. 2.24, p. 38).

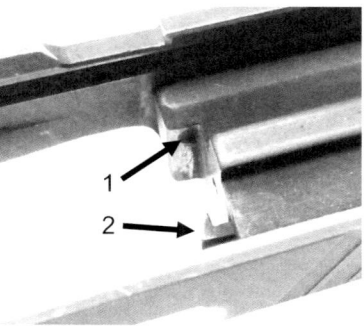

While the residual pressure drives the case out of the chamber,[206] the extractor claw remains inside the extractor groove of the case and pushes the base of the case sideways against the breech face guide block, on the opposing, non-extractor side of the breech face. This guide block improves precise positioning of the case's base for good ejection. If

Fig. 3.27: Groove-shaped breech face of a SIG SAUER P226, with half-moon cut (1) in the guide block opposite of extractor claw (2).

the positioning of the base is not perfect for each shot, the ejection pattern will keep changing. Failures to eject can be as a result of missing the case's head or not hitting the ejector properly.[207]

Basically, an extractor performs two functions: firstly, it pulls the cartridge out of the chamber when unloading a pistol manually, and secondly, it ensures dependable ejec-tion when firing. The extractor improves the reliability of a pistol, whereas the cartridge case is pushed out of the chamber by the residual propellant pressure rather than be-ing pulled out by the extractor.

On some guns, a small cut may be found on the guide block, opposite of the extractor claw. As the slide recoils, the rim of the case slides into this cut and consequently hits the ejector reliably providing good ejection (Fig. 3.27).

With long nitriding process times thin-walled areas will get brittle. That is, if the core of the metal does not have a sufficient amount of substrate, cracks are likely to be caused by the shocks introduced by every shot fired. Thin-walled spots are likely to be at the ejection port and at the breech face, especially at the cuts for the extractor and at the firing pin hole. The relatively small firing pin hole is not the cause of the crack sensitivity, but the striker channel behind the breech face, whose size and depth depends on the shape of the striker, and thus determines the wall thickness of the breech face (see Fig. 3.28). To achieve a sufficient du-

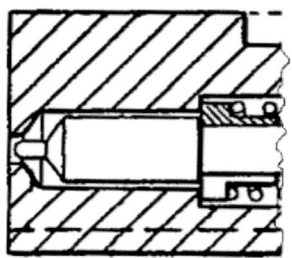

Fig. 3.28: Schematic view of striker channel behind and striker hole inside breech face.
(Patent AT 368,807, GLOCK, Fig. 5)

rability in the area of the firing pin hole, the wall thickness of the breech face should be as large as possible, and the diameter of the striker channel should be smaller than the diameter of the primer; so the primer cup can rest on a solid breech face during firing.

Patrick Sweeney made mention of "a curious admonition" by GLOCK, giving a warning about the use of "non-toxic" ammunition. The brisance of the replacement compounds in non-toxic primers is allegedly higher than that of its predecessors, which contained lead styphnate. Sweeney surmises, the breech face is thin around the Glock's firing pin slot and the effect of the greater brisance[208] of lead-free priming compounds is enough to cause peening or to chip the thin wall.[209]

A steeper gas pressure characteristic in the case's primer pocket will cause the primer cup to get pushed out of the primer pocket harder at the moment when the priming compound ignites. The primer cup slams back into the breech face, while the ignition of the propellant charge takes place. After the ignition of the charge, the case sets back at the full combustion pressure, thus seating the primer back in the pocket.

Sweeney foresees the issues that might arise if you are a police agency that uses non-toxic ammunition in large quantities, because the greater brisance of the non-toxic primer might affect the breech face.[210] Manufacturers should verify that the gas vent hole will do its job before the event of a ruptured breech face occurs – where the striker pokes through. Chances are that the striker and slide rear plate would be blown out rearward in this event. Sweeney recommends paying attention to the thin-walled breech face on GLOCK pistols when primers of high-powered ammunition slam back into it, or the gas jetting from pierced primers may cause the area around the firing pin slot to be eroded or even peened back.

Gas vent holes behind the breech face (Fig. 3.29) have been around since the introduction of the *Gewehr 98*, a bolt-action MAUSER rifle.[211] In the case of a pierced or ruptured primer, the propellant gas pressure will not get to a level where parts are blown toward the shooter. An equally important function besides this, is its function as a drain hole for fluids. The reason being that fluids are incompressible and might prevent the firing process with excess lubricant or presence of water inside the striker channel.

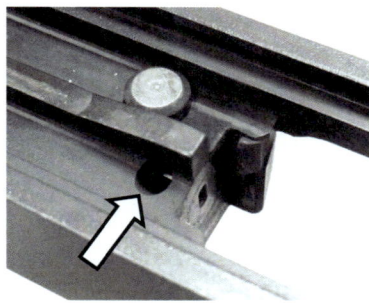

Fig. 3.29: Weep hole in the slide of a GLOCK *G19* pistol.

Manufacturing

Circular breech faces are manufactured by milling, while groove shaped breech faces are usually done by broaching or slotting. The use of a broach is better for internal machining types, where slots with small interior radii or chamfers, or sharp corners are to be cut. Using a broach demands that this tool can be pulled through the workpiece in the longitudinal direction. This process, however, requires the slide to be placed in a fixture on a broaching machine; which takes time and may lead to greater tolerances. This aids forensics investigations as broaching will leave tool marks on the breech face, which are in the same direction as the broach when pulled through the slot. These tool marks will then be transferred to the case's head when the cartridge is discharged.

In modern manufacturing, a groove shaped breech face can be machined using CNC machines by slotting, requiring only one set-up to machine the breech

Fig. 3.30: WALTHER *PPX* slide: Internal radii. Note radii on left and right side of breech face.

face and other milling operations. The slotting operation does put a considerable shock load onto the milling machine and causes some additional machining time. An alternative to this is to roughen out the barrel section including the breech face area with an end mill cutter in a single operation. This milling will produce generous inner radii (Fig. 3.30), which will be of the same dimension or slightly greater than the radius of the cutter itself. In other words, the sides of the breech face are far from the 90° groove cut, which serves as an ideal sideways rest for the base of the cartridge case.

Hence, this method is not applicable if conventional groove-shaped breech face blocks are required. However, the appeal of the new method lies in the design of the slide, where the extractor and the lateral guide block for the base of the cartridge are part of an insert, which is placed into the slide during assembly. The main body of this insert can be made of polymer and includes all functional components that were previously installed separately into a conventional slide.

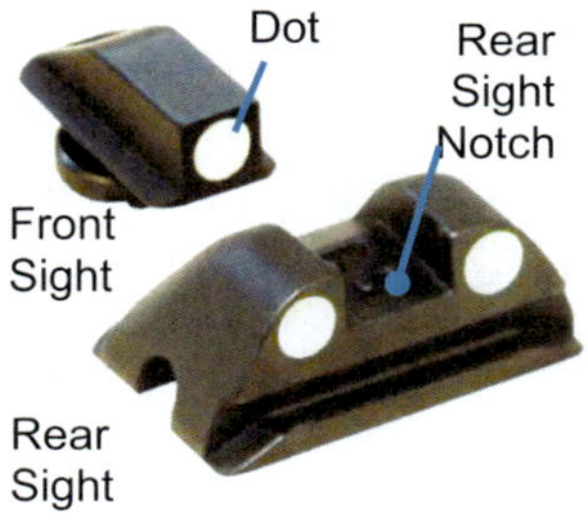

Fig. 3.31: Open post sight components.

(© WALTHER)

3.3.2 Sights

A set of open sights consist of a front sight and a rear sight. The cutout in the rear sight is the rear sight notch (Fig. 3.31).[212]

Markings are widely used to help aiming under low-light conditions. Traditionally, these markings are simple dots (Fig. 3.32), one on the front sight and two of them on the rear sight, and they may be made of self-luminous material.

Very common for self-luminous contrast markings in combination with light emitting materials is tritium, an isotope of hydrogen, represented with the symbol T, $^3_1$H or H-3. Because of its radioactivity, a free use of tritium is only allowed within the applicable country-specific activity.[213]

Fig. 3.32: Comparison: Tritium, SUPER-LUMI-NOVA, white dots (from above)

An alternative to active luminous dots are inactive markings, which make use of the physical effect of phosphorescence. They can be charged with a bright light, and the absorbed radiation is re-emitted at a lower intensity for up to several hours after the original excitation. Also widely used are simple white dots, which are manufactured by applying paint; they are the most cost-effective type among the markings considered here. Fig. 3.33 shows an overview of color markings, which are described in more detail below.

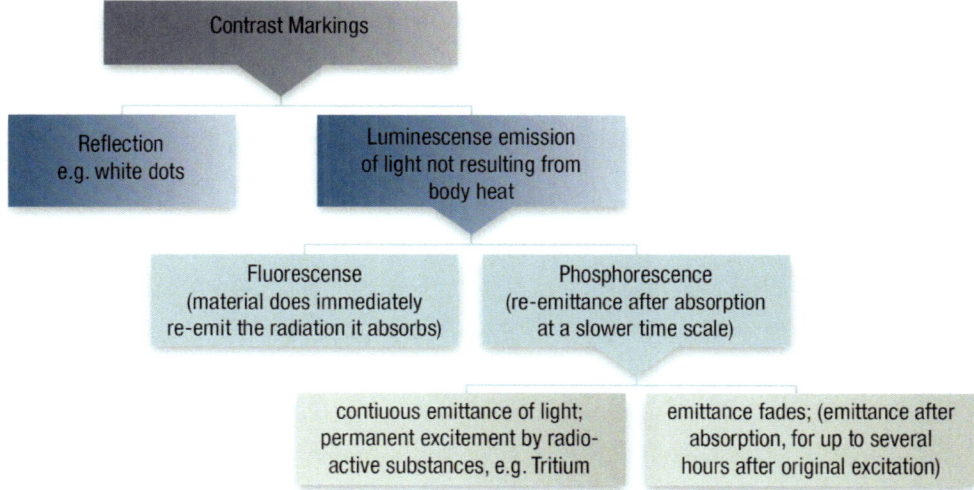

Fig. 3.33: Open post gun sight Contrast Markings.

Luminescence: the emission of light without thermal activity

The sun or the spiral-wound filament of an incandescent lamp emits light as a result of heating. However, luminescence is the emission of light by a substance in the absence of heat. Cold-body radiation emitters are made of luminescent materials and the associated radiation process is luminescence (light emission).[214] One of various forms of luminescence is light emission from any form of matter after the absorption of photons. If materials can absorb and re-emit photons, an activation by photons will lead to photoluminescence.

First, the electrons of the material are in their state of lowest energy. When an incoming photon of high energy collides with an atom or molecule, the photon transmits its energy to an orbital electron, exciting it to a higher energy level. Usually, this high-energy level takes only a short amount of time. After about 10 ns, the system relaxes and emits the excess energy in the form of light.[215]

Two important technical applications of luminescence are fluorescence and phosphorescence. In both cases cold-body radiation (luminescence) is caused by light, which can be either artificial light or sunlight, after which phosphorescent materials will re-emit for up to several hours, but fluorescent materials will not.[216]

Fluorescence: Form of Luminescence that ceases to emit light immediately
Fluorescent materials immediately re-emit the radiation they absorb. Fluorescence shifts energy present in the incident illumination from shorter wavelengths to longer wavelengths which might cause a fluorescent color to appear brighter than it could possibly be by reflection alone. An everyday example would be a highlighter pen, which takes advantage of this effect. Fluorescent materials cease to glow immediately when the radiation source stops, and will not continue to emit light.

Phosphorescence: Form of Luminescence that emits light for some time
Luminous pigments embedded in phosphorescent materials absorb radiation and will re-emit it within a few minutes or hours.[217] Afterglow pigments that re-emit for a longer period of time than thirty minutes with a luminance of at least 3 mcd/m² are called "photoluminescent" pigments.[218]

Energy in the form of daylight or artificial light stimulates the inorganic pigments of a substance and is released relatively slowly. The pigments consist of crystals, which include foreign atoms; for example, zinc sulfide and aluminum oxide, causing the glow-in-the-dark-effect. As soon as light stimulates excitation, the crystal's electron clouds are transferred to new configurations. The atoms then achieve an excited state of higher energy. When it relaxes, the energy difference will eventually be emitted in the form of light.

The more intense and longer the exposure to energy in the form of light, the higher the chances of atoms reaching an excited state. Once the exposure to light ends, many of the excited atoms will return to the lower energy state, causing an intense emission of light.[219] Later, fewer atoms left fall back gradually, and the glow slowly fades out. The total luminous intensity is determined by the number of light crystals as well as the excitation saturation, that is, the amount of materials involved and the maximum charge.

With an advantageous layout and appropriate conditions, the intensity of phosphorescent markings will be more visibly intense than that of tritium sights. Both make use of phosphorescent substances, but sights with pure phosphorescent markings carry more materials than tritium inserts. For this reason, during the first 60 minutes after exposure to natural light phosphorescence markings are perceived as very bright in dark environment, much brighter than tritium night sights. After two to three hours, the latter will appear to be brighter.

Applications of phosphorescence are seen in safety signs in buildings which are needed to be visible in the event of a power loss, watch dials and hands that glow in the dark. Phosphorescence is not based on radioactivity. The materials commonly used are non-toxic, can be charged an indefinite number of times and will not lose intensity over the years, which sets them apart from radioactive tritium sights.[220] Popular are SWISS SUPER-LUMINOVA® phosphorescent colors.[a] Their pigments are based on aluminum oxide crystals and deliver a higher light output than conventional pigments based on zinc sulfides.[221]

Phosphorescent color markings are luminous paints. The paint is placed in the cavities of front and rear sights, dried, and in a secondary operation covered with transparent paint. The transparent paint is a protection layer, used to protect the luminous paint from physical damage and moisture, because it can react chemically with water or humidity. In this reaction, a decomposition with formation of a layer of hydroxide on the surface will take place, that is, affected color markings will change their initial color to white.[222]

Luminous paint is destroyed by heat. A temperature of maximum 248 °F (120 °C) is acceptable temporarily; the use of ultrasonic cleaners with temperatures higher than 140 °F (60 °C) can damage the paint as well.

Apart from this, luminous paints will last as long as those markings on vintage wrist watches. Also, the activation and subsequent light emission process can be repeated continuously, that is, the material does not suffer any practical aging.

Excitation of Phosphorescent Material by Tritium
Another possible way to excite phosphorescent material is to use tritium instead of a light source. Tritium is an isotope of hydrogen, a light volatile gas, and also a low energy beta emitter, with a half-life of 12.3 years, which can be a potential radiation hazard agent when inhaled, ingested via food or water, or even absorbed through the skin.[223]

---

[a] SUPER-LUMINOVA is a registered trademark of NEMOTO & CO LTD, Japan.
  SWISS SUPER-LUMINOVA is a trademark of LUMINOVA AG SWITZERLAND.

For the application in gun sights, miniature vials (Fig. 3.34) are coated internally with a luminescent material, filled with the volatile gas bound in a polymer, and sealed. The beta decay of the tritium gas will excite the luminescent material, causing the emission of visible light (Fig. 3.35).

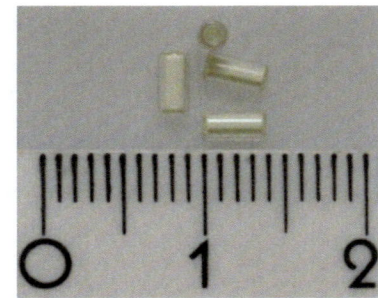

Fig. 3.34: Tritium vials

Both phosphorescent markings and tritium-activated sources make use of the same light emitting material. Tritium-activated devices have the advantage of emitting a consistent luminosity that does not fade under low-light conditions. However, being a radioactive material with a half-life of roughly 12 years, the intensity of tritium-based sources will gradually fade. Usually, the light intensity decreases within the first five years from 100 to 50 percent and later to 25 percent. The emission of light will end abruptly at a point when the excitement of the phosphorescent material is no longer sufficiently provided.[224]

When the light emission is no longer intense or when the dots are dead, the sights need to be replaced. The abrupt end of the emission occurs after about 10 years, and the sights need to be treated as nuclear waste and be disposed off according to radiation protection regulations.[225]

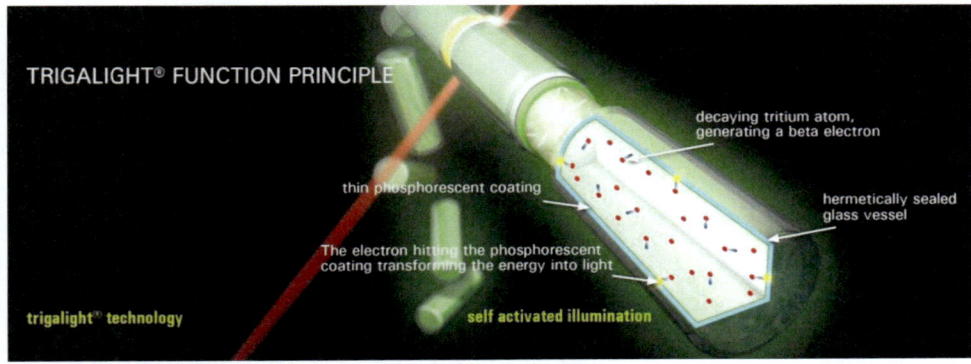

Fig. 3.35: Function principle of a TRIGALIGHT tritium vial.

(© MB-MICROTEC)

Tritium Night Sights

United States Nuclear Regulatory Commission (NRC) will neither constitute an unreasonable risk to the common defense or security or to the public health and safety, nor constitute a frivolous use of radioactive material. NRC generally considers that products proposed for distribution should be of some benefit or use to the public. Typically, the use of radioactive materials in toys, novelties such as fishing lures, and adornments have been considered to be of marginal benefit. Gun sights are exempted, provided the products have been initially transferred or distributed in accordance with a license issued to manufacture, process, produce or initially transfer. Also, users of the devices are instructed to return defective units and unwanted units to the distributor for disposal, and the cost of tritium gun sights is such that inadvertent or careless disposal is unlikely.[226]

Tritium night sights are officially referred to as *Tritium Illuminated Gunsights Containing Tritium Gas Sealed in Glass Vials*. This sort of sights can be made available in the United States only if they undergo a sealed source device safety evaluation prior to the issuance of a license according to U.S. Nuclear Regulatory Guide (NUREG) NUREG-1556 by NRC.

Applicants wishing to distribute tritium night sights must submit sufficient information concerning the product to demonstrate that the product will meet the safety criteria set forth for that type of product. At least five gun sights of each model are to be subjected to each of the following tests: radiation dose, labelling (with H-3 for tritium plus the name, the trade mark or license number of the manufacturer or dealer), chemical resistance, temperature resistance, temperature shock, vibration, pressure, penetration test, 60 drop tests and 5,000 rounds endurance test.[227]

3.4   Frame

Polymer injection molding technology revolutionized weapon design in the 1980s in a way rarely seen before by any manufacturing technique. Until then, pistol frames were made of steel or aluminum alloys by milling and other metal-cutting technologies. The aesthetics of such frames used to be rather simple and as a result of the time-consuming way of production, the production costs of the pistols were high. In 1990, a German-made SIG SAUER *P226* would sell for almost DM 1,600 in Germany. While the company that pioneered polymer frames, the Austrian company GLOCK, was able to offer its *G17* at about DM 1,200 (Table 3.1), which implies a 25 percent difference in price.

Table 3.1:   Price comparison, pistols, 1990 [228]

	BERETTA 92 F	GLOCK 17/19	HK P7 M13	SIG SAUER P226	WALTHER P88
Price [DM]	1,452	1,198	1,698	1,598	1,610
Price [€]*	742	613	868	817	823

* Exchange rate 1.95583 DM/€

Despite considerable cost pressure, gun manufacturers took a conservative approach toward the rise of polymer frames, and consumers in Europe perceived polymer frames as cheap. This prejudice might have originated when reports of one of the first handguns with a polymer frame appeared: the *VP70* pistol of HECKLER & KOCH (Fig. 3.36). The German gun magazine *Deutsches Waffen Journal* had published an article on the *VP70* in September 1972 where the author commented his opinion regarding the gun's aesthetics with the words, "it looked from a distance like a big plastic toy pistol." [229]

Fig. 3.36: HK *VP70*

Originally, manufacturers made pistol frames of steel. Efforts to use other materials, for the sake of an efficient production, began during the 1930s.[230] At that time, tests were carried out with frames made of aluminum alloys. It was not until the 1950s that frames produced from light alloy made their way into the market. The WALTHER *P1* pistol of the German Armed Forces already had a frame made of aluminum alloy in contrast to its predecessor *P38*, before the SIG SAUER *P225/P6* and WALTHER *P5* hit the market as newly designed pistols in the 1970s. The advantage for the user was in the weight reduction; the weight of a pistol with an aluminum alloy frame is approximately 100 to 200 grams (3.5 to 7 oz) less than the weight of guns with steel frames (see Appendix 12, p. 369).

A greater leap than just small grip panels and other polymer parts of this size started precisely in 1982 when the GLOCK *G17* pistol was introduced by the Austrian Armed Forces under the official designation *9 mm Pistole 80* and when this model was offered on the commercial market.[231] As the popularity of this pistol increased, and after GLOCK in 1985 founded a subsidiary company in the United States,[232] the rest of the firearms industry slowly started to realize the potential the polymer material had when used for essential weapon parts. At that time a hot debate on the "Glock" started with articles in *The Washington Post* and in several European news magazines, which said the gun could not be identified during X-ray inspections at airports. There were rumors that supposedly Muammar Gaddafi was interested in buying 100 *Glocks*.[233] Through these reports, the "Plastic Gun from Austria"[234] received attention from parts of the general public as well as the law enforcement community.

The trend towards polymers in the construction of firearms began well before the pistol designed by Gaston Glock. Already on the *MP38* [235] submachine gun, a large outside portion was made of Bakelite[a]. During the development of the American *M60* machine gun, outside polymer parts were used starting from 1951,[236] and the stock and handguard of the ARMALITE *AR15* were made of fiberglass plastic.[237] In the 1950s, Remington launched the *Nylon 66* rifle,[238] chambered in .22 Long Rifle, and in the 1970s STEYR designed and launched an

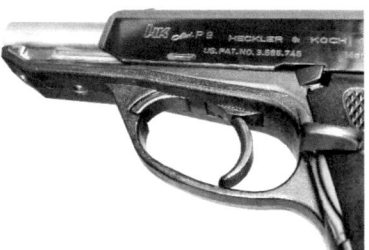

Fig. 3.37: Polymer insert on a HECKLER & KOCH *Mod. P9* pistol.

assault rifle with bullpup stock made of polymer. This STEYR *AUG* used two halves, which were permanently connected after molding by friction stir welding.[239] In the area of handguns, HECKLER & KOCH in 1970 made a first attempt with a polymer frame insert which formed the trigger guard and the front end of their *P9* pistol (Fig. 3.37).[240]

---

[a] According to *Wacker*, one *K98k* is known made in 1937 with a Bakelite stock. The rifle is marked with manufacturer code S/243, i.e. MAUSER-Werk Berlin-Borsigwalde. (Wacker, 1993 p. 62)

This was followed by the introduction of HK's *VP70* pistol in 1972.[241] It had a polymer frame, which carried a large metal insert (Fig. 3.38) molded into it.[242] The corresponding disclosure DE 1,728,251 by HECKLER & KOCH in 1968 explains a frame completely made of polymer, and the members to carry the trigger mechanism, slide rails, magazine release and metal parts to fix the barrel are pressed in or molded in.[243]

Fig. 3.38: Molded-in frame insert on a HK *VP70* pistol.

Gaston Glock consequently used the ideas provided by polymer injection molding technology further, by designing a one-piece molded pistol frame with just a limited number of metal parts molded into it; two slide rails (Fig. 3.39) and a serial number plate (Fig. 3.40). In short, the GLOCK patent differs from the one by HECKLER & KOCH. The disclosure DE 1,728,251 of the *VP70* was based on the idea of load-bearing functional parts which were molded into the polymer, avoiding the need to removably mount load-bearing parts.[244] The frame of the GLOCK, however, carries functional parts which are removable (see patent AT 368,807).[245]

Fig. 3.39: Molded in slide rails of a GLOCK *G19 Gen2* pistol.

Fig. 3.40: Metal plate with serial number on a GLOCK *G17* pistol.

Some publications about the GLOCK mention that it took 80-seconds cycle time to mold the frame of this pistol.[246] The method of molding in metal inserts used to be a common process until the end of the 20th century.

However, molding has the disadvantage of a slightly higher cycle time as a precise positioning of the components when inserting them into the injection molding tool requires some time, especially in cases where this is done by hand. Also, it possesses a certain risk of damage to the tool during closing if the tool collides with an incorrectly positioned insert. This risk can be reduced if the configuration of the pistol frame allows molding it without slide rails and uses minimum number of inserts in general. If integrally metal slide rails could be avoided, it would reduce the cycle time by 20 to 30 seconds. But there is a doubt, if the slide rails would be strong enough if they were made of polymer.

Nevertheless, there is another solution; a frame can be completely made of polymer without integrally rails, if the frame is designed to accept an interchangeable metal chassis with slide rails, which will then be inserted into the frame at the point of assembly.

There are currently two approaches of molded polymer frames without slide rails. On the one hand, frames in which a single relatively large chassis is installed and the slide rails are part of it (Fig. 3.41) and on the other hand, frames which carry more than one assembly (see Fig. 5.137, p. 296) similarly as in conventional frames made of metal, and where for example two of the inserts have slide rails.

Chassis, which carry all the functional parts, can be rather complex because they serve as a firing group housing with a variety of functional parts. As a consequence, their production can be challenging. They can be monolithic chassis, for example, machined out from solid aluminum alloy, or built-up, that is, they consist of several components which will be permanently connected as the chassis is made. An example of the latter is shown in Fig. 3.41; two sheet metal fittings interconnected by a MIM'ed

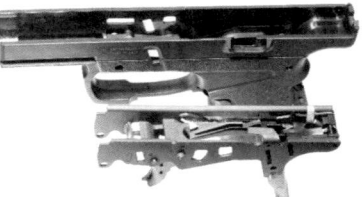

Fig. 3.41: Polymer module (top) and assembled chassis of a WALTHER *PPS* pistol.

locking block, where the permanent connection of parts is realized by brazing. The manufacturers are faced with the option of whether this effort is worthwhile or that the production of some simple, smaller function groups represents the better solution. Smaller assemblies might be less demanding in production. They are conversely independently located from each other in the frame and therefore the reliability of the gun depends on the sum of the manufacturing tolerances of the individual assemblies and in addition, the frame's dimensional tolerances.

The technical approach of the two is similar – they avoid integral inserts as the frame is being molded. However, the application of the Gun Control Act of 1968 affects them differently. Namely in regard to a frame made of a number of components, to distinguish which member of the components is the essential part representing the frame, this would ultimately be the member whose purchase requires a permission.

The frame or receiver is that part of a firearm which provides housing for the hammer, bolt or breechblock, and firing mechanism. According to this, in the first case of a one-piece frame insert, the chassis will be what is legally regarded as the receiver of the pistol, and in the case of a frame with several functional groups, the frame serves this purpose. Consequently, the removable one-piece chassis will carry a serial number and, on pistols manufactured in CIP member countries, a proof mark. Different requirements apply to the serial number around the world depending on the individual country. Basically, a legally valid serial number must be conspicuous and permanently inscribed on metal.

In cases where the frame carries more than one assembly with rails, the polymer shell is regarded as the pistol's frame. Thus, serial number and other markings (Fig. 3.42) are usually on an integral serial number plate. This insert has no mechanical function and, as a result, the cost of its production is at a minimum. If it is designed in a proper way, wrong insertion into the mold, tool collision, and damage are avoided. Consequently, the frame shell has no other molded-in inserts than the serial number plate,

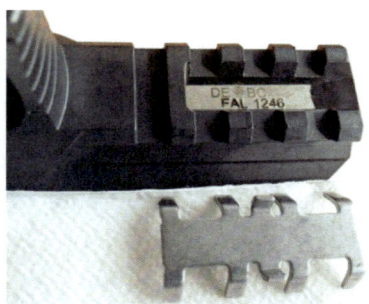

Fig. 3.42: Serial number plate: molded-in and separately (below).

can be molded with little effort, and despite the cost of two removable sub-assemblies with slide rails, it is an interesting solution from a manufacturing standpoint.

The weight of pistols with polymer frames is about 120 grams (4.2 oz) less than those with an aluminum alloy frame (see Appendix 12). This reduces the physical strain of people having to carry a gun, although a pistol on the duty belt of a police officer represents just one of many pieces of their personal equipment. However, with the weight of pistols being as low as it is today, a level is reached where malfunctions may occur when a polymer pistol is shot one-handed.

If the user applies a firm and solid gun grip on the pistol, he creates stability for shooting. If not, he has less control over recoil and the frame kicks back more than it is supposed to. Therefore, the slide cannot cycle with the necessary velocity relative to the frame, and as a result, failures to eject may occur. Another symptom that might be present is the slide short cycles, where the slide does not stay open on an empty magazine, reason being that it did not go fully back to the rear stop (i.e. underperformance). Moreover, failures to feed sometimes occur without proper pistol grip. Each one of these examples may indicate that something is wrong with the gun, but the root cause can also be related to a different part of the system in which its reliability depends on the gun, the ammunition, and the user.

In addition to the lower total weight, pistols with polymer frames have a further advantage; polymer can absorb shocks, and this can actually help to offset the recoil. First, it offers an advantage to the shooter in the sense that when the barrel gets uncoupled from the slide, and shortly thereafter, when the slide hits the rear stop inside of the frame, producing a perceived shock which will be noticed as not so bad as when compared to the one of a metal frame pistol. At the instance where the slide will bounce back from the rear stop with less energy, the recoil spring must get the job done to kick the slide with sufficient energy forward, so that the slide will feed a new cartridge from the magazine into the chamber, couple barrel and slide, and after all, to push the slide all the way to its forward stop; so that the slide must be pushed forward with clearly enough kinetic energy until it reliably reaches its forward stop.

Apart from the impending underperformance, the benefits of polymer frames are paying off for manufacturers and users. Rarely does one find similar ergonomic grips on today's metal frame pistols. Since the production of frames no longer involves metal cutting and other expensive manufacturing processes, but instead a low-cost injection-molding process is used, money invested one time only in constructive effort and tooling for a frame mold, leads to frames that can be produced in a series with consistent quality by injection molding. Some readers might remember a similar process that occurred with single-lens reflex cameras, where an evolution from square and purely technical product designs to ergonomic camera bodies of today could be witnessed.

The next step in improving polymer frames was taken with the introduction of the WALTHER *P99* pistol, where in addition to an improvement of the ergonomics, a pistol grip was introduced, that would allow an adjustment to the size of the user's hand by replaceable backstraps.[247] Other manufacturers followed, offering even richer customization possibilities, such as interchangeable grip panels in addition to backstraps (such as HECKLER & KOCH *P2000*).

In addition to the above benefits, polymer material proves to have an advantage when it comes to wear and tear. While regularly-toted handguns may show wear on external metal parts which is caused by abrasion, polymer frames retain their color because the color is embedded in the polymer material. Typically, the color tone of polymer raw-material is neutral and adding colorants allows a wide range of color tones (Fig. 3.43),

Fig. 3.43: Colored frame of a SCCY *CPX-1* pistol.

however colorants may change the material properties of the polymer. For example, shrinkage and thus the dimensional stability of the polymer could be different from that of a black polymer. Careful testing before release of series production therefore proves useful.

Instead of coloring by mixing colorants with the polymer prior to injection molding, the aesthetics of frames and other parts can be modified by applying an exterior coating using water transfer technology. This technology offers a wide variety of different colors as well as patterns (Fig. 3.44).

Fig. 3.44: WALTHER *PK380* frame with water transfer finish.

Changing the aesthetics of products is common in the United States. For manufacturers, it is a popular way to differentiate their products from those of competitors, but also add choice for the consumer. Product design can address existing market segments by themselves or in conjunction with other sales-political instruments, such as product quality, packaging or the product name, or may contribute to the formation of new segments.[248] For example, pistols in pink color, "tactical," or special make ups with colors and patterns reserved exclusively for the distribution through certain retail chains are standard practice. Usually, either completely modified guns or pistols with customized frames are offered.

Finally, a note on a useful detail: in some countries the law enforcement still uses lanyards to secure their pistols. Manufacturers consider this by including a lanyard loop on the frame (Fig. 3.45).

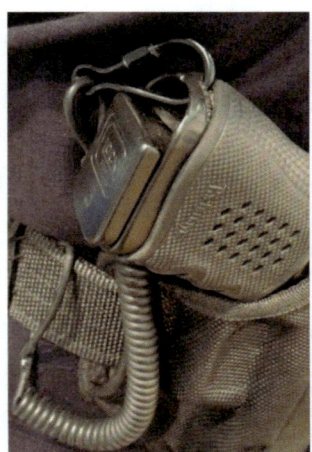  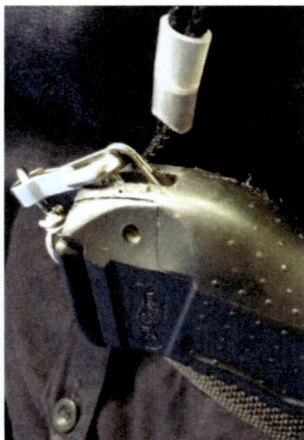

Fig. 3.45: Lanyard loop on pistol frames of GLOCK, SIG SAUER and WALTHER (f. l. t. r.).

3.5 Magazine Release

The ease of use of the magazine release might be of decisive importance during purchase. Users have a wide variety of demands with regard to size, shape, haptics, type of operation and even position of the magazine release. The latter is notable as there is an association with security in respect to the potential loss of a handgun's magazine. Magazine releases of a lateral push-button design on the left side of the pistol grip are extremely popular.[249]

Fig. 3.46: Ambidextrous paddle magazine release on a BERETTA *Pico* (see marking).

Sometimes, magazine releases are ambidextrous, or more recently are often on one side but reversible. Ambidextrous paddle style magazine releases are an alternative to the lateral push-button design. Usually, a pair of paddles that straddle the trigger guard are used. To release the magazine, either one can be pressed down. Paddle magazine releases hold the magazine reliably, because they do not get actuated by a lat- eral force. On the contrary, the popular push-button may inad- vertently release the magazine in cases where, for example, a

Fig. 3.47: Heel maga- zine release of an HK *P9* pistol.

seatbelt buckle presses against the holster or when a user carries the pistol inside the waistband. The still less common paddle release however requires the user to learn a different way of manipulating the magazine release if he is used to the more classic push-button. HECKLER & KOCH, WALTHER, and more recently BERETTA (Fig. 3.46) use a paddle mag release on some of their pistols.

In addition, there are magazine releases on the heel of the pistol grip (Fig. 3.47). This style can be found on early diminutive models as well as on pistols from the 1970s, such as the WALTHER *P5*, SIG SAUER *P6* and HECKLER & KOCH *P7* and *P9*. Actually, this is another example of an ambidextrous magazine release.

A heel release is activated with the support hand and the magazine will automatically slide into the user's hand, because his non-trigger hand is still at the mag release which is close beside the magazine well. This procedure keeps the user from dropping the magazine freely onto the ground. Which in turn saves the mag from hitting the ground or getting dropped into dirt, but it makes the mag change a cumbersome drill.[250] This is why the heel mag release is considered outdated.

3.6   Fire-control Mechanism

The means to initiate the discharge of a round is the *fire-control mechanism*. In hand-guns, it comprises of the *firing mechanism* and the *trigger mechanism*. The former provides the kinetic energy to the primer, and the latter is the means by which the firing mechanism is controlled.[251] Depending on the firing mechanism, handguns can be divided into hammer-fired or striker-fired and any of these may be in either uncocked or pre-cocked condition when a gun is carried.[252] Several essential functions are combined in the fire-control mechanism including: triggering, disconnecting, and, if applicable, cocking.

Firing mechanism, hammer-fired
Typically, the hammer on a hammer-fired pistol is part of the pistol's frame. The hammer is driven by the mainspring and, in most designs, this spring sits underneath the backstrap of the pistol's grip handle. The mainspring drives the hammer forward, which then hits the firing pin. As opposed to a striker, which is a spring-loaded firing pin directly propelled by the striker- or firing spring.[253] To keep things simple, we are going to refer to this striker spring as mainspring in this book.

   Hammer-fired mechanisms comprise of an internal or external hammer, the mainspring, the trigger with its return spring and drawbar or trigger bar, the sear spring, the sear and sometimes other connecting parts. What is called the sear is a part that retains the hammer in the cocked position. It rests in a notch in the hammer, under the tension of the mainspring. To initiate the firing of a cartridge, the trigger must be pulled. Pulling the trigger levers the sear from engagement with the sear notch in the hammer, the hammer then strikes and transfers its kinetic energy to the firing pin, which in turn transfers it to the primer cap.

   In many designs, the hammer is connected to the mainspring by a link, and it is either pulled or more commonly pushed by a hammer strut. The earliest and mechanically simplest fire-control mechanisms have a sear, which is part of the trigger. The hammer of such guns must be cocked by pulling it down to the point where the sear engages to hold the hammer. By pulling the trigger, the sear disengages and the hammer drops to cause a round to be fired. With semi-automatic guns, the discharge of a round starts the self-loading cycle, e.g. while the slide moves rearward, it will cock the hammer and eject the spent casing. When the slide reaches its rearward stop, the recoil spring will drive the slide forward to feed a cartridge from the magazine into the firing chamber.

Some hammer-fired pistols use a rebounding hammer where the hammer is automatically moved slightly rearward by a spring or cam after striking the firing pin. For example, hammer and hammer strut may be designed to be in a stable position to each other when at rest. This will keep the hammer a short distance away from the firing pin. As the trigger is being pulled to the rear, the hammer will be cocked starting from this resting position and will be released in normal manner at the end.

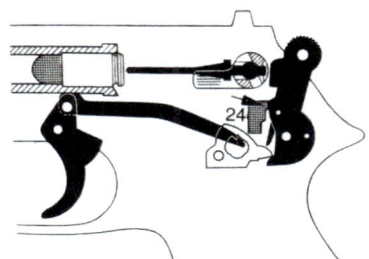

Fig. 3.48: Hammer block (24) on a WALTHER *PP*.

(© WALTHER)

As a consequence of its kinetic energy, the hammer will pass over the resting position until it comes in contact with the firing pin to discharge a round. Apart from the rebound feature, some other safety mechanisms are utilized on hammer-fired pistols. For example, a part of the drawbar engages to block the hammer, or a hammer block which isolates the hammer from the firing pin except when the trigger is pulled (Fig. 3.48). In firing mechanisms like that, the hammer will hit the firing pin during mechanical shock such as a jar, a blow or a drop, only if the safety device is getting disengaged at the same time.[254]

Firing mechanism, striker-fired

A striker is a spring-loaded firing pin that travels on an axis in-line with the primer cup. The striker and its spring are mounted into the slide, while some other parts of the firing mechanism are part of the pistol's frame (Fig. 3.49). Proponents of striker-fired actions opine the direct transfer of the mainspring's kinetic energy onto the striker is why less losses would

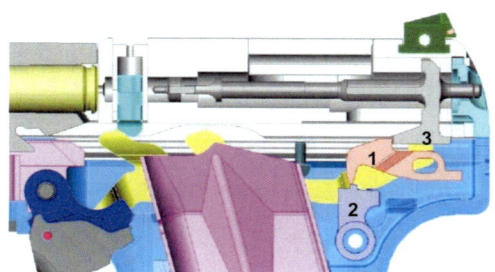

Fig. 3.49: SA trigger with sear (1), pawl (2) and striker (3).

(© WALTHER)

occur in the transmission of energy from the striker spring on the primer, and the lock time would be less than that on a hammer-fired pistol because fewer components have to be accelerated. In general, striker-fired firing trains have the reputation of being made of fewer parts than hammer-fired mechanisms.[255]

A striker-fired fire-control mechanism comprises a striker, a mainspring, the trigger bar and sometimes other parts such as springs, pins, pawl and sear (Fig. 3.50), the latter of which might be found between the trigger bar and the striker. The sear serves as the part which retains the striker, whereas the pawl keeps the sear engaged or releases it once the pawl is actuated as the trigger is being pulled.

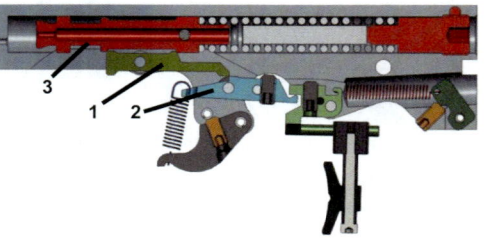

Fig. 3.50: Sear (1) and pawl (2) of a WALTHER *LG400* air rifle with striker (3).

(© WALTHER)

Pistols with firing mechanisms that are re-cocked when the slide travels rearward have more functional reserve as the slide is closing to battery. The work to cock the mainspring is already undertaken by the recoil, and the impetus of the recoil spring can then be used for other power consuming tasks, such as stripping a round from the magazine and feeding it into the chamber, and the proper locking of the slide.

Striker-fired pistols may lack this functional reserve if the firing mechanism is re-cocked as the slide is returning to battery. Yet all of this depends upon whether the striker spring bears on the receiver, as found on many early and simple blowback pocket pistols for low-powered ammunition types, or on a slide part. In the latter case, the striker spring's force will act in opposite direction to the recoil spring. Commonly, the force of the recoil spring decreases as the slide moves forward. At the same time the striker will get cocked, causing the mainspring's force to increase and act more and more against the forward motion of the slide – if the backstop for the striker spring is in the slide; which is a common feature of most modern striker-fired pistols.

Generally, all pistols can be considered to be a system of carefully designed springs, but particularly striker-fired pistols may have a tendency of not returning to battery while feeding a cartridge into the chamber if the slide's kinetic energy and hence the impetus of the recoil spring does not suffice. This might also be worsened if the shooter does not hold the pistol in the proper grip. To avoid this, pistols with weak recoil springs, or pistols for low-powered ammunition types are typically hammer-fired,[256] or if they are striker-fired, the design engineers take care that the rearward end of the striker spring bears on the receiver.

Striker-fired mechanisms do not have any external parts, and generally have the advantage of reduced dimensions[257] because they take advantage of the existing space above the magazine instead of bridging the distance between the primer and hammer by a passive firing pin. Depending on the recoil spring used, the force needed to retract the slide can be less than that on a hammer-fired pistol. On a hammer-fired pistol, the hammer will get cocked at the same time as the slide is racked whereas on most striker-fired pistols, the striker spring will get cocked when the slide is let go forward. This smooth cocking motion is handled by the recoil spring, and is done automatically without requiring any additional force to be applied by the user.

During the trials conducted by the German Army in trying to find a semi-automatic pistol, later to be known as *Pistole 08* designed by Georg Luger, this pistol's striker-fired action had been criticized. The firing mechanism would not indicate visibly if it was cocked. The Royal Prussian Rifle Commission had already objected in 1901 that there was no possibility of knowing whether the firing mechanism was cocked or not, a feature known with revolvers, where the cocked hammer would indicate the condition of the action.[258] Luger later modified his pistol in a way where at least the extractor would indicate if a round was in the chamber (patent 164,853, dated May 22, 1904).[259] During the early 1980s, GLOCK's design of a successful striker-fired pistol followed, and again it was a design without a cocking indicator.

SAVAGE pistols are a rare exception among striker-fired pistols. The main parts of the firing mechanism of these guns were in the slide and the action got cocked by means of a lever linked in the receiver while the slide moved rearward.[260] Because of this feature, the force of the recoil spring was used just for feeding and locking during the forward movement of the slide.

Trigger Mechanisms

The trigger mechanisms of modern pistols can be quantified into two types where the characteristics of the first shot is different from subsequent shots, and mechanisms with the same trigger characteristics for each round. On the latter, pistols with consistent trigger pull, the firing mechanism can be either uncocked or pre-cocked. Pre-cocked fire-control trains are sometimes referred to as pre-set trigger actions and can be separated into partially pre-cocked or fully cocked fire-control mechanisms (Fig. 3.51).

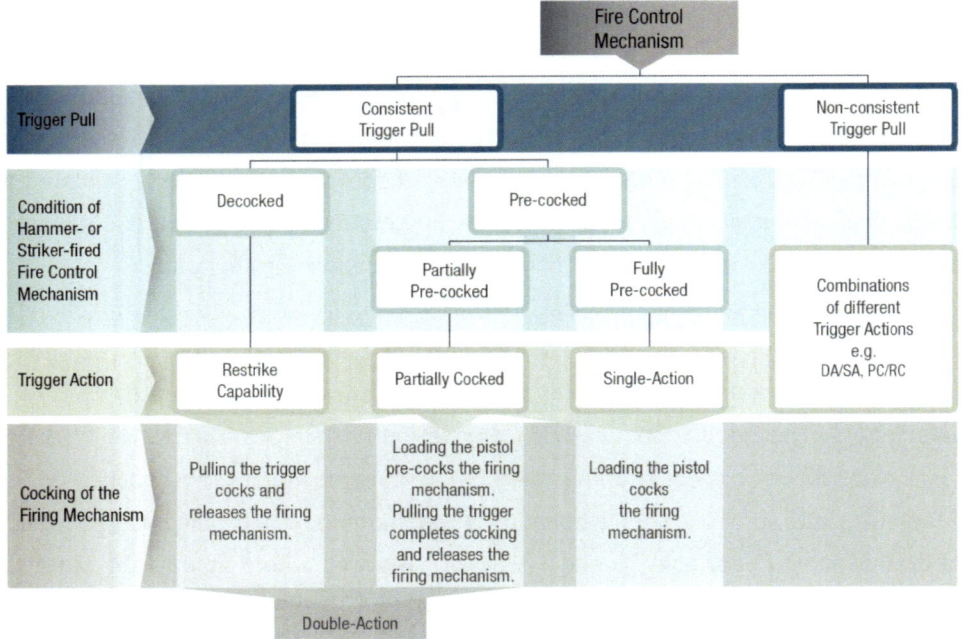

Fig. 3.51: Fire-control mechanisms of service pistols.

Single-Action Trigger

Single-action triggers are among the earliest trigger types, and date back to the 16th century when this innovation appeared in the Middle East, as the snaphance flintlock replaced the wheellock. In Europe, the first flintlocks were manufactured around 1630 and the technology was subsequently adapted for handguns. Flintlocks remained in service until about 1840.[261]

Fig. 3.52: Hammer of a percussion lock firing mechanism.

On these muzzle loaders the user had to draw back the hammer by hand for each shot.[262] This is why in Germany this action was called something which translates roughly as "manually-cocked hammers."[263] An improvement on the muzzle loaders took place in the second half of the 18th century with the event of impact-sensitive chemical compounds,[264] which culminated in the percussion lock (Fig. 3.52). The hammer used on the percussion lock mechanism was transferred to revolvers and pistols but still had to be cocked manually. This type of trigger action is called single-action (SA). It specifies a fully cocked firing mechanism, which releases the hammer or striker when the trigger is pulled.[265] Most of the revolvers that came out in the middle to the end of the 19th century were single-action.

The mechanical design of SA fire-control mechanism is not necessarily tied to a hammer. In the early days of self-loading pistols, they were also used in striker-fired guns, for example, in the first commercially successful automatic pistol in the world, the *C93* pistol[266] designed by Hugo Borchardt, as well as in the Luger pistol, which emerged from the *C93*. In the race of the firing mechanisms, the hammer-fired pistols lead. An outstanding example of this period in time are pistols based on the COLT *Model 1911* pistol, which are still popular today.

Regardless of whether a SA pistol is hammer or striker-fired, the gun can only be discharged upon cocking the firing mechanism prior to firing the first round. This can be achieved automatically, for example, when chambering a round as the user racks the slide. During the discharge, semi-automatic pistols use the reciprocating slide to cock the firing mechanism for the next shot. With their fully cocked firing mechanism, SA guns are commonly characterized by a short length of trigger travel, moderate trigger pull weight, and by a crisp trigger break at which the shooter feels a clear point of release with his finger on the trigger blade. Typically, a SA trigger is a two-stage trigger, where the first stage is the take-up and the second stage, where the sear releases the hammer or striker, is called the trigger break.

Another feature of SA triggers is the energy stored in the mainspring of a cocked firing mechanism. The energy must be sufficient for a dependable ignition of the next shot. However, an unintentional discharge may be caused by this condition, for example, in the case of a jar, a blow, a drop or vibration, if no safety is engaged at the same time.

One more trait, especially on a variety of striker-fired pistols, is that these pistols can be field-stripped[a] only when the firing mechanism is uncocked (see Fig. 3.53). On some of these pistols, the decocking

Fig. 3.53: Trigger bar and striker at the point of release shown on the example of a striker-fired pistol with partially cocked firing mechanism.

---

[a] Field-stripping a pistol: To break down or disassemble a pistol into its major components.

requires a separate manual operation, such as pulling the trigger, and a negligent discharge may be caused by this act. Therefore, an increasing number of manufacturers offers pistols in which their firing mechanism will automatically decock in a safe way as the user starts to strip them down. HECKLER & KOCH *VP9*, SPRINGFIELD *XD^M*, SIG SAUER *P320*, and RUGER *American Pistol* are examples of pistols which can be field-stripped without prior need to pull the trigger.

Worldwide, the term single-action is used for a trigger action that performs exclusively the one and only action of releasing the energy of the mainspring. Sometimes, the term single-action only (SAO) is used, especially when it comes to semi-automatic pistols, if there is the need to clearly differentiate between a pistol with a SA trigger and other pistols with a combination of SA and restrike capability.

Fig. 3.54: RUGER *American Pistol*, shown with Micro Red Dot Sight, slide open, magazine removed.

Restrike Capability

Invented in the 1830s,[267] this fire-control mechanism first gained popularity in revolvers as an alternative to single-action revolvers, which required drawing back the hammer by hand. Technically, restrike capability was relatively easy to achieve, as trigger and sear are arranged in direct distance from one another on a revolver. This allowed users to carry a revolver with hammer down, and to cock (action #1) and release (action #2) the hammer by squeezing the trigger. Hence, this trigger action was originally called double-action. Today, semi-autos with double-action triggers can either be hammer-fired or striker-fired.

Whether a gun's trigger has restrike capability can be found out by simply trying to pull the trigger repeatedly on a cleared gun without racking the slide and any other manual operation. If the hammer or striker falls every time the trigger is pulled, it is a trigger with restrike capability.[268]

With the advances of modern semi-automatic pistols, there has been a lot of confusion regarding the use of the term double-action. Today this particular feature of a trigger is often described as double strike capability or restrike capability instead.

Based on the history of trigger actions in the U.S., trigger actions are called after the trigger action of the first shot of a gun in carry condition. Traditionally, revolvers used to be single-action, followed later on by revolvers with both double-action and single-action, which were called double-action revolvers.[269] They were then followed by revolvers with shrouded hammer. With a bobbed off spur and the hammer being covered, these revolvers would not allow thumb cocking, hence every shot had to be taken by a double-action trigger and they were referred to as double-action only (DAO).[270]

The same applies to semi-automatics. A pistol with a DA/SA trigger is a "DA pistol", also sometimes known as traditional DA. They are double-action to single-action pistols, meaning DA trigger for the first shot, then the subsequent recoil action of the slide cocks the firing mechanism for the next shot. If a pistol trigger is consistently double action for every shot, then it is often called a double-action only (DAO).[271]

To find out what the official definition of the term double-action in the United States is, it is easiest to refer to the definition used by the National Institute of Justice (NIJ). The NIJ defines a double-action as "A mode of operation that permits a single pull of the trigger to cock and fire the pistol."[272] According to this definition, the term double-action just focuses on the two aforementioned actions, that is, cocking and releasing as the trigger is squeezed.

Remarkably, the NIJ definition of DA does not call out for a repeat strike capability of the trigger. It might be worth noting that it might cause some misunderstanding when communicating with Europeans because their understanding of double-action might strictly be connected with restrike capability.

As per the NIJ definition of DA, a partially cocked fire-control mechanism is regarded to be a DA trigger as well.

Partially Cocked Trigger

In the United States, partially cocked triggers lie within the definition of double-action triggers. The partially cocked fire-control mechanism is the most recent trigger action in our consideration. In the diagram Fig. 3.51 (page 96), it is located between single-action, i.e. fully cocked, and the firing mechanism with restrike capability, i.e. 0 percent cocked triggers. This trigger action can be any sort of partially cocked ranging from more than 0 percent to less than 100 percent, that is, it will neither be cocked at exactly 0 percent nor 100 percent. If however, after pulling the trigger the gun goes "click" and the slide does not cycle, the trigger of a now decocked "partially cocked fire-control mechanism" will not allow it to be released again by simply pulling on the trigger blade once more, just in the same manner as a single-action trigger would not work with a hammer down.

In many cases partially cocked firing mechanisms are found on striker-fired guns. In the very same year as mentioned in the model name of their Model 1907 pistol, ROTH/Krnka introduced a striker-fired pistol with partially cocked trigger.[273] It took how-ever until the 1980s for striker-fired pistols with partially cocked fire-control mechanism to become popular on a large scale. This happened with the market entrance of Gaston Glock's pistol in the 1980s.

NIJ's definition of "Striker Fire Action" is similar to what is already described above with regard to double-action.

> *A pistol design which employs an internal striker mechanism to detonate the primer. In operation, the pistol is normally in a partially cocked condition. Pulling the trigger com-pletes cocking the action, and then releases the striker mechanism to fire the pistol.*[274]

As it will be seen later, this definition is of importance when the Bureau of Alcohol, Tobacco, Firearms and Explosives (ATF) has to assess whether a gun may be im-ported into the United States (Chapter 4). The fire-control mechanism of GLOCK pistols as well as other makes of partially cocked pistols qualify as double-action triggers, which in turn has a positive effect when the model is evaluated for importation by ATF.[275]

A gun with partially cocked trigger can be verified by pulling the trigger on an empty chamber without racking the slide. Its trigger cannot be released a second time, but only after pre-cocking the firing mechanism by retracting the slide. Also, on some pis-tols, it will be noticeable on the index finger if squeezing the trigger completes cocking the action. This in turn is the salient characteristic between a partially cocked trigger and single-action; the latter releases at a more or less crisp point of release, without cocking the mainspring while squeezing the trigger.

On the one hand, when compared to triggers with restrike capability, the advantage of a partially cocked fire-control mechanism lies in the shorter trigger travel and a pos-sibly lighter trigger pull force. The latter depends however on the individual mechanical design of a pistol model's fire-control system.

On the other hand, the partially cocked mainspring may accidentally release the en-ergy stored in it when the weapon is being jarred, dropped or otherwise mishandled, which might deliver just enough energy to fire a round, if other safety elements acci-dentally become disengaged at the same time or if there are no safeties at all included in the design. The TR "Pistolen," that is, the law enforcement standards applied in Ger-many that sets performance standards for standard-issue sidearms, defines a limit until which a firing mechanism is regarded to be as save as an uncocked mechanism. The maximum is a 0.11 mm [0.004 in; PD] striker indent when a copper crusher gauge is used instead of a live primer.[276] Internationally, this subject receives little attention; quite in contrast to the issue of how to decock a pistol prior to field-stripping.

Some suggest that the combination of an operator's index finger and a gun's trigger should have only one function, which is to discharge a round. If the trigger was to fulfill two tasks, that is, discharging a round on the one hand, and on the other hand it is used during field-strip procedure, the latter might cause a negligent discharge during the disassembly process.[277]

Before we continue with the combinations of the three basic trigger actions, a brief overview of the features previously described should be reviewed:

Single-Action Trigger

The firing mechanism is fully cocked. Pulling the trigger releases it. It can be released once again only if the slide was cycled, either as a result of the auto-loading operation or if its cocked manually, e.g. during dry fire practice, such as racking the slide or thumb cocking a pistol with exposed hammer.

➢ Specific for this trigger action is in most cases a short trigger travel in combination with a low pull force.

Restrike Capability Trigger

The firing mechanism is not under spring tension. Trigger pull alone both cocks and releases the firing mechanism. This sequence can be carried out repeatedly without cycling the slide, i.e. in the event of a misfire or during dry fire practice.

➢ Specific for this trigger action is a fair amount of trigger travel and pull force.

Partially Cocked Trigger

The firing mechanism is neither uncocked (restrike capability) nor fully cocked (SA). Pulling the trigger completes cocking the mainspring, and then releases it (see restrike capability trigger). In the event of a misfire, the firing mechanism cannot be released again by pulling the trigger without further manual operation (see SA trigger).

➢ Specific for this trigger action is in most cases a short trigger travel during which the firing mechanism will get fully cocked and released. The pull force of pistol models with partially cocked trigger mechanism may vary depending on the mechanics of the respective fire-control mechanism.

Double-Action

The common feature of single-action, restrike capability and partially cocked trigger actions is that they offer a consistent pull weight for each shot. Historically, when service guns had SA triggers, it had always been requested to make the trigger safer to avoid unintentional discharges. It took until the beginning of the 20th century to find solutions to combine two different trigger actions.

Combinations of Trigger Actions

Traditionally, like in the days of muzzle loaders, guns with a fully cocked firing mechanism were regarded to be prone to unintentional discharge. The designers of semi-automatic pistols attempted to eliminate this risk by designing fire-control mechanisms which would allow to carry a gun safely, fire the first shot with a heavy trigger pull weight and for subsequent shots the trigger would be single-action. An early gun with two different types of trigger actions was the MANNLICHER Model 1894. The function of the trigger is literally similar to a revolver with restrike capability and singe-action, that is, when this semi-auto pistol is fired, the hammer does not automatically get cocked.[278]

An early pistol, with restrike capability for the first shot and single-action for subsequent shots like on many modern pistols, is based on the Austrian Patent No. 4430 of February 2, 1900, by the company G. ROTH, Vienna.[279] However, the internal mechanical parts of the trigger mechanism were on the side of this pistol, just like on a revolver.

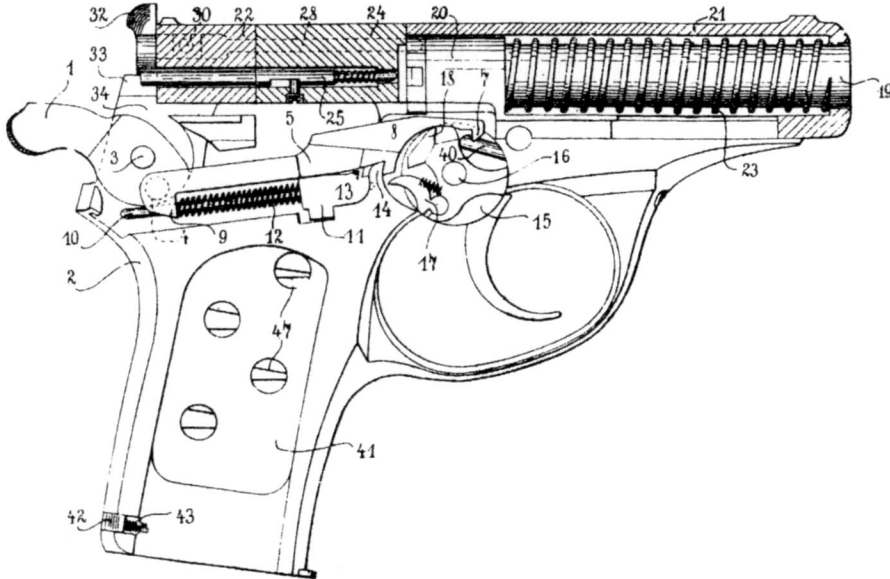

Fig. 3.55: Fig. 3 from Patent AT 36,387, Alois Tomischka.

In 1908, Alois Tomischka had then patented a pistol with restrike-capability/single-action trigger (*Little Tom*, Fig. 3.55). He claimed that an intermediate link between trigger blade and hammer is used to cock the hammer. This intermediate link was required, since removeable magazines got popular and the so needed magazine well inside the pistol's grip would put the trigger blade in a considerable distance to other parts of the firing mechanism.[280] Tomischka's design of a restrike-capability/SA trigger paved the way for trigger actions that are still used today.

Although, handguns of this kind are equipped with two trigger actions, the action will be named in the United States after the trigger action for the first shot of the gun in carry condition, which is mainly Double-Action (DA), sometimes also referred to as DA Auto, or Traditional DA.[281] This linguistic distinction is made to differentiate them from SA and DAO pistols.

In more recent times, the manufacturers also combined other trigger actions.[282] HECKLER & KOCH created its *Law Enforcement Module* (LEM) and the *P30 V0* pistol; SIG SAUER launched the *Double-Action/Kellermann* (DAK) trigger, and the TAURUS *PT 24/7 PRO* would have a combined trigger; whilst WALTHER offered its *P99Q* and *PPQ M3 / PPQ P3*. The standard trigger of any of these pistols is a pre-cocked firing mechanism, and they offer an additional action with restrike capability. The latter would allow the user to pull the trigger again and again in case of a failure to fire.

When combining two trigger actions in a pistol, engineers are striving to minimize the number of components of the fire-control mechanism and as a result, the cost. But in most cases combining two separate trigger mechanisms requires additional components. For example, the restrike trigger of a hammer-fired pistol might use the trigger bar to cock and release the hammer, but for single-action, a sear might be used to retain the hammer until the sear is levered out by the trigger bar.

In a design like this, it will be important to have separate release points, so that the hammer will not collide or get caught by the trigger bar's restrike functional face, when the hammer is released from the sear in SA; or when firing with restrike action, the trigger bar must release the hammer soon enough to keep the sear from engagement with the sear notch in the hammer. In both cases, the firing mechanism must transfer sufficient impact energy into the primer, even if its restrike action releases a little bit earlier than single-action and hence the mainspring will not get cocked as far.

The fire-control mechanism of the TARA *TM-9* pistol is different in its own way. For the first shot it uses a restrike trigger action, and for subsequent shots a partially cocked action. However, it does not switch automatically from one action to the other. This is initiated by the position of the trigger blade, and the trigger finger of the shooter. If the shooter leaves the finger on the trigger blade while the slide cycles (anything else would be almost impossible), the firing mechanism will be partially cocked. As soon as the shooter removes his finger from the trigger blade, the blade, trigger bar, and striker move forward, and this decocks the firing mechanism as a result.

Disconnector

As the slide cycles, the firing mechanism will be re-set. Before a hammer- or striker-firing mechanism can be re-set, the members which were needed to release the firing mechanism must get disengaged from each other as they would prevent the mechanism from getting back to its default position as long as the trigger is still under pressure by the firearm operator, hence keeping the firing mechanism in "released" condition.

The mechanism that interrupts what is holding the firing mechanism in released position is called the disconnector function. Without it, chances are that a self-loading gun might shoot full-auto. How much the latter is likely to happen depends on the individual design of the fire-control mechanism. For example, a hammer-fired pistol might not tend to shoot in multiples, as the hammer is held down and guided by the slide's stripper rail as the slide cycles, and as the slide closes, the hammer may not transfer enough kinetic energy to the firing pin in the event of a broken fire-control mechanism. But without the disconnector function, the firing mechanism would not get re-set and the discharge of another round by pulling the trigger might be impossible.

Before the next shot can be fired, the trigger mechanism must be re-set as well. The trigger is usually pushed forward by a spring. Releasing the pressure on the trigger blade will allow the trigger to return forward. At a certain point, this will re-engage the trigger mechanism attached to the trigger blade. What is called *reset* is the distance the trigger must travel forward from the trigger stop until the trigger mechanism is ready to release the firing mechanism again. The term "short stroke" would be used if the user fails to release the trigger far enough and trying to pull again without reset.[a] Usually, shooters acknowledge a tactile and audible trigger reset, which is at the same time short and smooth.

---

[a] *Firing on empty* would be used if the operator keeps trying to fire another shot on an already empty mag. One might also refer to it as *slidelock unawareness*.
A *dead trigger* is when the trigger goes all the way to the stop with nothing happening mechanically, e.g. the sear is not letting the striker drop.

The disconnection is functioned by the movement of the working parts as each shot is fired. Fire-control mechanisms of pistols using trigger bar and sear to release a hammer or striker normally use the recoiling slide to actuate the disconnector function. A recess under the slide will mechanically control the timing of when the trigger bar will get separated from the sear. This separation will allow the spring-loaded sear to now move independently of the trigger bar and to get re-engaged in the hammer or to block the striker again.

On hammer-fired actions, it is important to allow the sear enough time for a positive engagement with the sear notch in the hammer. To make sure enough time is given for the re-engagement, the slide rotates the hammer rearward past the engagement point of the sear. This is called over-travel, as opposed to the distance when the hammer snaps forward after the sear releases it, which is the hammer-fall. The combination of the both is called hammer travel.[283]

As we are defining terminology at this point: the disconnector would be that specific component which disengages the trigger mechanism from the firing mechanism.[284] The design of such may be any separate component or one that is actively involved in the release of a shot. Known solutions to function the disconnection are to utilize the rotation of the hammer, the movement of the hammer strut, levers, or parts of the receiver.[285]

The disconnector, as in most self-loading pistols, offers two functions. Firstly, it prevents more than one shot being fired with a single function of the trigger, and secondly it assists to avoid unintentional discharges. In most auto-loading pistols, the disconnector safeguards that the firing mechanism cannot be released for the next shot unless the slide has returned to battery, i.e. the cartridge case is fully supported by the chamber and the breech face, thus it works as a passive safety.[286]

Some striker-fired guns will release the striker by a direct interaction of trigger bar and striker, e.g. the sear is part of the trigger bar and no other intermediary parts are involved (Fig. 3.56). Squeezing the trigger causes the trigger bar to push the striker rearward. Also, the trigger bar is guided downward at a certain point and releases the striker. The trigger bar will initially remain in lowered position, and while the slide recoils, the rear end of the trigger bar must be raised again to positively engage with the striker as the slide closes.

After this cycle, the firing mechanism is partially cocked again. At the latest, the approach of the GLOCK pistol presents a technical solution, in which the disconnector moves transversely to the direction of the shot.

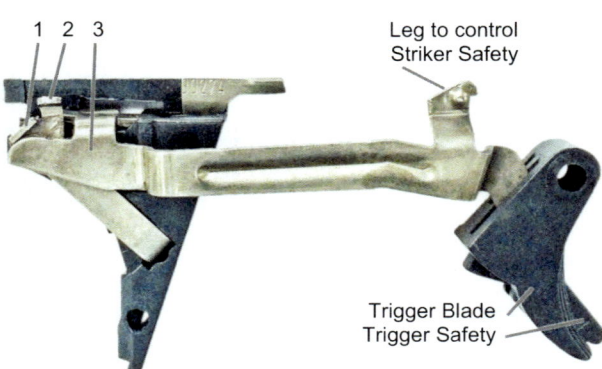

Fig. 3.56: Fire-control Mechanism of GLOCK pistols: The Trigger Bar (3) is cammed down at a Ramp (1). As the slide cycles, the Disconnector Leg (2) is pushed towards the center of the gun and at the same time, the Trigger Bar is raised (spring not shown).

Some fire-control mechanisms will allow the release of a shot, even though the slide is out-of-battery. A click might be heard when the firing mechanism releases, and it means this disconnector is not of the type which disrupts as long as the parts involved are not in correct position. Chances are that a design like this uses a striker safety instead to keep the gun from firing with the slide not fully closed. This idea, however, might work less reliably than a real disconnector.

In cases where a striker block is not fully engaged, the striker might cause a light tap on the primer, which might be just enough to initiate the primer. Depending upon how much energy is transferred into the primer, how far offset from the center the primer is being struck, and the thickness of the primer cup, all these could be enough to cause an ignition.

Please note: Do not try to dry fire a gun with the slide not fully closed. With slide out-of-battery, the proper control of the safety devices and the firing mechanism is not given and sensitive functional edges of parts involved can get damaged. This may cause a severe safety risk.

Primer Initiation: Kinetic Energy

The priming compound inside the primer cup has a crystal structure, and for an efficient ignition of the compound, it is important to break as many crystals as possible at one time. This is done when the nose of the striker transfers the striker's kinetic energy into the primer cup. In doing so, it deforms the primer cup, thus crushing the sensitive priming compound between the cup and the anvil (Fig. 3.57) to produce hot gaseous products and a shower of incandescent particles to ignite the powder charge.[287]

Note: We are going to refer to the tip of the striker, that is, the front end which is driven into the primer cup, as the nose of the striker or firing pin. The reason is that the front end should not be too pointy to avoid pierced primers.

Even though the initiation works with modern guns and ammunition with extreme reliability, failures to fire (Fig. 3.58) remain to be a severe issue, and whenever a failure to fire occurs, the question will be whether it was caused by the firearm or the ammo.

Fig. 3.57: Center fired ammo: The anvil can be part of the primer pocket (Berdan, left) or primer cup (Boxer).

Small arms and ammunition manufacturers in Germany assess striker energy using the copper crusher test. They will chamber a headspace gauge, which is in the shape of a cartridge case but instead of a primer cup, it can hold a copper crusher cylinder (Fig. 3.59). Then, the gun's trigger will be pulled. The firing pin strikes the copper crusher cylinder and indents it. An indent of 0.3 mm [0.012 in; PD][288] or more indicates that the firing mechanism of the tested pistol transfers enough kinetic energy (cf. A.7 Copper crusher indent, maximum diameter, p. 225ff, and Appendix 24). Hence, a failure to fire would be regarded to have been caused by the ammo.[289]

Fig. 3.58: Failure to fire: Indent after a one-time strike, 0.35 mm deep.

Some guns will reliably set off the primer, although they produce indents in copper crushers of less than 0.3 mm. This will be made possible, if the energy of the striker's nose still crushes enough crystals of the priming compound to initiate it.[290] Generally, the shape and size of the striker's nose and the impact velocity of the nose are crucial to break a sufficient number of crystals at one time to fire the cartridge. The measurement of the indent depth with copper crusher cylinders does not take into account

Fig. 3.59: Headspace gauge 19.15 and copper crusher cylinder, Type 5 × 7 mm.

these factors. However, the method provides comparison values that can be used as a reference for the impact energy of a particular pistol model.

In assessing the function of strikers or the impact energy transmitted by it, experiments with primed casings instead of copper crusher cylinders will not suffice. Measuring the depth of the indentation on a spent primer, e.g. on an empty primed casing used for this test, yields an incorrect result because the gas pressure caused by the initiated priming compound inside the primer pocket will blow back the primer cup against the nose of the firing pin. Non-reproducible marks of the firing pin nose will be the result, which will highly depend on the type of the primer, the thickness of the cup and the resulting gas pressure of the priming compound. The indent will be nicely deep on a used primer, but shallow on the event of a failure to fire or when a copper crusher is used.

Another criterion for the reliability of the ignition is the eccentricity of the strike in the primer cup. The TR "Pistolen" of the German authorities accepts 0.2 mm eccentricity, determined from the offset of the points of intersection of the centerlines of the firing pin dent and the central axis of the test cartridge.[291]

Striker Nose

An awareness of the graves of the topic of "firing pin nose" has been discussed since before the outbreak World War II. In the year 1938, Konrad Eilers writes with respect to hunting rifles:

> [...] Wrong shape or off-center positioning of firing pins may cause failures to fire, jammed bolts, or broken firing pins. The firing pin must neither be too pointed nor too blunt nosed. It must be driven into the primer without too much friction [...].[292]

The described friction can be tested by means of aforementioned copper crusher test. Also, mechanical defects such as burrs, scuffs and dirt can affect the strike-energy, and can be detected easily by comparing striker indent in copper crushers. However, this can only be evaluated correctly if the minimum requirement for indentation depth is known for the particular pistol model.

The shape and size[293] of the nose of the firing pin has great influence on the effectiveness of the firing pin; a large diameter and flattened nose can cover a larger area with regard to the anvil and thus break more crystals in the priming component, at the same time a large, flat nose may cause shallow indentations in the primer cup or copper crushers.

Fig. 3.60: Bulged primer: infamous prestage of a pierced primer.

Another criterion in the design of the firing pin nose is to avoid pierced primers: the diameter of the firing pin nose and thus the firing pin hole in the breech face must stay within reasonable dimensions because the primer might just tear, which is known as pierced primer (Fig. 3.60 and Fig. 3.61). Pierced primers pose a danger to the shooter because hot gases may be dis-charged under pressure from the weapon to the rear while the gun is held in line with the eyes and thus the face of the shooter.

Fig. 3.61: Pierced pri-mer, REMINGTON *Golden Saber* 147 gr JHP.

Allsop sums it up as follows:

- Excessive protrusion of the nose;
- Excessive striker energy; and
- A sharply pointed nose will produce pierced primers;
- The nose must remain in contact with the primer and must be adequately supported as the chamber pressure rises, to prevent the primer perforating and allowing high-pressure gas out into the firing pin hole in the bolt.[294]

If the firing pin hole in the breech face is too large, the primer cup is not properly supported and can possibly tear and will not provide its function for sealing against gas pressure towards the breech face and the firing pin hole. A diameter of about 1.6 mm at the nose of the firing pin has proven beneficial in practice. Exceptions are possible; for example, the WALTHER *P38* and *P5* pistols firing pin noses have a diameter of 2 mm. The shape of the *P38*'s firing pin nose is rounded and the one of the *P5* is flat; and since the introduction of the GLOCK pistol the lance-shaped striker nose in the approx-imate footprint of 0.8 × 2.5 mm (Fig. 3.62) has been existing. According to U.S. 4,825,744, filed 1988, the shape of a flattened triangular was chosen to "eliminate the problem of firing-pin tip breakage which plagues automatic pistols."[295]

The run-down test, a test for the determination of the impact sensitivity of the primers, uses a firing pin which front end is rounded into a hemispherical shape with a 1 mm [0.039 in; PD] radius. Under the conditions specified therein, a drop height of 350 mm [13.779 in; PD] is enough to drive the nose 0.3 mm [0.012 in; PD] deep into a copper crusher. This 0.3 mm are con-sidered to be the all fire condition, i.e. the functional limit at which primers must go off.[296]

Fig. 3.62: Lance-shaped striker nose of a GLOCK *G19* pistol.

When firing a cartridge, metal particles, e.g. metal particles torn off the primer by the firing pin nose, remain in the gun or are blown out of it. The particles can be caused, for example, if the nose of the striker is still in the indentation made in the primer and as the barrel is cammed down, the pin is pried out and drags across the edge of the indentation it has made.[297] In many cases, the material of the primer cup also floats around the striker nose and as the barrel is set in motion, the material is sheared off. Most guns with coupled barrel and slide group cam the barrel downward, which causes distinct shear marks as the primer material is forced out of the firing pin hole. Whether this shear mark occurs on a particular pistol model can be deter-mined if fired casings are inspected (Fig. 3.63).[298] If a firing pin drag or a shear mark is identified on the primer, chances are that metal particles will be found in the gun and the more obvious the striations are, the more parti-cles will be present in the gun.

Firing pin drag

Shear mark

Firing pin im-pression

Breech marks

Fig. 3.63: Striations and action marks.

In both cases, hammer and striker-fired pistols, it is important for the safety of the shooter that the nose of the striker does not protrude at the breech face dur-ing feeding. A protruding striker nose might cause a failure to feed, or even worse, an out-of-battery dis-charge. Numerous designs use a spring retracted striker or firing pin, to withdraw the nose. An excep-tion to these are MAKAROV and GLOCK pistols. Instead of a retracted striker, GLOCK uses a lance-shaped nose (Fig. 3.64). In case the nose would extend be-yond the breech face on a *Glock*, the base of the case could push the nose out of its way with no obstruction.[299]

Fig. 3.64: Protruding nose on a GLOCK *G19* pistol.

Retracted strikers are also used in many striker-fired pistols to withdraw the striker and set it back to a default position that is required to reset the firing mechanism and to push the striker behind the striker safety for a new engagement.

Modern pistols use inertial firing pins or strikers, that is, the nose of the striker does not protrude but is slightly recessed in the breech face, and the firing pin of a hammer-fired pistol would be shorter than the distance between the breech face and the hammer. In most cases, a retract spring is used to restrain the striker or firing pin.

Once kinetic energy is transferred to the striker or firing pin, it retains its new state of motion until it reaches the primer cup. A shorter firing pin helps to improve the safety of handguns, but the inertia of a firing pin held just by a retracting spring may also be a risk. When the slide slams abruptly into battery, the firing pin must not set off the primer. Whether a given pistol model is prone to slamfire, it can only be detected by testing and typically, a worst-case scenario would be where retract spring, striker block, etc. were removed from the gun before shooting live rounds.[300]

Fig. 3.65: Broken striker after dry fire endurance test.

Dry firing a handgun is extremely hard for the breech face and the striker, if the striker head smashes against the rear of the breech face at full speed. On any part of a striker where an abrupt change in its profile shape exists, heavy stresses will occur and might cause a great risk of breakage (Fig. 3.65). The failures shown in Fig. 3.65 and Fig. 3.66 were found after dry fire endurance testing.

Note: In the area of mechanical engineering in general – and especially in the design of guns – changes in cross-sectional areas are to be avoided and instead, smooth transition should be used to reduce the risk of cracks caused by stress risers.

Fig. 3.66: Cracked breech face after dry fire endurance test.

Summary Fire-control Mechanism

In summary, it can be said that the requirements of users range from rather safe triggers with long and heavy pull to very light ones with short reset. In the United States, the trend is to have generally easy to use triggers with a consistent mode of operation for each shot.[301] Ease of use is paramount in the U.S., as first-time buyers do not want to deal with technical features, while consumers in general prefer products that are convenient to use, particularly when confronted with evil.

3.7   Springs

Springs are commonly used to store mechanical energy, that is, to store mechanical work in the form of
elastic potential energy. A spring becomes deformed under the action of a force or torque and subsequently releases the stored energy with the
resumption of its initial form.[302] In gun design, a variety of different spring shapes are used: torsion
springs, tension coil springs, compression coil
springs and springs of non-cylindrical shape; a kind
of non-cylindrical shape would be used in magazines. In this chapter we are going to refer to helical
compression springs by the term "compression
spring."

Fig. 3.67: Small springs are wound
from spring steel on an automatic spring winding machine.

A spring and a mass attached to it forms a harmonic oscillator. In other words, the
spring with a member attached are normally in an equilibrium position and when displaced from its equilibrium position (pulse), a restoring force proportional to its displacement will occur which might cause harmonic motion (cf. Appendix 19, p. 418ff). A
compression spring that has experienced a pulse will first compress, and in doing so
develop a restoring force which pushes the member back to equilibrium position. At
this point, post-pulse oscillation can be experienced in various cases, in particular with
springs used in high-speed cyclical applications. Sudden shocks in combination with
high speeds can induce clashing and impact of adjacent coils, and for those springs
already carrying considerable loads, this post-pulse oscillation can cause extremely
high operating stresses inside the spring wire.[303]

Gun engineers are aware that compression springs under repetitive impact loading
conditions are highly stressed components. These springs absorb sudden shocks with
every round fired, e.g. recoil springs. Most users know little about factors affecting service life of springs or the scientific design effort required to give a reasonable longevity.

In material science, fatigue is the progressive weakening of a material caused by
cyclically applied loads. If the loads are above a certain threshold, the material will
fracture. This is called the fatigue limit of the material, and if it will operate indefinitely
without failure, this is referred to as endurance limit. This would be the standard verbiage in mechanical engineering.

When it comes to firearms, things are much different and everyone knows that gun parts can break. It is called low-cycle fatigue, where materials have a finite endurance for a certain type of load. Being aware of this, most law enforcement agencies demand a 10,000-round service life for the main parts of their service pistols. The main parts are typically considered to be the barrel, frame and slide.

Springs are considered to be expendable parts, for which normally a minimum life of 5,000 rounds is accepted.[304] Regardless of these theoretical considerations, the reliability of a spring can make the difference between life and death. A broken spring can lead to a catastrophic failure of the weapon and may put the operator in extreme danger if this malfunction occurs when he is face to face with a bad situation.

When a spring is made and then compressed the first time, the spring may get shorter. This depends on the stresses in the wire and it is referred to as "preset," or "removing the set." Once the spring takes this set, it will generally not take any significant length loss during operation. In case of an overstressed design, the spring will take a set over time, losing its length and load. When this happens with a spring in a firearm, it might not cause a life-threatening situation. In fact, malfunctions may happen more frequently, indicating the gun needs to be serviced.

Nevertheless, the setting phenomena is still a challenge not to be underestimated by gun designers. Dr. Karlheinz Walz recognized in 1944 when he wrote: "In the field of low-cycle fatigue caused by sudden shocks, the limited life cycle is less of an issue, as it would be the setting. Setting will influence and reduce the work performance of a spring." [305] This reduction of work performance can lead to malfunctioning of weapons. In the 1930s MAUSER had a study group, headed by Walz, set up to pursue a solution to improve the setting and the life cycle of springs.[306] Here, deflection vs. time diagrams gave helpful insight.[307] It was discovered the post-pulse oscillation caused the fracture of springs. As the gun and some of its functional parts move heavily during discharge, the springs get accelerated. They begin to oscillate and will still vibrate after the discharge. It was seen by Walz: "Long springs with small outside diameters are particularly vulnerable to setting. Setting is a plastic deformation of the spring and sudden shocks, such as happening in firearms, cause setting on a larger scale than would be experienced during a fatigue test."[308]

An important consideration in high-speed cyclic applications is the avoidance of waves of high inter-coil displacement travelling along the spring leading to high operating stresses, clashing and impact. In long springs subject to sudden shocks, an inter-coil displacement may occur, leading to clashing and impact of coils. This effect is referred to as surging. It can be observed by high-speed video (Fig. 3.68). The typical response for a recoil spring after a round has been fired would be the effect of post-pulse oscillation, where waves of high inter-coil displacement repeatedly travel back and forth along the spring before it finally comes to rest.

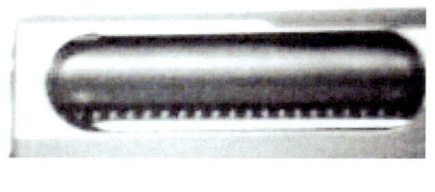
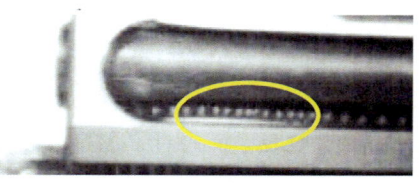
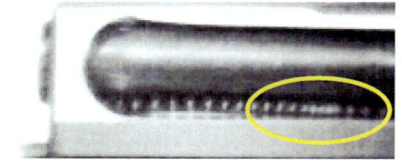

Fig. 3.68: Wave of high inter-coil displacement travels along the spring during recoil (shown at the occasion of an obstructed barrel test).

The team at MAUSER concluded that the service life of springs for fully automatic weapons would increase if they would find a way to diminish the post-pulse oscillation. And as surging arises from resonance between the operating and natural frequencies of the spring, MAUSER tried to dampen the vibrations of long springs. One of their approaches was to divide the spring into several shorter springs which changed the natural frequency of the spring system.

Another thought was to use springs arranged within one another, what equates to using two springs in parallel (Fig. 3.69). This approach, however, presented the disadvantage of a larger

Fig. 3.69: Concentric recoil springs of a RUGER LC9 pistol.

outer diameter, which can be problematic because of the typically limited design space envelope. Concentric springs can also damage one other as a consequence of changes in diameter and lateral deflection (buckling) when loads are applied. Concentric springs are typically arranged with each nested spring having an opposite hand helix to reduce the chances of one plane buckling.

Unlike MAUSER, small arms manufacturers in the Soviet Union had previously found a solution for springs under repetitive impact loading conditions and had already put it into service:[309] Boris Gabrielovich Shpitalny,[a] a Soviet weapon design engineer, had developed the multi-stranded coil spring (Fig. 3.70). A stranded spring was used from 1934 as a recoil spring in the Russian machine gun ShKAS[b].[310] This machine gun fires at a high cyclic rate of 1,500 to 1,800 rounds

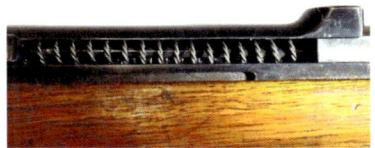

Fig. 3.70: Stranded coil spring in a SIMONOV *AVS-36* rifle.

per minute[311] and it was chambered in 7.62 × 54 mm R.[312] During the testing phase, this machine gun had to undergo some changes to improve its service life. The challenge in the development of automatic weapons is that each part can cause considerable problems without warning, which then has a major impact on the overall system. In this case, the most problematic part turned out to be the recoil spring. Despite all efforts being made, the original recoil spring was of insufficient strength and failed after 2,800 rounds or less. The engineers tried different types of steel, spring diameters, and wire diameters, but nothing helped. After a certain number of shots, they always had to stop shooting to change the recoil spring. Shpitalny later found an extraordinary solution. He proposed to make a stranded spring. Tests proved that such springs significantly improved the life cycle as compared to normal springs. Service life could be brought to the same level as that of other parts of the weapon.

According to a test report dated December 24, 1934, the *ShKAS* machine gun was test fired that day with a three-wire stranded spring. It was noted that the stranded spring was able to withstand 14,000 rounds, while a normal compression spring would last only 2,500 to 2,800 rounds under similar test conditions. In this case, the stranded spring increased the service life by a factor of 4.5 to 5.5.[313]

The Germans found out about this new spring design during the Spanish Civil War (July 17, 1936–April 1, 1939), when members of the volunteer Condor Legion[314] came across a stranded spring in a machine gun of a Soviet fighter aircraft.[315] The spring was made of three wires that were twisted around one other.[316]

As it was found at MAUSER, it proved to be very beneficial if stranded springs consist of three, five or seven wires. Besides the longer life cycle, another advantage of stranded springs is that their overall length can be shorter, or that they can produce

---

[a] English references sometime refer to the name also as Boris Gavriilovich Shpitalniy or Spitalny, whereas German references will refer to Boris Gawrilowitsch Schpitalny.

[b] *ShKAS*, short for <russ.>: aviatsionniy skorostrelniy pulemet sistemi Schpitalnogo/Kogaritskogo, <English>: Shpitalny-Komaritsky Aircraft High-Speed Machine Gun. (Chinn, 1952 p. 72) In English references the model name depends on how the name of the inventor is spelled: *ShKAS* or *Shkas*, or in German references *Schkas* or *SchKAS*.

high spring rates with relatively small overall dimensions, and MAUSER found out that the settings of stranded springs would be less than those on single-wire springs.[317] During the tests of the new springs, individual wires would still break. As it turned out, fractures were caused by a structural pre-damage when the wires were machine-twisted in production to form a strand. If an excessive bending moment was applied on the wire, a pre-damage of the spring wire occurred which would later result in a fracture at this spot.

Maintaining a constant angle of the wires – lay, that is, the angle in which the wires in a strand are laid in a helix – proved to be just as critical for stranded springs when they take high impact stresses and velocities.[318] As is now understood, the condition of the actual strand has a large influence on the wear of these springs. The orientation of the strand's helix must be the same as the spring's direction of wind, and the lay of three-stranded springs should be the equivalent of nine to twelve times of the wire diameter.[319] If the lay is too short, it can have a negative effect on the life cycle. In general, short life cycle or heavy setting of stranded springs are typical symptoms of a very short lay.[320]

Stranded springs are used whenever high-performance standards must be met. The strand acts in the same way as a single wire spring, and can damp migratory waves that traverse the spring under shock-loading. However, even if one strand breaks, the spring will still be functional and will continue working under stress.

The design considerations and calculations for stranded springs are the same as for parallel arranged springs; but as in any case in the design of guns, ultimately only endurance tests provide certainty if the performance requirements are met.[321] The failure of stranded springs is typically caused by friction between the individual wires, causing them to wear out. The damage on the surface expands increasingly and eventually leads to the breaking of a wire. More wire breaks will follow and eventually, the entire spring will fail.

The most common stranded springs use three wires. The duration between the rupture of the first wire and the complete failure of a three-wire spring is approximately 10 percent of the springs total life; for a seven-wire spring, it is between 20 to 40 percent.[322] Actually, stranded springs fail sooner than single-wire springs when purely oscillating loads are applied, but they are especially suited to repetitive impact loading conditions, under which they will have a 100 to 400 percent longer lifetime than standard springs.[323]

The length of a compression spring when under sufficient load to bring all coils into contact with adjacent coils, that is its solid height, can be a limiting factor when designing pistol recoil springs. Engineers face another challenge on top of that: the force needed to manually rack the slide. This maximum force is normally experienced when the slide is pulled back to the rear most limit of the slide movement, e.g. to engage the slide-lock. Some police might even specify a maximum slide racking force. The reason for this demand is that the pistols should be operable by any officer. For example, a maximum force of 100 N (circa 22.5 lbs) would be the standard in Germany,[324] while NIJ tends to 21 pounds[325] [93 N; PD]. The demand for an upper limit of more or less 100 N strongly constrains the engineers, because some pistols require a sufficiently high spring load to keep the slide reliably in battery, i.e. fully forward. This can be an issue with single-action or partially-cocked striker-fired pistols if they use a system where striker spring and recoil spring work in opposite directions. If the recoil spring's force was too light, the slide might get pushed rearward as the trigger is being pulled. Or when the gun is used under adverse conditions, for example when it gets bumped, the breechblock might get out of battery or extract the cartridge from the chamber, which then might get jammed in the ejection port. Hammer-fired pistols are not associated with this issue where mainspring and recoil spring work against each other; if we neglect other problems such as the friction which is caused by the hammer when getting pushed down on the underside of the slide as the slide cycles.

The demand for a preferably light slide racking force (e.g. F_2 = 100 N) in combination with a relatively heavy load F_1, that is, the force keeping the breech reliably in battery, can be addressed – within the limits of the material – by the use of flat-wire springs. Unlike the springs made of round wire, these have a rectangular cross-section (Fig. 3.71). Such recoil springs are normally of the kind that is wound to the spring axis, which means that the long cross-cutting side of the material is perpendicular to the spring's axis.[326] The advantage of flat wire springs is in their shorter solid height which is much less than

Fig. 3.71: Flat wire recoil spring.

that of round wire springs, giving it the required increase of travel distance.

In other words, if there are space limitations in a given design of a gun, a flat wire spring can be used and it will offer more active coils that will store and release more energy than in a round wire spring. A lower spring rate arises as a result of the greater number of active coils and thus the spring has a greater working capacity.[327] Accordingly, with a flat wire spring, and a given slide racking force F_2, and a deflection ($\delta_2 - \delta_1$) a heavier load F_1 can be realized than with a round wire spring (see Fig. 3.73, p. 122). In the following, a theoretical explanation is given.

The performance of a spring is characterized by the relationship between the deflections δ and the resultant spring forces F. For example, a recoil spring has a certain unloaded free length. Once installed in a pistol, the spring is compressed to δ_1 and the resulting spring force will be its locking force F_1. With the slide retracted all the way to the rear, the spring is compressed to δ_2 which will be providing a force F_2. The interdependence of a spring's force and compression can be shown in a force vs. deflection diagram.

The force-deflection curve in the diagram gives an idea of the spring's characteristics, and the rate of the spring is the gradient of the force vs. deflection curve. If the space for the spring inside the gun and thus the solid length is known, the number of coils can be determined and the spring rate R calculated. The following formula to calculate the spring rate applies to springs made of round wire:

$$R_r = 0.125 \frac{G_r\, d^4}{D_{mr}^3\, n_{ar}}$$

3.1[328]

where: R_r Spring rate of round wire spring [N/mm]
d Wire diameter [mm]
D_{mr} Mean coil diameter of round wire spring [mm]
G_r Shear modulus, round wire [N/mm²]
n_{ar} Number of active coils of round wire spring

The spring rate of flat wire springs is given by:

$$R_f = \frac{1}{\varepsilon} \frac{G_f\, h^2\, w^2}{D_{mf}^3\, n_{af}}$$

3.2[329]

where: R_f Spring rate of flat wire spring [N/mm]
D_{mf} Mean coil diameter of flat wire spring [mm]
G_f Shear modulus, flat wire [N/mm²]
h Height of flat wire's long edge, perpendicular to the spring's axis [mm]
n_{af} Number of active coils of flat wire spring
w Width of flat wire's short edge parallel to the spring's axis [mm]
ε Elasticity coefficient for h/w ratio

If we replace in Formula 3.2, h by $(h/w \cdot w)$, we end up with the following:

$$R_f = \frac{1}{\varepsilon} \frac{G_f \left(\frac{h}{w} w\right)^2 w^2}{D_{mf}^3 \, n_{af}}$$

3.3

Formula 3.3 can be transformed into:

$$R_f = \frac{(h/w)^2}{\varepsilon} \frac{G_f \, w^4}{D_{mf}^3 \, n_{af}}$$

3.4

Considering identical wire sizes and assuming equal conditions for round and flat wire, the ratio $h/w = 1$. In the case where the ratio $h/w = 1$ Meissner suggests to use an elasticity coefficient $\varepsilon = 5.59$ (see Table 3.2).[330]

Table 3.2: Shift of spring rate of flat wire springs vs. round wire springs under the assumption of $d = w$.

h/w ratio	1	2	3	4
Elasticity Coefficient ε	5.59	6.87	8.95	11.19
Factor, flat wire	0.179	0.582	1.01	1.43
Factor, round wire	0.125			
Shift of spring rate [a]	1.43	4.66	8.04	11.44

Therefore, the flat wire spring rate is:

$$R_f = 0.179 \frac{G_f \, w^4}{D_{mf}^3 \, n_{af}}$$

3.5

When comparing the factor 0.179 with the factor 0.125 in Formula 3.1, it may be realized that the spring rate for a flat wire spring material is 1.43 times more than the rate of a round wire spring ($0.179 \div 0.125 = 1.43$). As it can be seen from the figures shown in Table 3.2, the elasticity coefficient increases with an increasing h/w ratio. For $h/w = 2$ the elasticity coefficient is 6.87, for $h/w = 3$ it is 8.95, and for $h/w = 4$ it is 11.19.[331] Therefore, the spring ratio of flat wire springs is 4.66- to 11.44 times more for h/w-ratios ranging from 2 to 4.

Under the assumption of $d = w$ and the same number of active coils, the arithmetical comparison shows that square wire springs have a higher spring rate than round wire springs. If however, the flatter geometry of rectangular spring wire is used to get more

---

[a] Factor to anticipate additional coils of a flat wire spring to achieve the same spring rate as a round wire spring with $d = w$.

active coils in the space allotted, the spring rate of flat wire springs can be lower than the one of a round wire spring would be.

A factor for the required number of extra coils of a flat wire spring can be obtained if Formula 3.1 for R_r is identified with Formula 3.5 for R_f and solved for n_{af}. This derives the number of active coils n_{af}, with which a flat wire spring would have the same spring rate as a spring made of round wire:

$$n_{af} = n_{ar}\ \frac{8\, D_{mr}^3\, G_f \left(\frac{h}{w}\right)^2 w^4}{G_r\, d^4\, \varepsilon\, D_{mf}^3} \qquad\qquad 3.6$$

Under the assumption of equal values for the mean diameter D_m and the shear modulus G, Formula 3.6 will reduce to:

$$n_{af} = n_{ar}\ \frac{8 \left(\frac{h}{w}\right)^2 w^4}{d^4\, \varepsilon} \qquad\qquad 3.7$$

For the examples exhibited in Table 3.2, the shift of spring rate factors is confirmed. A flat wire spring with a ratio $h/w = 2$ would need to have 4.66 times more coils than a spring made of round wire of the same spring rate and material thickness to achieve the same spring rate.

In industrial practice, the spring rate of flat wire springs is controlled by further reducing the wire's width w, and optimizing other parameters, such as spring diameter or the free length L_0. However, the engineers have to stay within certain limits in designing compression springs. The space allotted governs the dimensional limits of a spring with regard to its acceptable solid height and clash allowance, i.e. when the slide is retracted, the recoil spring must not get compressed to solid.[a]

Professional mathematical calculation software for the optimization of the springs starts at this point. Among other effects, it will take into consideration the impact of the flat wire deforming into trapezoidal shape during coiling into a helix in the production of springs. The deformation on the inner edge (see Fig. 3.72) will cause a 20 percent increase in the springs solid length and around 10 percent shift in its spring rate.[332]

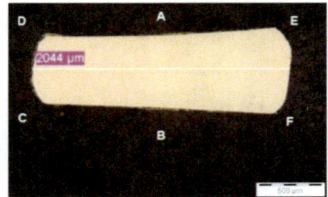

Fig. 3.72: Cross-sectional view; flat wire deformed into trapezoidal shape in area E-F.
(© BAUMANN FEDERN AG)

---

[a] The rearward travel of the slide must end at a well-defined stop at a point before the recoil spring is compressed to solid. If the solid length of a recoil spring would be used as a backstop, then the backstop dimension would be subject to the tolerances in the manufacture of the spring.

A comparison by the BAUMANN FEDERN AG is shown in Appendix 11 as an application-oriented example between flat wire and round wire recoil springs, assuming equal conditions: both recoil springs should have a maximum solid length of 24 mm, allow for a similar slide racking force F_2, and provide a maximum locking force F_1. As shown in the table, the recoil spring made of flat wire offers a locking force F_1 of 50 N, whereas the round wire spring would hold the slide with only 22 N in the closed position. The free length, L_0 of the flat wire spring is considerably longer, it has a larger number of coils, and the spring rate is about half as much as the round wire spring.

The force vs. deflection curves can be used to clearly show the influence of the spring rate on the slide locking force F_1 of recoil springs. The slope of the curve represents the spring rate R, which is the ratio of spring force F and deflection δ (Formula 3.8), if the displacement is shown on the axis of abscissae and the force is shown on the axis of ordinates.

$$R = \frac{F_1}{\delta_1} = \frac{F_2}{\delta_2} = \frac{F_2 - F_1}{\delta_2 - \delta_1} \qquad\qquad 3.8$$

where: R Spring Rate [N/mm]
F_1 Force, preloaded [N], e.g. slide fully forward
F_2 Force, fully loaded [N], e.g. slide fully retracted
δ_1 Deflection, preloaded [mm]
δ_2 Deflection, fully loaded [mm]

The higher the stiffness of a spring, the greater the slope of the force vs. deflection curve. The steeper the slope, the less the deflection of the spring at a given external load applied (compare deflection at point of intersection B_r and B_f at load F_{1f} in Fig. 3.73).

Force-deflection characteristics of compression springs are approximately linear, if they can work without friction; load and deflection behave proportional to each other. This constant rate was assumed in Fig. 3.73 for a graphical visualization and more clarity, indicating why a flat wire spring provides a greater slide locking force F_1 as a spring made of round wire. Two force-deflection curves are shown as an example: a curve "FW" for a flat wire spring with low slope and a steeper curve "RD" representing a round wire spring.

Starting from the intersection with the racking force $F_2 = 100$ N, the locking force F_1 for both springs can be determined by the intersection of deflection δ_1, that is, by considering the working range $\delta_2 - \delta_1$. As shown in the diagram, $F_{1f} > F_{1r}$, that is, the flat wire spring provides a greater locking force, due to their flatter curve in the diagram.

With a given racking force of $F_2 = 100\,\text{N}$ and identical working range $\delta_2 - \delta_1$, the flat wire spring will provide greater locking force F_1 than its round wire counterpart.

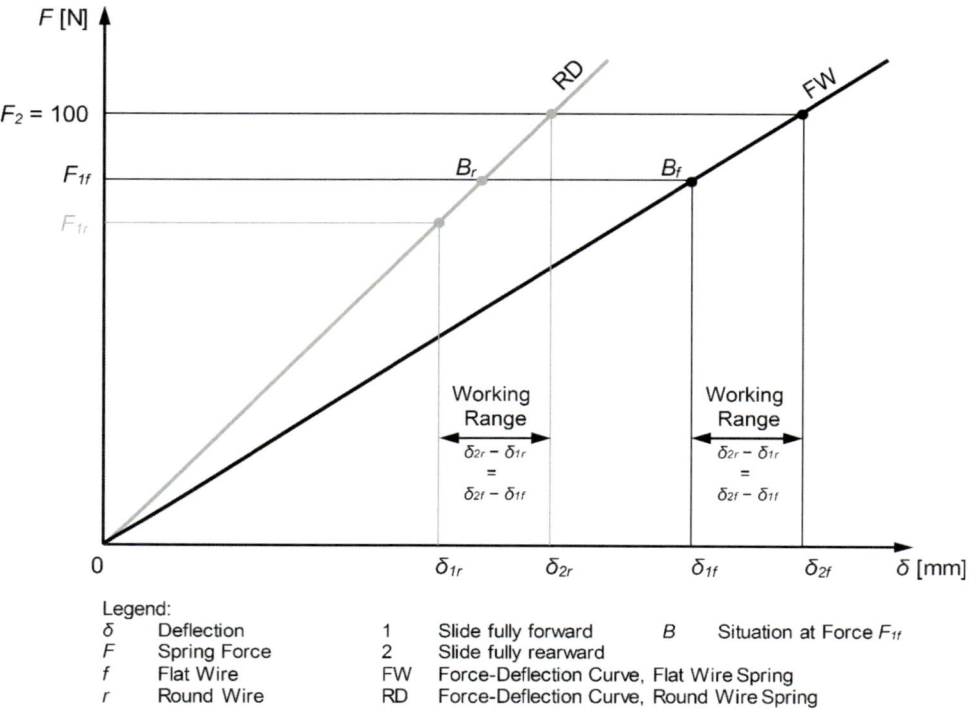

Fig. 3.73: Force vs. deflection curves; schematic comparison.

Flat wire recoil springs were already used at the beginning of the 20th century, for example, the ROTH STEYR Model 1907 pistol with partially cocked striker, or the hammer-fired STEYR Model 1912 pistol (Fig. 3.74). This type of recoil spring is still used today in GLOCK pistols as well as in pistols of various other manufacturers.

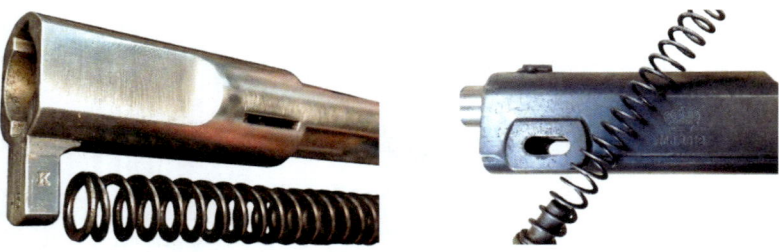

Fig. 3.74: Flat wire recoil spring: ROTH STEYR Model 1907 and STEYR Model 1912, both shown alongside the respective pistol's breech.

Presetting

When a compression spring is compressed and released, it is supposed to return to its original height, and the force supplied by the spring should remain constant. Whether or not set is an issue depends on the level of stress the spring undergoes versus the material's tensile strength. Compression springs may be categorized into three groups:[333]

1. Springs which set will be in a low range from 1 to 3 percent over time, even if they are compressed solid. They don't require an extra operation for removing set. Their torsional stress levels when compressed solid is not greater than 40 percent of the minimum tensile strength of the wire material.
2. Springs which can be compressed solid and which will have a permanent set from 5 to 10 percent after set has been initially removed. Their torsional stress levels when compressed solid is between approximately 40 and 56 percent of the minimum tensile strength of the material.
3. Springs which cannot be compressed solid without some further permanent set taking place because set cannot be completely removed in advance. They will be setting over time, that is, lose length and load anytime the maximum allowable spring deflection is exceeded. Their torsional stress levels when compressed solid is greater than approximately 56 percent of the minimum tensile strength of the material.

In the second case, one way for springs to stay within the specs is to make the spring initially a little bit too long and then to compress the spring in production all the way to solid. After the spring takes an initial set, it will be at the correct length to meet the drawing requirements. This process is referred to as "presetting," "removing the set" or sometimes "scragging," if it is carried out before a spring is put into service. At the point where it is done, it will reduce the stiffness of the spring and hence the loads.

Most springs are subject to some amount of permanent deformation during their life span, that is, relaxation or creep.[334] Permanent deformation from 1 to 10 percent are customary, and as experience teaches, flat wire recoil springs may show up to 15 percent relaxation and length loss in service.

Where post-pulse oscillation cannot be avoided, particularly when springs are used in high-speed cyclic applications such as weapons, setting of springs can sometimes not be avoided. Depending on design and space limitations, a more favorable material utilization through larger wire dimensions is rarely feasible, and leaves the engineers often with just the possibility to consider the influence of permanent set in the design of a spring. Without endurance testing, the permanent set can be simulated by speci-fying in the drawing that test samples must be compressed to solid for a certain period of time prior to measuring spring loads.[335]

In the manufacture of springs, the separate operation of presetting can lead to an improvement in the utilization of the material, when a more favorable stress distribution is formed after deliberately exceeding the material's elastic limit. Additional stresses are introduced during presetting and will cause this favorable effect, if a spring is de-signed to take torsional stress levels in the range from approximately 40 to 60 percent of the minimum tensile strength of the material when compressed to solid. This super-imposes the residual bending stresses produced by the coiling operation.[336]

The extra operation for removing set can take anything from short to long term period (24, 48, or 96 hours).[337] After releasing the springs internal stresses will remain, which have a positive effect on the stresses at the surface of the wire as the spring is de-flected; that is, the spring load will not decrease as much in the long run, because the residual stresses in the material will counteract the operating stress.[338] Thus, the pre-setting is carried out at the very end in the manufacturing process of springs. If another operation would be carried out on the springs after presetting, the positive effect would be lost. For example, the effect of presetting could be reduced by grinding the outside diameter of a spring.[339]

Rate of Fire

Fatigue failures and in particular the setting of springs often affect the functional relia-bility of pistols. Recoil springs and other slender springs, that is, long springs with small outside diameter, tend to surging which causes repetitive coil-to-coil contact, leading to impact and surface deterioration. How frequent coil-to-coil impacts will occur de-pends to a certain degree on how much time is remaining for the oscillating system between shots. In case of a low rate of fire, a relatively long period of time remains where waves of high inter-coil displacement can travel along the spring leading to high operating stresses, clashing and impact.

When the functional reliability of a gun is tested in an endurance test, the effect of the cyclic rate must be considered. A gun should not only provide a dependable func-tion during the requested service life, but it should still work effectively well when shot with muzzle up or down at the end of its life. Usually, this can only be found out when endurance tests are done, and thus if reproducible and meaningful test results are required, a specified rate of fire should be defined and used.

For example, NIJ Standard-0112.03 specifies a rate of fire in the range of two seconds per shot, up to no more than two shots per second, representing a rate of 30 to 120 rounds per minute (see p. 204f).[340]

If the limits for the cyclic rate and the magazine capacity are given, the maximum and minimum duration to empty one magazine can be calculated:

$$t_n = \frac{n}{K} \qquad\qquad\qquad 3.9$$

where: t_n Duration to shoot n rounds [s]
K Effective rate of fire [rounds/s]
n Magazine capacity [rounds]

According to the rate of fire specified in NIJ Standard-0112.03, a 12-round magazine would have to be emptied within 6 to 24 seconds.

3.8 Safety Devices

Safety devices can help to avoid unintentional discharges; however, they cannot prevent a shooter-induced discharge caused by the violation of basic safety rules, which would be called a negligent discharge. An unexpected discharge of a firearm beyond the control of the user, e.g. by a mechanical failure or malfunction, is referred to as an accidental discharge. Examples of the latter would be a cook-off, an ignition during feeding or a discharge when a gun is jarred (Fig. 3.75).

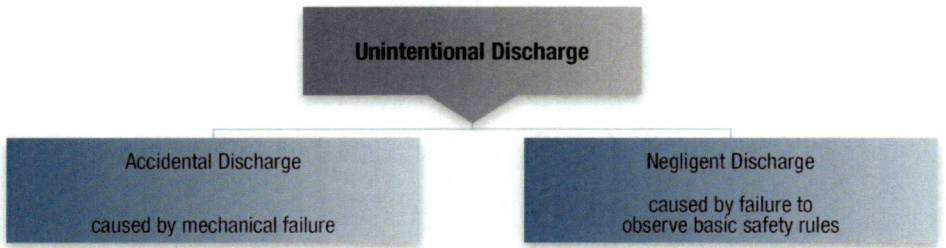

Fig. 3.75: Accidental Discharge vs. Negligent Discharge.

To avoid accidental discharges, a variety of features are included in pistols. Some are design features, while others are mechanical components, such as a disconnector. We've covered disconnectors already and will be focusing on safeties in this section.

With the rise of self-loading pistols, manual safeties became particularly acknowledged. Early single-action automatics were equipped with a safety switch, which would allow users to carry them "on-safe" and fully cocked. The most popular representative of this kind is probably the COLT *Model 1911* pistol and its derivatives, which are still carried "cocked and locked." During the draw, the thumb pushes the safety lever into "fire" position and the gun is ready to fire.

Later, when pistols with restrike-capability/single-action triggers became popular, the manual safety was not dropped, but additionally equipped with a de-cocking function. This can be confusing for some users, if they do not have the discipline to always carry a handgun in exactly the same condition. For example, when a decocked pistol with restrike-capability/SA trigger mechanism is carried on-safe and would not go off as the trigger is pulled in the event of an emergency, because the user did not flip the safety switch to off-safe.[341]

It took decades until the time was right for a simplification. At the time when U.S. police officers were carrying powerful revolvers, duty pistols in Germany were still chambered in .32 ACP. The pistols commonly had restrike-capability/single-action triggers and were equipped with manual safeties. This would change in the 1970s when the German authorities decided to put out a tender for a new service sidearm. The project group "specifications for handguns" (Pflichtenheft Faustfeuerwaffen) specified that manual safety switches would not be acceptable any longer for the new generation of service pistols. Instead, a duty pistol would have to be equipped with an automatically working firing pin safety which would allow to safely carry the pistol in ready-to-fire condition.[342] The design of the WALTHER *P5*, SIG SAUER *P6*, and HECKER & KOCH *P7* pistols are based on this demand. *Kaisertreu* already discussed the issue of readiness for a quick first shot in 1901 and explains why he rejects the term "automatic" when it comes to safety devices:

> *In question is actually [...] not the automatic safety, but rather the application of an automatic disengagement of safeties, which is necessary to have the gun ready to fire in no time.*
> *   Taken literally, an automatic disengagement of a safety – that is, without the manipulation by the firer – can never work!*[343]

Kaisertreu put his finger on it: "automatic" safeties do not work of their own accord, but through an action of the shooter. They can also affect the trigger pull, that is, increase the pull weight when they are worked by the movement of the trigger.

Generally, the safety engineering discipline distinguishes between measures to prevent an accident from happening, and other measures which will reduce the negative effects in case an accident actually happens, that is, active protection systems help to avoid the emergence of accidents, whereas passive safety elements minimize the damage and reduce the risk of injury or death at the very moment the incident occurs.[344] To create clarity about the terms related to safeties used in this chapter, terms will be explained, then grouped, and summed up in a chart.

Characteristics of Manual Safeties

Expressions, such as thumb safety, thumb operated safety, manually operated external safety device, safety switch or external safety are synonymous. *Positive safety* is used as a generic term; for example, in ATF Form 4590 (see Appendix 3).

The design of the manual safety is characterized by some sort of manual device on the outside of the gun, which takes a positive action of the shooter to engage or disengage it, for example to flip a lever from "on-safe" to "off-safe," and the safety remains in the respective position, which is known in English as "positive."

Manual safeties are associated with the field of active safety. They combine the following features:

- to be manipulated deliberately by the user
- positive safety, that is, the safety switch remains in the selected position
- external device

Characteristics of Automatic Safeties

In the U.S., the term "automatic"[345] is followed by an appropriate term which specifies the type of mechanical design, such as automatic firing pin block, automatic firing pin block safety, automatic striker safety, automatic grip safety, automatic magazine disconnect safety or automatic trigger safety. Automatic safeties fall into the category of active protection systems and are also referred to as active safeties. They combine the following features:

- as a round is to be discharged, the safety will get automatically disengaged
- operation at default: on-safe (i.e. engaged)
- external or internal device

Characteristics of Passive Safeties

A third type of pistol safeties is rarely mentioned in gun literature. They are fail-safe devices which work independently from the shooter and only under certain circumstances. An example would be a safety that allows the discharge of a round only if the slide is in closed position. This particular subject is being known since the advent of self-loading firearms: an example is the disconnect function which is also used in some pistols to make firing a round impossible with slide out-of-battery.

> *On the one hand it must be impossible to fire a shot by releasing the firing mechanism before the breech is completely closed, and on the other hand the fire-control mechanism must work reliably. Both requirements need to be axioms and so self-evident that this issue needs no further discussion.[346]*

Another safety device in weapons used to protect the user and acting independently of him, are drop safeties, or generally expressed as a redundant safety feature[347] or passive safety[348]. A passive safety is generally disengaged, that is, it does not affect the carry condition of a gun.

Three traits identify passive safeties:

- independent of the user
- operation at default: off-safe (i.e. disengaged)
- internal device

Fig. 3.76 shows a schematic classification of the above described mechanical working safeties by first dividing them with regard to their functional characteristics. In the overview, active safeties are shown on the left, that is, safeties which are operated through a manipulation of any kind by the shooter. Passive safeties are on the right. The latter are safeties which are supposed to prevent an accidental discharge independently from the user. The schematic's subsequent levels show the distinction after the operation at default, and with regard to the location of the device, whether it is an internal or external safety.

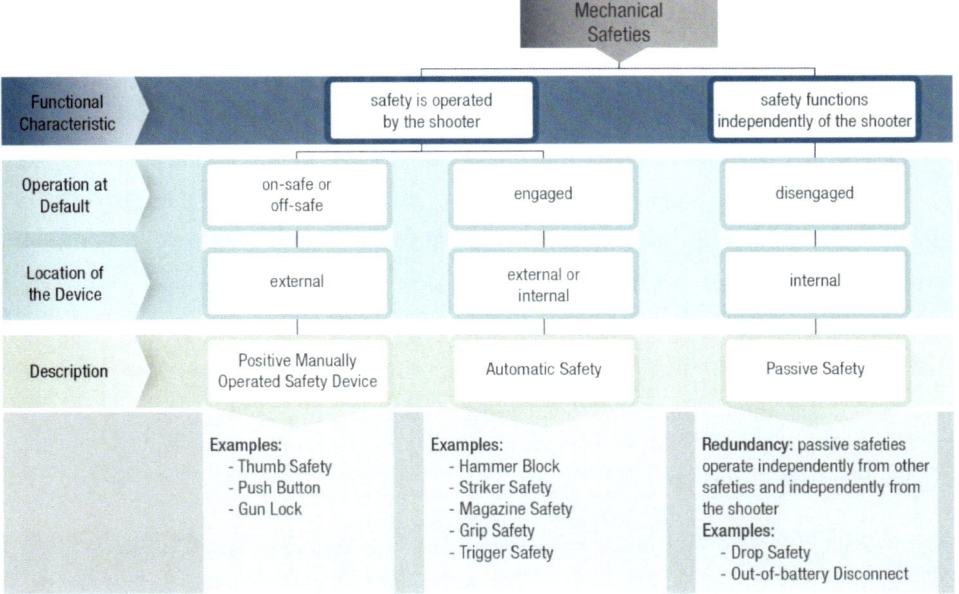

Fig. 3.76: Types of mechanical pistol safeties.

Design Types

Brukner wrote a still-current principle: "The ideal safety prevents the firing of a shot as long as a discharge is not intended by the shooter."[349] And, by preventing an unintentional discharge, safety devices protect the user! In the development of safeties, engineers face certain challenges; because often gun users demand a kind of active safety mechanism which will secure the weapon without involving any deliberate action by the shooter. Several of the design solutions for automatic safeties are linked to the movement of the trigger blade; and this is exactly why such safeties cannot effectively prevent unintentional discharges. These safeties are unable to do so because a gun is

129

intended to cause a round to be fired if the trigger is activated, that is, if the user places his finger on the trigger.

However, occasionally an interference with the trigger blade might happen without intention and may cause a discharge. For this reason, some users prefer to have a manual safety also on modern handguns. The technical details when designing a manual safety can be challenging for the engineer, for example, when a manual safety also has a decocking function. Placing the user in a dangerous situation must be avoided when he uses this decocking feature. In this case, the decocking should take place only after the safety is already engaged. The sequence should be: Off-Safe – On-Safe – Decocking. If the sequence would be different, decocking will take place without the manual safety engaged. The problems involved might be really in small details.

Generally, mechanical safeties affect individual components of the fire-control mechanism, that is, parts such as sear, pawl, hammer and firing pin or striker. The closer a safety device is to the primer, the more reliable this last resort can be in the event of a failure of the fire-control mechanism, where the safety would be the last element to avoid an unintentional discharge in this chain of elements. Regarding what kind of safety on a gun has to be used, manufacturers should focus on the requirements of their specific buyers. Police units usually prefer automatic safeties, whereas some military forces still have a tendency toward manual safeties.

3.8.1 Manual Safeties

Manual safeties were the standard type of firearm safeties for a long period of time: a thumb safety similar to what is known from the design of a Luger pistol or a COLT *M1911*, was frame- or slide-mounted, and the user had to manually rotate the safety lever from "Safe" to "Fire" or vice versa. Also, crossblock safeties were used on handguns (Fig. 3.77).

Fig. 3.77: Crossblock safety on an HK *VP70* pistol.

Manual safeties may block or deactivate the trigger, the trigger bar, or members of the firing mechanism, such as the sear, the hammer, the firing pin or striker. A deactivation is preferably advantageous, where the trigger gets uncoupled and may be pulled to the rear without getting blocked. Because, if a manual safety does not block the trigger mechanism or other parts involved when the gun is on-safe, the mechanics of the fire-control mechanism will be spared if the shooter pulls the trigger vigorously in a deadly-force encounter.

For importation into the United States, each pistol has to comply with ATF Form 4590. A positive, manually-operated safety device is a prerequisite. A safety switch on one side of the pistol would comply with this prerequisite. If, however, law enforcement agencies request a manual safety, then they often prefer ambidextrous designs to offer the same comfort for both, right- and left-handed users.

3.8.2   Automatic Safeties

A common type of automatic safety is the striker safety. A moveable member blocks the path of the striker. Pulling the trigger will cause the disengagement of the safety prior to releasing the firing mechanism (Fig. 3.78).

Fig. 3.78: HK *VP9* pivot-mounted striker safety

As the control of the plunger's movement is linked to the trigger mechanism, some manufacturers try to prevent an unintended movement of the trigger by securing the trigger blade (Fig. 3.79). For example, GLOCK pistols use a safety that is built into the trigger blade, which is commonly named *trigger safety or integrated trigger safety*.[350] Trigger safeties were first used at the end of the 19th century on IVER JOHNSON revolvers[351] and later on SAUER *M30* pistols[352].

Fig. 3.79: IVER JOHNSON, *1897, 2nd Model Safety Automatic Hammerless*, with integrated trigger safety.

(© *B.J. Martens*)

There are various formats of trigger safeties, some are obvious, others aren't. Very popular is the clearly recognizable bifurcated trigger assembly, in the style of a trigger within the trigger blade (see Fig. 3.81, p. 132). For users with short fingers, this type can be problematic, if he cannot bring his index finger around the tabbed trigger safety in front of the trigger blade to properly deactivate it. Other embodiments avoid this issue: The trigger on a SMITH & WESSON *M&P*, for example, is hinged in the middle, with the bottom half

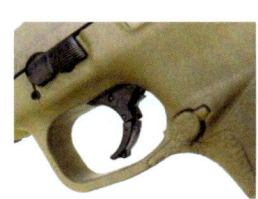

Fig. 3.80: Two-piece trigger blade; trigger safety of a S&W *M&P*.

of the trigger blade working as a trigger safety (Fig. 3.80). This design offers a "flat" trigger without tab.

Some users think the reason for a trigger safety is to avoid negligent discharges, for example, when a pistol is carried inside the waistband. But first of all, a trigger safety is an automatic safety. Depending on the constructive design, it can also serve as drop safety, which will block the trigger to prevent inertial discharge as the gun is grounded heavily, so the trigger bar which is attached to the trigger blade, will not disengage the striker block, and release the firing mechanism at the same time. This seems to work quite well in the case of the fire-control mechanism of GLOCK pistols; but trigger safeties cannot guarantee safety per se, as the firing mechanism might being jarred free when a weapon is dropped.

According to *Lugs*, trigger safeties are less reliable because they do not impinge on the firing mechanism and hence in some designs the firing mechanism can still get released even though the trigger safety is engaged.[353] An almost infinite number of different shocks is possible in practice. This is easy to grasp when we consider the total number of possibilities resulting from random combinations of drop angles, drop heights, impact surfaces, weather conditions etc. Mechanically-acting safeties may not work under all conditions, but only in those for which they were specially designed for.

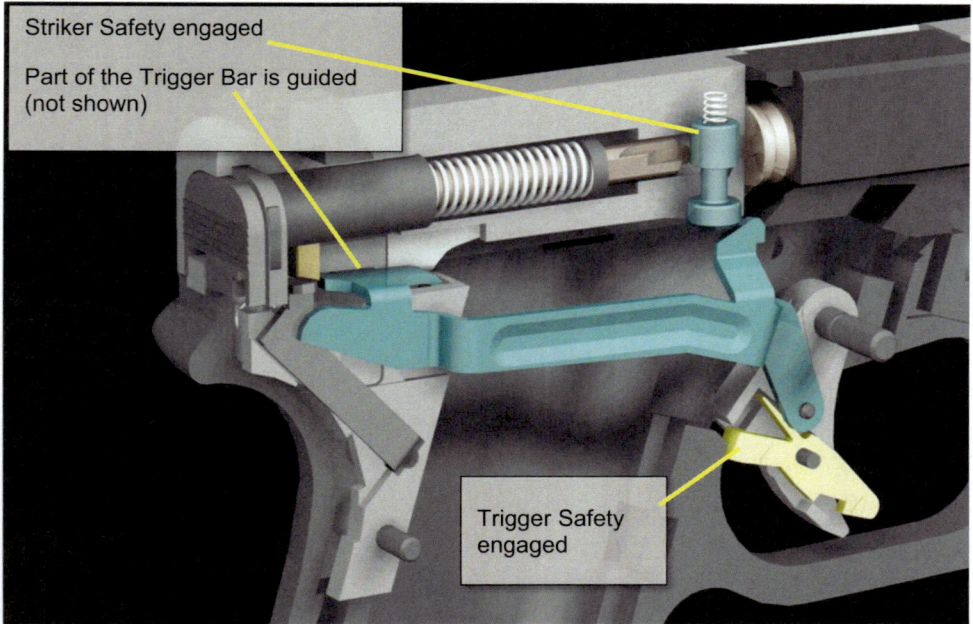

Striker Safety engaged

Part of the Trigger Bar is guided (not shown)

Trigger Safety engaged

Fig. 3.81: Safeties on GLOCK pistols. NB: Locking Block not shown.

(© GLOCK; comments PD)

Magazine disconnect safeties belong to the category of automatic safeties as well. Some states in the U.S. and also some police units in the United States require them. A magazine disconnect prevents a discharge as long as the magazine is removed from the pistol. Some purchasers think a police officer could use this feature to disable his firearm rapidly by pressing the magazine release and dropping the magazine.

Specifically when designing magazine disconnects, but generally when designing safeties, it is important to consider whether they are engaged with controls that can become damaged if a user under stress conditions does not remember to deactivate the safety while trying to pull the trigger with abnormally strong force. Before designing a safety feature such as a magazine disconnect, it should be discussed and specified with the specific client if the trigger must be blocked or if it is preferred that it may get pulled to the rear without causing a round to be fired when the magazine is not in the gun.

3.8.3 Passive Safeties

Passive safeties are used to protect the user and others in his vicinity at the instance when an accident is happening. They work independently from the user.

Disconnect Safety

Normally pistol designs with coupled barrel and slide will discharge a round only after the coupling is completed.[354] Most pistol designs will use an out-of-battery disconnect feature to accomplish this. Put in general terms a disconnect safety is a device that prevents the discharge of a round as long as the cartridge case is not fully supported.

Drop Safety

Technical features and function reliability of today's firearms, as well as the high level in the construction of the weapons, often misleads users to believe weapons are secure under all circumstances. A realistic examination of the facts cannot lead to such a conclusion; because the long life cycle of weapons and the wear and tear of components during the lifetime, as well as an almost infinite number of possible influences and conditions cannot be replicated to a full extent during the testing phase of a new weapon model, e.g. drop position, impact surface, ambient temperature, sensitivity of the primer, wear, and dirt.

To shield the user from an accidental discharge, well-defined standard tests under "laboratory conditions" are carried out. Usually test case conditions mimic real-world scenarios as closely as possible, e.g. a pistol would be dropped cocked with no manual safety applied, i.e. in the condition that it would be in, if it were dropped from a hand. Passive safeties therefore, are currently the design engineer's first choice when it comes to drop safety.

The reliability against an unintended discharge when a gun is dropped can be vital for the user, and during development it must be identified at an early stage which circumstances and components are likely to enable a contact of the striker nose with the primer. This will include all functional conditions of the weapon; taking into account all conditions of the trigger mechanism and setting conditions of the firing mechanism. Usually this requires a combination of several passive safeties to bring the complex processes under control when a gun is dropped at all kinds of different angles.

If a gun gets heavily grounded, its automatic safeties might become disengaged by their own inertia. In this case, successful solutions have proven to be passive type safeties, which are engaged by inertia as well. Their orientation is the same as the one of the automatic safety in question, but a passive safety will get engaged, for example it will block the striker, as the automatic safety is deactivated at the same time.

Example of a Striker Safety:

Very common are slides with a striker block safety (see Fig. 3.82, top). Basically, a spring holds a safety plunger down, which will in that case block the path of the striker. If the trigger is then pulled, the trigger bar cams the safety plunger upwards and out of the way for the striker.

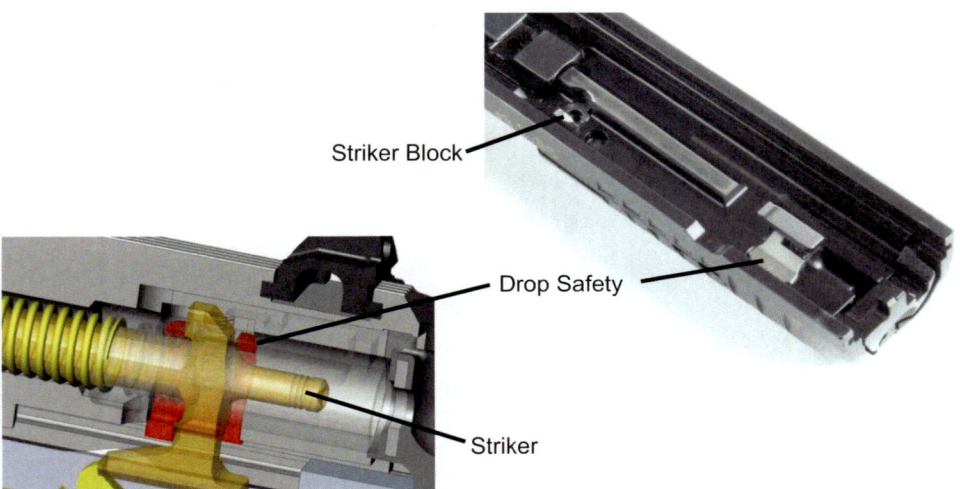

Striker Block

Drop Safety

Striker

Fig. 3.82: Striker Safety and Drop Safety of a WALTHER *P99Q* pistol.

(© WALTHER)

If a weapon of this type is dropped upside-down with the barrel in the horizontal position, the striker block could accidentally be deactivated as a consequence of its inertia, causing the risk of a discharge. This situation can be avoided by integrating a second, spring-loaded member (see "drop safety" in Fig. 3.82 and Fig. 3.83), which is not connected to the striker block but is moveable in the same direction. However, at the same time as the striker safety gets disengaged, the drop safety will start blocking the striker (Fig. 3.83).

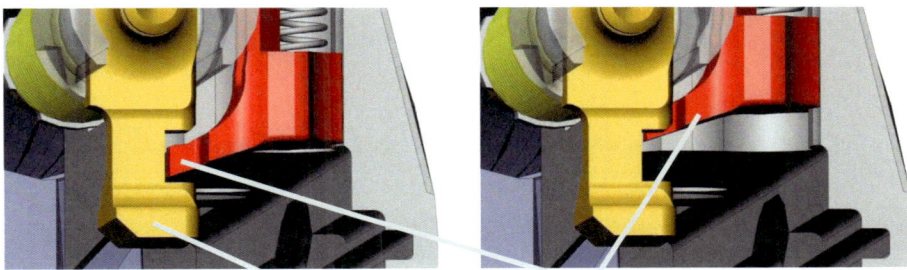

Striker   Drop Safety

Fig. 3.83: Drop safety: disengaged and engaged (right).

(© WALTHER)

Example of a drop safety integrated in the firing mechanism:
If a gun drops on its rear end, whereby the barrel is vertical with muzzle up, the trigger blade and the parts it is connected to, might be driven to the rear. A discharge may be caused by this because the trigger will just do what it is meant to do. It operates auto-matic safeties and triggers the firing mechanism. Here, a passive drop safety designed

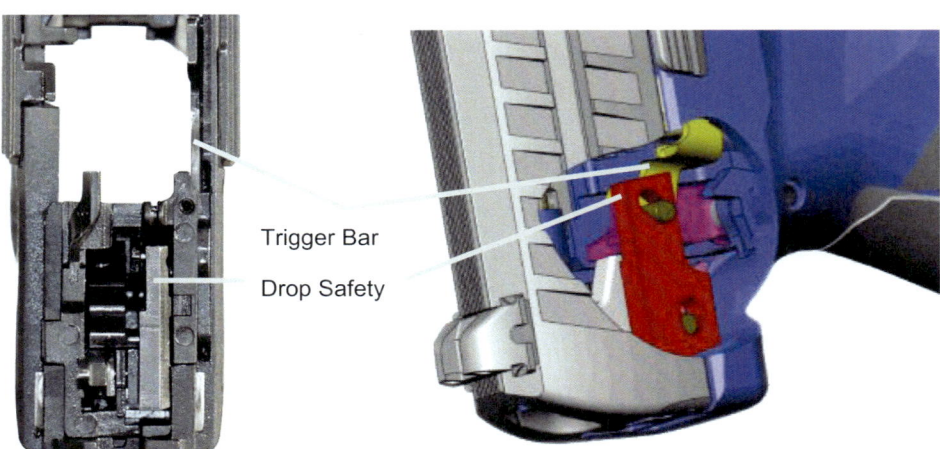

Trigger Bar

Drop Safety

Fig. 3.84: Drop Safety inside Firing Group Housing of a WALTHER *P99Q* pistol.

(© WALTHER)

as a piece of mass may be used to block the trigger bar. The drop safety can be inside the rear of the frame in order to block the end of the trigger bar when the weapon is heavily grounded with muzzle up (Fig. 3.84).

Passive safety devices as well as automatic safeties are normally the first choice of the law enforcement community. Reliably functioning passive devices however have the potential to replace trigger safeties and manual safeties. On the commercial market, some users might still prefer external safeties, either a manual safety, or a kind of automatic safety, which can be visually recognized or felt, because consumers might not be used to carrying a gun in a holster, or might keep their gun handily stored in a variety of different places and a discharge can be caused right at the instance when a "hot" gun is grabbed. People who act that carelessly might even consider a simple trigger safety to be useful avoiding a negligent discharge. Maybe this is the reason why trigger safeties will continue to be a design feature in the future.

3.9   Microstamping

Microstamping, ballistic imprinting and ballistic engraving are names for procedures to transfer an array of characters of a pistol on each cartridge case by imprinting when the firearm is fired. The idea is to provide markings that will allow the identification of the firearm or find out about the production lot by examining casings. For example, Hans-Peter Sigg disclosed DE 198,32,777 B4 on July 22, 1998, patent owner J. P. Sauer & Sohn. According to Sigg, raised and imprinted markings in the breech face, such as characters, letters, numbers, and similar marks, are used for identification purposes. J. P. Sauer & Sohn renounced the patent, after document RU 2,015,492 C1, filing date March 31, 1992, was revealed, in which the same method was disclosed to imprint markings on casings from gun parts which are in contact with the case.

Generally, marks are known, which enable the tracing of gun trafficking in crisis areas. The marks can be on casings or even on projectiles,[355] or in the context of forensic investigations, to identify the weapon a round has been discharged from. The latter is the reason why the State of California demands microstamping.[356] Since January 2010, California demands two microstamping arrays, which leave a firearm identification number (FIN) on the case. The FIN must contain the weapon model, manufacturer, and serial number of the gun. The implementation of this law, that is, the actual imprinting of a unique array on casings, is what weapon manufacturers regard to be not feasible from a production standpoint as well as with regard to the wear on the parts that would transfer the marks which would be exposed to heavy loads when they are in contact with the cartridge case.[357] Critics also say, the components on a gun needed for imprinting could be manipulated to bypass the FIN marking. Also, there is the possibility to leave casings with FINs, which were not fired from the weapon actually used at the crime scene.

First California's microstamping requirement was postponed. The requirement was under the condition that an accessible technology would be available to more than one manufacturer, meaning that the use of the technology is not legally protected, e.g. patented, in the United States, because the law should not penalize any manufacturer. This clause was necessary since one valid patent already existed, which consequently would have led to a preference of individual manufacturers.

> The microstamping process was invented 15 years ago by Todd Lizotte, a New Hampshire engineer who patented the process under the trademark NanoMark Technologies. Because the technology was available nowhere else, the Legislature required the attorney general to certify that it was available "to more than one (gun) manufacturer unencumbered by any patent restrictions."[358]

Todd E. Lizotte's patents US 6,886,284, US 7,111,423, and US 7,204,419 are known. In the first one, dated May 3, 2005, Lizotte disclosed the transfer of barcodes, signs, symbols and other marks onto casings, where firing chamber, breech face, extractor, ejector, or firing pin nose would transfer micro-markings. He also describes procedures on how to produce these markings, primarily by means of a laser. In patent US 7,111,423 of September 26, 2006, and US 7,204,419 dated April 17, 2007, Lizotte further discloses how to read off the marks on casings and even bullets. All patents have expired.

Patent protection for Todd E. Lizotte's US 6,886,284 and US 7,111,423 expired, because the patentee did not extend the patent protection.[359] Therefore California's requirement of Dangerous Weapon Control Law, § 12126 applies; that is, since May 17, 2013, new pistol models must be equipped with microstamping technology to get an approval for sale in California.[360]

Currently no weapon manufacturer offers any gun with microstamping technology.[361] The industry hopes for the positive outcome of the Second Amendment Foundation's (SAF) case in California,[362] Pena v. Lindley, a lawsuit challenging the state handgun roster requirements that include microstamping and magazine disconnects. Attorneys for GLOCK, INC. have filed an amicus curiae brief supporting the SAF's case (see Appendix 7). According to the brief filed, California's requirements constitute an unreasonable burden for producers and consumers. Neither GLOCK pistols nor any other manufacturer's pistols in full production can comply with the microstamping mandate, and the newest generation of GLOCK pistols would not be allowed for sale in California to private individuals.

The requirement for magazine disconnects is seen as a similar burden: GLOCK pistols come without a magazine disconnect, just like most of the handguns available on the market, and an overwhelming majority of the police require pistols that do not have a magazine disconnect feature. Therefore, the claim constitutes magazine disconnects are a burden for the manufacturers. The outcome of the case is still open. The SAF relies on the support from GLOCK in the fight against microstamping and magazine disconnects.

3.10 Electronics

There is a considerable range of accessory items which can be added externally to handguns, for example electronic devices that can be attached to an accessory rail or installed on the slide. The market is dominated by tactical flashlights and aiming devices, such as laser aiming devices or carry optics (slide mounted electronic optics, for example, often called micro red dots or reflex sights); whereas the handguns themselves are still strictly mechanically operated devices if we ignore electronic trigger mechanisms on sporting firearms.

In contrast, the evolution of integrated, advanced safety technology started in the 1990s. This technology is often tied to smart guns, or firearms that "utilize integrated components that exclusively permit an authorized user or set of users to operate or fire the gun and automatically deactivate it under a set of specified circumstances, reducing the chances of accidental of purposeful use by an unauthorized user." [363] Such user-authorized handguns are being controversially discussed in the United States and some users regard them as an immature technology. [364] The National Shooting Sports Foundation (NSSF) conducted a survey in 2013 asking more than 1,200 Americans about smart guns. Eighty-four percent of the respondents thought that the technology was not reliable, and 60 percent said they would not buy any gun with smart-gun technology. [365] Apart from technical issues, there are concerns that legislation could use the technology to control the possession of arms.

In California for example, the state has not only one of the strictest gun laws in the United States, but this state also represents one of the largest handgun markets in America (see Appendix 10, Table A10.10). A state senate bill was suggested in 2013 to make owner-authorized handguns mandatory, if two owner-authorized handguns have been placed on the roster. A two-year lead time would have applied from the date that the second handgun was placed on the roster. Starting from that date, only owner-authorized handguns would have had a chance to be accepted for sale in California.

This California Senate Bill SB-293[366] introduced by State Senator Mark DeSaulnier (D-7th State Senate District) also included some specifications for the scope of the user ID, for example, the time from first contact to use recognition and firearm enablement should have not been more than 0.5 seconds, and if the power was down, the firearm should have been inoperable.[367]

By end of 2014, the bill was defeated in the California Senate, although a first supplier of an owner-authorized handgun had already started selling its pistol in California; in 2013, the German manufacturer ARMATIX had received the approval for its *iP1* pistol chambered in .22 Long Rifle.[368] On October 25, 2013, the ARMATIX *iP1* pistol was placed on the California roster, the list of California-approved weapons. The pistol retailed in California at $1,399 U.S. (as of 2014). The identification of the user was done via a wristwatch, which could be obtained for an additional $399 U.S.[369]

Besides the concerns about gun control under the legislature, the question is how a smart gun should perform if the electronics break down. For the consumer it is clear: weapons with smart-gun technology must function as safely and flawlessly as mechanically functioning firearms.[370] NIJ has addressed this issue now. In 2016 NIJ published "baseline specifications," which outline requirements for any LE firearms equipped with gun safety technology. It specifies, "if the security device malfunctions, it shall default to a state to allow the pistol to fire." When integrating smart gun technology into a firearm's design the task is to not compromise the reliability, durability, and accuracy that officers expect from their service weapons.[371]

When the National Institute of Justice held a workshop in 2013, the research activities in the field of smart guns in the years from 1994 to 2006 were referred to as "early R&D," and the period from about 2006 until then as "existing and emerging gun safety technology and smart guns today." Research in the area of smart guns began in the mid of the 1990s, after NIJ started to fund research activities. Until 2013, over $12.6 million U.S. were funded by NIJ and the Bureau of Justice Assistance on research and

development projects to investigate different technologies and develop functional prototypes of handguns with electronic safety mechanisms built in that would prevent anyone other than an authorized user from firing it. For example, NIJ granted
$500,079 U.S. to COLT in 1997, in the years from 2000 to 2006, $2,606,156 to FNH
USA, and in the period from 2000 to 2005, $3,673,361 to SMITH & WESSON. NIJ gave
over $1.1M in a number of smaller awards to various smaller firms, including iGUN
TECHNOLOGY CORPORATION and NIJ provided $ 2,515,475 U.S. to the New Jersey In-
stitute of Technology.[372]

A survey published by National Academies Press in 2005 entitled *Committee on
User-Authorized Handguns, National Academy of Engineering, Technological Options
for User-Authorized Handguns: A Technology-Readiness Assessment*, focuses on two
types of handgun owners: (1) people responsible for public safety and (2) people concerned with personal safety and handgun misuse, particularly by children, in the
home.[373] Typically, police officers are the prime example in publications on smart guns,
because cases are known where officers have been killed on duty with their own gun
by unauthorized users.[374]

The SANDIA report of 1996 notes the introduction of electronics to weapons was
needed in the United States. It explains, the number one concern among officers is the
reliability of the smart gun technology and the addition of it must significantly reduce
the reliability of the firearm system compared to existing firearms.[375] The paper was
followed by a 2001 update. However more up to date than the SANDIA report is the
NIJ research report by Mark Greene, *A Review of Gun Safety Technologies* published
in June 2013 by NIJ. This paper examines firearms safety technologies. It takes a look
at existing technology as well as technologies to come, their availability and the benefits they might provide. In his review the researcher provides a current and comprehensive overview of the state of the art. Below is a summary of Greene's report.

The 2013 review concedes the discharge of a round is the intended function of a
firearm. The logic is that a discharge can be prevented by safety features, and the
purpose of safety features in general is to mitigate the risks associated with the use or
misuse of a firearm. A benefit of advanced safety features can be that the operation of
an illegally possessed firearm for criminal purposes or against law enforcement can be
eliminated and maybe even the illegal acquisition of firearms. But the reputation of
modern safety technology is that it will cause the firearm to malfunction during a situation, at the critical instant when a life-or-death situation occurs and the reliability in

general has been cited as a concern for users.[376] If a malfunction occurs in the system of the firing mechanism, ammunition, safety mechanism, and other components, the following four scenarios can be the result:[377]

1. The gun fires when the trigger is pulled and this is the desired result (true positive);
2. The gun fires when the trigger is pulled and this is not the desired result (false positive);
3. The gun does not fire when the trigger is pulled and this is the desired result (true negative); and
4. The gun does not fire when the trigger is pulled and this is not the desired result (false negative).

Greene emphasizes these four results would apply to weapons in general, regardless of whether they are equipped with an advanced safety technology or a traditional mechanically operated safety device.

3.10.1 User Authorization Technologies

All technologies Greene looked at concerned user authorization, and technical implementation took place by integrating new technology into weapons. They would combine an authentication mechanism which actuates a blocking mechanism. The length of time for this process should be shorter than for traditional weapons. As soon as the user was identified and recognized as authorized, typically an electronic circuit would produce a physical change such as removing a mechanical block. Blocking mechanisms would include solenoids, motors and piezoelectric which were used as actuators.[378]

Generally, radio frequency identification or biometric systems were found to be used to identify the operator; whereby the identification of the user was actually based on two techniques: on the detection of a device (token-based technologies), or on the detection of body characteristics (biometric technologies).[379]

Token-Based Technologies
A token is an electronic component, which provides information that will be evaluated by the recipient of this information. In token-based systems the user has to remember to have the token on him, or if the token was stolen, it can be used to authorize their associated firearm. A token implanted into an authorized user are an alternative to this scenario.

The token can exchange information in three ways: (1) via radio frequency identification (radio frequency electromagnetic fields), (2) via ultrasonic signals, or (3) magnetism.

Token: Radio Frequency Identification

Radio Frequency Identification (RFID) is the wireless transmission of data from a tracking tag to a receiver using electromagnetic fields. The tags used can be active or passive; the former require a battery, while the latter do not. The receiver is usually embedded in the weapon. Information is stored on the tag, which might be readable at a distance of several meters away. As for electromagnetic waves, a visual contact between the transmitter and receiver is not needed, hence the user can wear gloves.

The disadvantage of RFID technology is that it can be influenced by interference from electromagnetic waves.[380]

Token: Ultrasonic Technologies and Magnetic Technologies

A user ID via an ultrasonic based token or magnetic fields is rarely used. For a user ID by means of ultrasonic based token, the token is worn on the body of the user and emits an ultrasonic sound which is too high for humans to hear. In the case of a magnetic token, the magnetic field of a permanent magnet may be used to move a blocking mechanism in the gun, while the magnet is at close distance to the weapon.[381]

Biometric Technologies

Biometric systems utilize unique features of individuals as the way of identification. Usually fingerprint, palm print, voice, face or vein pattern serves as a unique feature of identification.

An electronic sensor is optimized to collect the biometric and compare it to those stored in computer memory.

Biometric Technologies: Fingerprint Technologies and Palm Print Technologies

An image of the papillary lines is scanned at a place accessible with no conscious effort by the user, and then compared with a database of authorized users. If a match is found, the firearm gets unlocked.[382]

Biometric Technologies: Dynamic- or Static Grip Technologies

How the user grasps a pistol grip, with what force, where he places his fingers, and the hand geometry allow his identification. Distinction is made between static and dynamic grip recognition, depending on whether the detection is done once, or by means of the change in behavioral characteristics over a duration of time.[383]

Biometric Technologies: Optical Technologies

Optical methods may rely on spectroscopic detection features. Suitable for slight vari-ances in the color or pattern, such as vein pattern recognition in the palm of the hand.[384]

Greene's research identifies at least three products which he describes as commer-cializable or pre-production. Among them are the ARMATIX *iP1* pistol, which is activated by RFID, the latter is worn like a wristwatch (Section 3.10.2, p. 144), KODIAK INDUSTRIES with *Intelligun*, a gun with fingerprint-based locking system installed on a Model 1911-style pistol, and iGUN TECHNOLOGY CORPORATION with a shotgun, where the user wears a finger ring with passive RFID tag.[385]

The 2013 study also mentions SMITH & WESSON. This company explored a variety of methods of authentication and investigated new ways in weapons technology that could move within reach if an electrical power supply was included in the design of a gun. In this context, SMITH & WESSON tested electronic firing and fired over 60,000 rounds with prototype firearms with no limitations related to reliability or power sources.[386]

Greene's conclusion: Pulling the trigger to make the gun fire is the intended function of the product; and safety features are intended to mitigate the risks with the use of firearms. Specific smart gun requirements for potential users did not exist at the date of his writing. He concludes, these requirements are needed by the manufacturers as orientation for their research and development projects. The report suggests test pro-tocols need to be developed, which could be used to test smart gun systems under different operating conditions on a uniform basis. Greene emphasizes, the tests should also include the human factor under different stated conditions. As a result, information about the behavior of the system under the influence of the human factor could be obtained if the tests under different conditions would be reviewed. This would help to distinguish reliability from related concepts such as usability, durability, maintainability, and proficiency.[387]

3.10.2 Firearm Accessories and Current Applications

Currently, a trend towards shot counters, as well as other useful electronics offered by a few manufacturers can be seen. They are accessories which do not interfere with the mechanics of firearms.

ARMATIX

ARMATIX filed a request for opening of the insolvency proceedings, ref. 1508 IN 1215/15 in 2015. Before this, the mission of ARMATIX GMBH in Germany used to be, according to its own information, "to raise the bar with regard to the use of handguns, revolvers, long barreled weapons and rifles to set new standards across the board." [388] ARMATIX is known for external electronic gun locks, electronic compartment systems for the

Fig. 3.85: ARMATIX
Smart System iP1
(© ARMATIX)

safe storage of weapons, and a token based *iP1* .22 caliber pistol. The activation of the pistol via RFID and a wrist watch which includes an active tag (Fig. 3.85), where the user would identify himself by PIN code. The handgun will get deactivated if it is more than forty centimeters away from the watch, explains ARMATIX.[389]

On October 19, 2016, Computerworld.com reported that gunmaker ARMATIX, under CEO Wolfgang Tweraser, planned to have a 9 mm *iP9* on the U.S. market by mid-2017, which was supposed to retail for $1,365 U.S.

BERETTA *GunPod*

BERETTA offers the battery-powered, detachable computer *GunPod* (Fig. 3.86) at a price of $215 U.S. for the semi-automatic shotgun *A400*: an optional accessory, which can be subsequently mounted instead of the pistol grip cap of select shotguns. The computer can count the number of shots, can be used as a thermometer for ambient temperatures from −25 °C to +70 °C, and finds a numeric comparison value for the recoil energy of the last shell which was fired. According to the manufacturer, the shot counter works without a battery, and the device will generally serve to

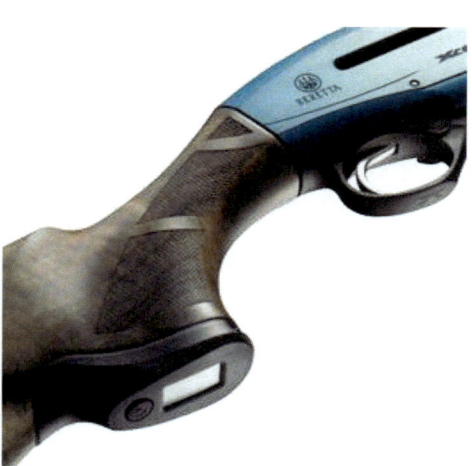

Fig. 3.86: BERETTA *A400* with *GunPod* housed in pistol grip cap.
(© BERETTA)

determine the best conditions for the performance of the shooter and weapon.[390] At this time BERETTA offers *GunPod 2*, a new generation that can be hooked up to smartphones.

BERETTA *i-PROTECT*

BERETTA offers one of the most complex electronic system in this survey. The i-PRO-TECT system is intended for police use and incorporates the following applications: a system for mission control centers (Intellicore), user support software application (O.d.i.n.o) by tablet computer (Mercurino and Mercurino Metropolitan Police), clothing (BZero), and the BERETTA *PX4i* storm pistol (Fig. 3.87).[391]

Fig. 3.87: BERETTA *i-Protect*, schematic.

(© BERETTA)

The handgun is based on the BERETTA *PX4* pistol, which is equipped with sensors and electronics for duty use of law enforcement personnel. The electronics exchanges information with the i-PROTECT-system. The sensors of the weapon monitors the number of cartridges in the magazine, whether a cartridge is in the firing chamber, note the position the slide, determine whether the firing mechanism is cocked, and even note the position of the manual safety. The manufacturer also integrated an accelerometer and a radio transmitter in the firearm. With this radio transmitter, the gun stays connected to the control center in real time, transmits audio and video signals and GPS data. Standard maintenance information of the weapon is collected as well, passed on, and evaluated for logistics. The communication interface is the i-PROTECT Black Box (Fig. 3.88). The Black Box can be detached from the pistol at any time, that is, it is a separate accessory item with integrated flashlight, and sits on the accessory rail. The handgun itself works purely mechanical and may be used without this communication interface.

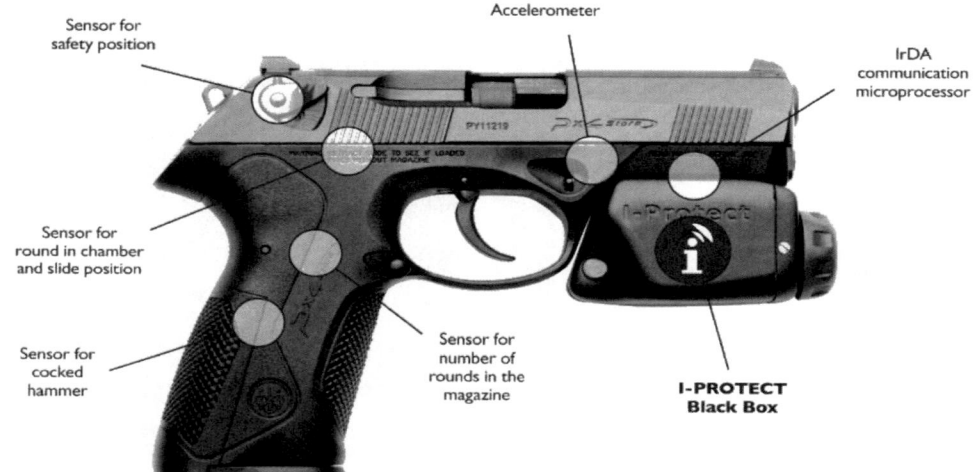

Fig. 3.88: BERETTA *Px4i* pistol.

(© BERETTA)

Body worn sensors (BZero) monitor heart rate, breathing and body position of the officer. In combination with the gun they form a system which offers additional security and assistance. The system reports biometric and other data to track athletic performance and can send an emergency signal with GPS data to the control center, e.g. if the sensors detect that an officer is down. This allows the control center to react accordingly and send in first responders to the person in distress. At the same time, the control center will stay informed, and note the condition of the body. According to BERETTA, the transmission of the signal will work reliably even in tunnels or shielded environment. This is done by the simultaneous use of different ways of transmission. The two-way communication of the system is in encrypted form.[392]

FN HERSTAL *SmartCore*

FN HERSTAL's *SmartCore®* is a shot counter inside the grip handle of rifles, which communicates via radio signal. According to the manufacturer, the shot counter contributes to the reduction of maintenance costs, as well as to optimize the availability period of the respective weapon because service work can be planned. The information it monitors includes jars, as well as the number of shots, with blank, live and dry firing being discriminated. It's a device for military rifles, which is why FN also points out the advantage of a real-time data transmission to a mission control center when the gun is being fired.

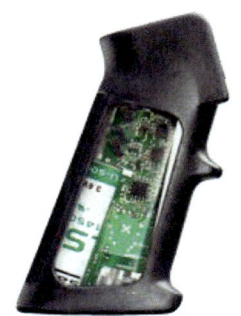

Fig. 3.89:
FN *SmartCore*
(© FN HERSTAL)

The new generation of the shot counter generates its own power. The new *SmartCore®* is a no-battery device. It is located inside the lower receiver of an FN assault rifle and uses the rearward movement of the moving parts to generate power. Also new is the ability to transmit data wirelessly now at short-range and only upon request, ensuring it is undetectable and there is no unintentional emission.

SECUBIT *GSC*

The company, SECUBIT LTD. offers the gun shot counter (*GSC™*), as an aftermarket accessory. The weight of its heart is 3 g and can either be attached directly to firearms or that the manufacturer uses this core in a variety of adapter housings. Depending on the housing, the shot counter can be placed on the picatinny accessory rail[393] or perhaps in the cavity behind the magazine well of GLOCK pistols *G17* and *G19* (Fig. 3.90). The weight of the shot counter intended for picatinny rails is 24 g, for GLOCK pistols 12 g.

Fig. 3.90: SECUBIT
GSC™ for GLOCK
G17 and *G19*.
(© SECUBIT)

SECUBIT describes the performance of the system in connection with its application in a long gun. Here, the shot counter should be put on the upper picatinny rail, closely to the rear sight, and it will be fixed with a screw. The shot counter can record the data of at least 500,000 shots, where the accuracy of detection would amount to over 95 percent.

After the data is transferred via USB interface from the gun shot counter to a standard PC, the data can be processed and analyzed with a software from SECUBIT. This software can deliver information on how the user and the gun performs. It can also determine information related to the endurance of the firearm or the parts that may need maintenance.[394]

The new generation of SECUBIT's shot counter *WL™* is smartphone ready and discriminates between life fire, cocking the gun, and other events that should be monitored.

TASER

This American company, originally established in 1993, and now under new name AXON, coined a niche market: the TASER name has become synonymous with less than lethal weapons. The company's long-term vision is making the bullet obsolete. Their TASER can be carried and handled just like a pistol. It discharges arrow pairs, which remain connected to the control unit by electric wires. Initially, a five second continuous electrical current is transferred to the struck target, which, for example, will cause a person to lose control over his muscles and drop. After those first five seconds, the police officer could resubmit current pulses that persist from now on for the duration of actuation of the trigger. Another pair of arrows from the electronic control unit could be fired if required.

Fig. 3.91: TASER *X2*

The electronic control unit performs a self-check for determining operational readiness and records the date and time of the shot. On the back, a small display (Fig. 3.92) shows information about battery life and power supply. At the business end are two laser diodes, each one zeroed in for the respective pair of arrows to enable a precise aiming. An additional module is available as an accessory for up to 1.5 hours of audio/video recording.[395]

Fig. 3.92: Display of a TASER *X2*.

TASER INTERNATIONAL in 2017 changed the company's name to reflect the evolution of their company from a weapon manufacturer to a full solutions provider of cloud and mobile software, connected devices, wearable cameras and artificial intelligence. All of which now fall under the AXON brand.

Radio Frequency Identification

HECKLER & KOCH disclosed the idea for an electronic ID tag which is to be attached at a hidden location of a weapon (US 8,171,665 B2[396], dated December 6, 2004). Objective of the invention is to provide a hidden serial number which could be determined in cases where the original serial number was tampered.

Fig. 3.93: Frame with transponder (covered with adhesive) and separate transponder (glass encapsulated).

Electronic ID tags are being used to automatically identify products, people and animals in applications that range from animal identification, access control, sports timing, industrial automation, event management, and asset tracking to logistics. On firearms, ID tags are being used ever more frequently to reduce logistic effort: the memory of an RFID tag on a gun will store the serial number of the firearm and maybe other data, which is then transmitted wireless from the tag to a reader at short range (Fig. 3.93).

ID tags can help to reduce the time that may be needed e.g. for an incoming inspection, where normally the gun boxes have to be opened, a safety check carried out, and then the serial number would be identified visually. In this process, not only the time required is considerable, but has also the potential for errors while reading the serial number of the firearm and manually transferring this information in a computer system.

KEMA TECHNOLOGIES offers storage systems for weapons, which can identify data stored on ID tags and thus monitor whether a gun was placed back in a gun compartment. A possible application would be a security company, where each duty gun has to be placed back in a locker at the end of the shift. The storage system will recognize whether a gun was placed back and which gun specifically. The advantage of this system lies in the avoidance of bureaucracy at the end of a work shift, as the inspection and registration of guns takes time and is costly; for certain professionals who may invest in this storage system it can pay for itself.

3.10.3 Summary "Electronics"

Since the mid-1990s, numerous companies have worked on personalized firearms. The objective of these efforts is to avoid unapproved use of weapons or the problem of police firearm takeaways by adversaries and how carrying personalized firearms could prevent officers' injury or death when their own or another officer's firearm is being pointed against them by dangerous individuals. The smart gun technology polarizes; some condemn it because when designed in the appropriate way, the technology could be applied to control gun ownership and various gun rights advocacy groups are opined it will be encroaching on gun owner's rights.

Even though NIJ took a bold step in publishing *Baseline Specifications for Law Enforcement Service Pistols with Security Technology*, widely accepted and clear specifications of the performance requirements are yet to arrive, as well as a standard on how the reliability of weapons with smart-gun technology is to be tested. An important question in this regard is, whether the failure of the technology may or may not inhibit firing of the weapon. NIJ has answered this by stating: "If the security device malfunctions, it shall default to a state to allow the pistol to fire."[397] As it's been shown earlier, there is no doubt for civilian users: the system must function flawlessly, and at any point in time. *Günther* may be cited at this point. In 1900, he describes the understandable disappointment of the user when a malfunction occurs. His words are timeless:

> How will you help yourself in urgent moments, when the mechanism [...] suddenly fails? The most expensive and sophisticated gun turns out to be an old and useless piece of iron, which pushes even harder on the mind of its unfortunate owner the more he expects to get out of the precious product.[398]

In contrast to smart guns, electronic accessories such as reflex sights are becoming increasingly popular. Firearms are certainly part of some areas of daily life, and it may take slightly longer until consumers accept the idea of new comfort and new technology. The consumer behavior may depend on which generation they belong to. Younger buyers grow up in an environment with computers and electronics (Digital Natives), and are more open toward embracing new technologies.

3.11   Summary "Pistol Technology"

The most popular operating method of semi-automatic pistols is based on the more than 100-year-old principle of recoil-actuation. The remaining technology on self-loading handguns has continuously improved in small steps over time and is now at a very high level, but as we have seen, the improvements in recent times focused mainly on two areas: trigger mechanisms and safeties. Apart from these, no significant progress has been seen in pistol technology over the last few decades. Instead, a shift from technical progress to simplified but nevertheless advanced production techniques in the manufacture of handguns has taken place since the 1980s, which has allowed a more efficient production.

In electronics, a slight trend is emerging. Some users vehemently reject electronics, but despite all the negative bias, progress in this area can be expected, which will be accompanied by the proliferation of electronics in other parts of our everyday life. The emergence of the use of electronics for firearm accessories is noticeable and the next step towards the integration of electronics into the guns themselves seems feasible.

Maybe manufacturers will start to integrate electronics only where it will not interfere with the firing of a shot, because confidence in the new technology plays an essential role in the purchasing decision of consumers in the future. Also possible would be the introduction of electronic safety systems in areas where the users and people in their vicinity should be protected against an unintentional discharge, e.g. in training and in a classroom setting.

Apart from this, polymer frames, safe field-stripping procedures, ease of use and consistent trigger pull for each shot are standard these days. Generally, despite all the progress, technical features still turn out to be demanding and their mastery requires technical knowledge and a high level of skills, which is why users should continuously hone their firearms skills through regular training. The training regimen should include troubleshooting, because regardless of technical improvements, when using guns, malfunctions may occur in this process which is best described as a cycle of operation.

A working cycle of the operation consists of two strokes: Recoil and counter-recoil. Fig. 3.94 shows the shot cycle of most semi-automatic pistols. It is explained below based on the example of a short-recoil-operated pistol with coupled barrel and slide. The cycle is divided into quadrants: gun is fired, recoil, counter-recoil, and trigger reset. The shot cycle starts with the loading of the pistol by the user, which is illustrated outside the cycle, at 12 o'clock.

The discharge of a round described in the first quadrant is initiated by the operator when they pull the trigger, which disengages automatic safeties before the firing mechanism is released. This is followed by the ignition, which represents the transition onwards to the second quadrant of the cycle. As the coupled barrel and slide group recoils, the disconnector gets activated. Afterwards the barrel is uncoupled from the slide, and its motion is terminated. While the slide continues its rearward movement under its momentum, the spent case is ejected. The rearward movement of the slide compresses the recoil spring. Full recoil stroke is accomplished when the slide strikes the backstop.

On completion of the rearward stroke the recoil spring is fully tensioned. In the third quadrant the stored energy in the spring then starts the slide forward. As the slide is forced forward, it feeds a cartridge from the magazine into the chamber. Also, the coupling of barrel and slide is accomplished, the coupling will be safeguarded, and the disconnect function ends shortly before the coupled barrel and slide group reaches the front stop. At the end of the third quadrant, the gun is loaded, but the firer still has their finger on the trigger blade, which is why the trigger is in the rear position. Before the shooter can fire another round, they must release the trigger. The reset function is completed once the trigger mechanism re-engages into the firing mechanism. At this point, it is possible to pull the trigger again to fire the next shot. If the operator decides to let the trigger go to the forward-most position instead, the automatic safeties are engaged. With the user's finger off the trigger and outside the trigger guard, they now hold in their hand the gun ready to fire the next shot and with automatic safeties engaged. The latter is what is shown in the fourth quadrant.

The sequence of events ends abruptly when the ammunition supply is exhausted: the empty magazine's follower causes the slide stop to engage the slide, just after the slide hit the rear stop. This function is shown outside of the cycle, at six o'clock.

Shown in the inner circle of the diagram is the cocking of the firing mechanism. This mechanism may get cocked either during recoil or counter-recoil, mostly depending on if the pistol is hammer- or striker-fired.

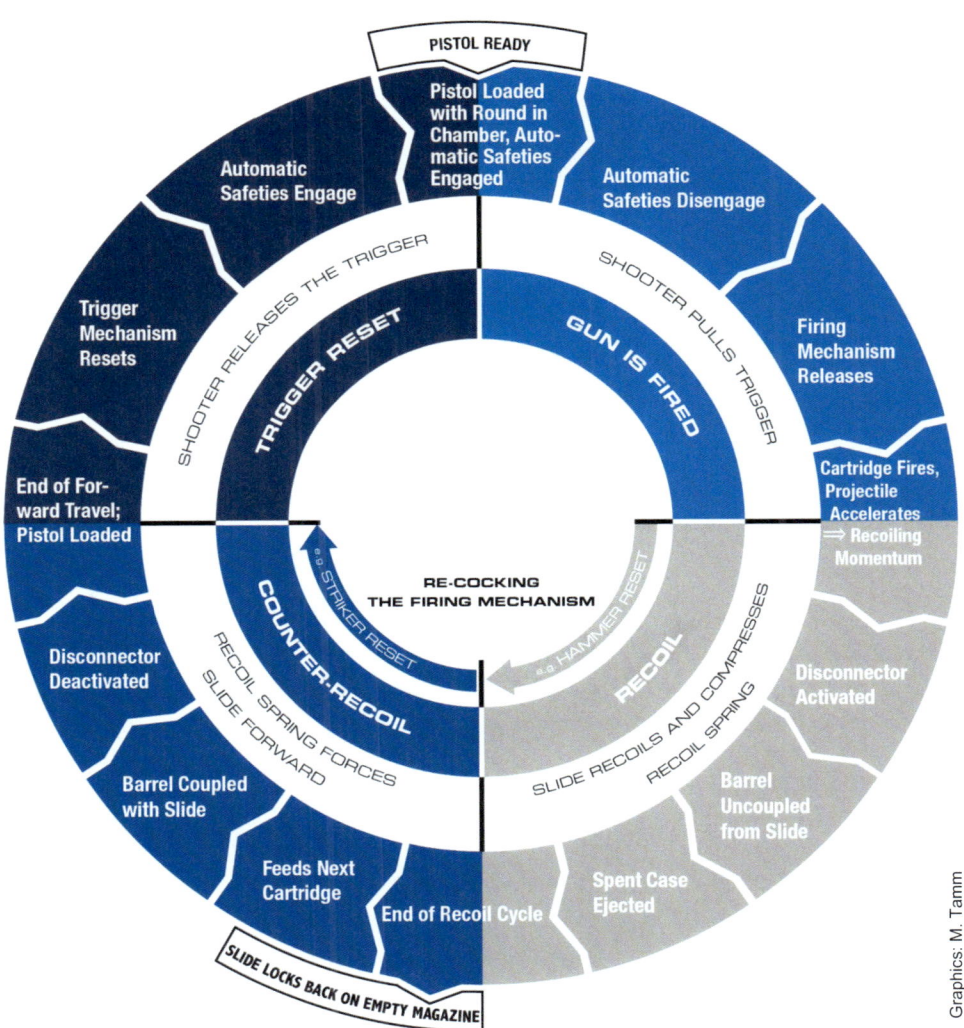

Fig. 3.94: Schematic of the shot cycle of semi-automatic pistols.

Graphics: M. Tamm

153

Fig. 3.95: The RUGER-57 uses 5.7 × 28 ammo and offers 20-round mag capacity.

(© RUGER)

4   U.S. Regulations and Legal Requirements

In general, the National Firearms Act of 1934 (26 USC Chapter 53) and the Gun Control Act of 1968 (18 USC Chapter 44) serves as the cornerstone for firearms regulations on manufacturers in the USA (federal legislation), supplemented by state regulations implemented by the individual federal states, the laws of Washington D.C., and U.S. territories. The individual states are entitled to enact additional laws (state legislation), which may vary considerably and are independent of federal firearms laws.

State-level legislation may be broader or more limited in scope than that of the federal laws; when state and federal laws are not in agreement, federal law prevails. As a result, firearm manufacturers must consider a wide range of features to be incorporated within their handguns, in order to ensure their handguns are compliant for purchase throughout the U.S.

Fig. 4.20 on page 215, depicts a chart that provides an overview of the contents contained herein in this chapter. We stretch an arc beginning with core U.S. federal laws, primarily in regard to assessing the importation of handguns. We then move on to restrictive state laws of selected federal states, varying requirements for police guns, testing in accordance with SAAMI specs, as well as considering IPSC handgun rules.

Individual state laws are not intended to be discussed, instead, we take a closer look at the requirements imposed on handguns by four restrictive states: California, Massachusetts, Maryland and New York. Pistols that are specifically-designed to comply with the strict requirements presented by these four states can easily be modified in accordance with the specific requirements of the remaining states. This can be done without incurring additional costs to the manufacturers, as the requirements imposed by the said four states incorporate the basic requirements into the design of a pistol model, thus investment of any additional engineering is no longer required.

Furthermore, we are only concerned as to how the regulations apply to new semiautos that manufacturers would like to develop and introduce into the market. Based on the previous statement, the contents of this chapter will be of main interest to anyone intending to import pistols into the United States.

Manufacturers who produce handguns outside of the U.S. need to consider compliance with non-U.S. gun laws and regulations, export laws of their respective country, as well as other legal requirements such as proof laws, including CIP standards. In the latter sections, we will focus extensively on the requirements that apply to the U.S. market alone.

4.1   Review of the Historical Development of the U.S. Firearms Act

In the United States the terms "Junk Guns" and "Saturday Night Specials"[399] are used interchangeably to describe handguns that are made from inferior materials or perceived as poor quality and designed with the main aim of cutting manufacturing costs while increasing availability.

The term "Junk Guns" and its designated type of weapon became more widespread with the passing of the GCA of 1968 and grew in popularity in the early 1980s with a surge in production by U.S. companies. The producers of these inexpensive handguns soon became known as "Ring of Fire" companies; which comprised of a small group of Los-Angeles-based gun manufacturers.

The Ring of Fire companies increased production steadily and achieved a market share of approximately 33 percent of U.S.-made handguns at the beginning of the 1990s.[400] The increase in market share was accompanied by an increase in crime rate, which began in the 1960s as a result of demographic and social changes and continued until the early 1990s.[401] In 1991, guns were involved in over two-thirds of all homicide cases, 80 percent of which were carried out with handguns.[402] Typically, cheap U.S.-made revolvers and pistols were used in the commission of crime.[403]

Weapons experts criticized Junk Guns because of their moderate technical features, in particular the lack of safety features, low reliability and inferior quality. By reason of their poor design, low level of accuracy and unreliability these handguns were considered as inadequate for self-protection and sporting use in wide circles.[404] The State of California responded to this public safety threat by adopting safety standards for handguns in 1999, and by 2003, five of the six original Ring of Fire companies had declared bankruptcy; but the market had swallowed too many of their guns. In five out of 10 crimes committed, guns from Ring of Fire manufacturers were reportedly used.[405] To reduce further widespread of Junk Guns, some states introduced gun laws, which went beyond existing federal laws.

The Gun Control Act of 1968[406] was passed into law, after gun sales had quadrupled in the years from 1962 to 1968[407], while handgun sales had risen from 1.7 million to 2.5 million in the year 1967 alone.[408] In addition, the crime rate had increased, with homicides taking of 55 percent in 1964.[409] Across the nation it was possible to buy a gun in stores with little in the way of a background check. Only a few states enforced a minimum age to purchase a handgun.[410]

The country also saw several high-profile assassinations of public figures, and famous victims included President John F. Kennedy in 1963, followed by his brother Senator Robert Kennedy and Civil Rights Leader Dr. Martin Luther King, Jr. in 1968.

At the same time, there were protests across the United States in college campuses and in cities across the nation against America's war in Vietnam. Thus, the time seemed right for a new gun control law, and by passing the *Gun Control Act of 1968*, the United States introduced a federal law that would control gun purchases, limit the sales of guns and regulate interstate commerce in firearms by generally prohibiting interstate firearms transfers except among licensed manufacturers, dealers and importers.

The Gun Control Act of 1968 would basically exclude the importation of weapons into the United States by creating what is known as the "sporting purposes" standard for imported firearms,[411] which implies that handguns can only be imported if they are particularly suitable for sporting purposes.[412]

The Gun Control Act of 1968 is part of the Code of Laws of the United States of America; the latter is often referred to as United States Code or USC as an acronym. The United States Code consists of 54 numbered "titles." Title 18 – Crimes and Criminal Procedure; and Appendix, as well as Title 26 – Internal Revenue Code, contain gun laws. Wherever applicable, the Titles are divided into sub-titles, e.g. Title 26.[413] The full name of the Gun Control Act of 1968 is: Chapter 44 of Title 18 of the United States Code.

4.2 Federal Firearms Regulations

The Gun Control Act of 1968 (GCA) prohibits, in principle, the import of firearms, firearm frames or receivers, barrels, and ammunition.[414] Exceptions are listed in GCA, 18 USC § 925(d), i.e. the Attorney General shall authorize a firearm or ammunition to be imported or brought into the United States; or any possession thereof if the firearm or ammunition falls under the following cases:

1. Is being imported or brought in for scientific or research purposes, or is for use in connection with competition or training pursuant to chapter 401 of title 10;
2. Is an unserviceable firearm, other than a machinegun as defined in section 5845(b) of the Internal Revenue Code of 1986 (not readily restorable to firing condition), imported or brought in as a curio or museum piece;

3. Is of a type that does not fall within the definition of a firearm as defined in section 5845(a) of the Internal Revenue Code of 1986 and is generally recognized as particularly suitable for or readily adaptable to sporting purposes, excluding surplus military firearms, except in any case where the Attorney General has not authorized the importation of the firearm pursuant to this paragraph, it shall be unlawful to import any frame, receiver, or barrel of such firearm which would be prohibited if assembled; or
4. Was previously taken out of the United States or a possession by the person who is bringing in the firearm or ammunition.

The Attorney General shall permit the conditional importation or bringing in of a firearm or ammunition for examination and testing in connection with the making of a determination as to whether the importation or bringing in of such firearm or ammunition will be allowed under this subsection.

Please note, Chapter 401, Title 10 referred to in § 925(d)(1) applies for training performed by members of the U.S. Armed Forces.

According to section 5845(a)[415] referred to in § 925(d)(3), the following weapons are excluded from importation:

1. A shotgun having a barrel or barrels of less than 18 in [457.5 mm; PD] in length;
2. A weapon made from a shotgun if such weapon as modified has an overall length of less than 26 in [660.4 mm; PD] or a barrel or barrels of less than 18 in [457.5 mm; PD] in length;
3. A rifle having a barrel or barrels of less than 16 in [406.4 mm; PD] in length;
4. A weapon made from a rifle if such weapon as modified has an overall length of less than 26 in [660.4 mm; PD] or a barrel or barrels of less than 16 in [406.4 mm; PD] in length;
5. Any other weapon, as defined in subsection 5845(e);
6. A machinegun;
7. Any silencer (as defined in section 921 of title 18, United States Code); and
8. A destructive device.

The term "firearm" shall not include an antique firearm or any device (other than a machinegun or destructive device) which, although designed as a weapon, the Secretary finds by reason of the date of its manufacture, value, design, and other characteristics is primarily a collector's item and less likely to be used as a weapon. According to §5845(g), the term "antique firearm" refers to any firearm not designed or redesigned for using rim fire or conventional center fire ignition with fixed ammunition and manufactured on or before 1898 (including any matchlock, flintlock, percussion cap, or similar type of ignition system or replica thereof, manufactured before or after the year 1898) and also any firearm using fixed ammunition manufactured on or before 1898, for which ammunition is no longer manufactured in the United States and is not readily available in the ordinary channels of commercial trade.

According to §5845(b), the term "machinegun" refers to any weapon that shoots, is designed to shoot, or can be readily restored to shoot automatically more than one shot without manual reloading by a single function of the trigger. The term shall also include the frame or receiver of any such weapon, any part designed and intended solely and exclusively, or combination thereof for use in converting a weapon into a machinegun, and any combination of parts from which a machinegun can be assembled if such parts are in the possession or under the control of a person.

The term, "any other weapon" in § 5845(e)(5) refers to any weapon or device capable of being concealed on a person from which a shot can be discharged through the engagement of an explosive, a pistol or revolver having a barrel with a smooth bore designed or redesigned to fire a fixed shotgun shell, weapons with combination shotgun and rifle barrels 12 in [304.8 mm; PD] or more, less than 18 in [457.2 mm; PD] in length, from which only a single discharge can be made from either barrel without manual reloading, and shall include any such weapon which may be readily restored to fire. Such term shall not include a pistol or a revolver having a rifled bore, or rifled bores, or weapons designed, made, or intended to be fired from the shoulder and not capable of firing fixed ammunition.

The term "destructive device" in § 5845(a)(8) refers to (1) any explosive, incendiary, or poison gas (A) bomb, (B) grenade, (C) rocket having a propellant charge of more than 4 ounces [113 g; PD], (D) missile having an explosive or incendiary charge of more than one quarter ounce [7.1 g; PD], (E) mine, or (F) similar device; (2) any type of weapon by whatever name known, which will or may be readily converted to expel a projectile by the action of an explosive or other propellant, the barrel or barrels of which have a bore of more than 0.5 in [12.7 mm; PD] in diameter, except a shotgun or shotgun shell, which the Secretary generally recognizes as particularly suitable for sporting purposes; and (3) any combination of parts, either designed or intended for use in converting any device into a destructive device as defined in subparagraphs (1) and (2) and from which a destructive device may be readily assembled.

The term "destructive device" shall not include any device which is neither designed nor redesigned for use as a weapon; any device, although originally designed for use as a weapon, which is redesigned for use as a signaling, pyrotechnic, line throwing, safety, or similar device; surplus ordnance sold, loaned, or given by the Secretary of the Army pursuant to the provisions of section 4684(2), 4685, or 4686 of Title 10 of the United States Code; or any other device which the Secretary finds is not likely to be used as a weapon, or is an antique or rifle that the owner intends to use solely for sporting purposes.[416]

As we have just seen, § 5845(f)(2) limits the caliber of barrels to 0.5 in. This explains why manufacturers of revolvers, e.g. the S&W *Model S&W500* revolver, despite the success of this big bore revolver; can not offer a handgun of any caliber exceeding 0.5 in.

The category of destructive device had a direct influence on the development of the *Gyrojet* pistol. The inventors, Dr. Robert Mainhardt and Dr. Arthur Biehl joined forces to form their company MBAssociates, and had already begun development in 1960 with the project and decided to increase the caliber of the rocket projectiles from 0.49 to 0.51 in. With the introduction of the Gun Control Act of 1968, weapons and ammunition were now classified as destructive devices, which required MBAssociates to react and reduce the caliber to 0.49 in.

GCA § 925(d)(3), Sporting Purposes (Factoring Criteria for Weapons)
What is called "sporting purposes" of GCA § 925(d)(3), i.e. a type of firearm that is generally recognized as particularly suitable for or readily adaptable to sporting purposes, is the lawful foundation for the evaluation of handguns carried out by the Bureau of Alcohol, Tobacco, Firearms and Explosives (ATF) Firearms Technology Industry Services Branch (FTISB). FTISB evaluates round about 300 handgun models per year.

The ATF is the primary agency responsible for enforcing Federal firearms laws with its headquarters based in Washington, DC.[417] Its mission is defined as follows:

> ATF protects the public from crimes involving firearms, explosives, arson, and the diversion of alcohol and tobacco products; regulates lawful commerce in firearms and explosives; and provides worldwide support to law enforcement, public safety, and industry partners.[418]

However, prior to the Gun Control Act of 1968, criteria had to be created initially for the assessment of "suitable for sporting purposes." In 1968, then Secretary of the Treasury Henry H. Fowler founded the Firearms Evaluation Panel, in order to provide guidance in determining an import standard and to identify which firearms met this standard for importation into the United States. The panel focused its attention on handguns and recommended the adoption of a factoring criteria to evaluate certain types of handguns.[419]

The resulting ATF Form 4590, *Factoring Criteria for Weapons*, is a one-page worksheet utilized by ATF's FTISB to calculate a numeric score that is used in determining whether a certain handgun may be approved legally for importation into the United States. Form 4590 establishes certain standards and prerequisites for imported handguns.[420] It mandates minimum size and weight requirements for handguns under consideration for importation.

Each submitted sample can accrue additional points based on the cumulative evaluation of its configuration, design and enhanced safety features that further contribute to the overall sporting and safety characteristics. When handguns are evaluated, pistols must obtain a numeric value of 75 points, revolvers 45 points, before they are considered and approved for importation. The ATF Form 4590 Factoring Criteria for Weapons represents the first and possibly most important hurdle for non-U.S. manufacturers of handguns for importation of a new handgun model into the United States.

Generally, U.S.-made firearms are not subject to the factoring criteria as long as they remain within the shores of the United States. However, if a U.S.-made handgun were to be exported, it would be subject to the factoring criteria before it could be imported back into the United States. An exception to this rule is provided if the firearm was previously taken out of the United States or a possession by the person who is bringing it back to the U.S. In such scenario, the sporting purposes test does not apply.

An ATF Form 4590 evaluation is limited to complete weapons only, and does not apply to actions, receivers or frames.[421] Firearm receivers and frames may be evaluated by the FTISB for purposes other than those outlined within this chapter. The evaluation period for a complete firearm is usually 90 days. However, before an evaluation can begin, an estimated 90 days must be considered upfront for the conditional import permit application, which starts the process. Once a favorable FTISB approval is received, another 90 days is required for the import permit approval of series weapons. In total, a lead time of approximately 270 days should be considered until the series delivery on a new product can begin.

Before a weapon sample can be submitted to the ATF's FTISB, the U.S. importer must first apply for a temporary import permit by means of *conditional importation*[422]

as defined in U.S. 925(d)(4). The Attorney General shall permit the conditional importation or bringing in of a firearm or ammunition for examination and testing in connection with the making of a determination as to whether the importation or bringing in of
such firearm or ammunition will be allowed under this subsection. 18 USC § 925(d)
and 27 CFR 478.116 authorize the conditional importation of a firearm or ammunition
for the purpose of examination and testing by ATF in determination as to whether the
importation of the firearm or ammunition will be allowed under this section.

An ATF Form 6, *Application and Permit for Importation of Firearms, Ammunition and
Implements of War*, is used to initiate the importation. A Form 6 that is conditionally
approved, directs the U.S. Customs and Border Protection (CBP) to deliver the firearm
or ammunition from the port of entry to ATF. If upon completion of ATF's examination,
the firearm or ammunition is determined to be importable, the firearm or ammunition
will be returned to the importer or their broker. If the firearm or ammunition is otherwise
found to be unsuitable for importation, the firearm or ammunition must be exported
back to its origin by the importer, pursuant to an export license from the Department of
State or Department of Commerce, abandoned to ATF for destruction, or ATF will seek
forfeiture.

Once the importer receives the weapon sample, it must be forwarded promptly to
the ATF for evaluation in accordance with the requirements of ATF Form 4590.

4.2.1 Factoring Criteria for Weapons

Each and every handgun imported into the United States is subject to the factoring
criteria of evaluation by the ATF Firearms Technology Branch, even if it is a copy of a
U.S.-made handgun or considered a *curio or relic* (C&R).[423] The only exemptions from
evaluation are weapons manufactured in the United States alone. U.S.-made long- or
handguns that have been exported from the U.S. and later re-imported back into the
U.S., as long as the original export is not restricted by Arms Export Control Act, the
Foreign Assistance Act or additional internal U.S. politics, must be evaluated before
re-importation; in the case of pistols and revolvers to ensure compliance with ATF Form
4590. If the firearm is classified according to 18 USC § 925(e)(2) as a curio or relic, an
exemption for the import can be granted if necessary. 27 CFR § 478.11 defines curio
or relic firearms as:

1. Firearms which were manufactured at least 50 years ago, but not replicas thereof;
2. Firearms, which are intended to remain in museums;
3. Valuable and historically significant weapons.[424]

For any surplus firearm, the same rules would apply for this type of weapon and in general: an import of these weapons into the United States is not possible unless they are updated.

ATF Form 4590 (5330.5) "Factoring Criteria for Weapons"

ATF Form 4590, *Factoring Criteria for Weapons*, distinguishes between pistols and revolvers (see Appendix 3), listing the criteria for pistols on the left side and revolvers on the right. In carrying out an assessment, the weapon is first checked with regard to construction-related prerequisites and then individual characteristics are evaluated and rated with scores as proscribed within the ATF Form 4590 with respect to the following criteria: size, weight, material, caliber, safety devices and technical features. In accordance with assessment of suitability for sporting purposes, a handgun is awarded more points based if it's large, heavy, powerful and of quality material.

We are going to have a close look at pistols first and then revolvers.

Factoring Criteria for Pistols (ATF Form 4590)

Pistols must be awarded at least 75 points in the evaluation process to be approved for importation. Two prerequisites must be met regardless of the qualifying score:

1. The pistol must have a positive manually operated safety device.
2. The combined length and height must not be less than 10 in [254 mm; PD] with the height (right angle measurement to barrel without magazine or extension) being at least 4 in [101.6 mm; PD] and the length being at least 6 in [152.4 mm; PD].

The first prerequisite calls for a manual safety, that is a safety device that is manually operated by the user. Acceptable is a classic safety switch, which is intended to be operated by the user from "safe" to "fire", similarly to the manual safety found on a *Model 1911* pistol. Alternatively, automatically-acting safeties are accepted if they act directly on the firing pin or striker. Trigger safeties do not meet the definition of the first prerequisite (see Appendix 4).

In regard to the second prerequisite, pistols manufactured in the United States can have smaller dimensions than pistols imported from abroad. A comparison between a German-made WALTHER *PPS* pistol and a RUGER *LCP* manufactured in the United States may serve as an example (Fig. 4.1); the combined length and height of the WALTHER *PPS* is 10.7 in (LCP: 8.7 in), with a length of 6.3 in (*LCP*: 5.1 in) and a height of 4.4 in (LCP: 3.5 in). The *LCP* is considerably smaller than the *PPS* and

Fig. 4.1: Size comparison RUGER *LCP* vs. WALTHER *PPS*.

too small to satisfy the second prerequisite, thus would not be considered for importation, if it were made outside the United States. For gun manufacturers that intend to sell deep-concealed carry guns in the U.S., the best option is to set up their manufacturing in the USA.

After both prerequisites have been evaluated and found to be met, the pistol's individual characteristics will then be evaluated and scored accordingly. During the evaluation process, only full points will be awarded, and the partial points evaded. The evaluation starts with the pistol's overall length.

Overall Length

Starting with a minimum length of 6 in [152.4 mm; PD], pistols are awarded one point for each one quarter of an inch [6.35 mm; PD] of additional length.

Frame Construction

A frame constructed from investment cast or forged steel is awarded 15 points. Frame constructions consisting of investment cast alloy or forged HTS alloy (i.e. High Tensile Strength non-ferrous alloy), are awarded 20 points.

Non-ferrous metals are any metals and metal alloys in which iron is not included as a main element. Non-ferrous alloys must have a tensile strength of 50,000 psi [345 N/mm²; PD] or greater. When submitting a firearm comprising of an alloy frame, information regarding the type of alloy used must be submitted in English (see Appendix 5).

A pistol model with polymer frame will only be awarded points for frame construction, if it uses a metal chassis which carries the fire-control unit and the metal used for the chassis meets the above-mentioned criteria (see p. 87). Polymer pistol frames with integrally formed slide rails do not meet this criterion and will not be awarded any point, even if the rails are made of metal.

Weapon Weight

The pistol with one unloaded magazine will be awarded one point per ounce [28.35 grams; PD].

Caliber

Pistols chambered in .22 Short and .25 Auto are not awarded any points. Pistols chambered in calibers .22 LR and 7.65 mm [.32 in; PD] to .380 Auto [.380 ACP; PD] will be awarded three points, 9 mm Parabellum [9 mm Luger, 9 mm × 19; PD] and over, will be awarded ten points.

Safety Features

Five points are given for a locked breech mechanism.

A loaded chamber indicator is also awarded five points. Acceptable are viewports, mechanical indicators and other devices that fulfill the same task.

Three points are given for a grip safety, five points for a magazine safety, and a firing pin block or lock gets ten points.

Miscellaneous Equipment

In this section, the pistol will be awarded two points if equipped with an external hammer and 10 points if equipped with a double-action trigger. Double-action is defined as a mode of operation that permits a single pull of the trigger to cock and fire the pistol. Restrike capability is not required within the scope of the definition.

Target sights are awarded five points, that is drift adjustable sights with contrast markings. If such a target sight is click-adjustable for windage and/or elevation, 10 points will be awarded.

Target grips are awarded five and target triggers two points. A target grip suitable for sporting purposes might for example offer a thumb rest, and a target trigger is vertically grooved to provide more positive trigger control.

Factoring Criteria for Revolvers (ATF Form 4590)

Revolvers must fulfill three prerequisites. They must pass a safety test, and an overall frame length of 4.5 in [114.3 mm; PD] minimum with conventional grips, not measured diagonal, and must have a barrel length of at least 3 in [76.2 mm; PD].

The safety test of the first prerequisite focuses on accidental discharges. A double-action revolver must have a safety feature which automatically (or in a single-action revolver by manual operation) causes the hammer to retract to a point where the firing pin does not rest upon the primer of the cartridge. The safety device must withstand the impact of a weight equal to the weight of the revolver, dropping from a distance of 36 in [914 mm; PD] in a line parallel to the barrel upon the rear of the hammer spur, a total of five times.

After the three prerequisites have been evaluated and met, the revolver's individual characteristics will then be evaluated. With the exception for barrel length only full points will be awarded and a total of at least 45 points must be achieved.

The evaluation starts with the revolver's barrel length, and it is measured from the muzzle to the front face of the cylinder. Starting from a minimum length of 4 in [101.6 mm; PD], revolvers will get one half a point for each additional full multiple of one quarter inch [6.35 mm; PD].

Under frame construction, revolvers will achieve 15 points for an investment cast or forged steel frame. Frames consisting of investment cast alloy or forged HTS alloy, are awarded 20 points. Non-ferrous alloys must have a tensile strength of 50,000 psi [345 N/mm²; PD] or greater (see Appendix 5). One point is awarded per ounce [28.35 g; PD] of the weight of the unloaded revolver.

Revolvers chambered in .22 Short to .25 ACP are awarded zero points, chambered in .22 LR and .30 to .38 S&W accrue three points, .38 Special, four points, and .357 Mag and over, five points.

Under Miscellaneous Equipment drift or click adjustable target sights are listed with five points, and target grips with additional five points. A target hammer in combination with a target trigger will also earn five additional points to the total score.

ATF requirements can be summarized as follows:

- Safety device
- Minimum dimensions
- Barrel length (Revolver)
- Suitable for sporting purposes (Factoring Criteria for Weapons, ATF Form 4590)

Fig. 4.2: Summary of ATF requirements.

4.2.2   Marking Requirements

Any weapon, including a starter gun, which will or is designed to or may readily be converted to expel a projectile by the action of an explosive including the frame or receiver of any such weapon; is defined as a firearm under the Gun Control Act of 1968. The term firearm, does not include any antique firearm which was manufactured on or before 1898.

The Gun Control Act of 1968, Title 18, United States Code, section 923(i), and the National Firearms Act of 1934, Title 26, USC, section 5842(a), requires all licensed importers and manufacturers to identify each firearm imported or manufactured by means of a serial number engraved or cast on the frame or receiver of the weapon, in such a manner as prescribed by Federal Regulation 27 CFR 478.92(a) and 479.102(a); in addition to additional markings that must be legibly placed on the firearm.

The unique marks of identification on firearms serve several purposes. First, the marks are used by Federal firearms licensees to effectively track firearm inventories and maintain all required records. Second, the marks enable law enforcement agencies to trace specific firearms used in crimes directly from the manufacturer or importer to individual buyers, and to identify firearms that have been reportedly lost or stolen. Furthermore, unique marks help determine if firearms used in a crime have travelled in interstate or foreign commerce.

The complete list of required markings is comprised of a serial number and "additional information," which includes the following: [425]

- The model, if such designation has been made,
- The caliber or gauge,
- Importer's name (or recognized abbreviation) and also, when applicable, the name of the foreign manufacturer,
- In the case of an U.S. made firearm, the city and state (or recognized abbreviation thereof) where the manufacturer maintains his place of business; and,
- In the case of an imported firearm, the name of the country in which it was manufactured and the city and state (or recognized abbreviation thereof) where the importer maintains his place of business.

Additional Information must include:

- Engraved, cast, stamped (impressed) or placed on the frame, receiver, or barrel;
- Must be conspicuous;
- Must be placed in a manner not susceptible of being readily obliterated, altered, or removed;
- The engraving, casting, or stamping (impressing) of additional information must be to a minimum depth of 0.003 in [0.08 mm; PD].

The serial number:

- Must be an individual serial number;
- Must be engraved, cast stamped (impressed) or placed on the frame or receiver;
- Must be conspicuous (i.e. it must be placed in such a manner as to be wholly unobstructed from plain view);
- Must not duplicate any serial number placed by the same manufacturer on any other firearm;
- Must be placed in a manner not susceptible of being readily obliterated, altered, or removed;
- The engraving, casting, or stamping (impressing) of the serial number must be to a minimum depth of 0.003 in [0.08 mm; PD] and in a print size no smaller than $^{1}/_{16}$ in [1.6 mm; PD]. The depth of all markings required will be measured from the flat surface of the metal and not the peaks or ridges. The height of serial numbers will be measured as the distance between the latitudinal ends of the character impression bottoms (bases);[426]
- The serial number must be placed on a metal part.[427]

Wherever the serial number is shown its notation must be identical, even as it appears on gun parts or labels. ATF does not accept, for example, to improve the readability of the serial numbers on labels or packaging by inserting spaces between letters and numbers, whereas the serial number on the firearm would not contain such characteristics.

It is important to comprehend that any of the above markings must be indented and may not be embossed. This is important for manufacturers who intend to keep the cost of production low. Manufacturers of injection molded parts tend to place most of the required markings on the frame (Fig. 4.3), because the markings can be included in the

Fig. 4.3: Importer marking, indented.

mold, which avoids secondary operations at a later step in the manufacturing process. Toolmakers might prefer embossed markings because it's easier to include in the tool, but embossed markings do not meet the legal requirements of engraved, cast and stamped (impressed).

The licensed importer of firearms is responsible for the compliance of the markings placed on the weapons. The importer must ensure that the required markings appear on each firearm. If the manufacturer did not mark the firearm, the importer must place the required markings on the firearm within 15 days of their release from Customs and Border Protection custody.

Alternate means of identification

The Director of the ATF may authorize other means of identification upon receipt of an *Application for Alternate Means of Identification of Firearm(s) (Marking Variance) ATF Form 3311.4* application submitted electronically by a licensed manufacturer or licensed importer of firearms, showing that such other identification is reasonable and will not hinder the effective administration of this part.[428]

Multiple identification markings may be confusing to law enforcement agencies and potentially hinder effective tracing of firearms used in crimes. Therefore, ATF finds that the other means to identify firearms specified in this ruling are reasonable and will not hinder the effective administration of the firearms regulations.

Regardless of the legal requirements, a trend towards two-dimensional barcodes can be found for the labeling of all kinds of equipment (e.g. Data Matrix code or symbol, see MIL-STD-130). What is used must be a permanently affixed marking, which includes serial number, part number and, if necessary, further information. For example, Unique Identification marking (UID marking) or Item Unique Identification (IUID), which is a part of the compliance process mandated by the U.S. Department of Defense (DOD).

The DOD has also made it mandatory for all federal contractors to utilize UID-marking on government-furnished military, non-military, and more by U.S. federal authorities carrying firearms.

Identification of Spare Parts

Firearms, including frames, receivers, and firearm barrels, can only be imported into the U.S. by a licensed importer, if the Director of the ATF has authorized the importation of such items.

A firearm frame or receiver that is not a component part of a complete weapon at the time it is sold, shipped, or otherwise disposed of by the importer must be marked as required by 27 CFR § 478.92(a)(1), that is, the same way as it is required for a fire-arm.[429]

Thus, the term frame or receiver in regard to handguns would be that part of a firearm that provides housing for the hammer, bolt or breechblock, firing mechanism; and in the case of a revolver, the frame would be usually threaded at its forward portion to receive the barrel.

Required identification data pursuant to 27 CFR § 478.92 must be placed on each firearm within 15 days of release from Customs custody in case the firearm did not initially bear the correct markings.[430] Required identification markings called for in § 478.112(d)(2) refers to firearms, including firearms and firearm frames or receivers. Other parts, such as barrels, slides and other spares require no importer identification markings.

4.2.3 Semi-Automatic Assault Weapons

The following provisions of the GCA were repealed when the semi-automatic assault weapon and large capacity ammunition feeding device bans sunset in 2004. We are going to mention it, and that is because some States (e.g. CA, MA, NY) still refer to it:

18 USC § 921(a)(30)(C) established that a pistol is also within the definition of an assault weapon,[431] if it possesses the ability to accept a detachable magazine and has at least two of the following:[432]

(i)   An ammunition magazine that attaches to the pistol outside of the pistol grip;
(ii) A threaded barrel capable of accepting a barrel extender, flash suppressor, forward handgrip, or silencer;
(iii) A shroud that is attached to, or partially or completely encircles, the barrel and that permits the shooter to hold the firearm with the non-trigger hand without being burned;
(iv) A manufactured weight of 50 ounces [1418 g; PD] or more when the pistol is unloaded; and
(v) A semi-automatic version of an automatic firearm.

4.2.4 Undetectable Firearms Act

Title 18 USC § 922(p), also known as Undetectable Firearms Act of 1988, was set in place to ensure weapons can be detected by conventional screening technologies, such as those used in airports. With the advent of 3D printing where it is possible to produce polymer weapons at home, the content of the Undetectable Firearms Act is more pronounced than ever before; even though this Act was introduced a few decades ago, after the GLOCK pistol had entered the market.[433]

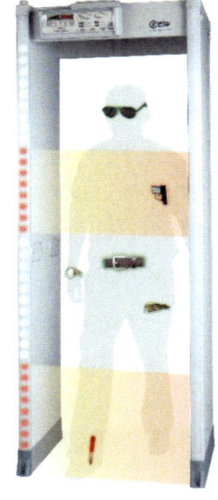

Fig. 4.4: Walk-Through Metal Detector.
(© CEIA GMBH)

With the launch of the GLOCK pistol, and rumors that this gun would be invisible to metal detectors at airports, firearms whose major components would not appear on the screen in the usual manner during passenger security checks were banned. Details were specified in the Undetectable Firearms Act of 1988. The original Act had a 10-year sunset clause. Congress keeps renewing it and there are intentions to modernize it to reflect 3D printed guns.

The Undetectable Firearms Act makes it illegal for any person to manufacture, import, sell, ship, deliver, possess, transfer, or receive any firearm that, after removal of frame, grips, stocks, and magazines, is not as detectable as the Security Exemplar, by walk-through metal detectors calibrated and operated to detect the Security Exemplar. The term "Security Exemplar" means an object, to be fabricated at the direction of the Attorney General, that is constructed of 3.7 ounces [105 g; PD] of material type 17-4 PH stainless steel in a shape resembling a handgun; and suitable for testing and calibrating metal detectors.[434]

Also unlawful is any major firearm component (i.e. barrel, slide or cylinder, or frame or receiver), if it does not generate an image that accurately depicts the shape of the component when subjected to inspection by the types of X-ray machines commonly used at airports. Barium sulfate or other compounds may be used in the fabrication of the component to deliver a better image during inspection by X-ray machines.[435]

4.3   State Legislations of Selected States

As previously mentioned, the increasing availability of Junk Guns, as well as the attempts at tightening gun laws, has led more states to introduce and dictate certain legal design and safety standards for handguns.

As a means to understand the topic and the complexity of mandatory requirements which are to be met by gun manufacturers, we are going to have a look at some of the requirements from seven states in particular, including California, Hawaii, Illinois, Maryland, Massachusetts, Minnesota and New York.

However, it is important to note that different laws and varying requirements, apply in each of these states. Usually the state laws involve claims of drop safety, functional reliability, material properties and safety features, which need to be evaluated and approved by the respective states before the sale of new products is considered lawful. Often this is accompanied by a test performed by an independent testing authority.

Minimum requirements for drop safety and reliability, as well as special safety devices, exist in California, Massachusetts and New York. Minimum requirements for the material are required in Hawaii, Illinois, Massachusetts, Minnesota and New York. States that maintain a list of approved weapons for sales (roster) are California, Maryland and Massachusetts.

There is no federal law that requires certain safety standards for guns manufactured in the United States. The same seven states (CA, HI, IL, MA, MD, MN and NY) each have laws which contain certain minimum requirements to make the use of weapons safe for the users and to limit the spread of Junk Guns.[436]

In California, Massachusetts and New York extensive requirements apply to construction and safety standards for handguns. Construction and safety standards were introduced to promote the development of high-quality handguns, as well as to curb the spread of weapons prone to failure, fire and other malfunctions. Examples of constructive standards compliance are: drop tests, reliability tests or a minimum melting point for the material used. Reliability tests will prove whether a weapon remains operational during a defined number of rounds fired, or tends to malfunction.

Typically, 600 rounds will be fired and the handgun is then checked for defects and cracks. Drop tests are done to test weapons for potential risk of an accidental discharge. These tests are observed by dropping weapons from a given height on a defined impact surface.

Hawaii, Illinois, Massachusetts, Minnesota, and New York consider the material's melting point for evaluation. The materials used shall provide sufficient security if a weapon reaches temperatures that could occur under high endurance. These states demand minimum temperatures for the melting point in the range 427 °C to 538 °C (800 °F to 1000 °F). Massachusetts and Minnesota additionally require proof of material density and tensile strength.

In California, Massachusetts and New York, handguns are regarded to be unsafe if they do not have a safety feature that protects the user from an accidental discharge. These states demand, for example, loaded chamber indicators that allow the user to determine if a round is chambered, safeties that prevent unintentional discharges, and magazine disconnect safeties to prevent the gun from firing after the magazine has been removed.

Since 2007, every centerfire semi-automatic pistol with removable magazine is considered to be an unsafe handgun in California if it is not equipped with a loaded chamber indicator and a magazine disconnect feature. Also, microstamping is mandatory in California for new firearm petitions submitted after May 17, 2013. The District of Columbia had planned to follow in 2016, but has continued to delay implementation. By now, the District prohibited any licensed dealer from selling or offering for sale any semiautomatic pistol manufactured on or after January 1, 2018, that is not microstamp-ready.

In California, Massachusetts and Maryland, only handguns that may be transferred are published on a list of weapons approved for sale within each state. Weapons listed on the rosters of California and Massachusetts have successfully completed a qualification test, which is carried out by an independent testing body in accordance with the specifications of the respective state. In Maryland, a Committee decides if a weapon is to be published on the roster; additional testing is not required.

Apart from this, magazine capacity is a constant and recurring theme that tends to be a popular topic of discussion, particularly after mass shootings.[437] The states of California, Colorado, Connecticut, Hawaii, Maryland, Massachusetts, New Jersey and New York, all impose different limits on magazine capacity, which ranges from 7 to 15 rounds.

For example, the permitted capacity of a magazine is 10 rounds in New York. A European could be puzzled to note that, not too long ago, New York allowed the 10-

shot magazines to be sold, but users were restricted from loading more than seven rounds.

Magazine capacity must be permanently limited, but States can set different standards when assessing the embodiment of this limit (Fig. 4.5). For example, Connecticut and California do not accept stakings in the magazine body, which would block the movement of the follower from the tenth cartridge, because a staking could be removed later.

Europeans might also be confused, as to why it is officially legal in all 50 states to carry concealed handguns (42 states require a license, 8 states do not), open carrying in some states (45 states allow, five states do not).[438] Or the fact that a verification of the purchaser (i.e. background check) for the purchase of a weapon in a gun store is required, while individual private sales are handled without any permission or validation depending upon state laws. The latter is often referred to as the gun show loophole or private sale loophole.[439]

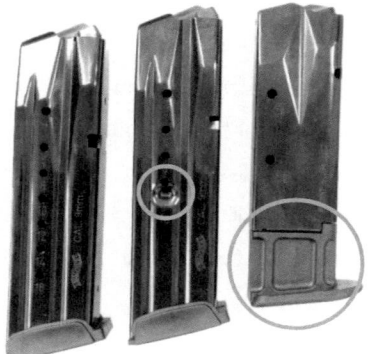

Fig. 4.5: Magazine with standard capacity vs. limited to ten rounds by a staking or shortening the magazine body (f. l. t. r.).

4.3.1 California

In California (CA) the term "firearm" is defined as a rifle, shotgun, revolver, pistol, or any other device designed to be used as a weapon from which a projectile is expelled by force of any explosion or other form of combustion. The term *firearm* includes the frame or receiver of any such weapon,[440] which means pistol frames are legally viewed the same way as a firearm. It is regarded to be the main part of the pistol and is subject of import regulations.

A *handgun* is any pistol, revolver, or other firearm capable of being concealed upon the person carrying it and has a barrel length of less than 16 in [406 mm; PD].[441] The pistol's barrel length is defined as the internal length of the barrel as measured from the face of the closed breech when it is unloaded, to the forward face of the end of the barrel.[442] The term handgun also applies to any device that has a barrel length of 16 in or more, and is designed to be interchanged with a barrel less than 16 in.[443]

No handgun can be manufactured or sold to the public unless it is of a design and model that has passed required safety and functionality tests and is approved for publication in the Department of Justice's (DOJ) official list of handguns certified as safe for sale in California. Any person who manufactures, sells or imports into California for sale, gift, or loan of an unsafe handgun is guilty of a misdemeanor.[444]

The CA Department of Justice compiles, publishes, and thereafter maintains a roster listing all pistols, revolvers and other firearms capable of being concealed upon a person and have been tested by a certified testing laboratory (Fig. 4.6), determined not to be unsafe handguns and may be sold in California. Within the firearms industry this is also known as "California Compliance."[445] The roster includes for each firearm: manufacturer, model number, model name and further information, if the CA Department of Justice considers this information necessary to uniquely identify a particular model of weapon.[446]

The current list of handguns certified as safe for sale in California is available on the Bureau of Firearms website at http://www.ag.ca.gov/firearms/certlist.htm.

The firearms safety test does not apply to firearms listed as C&R firearms, as defined in section 478.11 of Title 27 of the Code of Federal Regulations. Pistols designed expressly for use in the Olympics and target shooting events are also exempted.[447]

It might be interesting to note that any person who for commercial purposes purchases, sells, manufacturers, ships, transports, distributes or receives a firearm where the coloration of the entire exterior surface of the firearm is bright orange or bright

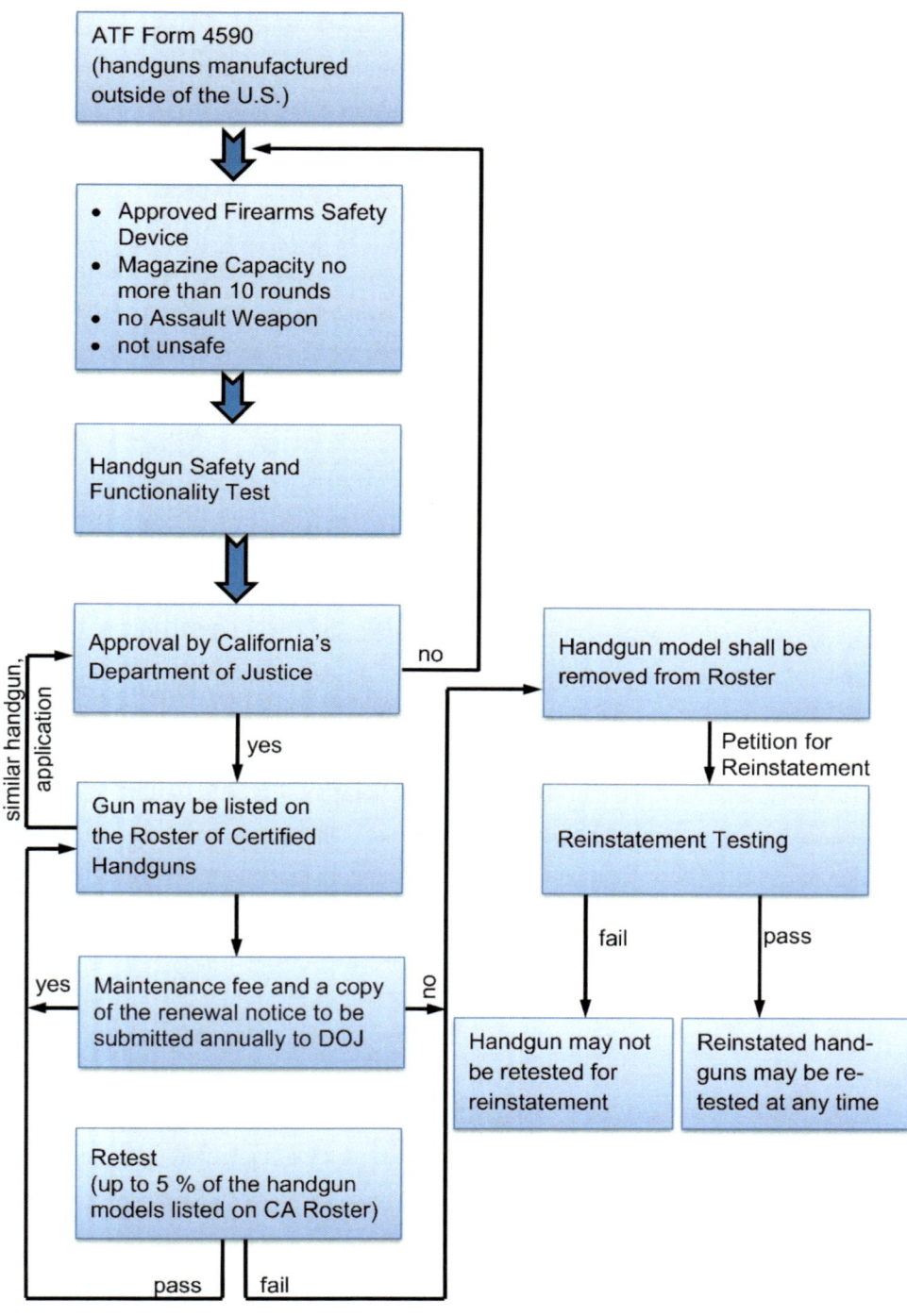

Fig. 4.6: California Compliance, schematic

green, either singly, in combination, or as the predominant color in combination with other colors in any pattern, is liable for a civil fine.[448]

It is punishable by imprisonment to manufacture or cause to be manufactured, import into CA, to keep for sale, or offer or expose for sale, or to give, lend, or possess any cane gun or wallet gun, any undetectable firearm, any firearm which is not immediately recognizable as a firearm, any camouflaging firearm container etc.[449] As used in this context, an "undetectable firearm" means any weapon which meets one of the following requirements:[450]

a) When, after removal of grips, stocks, and magazines, it is not as detectable as the Security Exemplar, by walk-through metal detectors calibrated and operated to detect the Security Exemplar.
b) When any major component of which, when subjected to inspection by the types of X-ray machines commonly used at airports, does not generate an image that accurately depicts the shape of the component. Barium sulfate or other compounds may be used in the fabrication of the component.
c) For purposes of this paragraph, the terms "firearm," "major component," and "Security Exemplar" have the same meanings as those terms are defined in section 922 of Title 18 of the United States Code.
   All firearm detection equipment newly installed in non-federal public buildings in this state shall be of a type identified by either the United States Attorney General, the Secretary of Transportation, or the Secretary of the Treasury, as appropriate, as available state-of-the-art equipment capable of detecting an undetectable firearm, as defined, while distinguishing innocuous metal objects likely to be carried on one's person sufficient for reasonable passage of the public.

Basically five provisions need to be considered for the sale of handguns in California:

1. Handguns must be accompanied by a DOJ-approved firearms safety device.
2. Large-Capacity Magazines are not legal.
3. Handguns must not fall under the provision of Assault Weapons.
4. Handguns must pass the Safety and Functionality Tests.
5. Handguns must be certified for sale in CA by the Department of Justice (listed on the *Roster of Handguns Certified for Sale*).

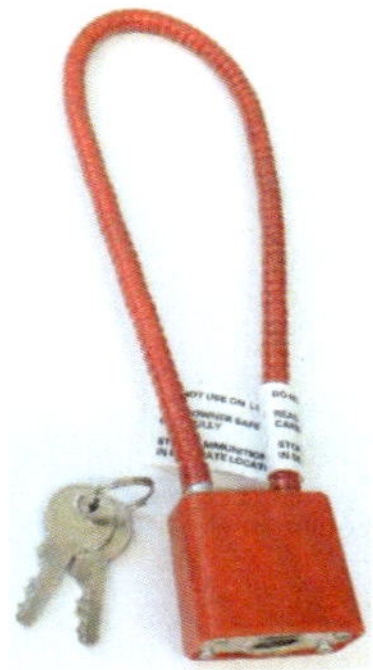

Fig. 4.7: Safety device, e.g. cable lock.

Note on 1. Firearms Safety Device

Firearms safety device means a device other than a gun safe that locks and is designed to prevent children and unauthorized users from firing a firearm. The device may be installed on a firearm, be incorporated into the design of the firearm, or prevent access to the firearm.[451]

Each firearm sold, transferred, or manufactured in California must be accompanied with a firearm safety device approved by the Department of Justice and identified as appropriate for that firearm. The Department of Justice is required to compile and publish a listing of all of the safety devices that have been determined to meet the department's standards. A list of these devices is available on the Department of Justice Bureau of Firearms website at http://www.ag.ca.gov/firearms/fsdcertlist.htm.[452]

Note on 2. Large-Capacity Magazine
It is unlawful for any person to manufacture, cause to be manufactured, import into California, keep for sale, or offer or expose for sale, or give or lend, any large-capacity magazine.[453] A large-capacity magazine means any ammunition feeding device with the capacity to accept more than 10 rounds, but shall not be construed to include a feeding device that has been permanently altered, so that it cannot accommodate more than 10 rounds, a tubular magazine that is contained in a lever-action firearm, or a .22 caliber tube ammunition feeding device.[454]

Note on 3. Assault Weapon
A pistol is no longer a handgun but is classified as an assault weapon, if this semi-automatic pistol has the capacity to accept a detachable magazine and any one of the following:[455]

- A threaded barrel, capable of accepting a flash suppressor, forward hand-grip, or silencer.
- A second handgrip.
- A shroud that is attached to, or partially or completely encircles, the barrel that allows the bearer to fire the weapon without burning his or her hand, except a slide that encloses the barrel.
- The capacity to accept a detachable magazine at some location outside of the pistol grip.

Also, any semi-automatic pistol with a fixed magazine that has the capacity to accept more than 10 rounds is classified to be an assault weapon.[456]

Note on 4. Safety and Functionality Test
The headline of *Dangerous Weapons Control Law* Chapter 1.3 is "Unsafe Handguns."[457] In California handguns can only be sold if they are not unsafe for the user. According to section 12126, a handgun is not unsafe if it meets the following criteria:[458]

a) Revolver

It must have a safety device that, either automatically in the case of a double-action firing mechanism, or by manual operation in the case of a single-action firing mechanism, cause the hammer to retract to a point where the firing pin does not rest upon the primer of the cartridge. Revolvers must also meet the firing requirement for handguns and pass the drop safety requirement for handguns.

b) Pistol

1. It must have a positive manually operated safety device, as determined by standards relating to imported guns (see ATF Form 4590).
2. It must meet the firing requirement for handguns.[459]
3. It must meet the drop safety requirement for handguns.[460]
4. Centerfire semi-automatic pistols that are not already listed on the roster must have both a chamber load indicator and if it has a detachable magazine, a magazine disconnect mechanism.
5. Any rimfire semi-automatic pistol that is not already listed on the roster must have a magazine disconnect mechanism, if it has a detachable magazine.
6. All semi-automatic pistols that are not already listed on the roster must be equipped with a microscopic array of characters that identify the make, model, and serial number of the pistol, etched or otherwise imprinted in two or more places on the interior surface or internal working parts of the pistol, and that are transferred by imprinting on each cartridge case when the firearm is fired.

The requirement for microstamping applies to semi-automatic pistols only, and this is because revolvers typically will not automatically eject expended casings when they are fired, and hence their casings are less likely to be found on crime scenes. Following its adoption, implementation of the microstamping requirement was postponed until the Department of Justice could certify that microstamping technology is available to more than one manufacturer and it was not encumbered by a patent. On May 17, 2013, it was certified that microstamping technology was unencumbered by any patent.

The Attorney General may also approve a method of equal or greater reliability and effectiveness in identifying the specific serial number of a firearm from spent cartridge casings discharged by that firearm than that which is described above.

The microscopic array of characters required shall not be considered the name of the maker, model, manufacturer's number, or other mark of identification, including any distinguishing number or mark assigned by the Department of Justice, within the interpretations of Sections 12090 and 12094.

Pistols with Smart-Gun-Technology or pistols equipped with user recognition are not exempt from microstamping. BTW: The ARMATIX *iP1* was approved on May 17, 2013, and listed on the roster before the enforcement of the microstamping requirement.

A manufacturer's handguns, which are similar to other handguns by the same manufacturer already listed on the Roster shall be deemed to be not unsafe handguns, if the similar handgun differs only in one or more of the following features:[461]

a) Finishing including but not limited to bluing, chrome-plating, oiling or engraving.
b) The material from which the grips are made.
c) The shape or texture of the grips, so long as the difference in grip shape or texture does not in any way alter the dimensions, material, linkage, or functioning of the magazine well, the barrel, chamber, or any of the components of the firing mechanism of the firearm.
d) Any other purely cosmetic feature that does not in any way alter the dimensions, material, linkage, or functioning of the magazine well, the barrel, chamber, or any of the components of the firing mechanism of the firearm.

Similar handguns are not automatically exempted from the certification. If a manufacturer wants to have a similar firearm listed on the roster, he has to provide to the Department of Justice all of the following:

1. The model designation of the listed firearm.
2. The model designation of each similar firearm that the manufacturer seeks to have listed.
3. A statement, under oath, that each similar firearm for which listing is sought, differs from the listed firearm only in one or more of the ways identified above from a) to d) and is in all other respects identical to the listed firearm.[462]

The Department of Justice may, in its discretion and at any time, require a manufacturer to provide to the department any model for which listing is sought, to determine whether the model complies with the requirements.

Note on 5. Certified for sale in CA by the Department of Justice

California's Dangerous Weapons Control Law, section 12131, describes the *Roster of Certified Handguns*,[463] e.g. the list of not unsafe pistols, revolvers, and other firearms capable of being concealed upon the person, certified after January 1, 2001, to be sold in California. The listing will be valid for one year from the date the model was added to the Roster. It must be renewed prior to the expiration of the handgun model listing and an annual listing fee must be paid.[464]

For renewal of a listing, the DOJ will mail a renewal notice to the manufacturer/importer or other responsible person 60 days prior to the expiration of the handgun model listing. The person wishing to renew the listing will have to submit to the DOJ a copy of the renewal notice with the annual maintenance fee, and once these requirements are met and the request has been processed, the DOJ will then send a notification that the listing has been renewed.

If the manufacturer/importer or person responsible fails to comply with these renewal requirements, the handgun model listing shall expire by operation of law at midnight on the date of expiration of the listing and the model will be removed from the Roster.[465] A handgun model can be removed from the Roster if the annual maintenance fee is not paid or if it is determined that the handgun is in fact unsafe based upon further testing, or if it is determined that the handgun model submitted for testing was modified in any way from those that were sold at the time the certification was granted.

If in any way the dimensions, material, linkage, or functioning of the magazine well, the barrel, chamber, or any of the components of the firing mechanism of the firearm were altered, the model will be removed from the roster.[466]

Manufacturers should take note that California is very strict about alterations. Manufacturers who make minor changes to their products as new technology or manufacturing processes become available are required to have these models approved through the same process. If they are not accepted as "similar handgun" of an already listed model, they will be treated as a new handgun model and must comply with the current requirements at the time of petition before it can be listed on the Roster.[467]

SMITH & WESSON, for example, was going to remove their *M&P* pistol from the California market, after "performance enhancements" had been included in the design of the pistol.[468] Actually, this should not influence companies such as SMITH & WESSON and RUGER that much, as long as they maintain a portfolio of California-compliant revolvers. However, after being removed from the roster, the only way an *M&P* could be sold in CA is if it now passes the microstamping requirement, and the problem associated with that is the microstamping process itself, which is not a viable technology. No firearm manufacturer is currently producing a handgun with microstamping.

Section 12131 (c) explains that the Attorney General may annually retest up to 5 percent of the handgun models that are listed on the roster.[469] This retest should not be underestimated by manufacturers.

For the retest, three samples of the handgun model will be obtained from retail or wholesale sources, or both. The Attorney General shall select the certified laboratory to be used for the retesting. The ammunition used for the retesting shall be of a type recommended by the manufacturer in the user manual for the handgun. If the user manual for the handgun model lists no ammunition recommendations, the Attorney

General shall select the ammunition to be used for the retesting. The ammunition shall be of the proper caliber for the handgun, commercially available, and in new condition.

The retest is to be conducted in the same manner as for the handgun safety and functionality test prescribed, which means they have to pass the "firing requirement for handguns" – as described in section 12127 – and the "drop safety requirement for handguns" – as described in section 12128. If the handgun model fails retesting, the Attorney General will remove the handgun model from the roster. Reinstatement testing may be carried out only once.[470]

Reinstatement Testing of handguns which failed retest
Any handgun model that fails retesting will be removed from the roster.[471] A handgun model removed from the roster may be reinstated back on the roster if the manufacturer petitions the Attorney General for reinstatement of the handgun model and pays the Department of Justice for all of the costs related to the reinstatement testing of the handgun model, including the purchase price of the handguns, prior to reinstatement testing. For the reinstatement testing, the Attorney General shall obtain from retail or wholesale sources, or both, three samples of the handgun model to be retested and he shall select a certified laboratory to be used for the retesting.

The ammunition used for the reinstatement testing shall be of a type recommended by the manufacturer in the user manual for the handgun. If the user manual for the handgun model makes no ammunition recommendation, the Attorney General shall select the ammunition to be used for the reinstatement testing. The ammunition shall be of the proper caliber for the handgun, commercially available, and in new condition. Three handgun samples shall be tested and must pass the "firing requirement for handguns" – as described in section 12127 – and the "drop safety requirement for handguns" – as described in section 12128. The handgun samples shall be tested only once for reinstatement. If the sample fails, it may not be retested. If the handgun model successfully passes testing for reinstatement, and if the manufacturer of the handgun is otherwise in compliance with the provisions, the Attorney General shall reinstate the handgun model on the roster. Before, the manufacturer shall provide the Attorney General with the complete testing history for the handgun model. It is important to note that the Attorney General may, at any time, further retest any handgun model that has been reinstated back to the roster.[472]

Safety and Functionality Test Procedure

The testing will be carried out by a DOJ-certified laboratory. Three handguns of each model to be tested shall be submitted to the DOJ-certified laboratory. The handguns submitted for testing shall not be modified in any way from those that would be sold if certification is granted. If it is determined by the DOJ that the handguns submitted for testing are modified in any way from those that are being sold after certification has been granted, that model will be immediately removed from the Roster of Certified Handguns.

Other than the DOJ, only the manufacturer/importer of a handgun model is authorized to submit the handgun model to a DOJ-certified laboratory for testing. Manufacturers/Importers may supply any information that they believe may be needed by the laboratory for proper and safe operation of the handgun. The following information shall be supplied in the English language with each handgun model submitted for testing:[473]

1) Instructions for field disassembly/assembly and diagram(s) identifying all parts.
2) Cleaning instructions. These may be different from and in addition to the instructions that are provided when the handgun model is sold.
3) A description of each safety feature designed into the handgun, how each safety feature is intended to function, and for those under shooter control, how the shooter should operate (activate/deactivate) each safety feature.
4) A statement regarding the ammunition the manufacturer/importer markets and/or recommends that the handgun being tested is designed to handle. This may also include information on ammunition known to be beyond the design limits of the handgun and/or known not to function in the handgun.
5) Microstamping: A statement by the manufacturer indicating that for each handgun of the make and model of semi-automatic pistol submitted for testing:
   i.   The pistol is designed and equipped with a Firearms Identification Number (FIN) etched or otherwise imprinted in two or more places on the interior surface or internal working parts of the pistol, and are transferred by imprinting on each cartridge case expended from the pistol when the pistol is fired; and
   ii. The pistol's complete FIN can be identified from the one or more etchings on each cartridge casing.
6) The FIN for each handgun of the make and model of semi-automatic pistol to be tested. The FIN shall also be displayed or recorded on the manufacturer's packaging of any semi-automatic pistol which is manufactured, caused to be manufactured, imported into the state for sale, kept for sale, offered or exposed for sale, given, or lent in the state and subject to the

microstamping requirement set forth in Penal Code section 31910, subdivision (b)(7). The FIN must be clearly marked as the FIN wherever the serial number of the pistol is displayed or recorded on the packaging of such a pistol.

The manufacturer/importer shall be allowed, but not required, to provide the standard ammunition to be used during the firing test provided that, if applicable, it is the more powerful cartridge that is marketed/recommended by the manufacturer/importer. The manufacturer/importer shall be allowed to inspect any laboratory-supplied standard ammunition before testing begins. The manufacturer/importer or DOJ-certified laboratory shall indicate the ammunition lot number on the Compliance Test Report. Notwithstanding the above, the DOJ may allow a handgun to be tested with newly-designed non-standard ammunition that is not yet "available for purchase at consumer-level retail outlets." Any such ammunition shall be commercially produced and factory loaded.[474]

Safety and Functionality Test Procedure: General Information
The handguns may not be refined or modified in any way from those that would be made available for retail sales if certification is granted. The magazines of a tested pistol must be identical to those that would be provided with the pistol to a retail customer.[475]

The only persons allowed to conduct handgun testing are authorized staff of the DOJ-certified laboratory. In addition to this staff, representatives of the manufacturer/importer and/or the DOJ shall be allowed to be present during testing. Any such representative(s) shall not participate in the testing. However, if deemed necessary by the staff of the DOJ-certified laboratory, representative(s) of the manufacturer/importer may be asked to provide advice and/or guidance regarding the characteristics, handling, and/or operation of the handgun.[476]

The Safety and Functionality Test comprises of the inspection of the following: positive manually operated safety device; chamber load indicator; magazine disconnect mechanism; the firing requirement for handguns and the drop safety requirement for handguns.

Requirement Testing: Safety Device
Prior to the commencement of the **Safety and Functionality Test**, the DOJ-certified laboratory shall determine whether the revolver does have a safety device that either automatically in the case of a double-action firing mechanism, or by manual operation

in the case of a single-action firing mechanism, causes the hammer to retract to a point where the firing pin does not rest upon the primer of the cartridge,[477] or in the case of a pistol, if it does have a positive manually operated safety device, as determined by standards relating to imported guns promulgated by the ATF (see ATF Form 4590).[478]

Handguns that have passed the *factoring criteria for weapons* automatically qualify. Hence, this safety device requirement is something U.S. manufacturers must take into consideration as well, while the rest of the world will already include a safety device into the design to pass the factoring criteria for importation into the USA.

If the DOJ-certified laboratory needs guidance in making this determination, the description of the safety features and other information supplied by the manufacturer/importer should be consulted first. If the DOJ-certified laboratory is still not able to make this determination, they should contact the manufacturer/importer for additional information. Any additional information received from the manufacturer/importer shall be included with the other information initially submitted.[479]

If the DOJ-certified laboratory is still uncertain whether a positive manually operated safety device is present on a pistol even after it received additional information, the firing and drop tests should be performed. If the pistol passes these tests, the laboratory should submit the pistol to the DOJ with a letter explaining the steps taken to determine whether the positive manually operated safety device is present. The laboratory must indicate its preliminary decision regarding the positive manually operated safety device. The letter should also include any information that would support the position taken by the laboratory. This includes a description of the positive manually operated safety device(s) incorporated into the pistol's design and an explanation of how this design replicates the positive manually operated safety device of a pistol design that has already been determined to meet the standards promulgated by the ATF. The DOJ will use this information to determine whether the pistol can be sold in California.[480]

The DOJ-certified laboratory shall conduct the required testing of a centerfire semi-automatic pistol only after ascertaining the firearm has a functioning chamber load indicator and a functioning magazine disconnect mechanism and that it complies with the microstamping requirement for semi-automatic pistols.[481]

Requirement Testing: Chamber Load Indicator

A functioning chamber load indicator (e.g. Fig. 4.8) must be visible from a distance of at least 24 in [610 mm; PD] and only when there is a round in the chamber. Explanatory text and/or graphics either incorporated within the chamber load indicator or adjacent to the chamber load indicator is/are permanently displayed by engraving, stamping, etching, molding, casting, or other means of permanent marking (Fig. 4.9). Each letter of explanatory text must have a minimum height of $1/16$ in [1.6 mm; PD]. The explanatory text and/or graphics shall be of a distinct visual contrast to that of the firearm.

Fig. 4.8: Chamber Load Indicator of a RUGER *LC9* Pistol.

The "loaded" indication, that portion of the chamber load indicator that visually indicates there is a round in the chamber, shall be of a distinct color contrast to the firearm.

The text and/or graphics and the "loaded" indication together inform a reasonably foreseeable adult user of the pistol that a round is in the chamber without requiring the user to refer to a user's manual or any other resource other than the pistol itself.[482]

Fig. 4.9: Chamber Load Indicator (red pin) and text found on a FMK *9C1 G2* pistol.

Requirement Testing: Magazine Disconnect Mechanism

All centerfire semi-automatic pistols with detachable magazines must have a magazine disconnect mechanism.[483] This will be tested by pulling the trigger. A functioning magazine disconnect mechanism must prevent the ammunition primer from being struck with a pull of the trigger or attempted pull of the trigger whenever a detachable magazine is not inserted in the pistol.[484]

Requirement Testing: "Firing Requirement for Handguns"

The "firing requirement for handguns" is the first test to be undertaken by a DOJ-certified laboratory.[485] How a pistol will have to perform to meet the firing requirement for handguns is described in California Penal Code section 31905, 11 CCR § 4060 (f), and in CA Dangerous Weapons Control Law, section 12127.

The "firing requirement for handguns" is comprised of a 600-round firing test in which the manufacturer provides three handguns of the make and model for which certification is sought to a DOJ-Certified independent testing laboratory.[486] These handguns may not be refined or modified in any way from those that would be made available for retail sale, if certification is granted. The magazines of a tested pistol shall be identical to those that would be provided with the pistol to a retail customer.

Prior to conducting the firing test of a semi-automatic pistol, the laboratory shall fire each pistol two times. After firing the pistol two times, the laboratory shall collect the two cartridge casings expended from that pistol, store the casings in a container labeled with the Firearms Identification Number of the pistol from which they were expended and indicating that the two cartridges were expended immediately preceding the firing test, and retain the casings for possible future analysis at the conclusion of the 600 rounds firing test.[487]

Immediately, as time permits, after the successful completion of the 600-round firing test, the laboratory shall fire each of the three semi-automatic pistols two additional times. After firing the pistol two additional times, the laboratory shall collect the two cartridge casings expended from that pistol and store them in a container labeled with the FIN of the pistol, indicating that the two cases were expended immediately following the firing test, and keeping them separate and apart from the cases expended and collected from the same pistol prior to conducting the 600-round firing test.[488]

In order to verify compliance with the microstamping requirement for semi-automatic pistols set forth in Penal Code section 31910, subdivision (b)(7), the laboratory shall use the following procedures and criteria to examine the cartridge casings collected from each tested semi-automatic pistol to determine whether a FIN was transferred by imprinting onto each cartridge case when the pistol was fired.[489]

1. Using a stereo zoom microscope (with a low magnification of 25× or less, and a high magnification of at least 60×; the microscope must also be equipped with a ring-light for illumination and with polarizing filters to aid in reflection control and should have a digital camera with the ability to capture digital images sufficient to adequately document the markings made on the cartridge cases by the microstamp.[490] The DOJ-certified laboratory shall examine each of the cartridge casings collected prior to and after the 600 round firing test to verify that the pistol has transferred an imprint or etching in at least two places on each cartridge casing. So long as the pistol's complete FIN can be identified from the one or more etchings on each cartridge casing, the pistol will meet the microstamping requirements.

2. The laboratory shall take digital photographs sufficient to adequately document the markings made on the cartridge cases by the microstamp.
3. The laboratory shall repeat the examination process described above for each set of cartridge casings expended from all three tested semi-automatic pistols. If each cartridge casing from each set of expended cartridge casings satisfies paragraph (1) above, then the laboratory shall certify that the model of semi-automatic pistol complies with the microstamping requirement.

The laboratory shall fire 600 rounds from each gun, take a pause after each series of 50 rounds has been fired for 5 to 10 minutes to allow the weapon to cool, then pausing after each series of 100 rounds has been fired as well to tighten any loose screws and clean the gun in accordance with the manufacturer's instructions, and taking a pause as needed to refill the empty magazine or cylinder to capacity before continuing. The ammunition used shall be of the type recommended by the handgun manufacturer in the user manual, or if none is recommended, any standard ammunition of the correct caliber in new condition that is commercially available.[491]

If the manufacturer/importer markets and/or recommends that the handgun model is designed to handle multiple cartridges, the standard ammunition used during the firing test shall be the more powerful marketed/recommended cartridge. However, the Laboratory shall not use any standard ammunition known to be beyond the design limits of the handgun and/or known not to function in the handgun.[492]

If a pistol has multiple chambers, the 600 rounds shall be evenly apportioned between the chambers.[493]

The DOJ-certified laboratory shall then determine whether there is any crack or breakage of an operating part of the handgun that increases the risk of injury to the user.[494]

A total number of six malfunctions per handgun is acceptable. For the purposes of determining whether a handgun passes the "firing requirement for handguns," "malfunctions" include any failure to operate as designed and also include failure of a pistol's slide to remain open after a manufacturer-approved magazine has been expended, provided that the handgun was designed by the manufacturer to remain open.[495] For example, the following malfunctions will be counted: failure to properly feed, fire, eject a round or failure of a pistol to accept or eject the magazine.[496]

A handgun shall pass the firing test if each of the three test guns meets both of the following:

1. Fires the first 20 rounds without a malfunction that is not due to ammunition that fails to detonate.[497]
2. Fires the full 600 rounds with no more than six malfunctions that are not due to ammunition that fails to detonate and without any crack or breakage of an operating part of the handgun that increases the risk of injury to the user.[498]

If a pistol does not meet the requirements that apply for the first 20 or the full 600 rounds as a result of ammunition that fails to detonate, the pistol shall be retested from the beginning of the "firing requirement for handguns" test. A new pistol may be submitted for the test to replace the pistol that failed.[499]

Should a handgun fail the 600-round firing test, three handguns of that make and model must be re-submitted for the firing test. Handguns that do not pass the "firing requirements for handguns" test may not be submitted for the "drop safety requirement for handguns" testing.[500]

Requirement Testing: "Drop Safety Requirement for Handguns"
The "drop safety requirement for handguns" means that at the conclusion of the firing requirements for handguns described previously, the same certified independent testing laboratory shall subject the same three handguns of the make and model for which certification is sought, to the following test:[501]

A primed case (no powder or projectile) shall be inserted into the chamber. The slide shall be released, allowing it to move forward under the impetus[a] of the recoil spring, and an empty magazine shall be inserted. The pistol shall be placed in a drop fixture capable of dropping the pistol from a drop height of 1 m + 1 cm (39.4 + 0.4 in) onto the largest side of a slab of solid concrete having minimum dimensions of 7.5 × 15 × 15 cm (3 × 6 × 6 in).

The drop distance shall be measured from the lowermost portion of the weapon to the top surface of the slab. The required concrete slab shall rest upon a firm surface and the face of the slab shall be perpendicular to the direction of the drop. If a handgun

---

[a] Note: The seating point for straight walled cartridge cases is where the case mouth seats at the chamber's front end. Inserting a primed casing into the chamber and allowing the slide or extractor to hit that chambered casing under the impetus of the recoil spring might push the casing over the seating point and partially into the barrel's leade, causing the head of the casing, i.e. the primer, to be seated deeper in the chamber as normally a cartridge would be, which was fed from the magazine into the barrel's chamber and hence cause an indentation where a magazine fed cartridge would have been already fired. Maybe the DOJ-test tries to check the inertia of the firing pin at the same time. However, experience shows that, drop test and inertia test should be considered as two separate tests and in any case for each new test, a new primed casing should be used. If an inertia test of the firing pin was to be conducted, any part should be removed from the slide which would reduce the impetus of the slide, e.g. the extractor.

has an exposed hammer, the hammer shall be fully cocked during each drop test. When dropped, the handgun shall initially strike the face of the required concrete slab and then come to rest without interference.[502]

The weapon shall be dropped in the condition that it would be in if it were dropped from a hand (cocked with no manual safety applied). If the design of a pistol is such that upon leaving the hand, a "safety" is automatically applied by the pistol, this feature shall not be defeated.

The weapon shall be dropped from a fixture and not from the hand. An approved drop fixture is a short piece of string with the weapon attached at one end and the other end held in an air vise[a] until the drop is initiated. The following seven drops shall be performed:

1. Normal firing position with barrel horizontal.
2. Upside down with barrel horizontal.
3. On grip with barrel vertical.
4. On muzzle with barrel vertical.
5. On the right or left side with barrel horizontal.
6. On the remaining side with barrel horizontal.
7. If there is an exposed hammer or striker, on the rearmost point of that device, otherwise on the rearmost point of the weapon.

The primer shall be examined for indentations after each drop. If indentations are present, a fresh primed case shall be used for the next drop.

The handgun shall pass this test if each of the three test guns does not fire the primer.[503]

The primed cases used during the drop test shall be produced by the ammunition manufacturer and made from the same cases and primers as the standard ammunition that is used during the 600-round firing test.[504]

If a pistol has multiple chambers and/or firing pins[b], then for each of the drop tests a primed case will be placed in each chamber. If the hammer or firing pin alternates between chambers, the pistol will be dropped once for each hammer or firing pin position.[505]

---

[a] Note: Electromagnets have proven to work not satisfactorily to hold and release weapons in drop tests. As a consequence of the magnetic remanence an accidental turn of the test sample may occur while the magnet is turned off.

[b] California Code of Regulations uses the term *firing pin* and does not mention *striker*. *Johnston* uses the following definition: "**Firing Pin.** That part activated by a hammer to strike the primer to fire the cartridge." (Johnston, et al., 2010 p. 1197) and: "**Striker.** A firing pin released directly by the sear to go forward under its own mainspring. A striker uses no hammer." (Johnston, et al., 2010 p. 1200)

Minimal damage, such as broken grips or sights, can and will occur during the course of the drop testing. Damage and/or breakage that affects the overall dimensions of the handgun shall be repaired prior to continuing the drop tests. After each of the first five drop tests, the laboratory shall determine whether the handgun has been rendered incapable of firing a primed case prior to conducting the next drop test. If so, the handgun model shall either be repaired, or the test shall be stopped and three new handguns must be submitted for testing, beginning with the "firing requirement for handguns."[506]

After examining the primed case for indentations after each drop test, each primed case shall be fired to determine whether the primer was functional. If not, the drop test shall be repeated with a new primed case. A new primed case will be used for the next drop test, or in the case of multiple chambers, for each chamber a new primed case will be used.[507]

Should a handgun fail the "drop safety requirement for handguns," or be found incapable of firing a primed case, three new handguns of that make and model must be submitted for testing, beginning with the "firing requirements for handguns" test.[508]

The DOJ-certified laboratory shall report a handgun to the DOJ as "not unsafe" only if it has passed the required testing, has been found to comply with the microstamping requirement for semi-automatic pistols, if applicable, and the laboratory has confirmed that any chamber load indicator and magazine disconnect continues to function upon completion of the required testing.[509]

Figure 4.10 shows a summary of California's provisions.

California Provisions

- DOJ-approved Firearms Safety Device
- Magazine capacity
- Assault Weapon Ban
- Positive manually operated Safety Device as determined by ATF Form 4590
- Firing Requirement for Handguns
- Drop Safety Requirement for Handguns
- Chamber Load Indicator (centerfire pistols)
- Magazine Disconnect Mechanism
- Microstamping
- Roster of Certified Handguns
- Detectable by walk-through metal detector and accurate image on X-ray machines

---

Fig. 4.10: California's provisions in a nutshell.

4.3.2   Commonwealth of Massachusetts

In Massachusetts (MA), "handgun" is defined as a weapon designed to be fired by the use of a single hand from which may be fired or ejected one or more solid projectiles propelled via a chemical ignition, and which has a smooth bore with a barrel less than 18 in [457 mm; PD] long, a smooth bore and an overall weapon length of less than 26 in [660 mm; PD], or a rifled bore with a barrel less than 16 in [406 mm; PD].[510]

Large capacity magazines are banned, i.e. any "large capacity feeding device," such as a fixed or detachable magazine, box, drum, feed strip or similar device capable of accepting, or that can be readily converted to accept, more than ten rounds of ammunition.[511]

Illegal is any weapon that is constructed in a shape that does not resemble a handgun, short-barreled rifle or short-barreled shotgun including, but not limited to, covert weapons that resemble key-chains, pens, cigarette-lighters or cigarette-packages; or not detectable as a weapon or potential weapon by X-ray machines commonly used at airports or walk-through metal detectors.[512]

Under the Commonwealth's statutory definition, "Copies" or "Duplicates" of the enumerated weapons in MGL c. 140, § 121, such as the COLT *AR-15*, and other weapons with certain features as identified formerly in 18 USC § 921(a)(30), are assault weapons and therefore banned in Massachusetts. On July 20, 2016, the Massachusetts Office of the Attorney General issued an Enforcement Notice to provide a framework for understanding the definition of "assault weapon" contained in MGL c. 140, § 121. This notice also made clear that the sale, transfer, or possession of an "assault weapon" as defined in section 121, is still unlawful in MA even though section 121 refers to the former: 18 USC § 921(a)(30), which contained a sunset clause repealing § 921(a)(30) on September 13, 2004.

A semi-automatic pistol is a semi-automatic assault weapon banned under Massachusetts law, if it has an ability to accept a detachable magazine and has at least two of the following:[513]

- An ammunition magazine that attaches to the pistol outside of the pistol grip.
- A threaded barrel capable of accepting a barrel extender, flash suppressor, forward handgrip, or silencer.
- A shroud that is attached to, or partially or completely encircles, the barrel and that permits the shooter to hold the firearm with the non-trigger hand without being burned.

- A manufactured weight of 50 ounces [1418 g; PD] or more when the pistol is unloaded.
- A semi-automatic version of an automatic firearm.

Tamper-Resistant Serial Number

All firearms must bear serial numbers permanently inscribed on a visible metal area, and the manufacturer of said firearm must keep records of the serial numbers and the dealer, distributor or person to whom the firearm was sold or delivered.[514] Remember: The pistol's frame is regarded to be the firearm in the U.S. (see Chapter 4.2.2. Marking Requirements). Additionally, a tamper-resistant serial number must be included on handguns, which typically is a hidden serial number. It is a violation of state law to transfer any handgun on which the serial number has been placed solely in a location on the handgun that results in the number's susceptibility to eradication. A serial number shall be deemed not susceptible to eradication if it is placed on the interior of the handgun and the handgun-purveyor provides information regarding the location of the interior serial number to the Office of the Attorney General and other law enforcement officials upon request, or it is placed on the exterior of the handgun in a way that is not visible to the un-aided eye (Fig. 4.11), but is visible with the aid of an infrared detector or other device, and the handgun-purveyor provides information regarding the location of the non-visible serial number or any method by which this number can be made viewable to the Office of the Attorney General and other law enforcement officials upon request.[515]

Fig. 4.11: DataMatrix-Code on a frame of a REMINGTON *R51*.

Inferior Materials

Handguns made from inferior material can not be offered in Massachusetts. Inferior is any handgun that:[516]

1) has a frame, barrel, cylinder, slide or breechblock:
   a. composed of any metal having a melting point of less than 900 °F [482 °C; PD],
   b. composed of any material having an ultimate tensile strength of less than 55,000 pounds per square inch [379 N/mm²; PD], or
   c. composed of any powdered metal having a density of less than 7.5 g/cm³; or

2) is prone to repeated firing based on a single pull of the trigger, explosion of the handgun during firing with standard ammunition, or prone to accidental discharge.

Handguns made from inferior material may be offered in Massachusetts, if they satisfy the Handgun Performance Test. The Attorney General may require that the handgun-purveyor, or the entity testing the make and model in question on behalf of the hand-gun-purveyor, provide a sworn certification verifying that the make and model met the performance requirements. At the Attorney General's discretion and within 60 days notice the office may require that any such test be performed again by an independent testing entity chosen by the office, upon three test guns of the make and model purchased at retail. In such a case, the prior certification shall be prospectively invalid at the conclusion of the notice period and the make and model in question may henceforth only meet the Make and Model Performance Requirements by obtaining a certification from the independent tester. A handgun-purveyor may resubmit a make and model to the independent tester for testing, unlimited number of times.[517]

Childproofing and Safety Devices

With the exception of guns solely designed and sold for formal target shooting competition, and which are listed on the *Formal Target Shooting Firearms Roster*, special requirements apply in Massachusetts regarding childproofing and safety devices.

Fig. 4.12: WALTHER *Q5 Match*, target pistol in 9 mm Luger.

Any firearm sold in MA must include or incorporate a safety device designed to prevent the discharge of the weapon by unauthorized users and approved by the Colonel of State Police. Accepted safety devices include, for example, mechanical locks or devices designed to recognize and authorize or otherwise allow the firearm to be discharged only by its owner or authorized user, by solenoid use-limitation devices, key-activated or combination trigger or handle locks, radio frequency tags, automated fingerprint identification systems or voice recognition, provided that such device is commercially available.[518]

Childproofing is based on the abilities of an average five-year-old child. Handguns must contain a mechanism that effectively precludes an average five-year-old from operating the handgun when it is ready to fire; such mechanisms shall include, but are not limited to: raising trigger resistance to at least a 10 lbs pull [44.5 N; PD] or altering the firing mechanism so that an average five-year-old's hands are too small to operate the handgun, or requiring a series of multiple motions in order to fire the handgun.[519] Handguns equipped with a hammer deactivation device meet this criterion.[520]

Loaded Chamber Indicator or Magazine Disconnect

Handguns that have a mechanism to load cartridges via a magazine must either contain a load indicator or a magazine safety disconnect.[521]

Approved Weapon Roster

The Approved Weapon Roster[522] is a list of firearms approved by the Secretary of Public Safety and Security which meet or exceed the testing criteria as outlined in MGL c. 140, § 123 clauses 18, 19, 20, and 21. The said clauses do not apply to target guns. The testing criteria is explained below under *Handgun Performance Test[523]*.

An approved Firearm is defined as a firearm make and model that passed the testing requirements of MGL c. 140, § 123 and was subsequently approved by the Secretary of Public Safety. Included are those firearms listed on the current Approved Firearms Roster and those firearms approved by the Secretary of Public Safety that will be included on the next published Approved Firearms Roster.

A firearm may be considered for placement on the Approved Weapon Roster only after the Secretary has received a final test report from an approved independent testing laboratory certifying that the specified firearm make and model has successfully completed all testing requirements in compliance with MGL c. 140, § 123 and 501 CMR 7.00, or the Secretary has determined that the firearm is the functional equivalent of a previously approved firearm, or has been tested by another state which has identical testing requirements of the Commonwealth, pursuant to 501 CMR 7.04.

Modifications to this roster are likely to occur periodically, and licensees and law enforcement personnel should always utilize the most recent roster for the purpose of determining statutory compliance. The Approved Firearms Roster posted on the website of the Executive Office of Public Safety and Security (www.mass.gov/EOPSS) will contain the most recently approved models.[524]

Handgun Performance Test

In order to qualify to be listed on the Approved Weapon Roster, a sample of three handguns in new condition must all pass the following test: Each of the three samples shall fire 600 rounds, stopping every 100 rounds to tighten any loose screws and clean the gun if required by the cleaning schedule in the user manual, and as needed to refill the empty magazine or cylinder to capacity before continuing. For any firearm that is loaded in a manner other than via a detachable magazine, the tester shall also pause every 50 rounds for 10 minutes.

The ammunition used shall be the type recommended by the firearm manufacturer in its user manual or, if none is recommended, any standard ammunition of the correct caliber and in new condition. A firearm shall pass this test if it fires the first 20 rounds without a malfunction, fires the full 600 rounds with no more than six malfunctions and completes the test without any crack or breakage of an operating part of the firearm.

The term "crack" or "breakage" shall not include a crack or breakage that does not increase the potential of injury to the user.

For purposes of evaluating the results of this test, "malfunction" shall mean any failure to feed, chamber, fire, extract or eject a round or any failure to accept or eject a magazine or any other failure which prevents the firearm, without manual intervention beyond that needed for routine firing and periodic reloading, from firing the chambered round or moving a new round into position so that the firearm is capable of firing the new round properly. "Malfunction" shall not include a misfire caused by a faulty cartridge, or the primer, which fails to detonate when properly struck by the firearm's firing mechanism.[525]

Accidental Discharge Test

Other than a negligent discharge, which is caused by the shooter, an accidental discharge happens for an event such as when a gun is dropped. Handguns will only be listed on the Approved Weapon Roster, if they pass a drop test.[526]

A sample of five firearms in new condition will be drop-tested and none of them must discharge during the test. Each of the five sample firearms shall be test loaded and set so that the firearm is in a condition such that pulling the trigger and taking any action that must simultaneously accompany the pulling of the trigger as part of the firing procedure would fire the handgun. Simultaneously the handgun is dropped onto a solid slab of concrete at a height of one meter [3 ft 3.37 in; PD] from each of the following positions:

- Normal firing position;
- Upside down;
- On grip;
- On the muzzle;
- On either side; and

- On the exposed hammer or striker or, if there is no exposed hammer or striker, the rearmost part of the firearm.

If the firearm is designed so that its hammer or striker may be set in other positions, each sample firearm shall be tested as above with the hammer or striker in each of such position but otherwise in such condition that pulling the trigger, and taking any action that must simultaneously accompany the pulling of the trigger as part of the firing procedure, would fire the firearm. Alternatively, the tester may use additional sample firearms of the same make and model, in a similar condition, for the test of each of these hammer or striker settings.[527]

Inferior Make

A handgun cannot be transferred in Massachusetts, if such firearm is prone to firing more than once per pull of the trigger, or explosion during firing.[528]

Accuracy disclosure when barrel length is shorter than three inches

It is considered an unfair or deceptive practice for a handgun-purveyor to transfer to a customer located within the Commonwealth, a handgun that has a barrel shorter than 3 in [76.2 mm; PD], unless the handgun-purveyor discloses the limits of the accuracy of the make and model of handgun for sale by providing in writing to the customer, prior to the transaction, the make and model's average group diameter test result at 7, 14, and 21 yards [6.4, 12.8, and 19.2 meters; PD].[529]

The purpose of this clause, "average group diameter test result" shall serve and refers to the arithmetic mean of three separate trials, each performed as follows on a different sample firearm in new condition of the make and model being examined.

Each firearm shall fire five rounds at a target from a set distance and the largest spread in inches between the centers of any of the holes made in a test target shall be measured and recorded. This procedure shall be repeated two more times on the firearm. The arithmetic mean of each of the three recorded results shall be considered to be the result of the trial for that particular sample firearm. The ammunition used shall be the type recommended by the firearm manufacturer in its user manual or, if none is recommended, any standard ammunition of the correct caliber in new condition.[530]

Safety Warning/Disclosures

It shall be an unfair or deceptive practice for a handgun-purveyor to transfer or offer to transfer to any customer located within the Commonwealth, any handgun unless that handgun is accompanied by the following warning, provided on a separate sheet of paper included within the package enclosing the gun, which, in at least 12-point type [4.23 mm; PD], states the following:

> *WARNING FROM THE MASSACHUSETTS ATTORNEY GENERAL: This handgun is not equipped with a device that fully blocks use by unauthorized users. More than 200,000 firearms like this one are stolen from their owners every year in the United States. In addition, there are more than a thousand suicides each year by younger children and teenagers who get access to firearms. Hundreds more die from accidental discharge. It is likely that many more children sustain serious wounds, or inflict such wounds accidentally on others. In order to limit the chance of such misuse, it is imperative that you keep this weapon locked in a secure place and take other steps necessary to limit the possibility of theft or accident. Failure to take reasonable preventive steps may result in innocent lives being lost, and in some circumstances may result in your liability for these deaths.[531]*

Failure to include this warning in the package enclosing the gun shall not be a violation if the handgun in question complies with 940 CMR 16.05(1) by means of a built-in passive use-limitation device, including but not limited to a non-detachable solenoid use-limitation device.[532]

It is an unfair or deceptive practice for a handgun-purveyor to transfer a handgun directly to a retail customer located in Massachusetts without demonstrating how to load, unload, and safely store the handgun, and how to engage and disengage all safety devices on the handgun. This shall include an explanation of the circumstances for which the safety devices are designed to prevent the firing of the handgun. The handgun-purveyor shall also note for the retail customer, the absence, if any, of the following: a load indicator, a magazine safety disconnect or an internal safety.[533]

Massachusetts Provisions

- Magazine capacity
- Detectable as a weapon by X-ray machines or walk-through metal detectors
- Assault Weapon Ban
- Tamper-Resistant Serial Number
- High Quality Material or Performance Test (optional)
- Childproof
- Safety Device
- Loaded Chamber Indicator or Magazine Disconnect
- Approved Weapons Roster
- Handgun Performance Test
- Accidental Discharge Test
- Reliability
- Accuracy Disclosure for barrel length shorter than 3 in
- Safety Warning/Disclosures

Fig. 4.13: Massachusetts' Provisions in a nutshell.

4.3.3 Maryland

In Maryland (MD) "handgun" is defined as a firearm with a barrel less than 16 in [406 mm; PD] in length, and includes signal, starter, and blank pistols.[534] Any handgun manufactured after January 1, 1985, that is not included on the Maryland handgun roster may not be offered for sale in MD.[535] Maryland's handgun roster is a list of authorized handguns that are useful for legitimate sporting, self-protection, or law enforcement purposes. The handgun roster is compiled and maintained by the Handgun Roster Board in the Maryland Department of State Police.[536]

The Board will consider carefully each of the following characteristics of a handgun without placing undue weight on any one characteristic in determining whether any handgun should be placed on the handgun roster:[537]

1. Concealability;
2. Ballistic accuracy;
3. Weight;
4. Quality of materials;
5. Quality of manufacture;
6. Reliability as to safety;
7. Caliber;
8. Detectability by the standard security equipment that is commonly used at an airport or courthouse and that is approved by the Federal Aviation Administration for use at airports in the United States; and
9. Utility for legitimate sporting activities, self-protection, or law enforcement.

The Board may place a handgun on the handgun roster at its own discretion or the Board shall place a handgun on the handgun roster after the successful petition of any person, unless a court, after all appeals are exhausted, has made a finding that the decision of the Board shall be affirmed. A petition to place a handgun on the handgun roster shall be submitted to the Board in writing in the form and manner that the Board requires and the petitioner has the burden of proving to the Board that the handgun should be placed on the handgun roster.[538]

Within 45 days after receipt of a petition to place a handgun on the handgun roster, the Board shall either deny the petition in writing, stating the reasons for denial; or approve the petition and publish a description of the handgun in the Maryland Register, including notice that any objection to the handgun's inclusion on the handgun roster shall be filed with the Board within 30 days. If the Board fails to deny or approve a petition within 45 days, the petition shall be considered denied.[539]

A person may not manufacture, sell, offer for sale, purchase, receive, or transfer a detachable magazine that has a capacity of more than 10 rounds of ammunition.[540] And in Maryland, a handgun is regarded to be an assault weapon if it is a semi-auto pistol that can accept a detachable magazine and has any of the following:

- A threaded barrel, capable of accepting a flash suppressor, forward handgrip, or silencer.
- A second handgrip.
- A shroud that is attached to or that partially or completely encircles the barrel, except for a slide that encloses the barrel, and that allows the bearer to fire the weapon without burning the bearer's hand.
- The capacity to accept a detachable magazine outside the pistol grip.

Also, a semi-automatic pistol with a fixed magazine that can accept more than the 10 rounds is an assault weapon.[541]

A dealer may not sell a handgun manufactured on or before December 31, 2002, without an external safety lock. If a handgun was manufactured after the year 2002, it may only be sold if it has an integrated mechanical safety device.[542] "Integrated mechanical safety device" means, a disabling or locking device that is built into a handgun, and designed to prevent the handgun from being discharged unless the device has been deactivated.[543]

The Handgun Roster Board will annually review the status of personalized handgun technology; and on or before July 1 report its findings to the Governor and to the General Assembly. In reviewing the status of personalized handgun technology, the Handgun Roster Board shall consider the number and variety of models and calibers of personalized handguns that are available for sale, and each study, analysis, or other evaluation of personalized handguns conducted or commissioned by the National Institute of Justice, a federal, state, or local law enforcement laboratory, or any other entity with an expertise in handgun technology, and any other information that the Handgun Roster Board considers relevant.[544]

Maryland Provisions

- Handgun Roster
- Detectability by standard security equipment
- Utility for legitimate sporting activities or self-protection
- Magazine capacity
- Assault Weapon Ban
- Integrated Mechanical Safety Device

Fig. 4.14: Maryland's Provisions in a nutshell.

4.3.4 New York

In New York (NY), the term "firearm" means any pistol or revolver, a shotgun having one or more barrels less than 18 in [457 mm; PD] in length, a rifle having one or more barrels less than 16 in [406 mm; PD] in length, any weapon made from a shotgun or rifle whether by alteration, modification, or otherwise if such weapon has an overall length of less than 26 in [660 mm; PD], or an assault weapon. In New York, "firearm" does not include an antique firearm.[545]

"Semi-automatic" is defined as any repeating rifle, shotgun or pistol, regardless of barrel or overall length, which utilizes a portion of the energy of a firing cartridge or shell to extract the fired cartridge case or spent shell and chamber the next round, and which requires a separate pull of the trigger to fire each cartridge or shell.[546]

No person, firm or corporation engaged in the retail business of selling firearms, shall sell, deliver or transfer any firearm to another person unless the transferee is provided at the time of sale, delivery or transfer with a gun locking device and a label containing the following verbiage:[547]

> The use of a locking device or safety lock is only one aspect of responsible firearm storage. For increased safety firearms should be stored unloaded and locked in a location that is both separate from their ammunition and inaccessible to children and any other unauthorized person.

This label must be either affixed to the firearm or placed in the container in which the gun is sold, delivered or transferred.

The term "gun locking device" is defined as an integrated design feature or an attachable accessory that is resistant to tampering and is effective in preventing the discharge of such firearm by a person who does not have access to the key, combination or other mechanism used to disengage the device.[548]

Under the new law in the State of New York, a semi-automatic pistol is an assault weapon when it is semi-automatic and has one or more of the following military characteristics:[549]

- Folding or Telescoping Stock.
- Thumbhole Stock.
- Second Handgrip or Protruding Grip that can be held by non-trigger hand.
- Capacity to accept an ammunition magazine that attaches to the pistol outside the pistol grip.
- A threaded barrel capable of accepting a barrel extender, flash suppressor, forward hand grip or silencer.

- A shroud that is attached to, or partially or completely encircles, the barrel and that permits the shooter to hold the firearm with the non-trigger hand without being burned.
- A manufactured weight of fifty ounces or more when the pistol is unloaded.

"Large capacity ammunition feeding device" is defined as a magazine, belt, drum, feed strip, or similar device, manufactured after September 13, 1994, and has a capacity of more than 10 rounds of ammunition. Large capacity ammunition feeding devices are banned in the State of New York. In addition, New York compliant magazines must be limited in a way that they cannot be readily restored or converted to accept more than 10 rounds of ammunition. The limitation does not apply to an attached tubular device designed to accept, and capable of operating only with, .22 caliber rimfire ammunition.[550]

New York Provisions

- Gun Locking Device
- Assault Weapon Ban
- Magazine capacity

Fig. 4.15: New York's Provisions in a nutshell.

4.4   National Institute of Justice Performance Requirements

NIJ Standard-0112.03, *Autoloading Pistols for Police Officers*,[551] is a technical document that specifies performance requirements and test methods for the kind of pistols typically carried by law enforcement personnel as their duty weapons. Purchasers can use the test methods described in this NIJ Standard to determine whether a particular pistol model meets the essential requirements, or they may have the tests conducted on their behalf by a qualified testing laboratory. Procurement officials may also refer to this standard in their purchasing documents and require that equipment offered for purchase meet the requirements. Compliance with the requirements of the standard may be attested to by an independent laboratory or guaranteed by the vendor. The NIJ Standard is designed as a procurement aid and provides precise and detailed test methods.[552]

In most cases the so termed *conformance testing* is carried out by a qualified testing laboratory. Typically, this will be H.P. WHITE LABORATORY, INC. based in Maryland. Costs will be in the neighborhood of $3500 U.S. plus additional costs for ammunition.

The following minimum information must be supplied in the English language by the manufacturer:

a) Instructions for field disassembly/assembly and diagram(s) identifying all parts.
b) Cleaning instructions.
c) A description of each safety feature designed into the pistol, how each safety feature is intended to function, and for those under shooter control, how the shooter should operate (activate/deactivate) each safety feature.
d) A statement on ammunition known to be beyond the design limits of the pistol (e.g., +P ammunition[a] in a pistol not designed to handle +P ammunition) and/or known not to function in the pistol.
e) A statement identifying how a parts list may be obtained.

Manufacturers may supply any other information that they believe may be needed by the user for proper and safe operation of their handguns.[553]

---

[a] This ammunition is loaded to a higher pressure. In general, the pressure is higher than the SAAMI Maximum Average Pressure ($p_{T\,max}$) specifications, and is indicated by a "+P" marking on the case headstamp. Official +P pressures are established by SAAMI for certain cartridges. SAAMI Z299.3-2015, p. 17, specifies for the 9 mm Luger cartridge a $p_{T\,max}$ of 35,000 psi (241.3 MPa) and for 9 mm Luger +P 38,500 psi (265.5 MPa), i.e. the +P pressure is approximately 10% higher than the Maximum Average Pressure. – CIP specifies $p_{T\,max}$ = 235.0 MPa for the 9 mm Luger cartridge.
    Note: The "+P" designation is not currently used by SAAMI, but by some manufacturers to designate loads that exceed the +P SAAMI specification and may reach a gas pressure of 9 mm Luger CIP proof test cartridges.

Two representative samples of each pistol model to be tested are required. The samples can be selected at random from a current purchase lot for acceptance testing, or alternatively, two test pistols can be supplied by the manufacturer for qualification compliance testing separately from a purchase lot, in which case they shall be selected randomly from the current production.[554]

Hammer travel, surfaces and particles are the three parts of the visual inspection. If pistols are equipped with an external hammer, it will be cocked to the single-action full-cock position, if the weapon will fire on single-action. In doing so, it will be verified that there is perceptible travel past this position. This is followed by an examination of the pistols where any shavings or filings that should not be inside the pistol will be noted. Also noted will be any chips, scratches, sharp edges, burrs, or rust spots. There shall be no sharp edges or corners that could cut the shooter's hand while firing or during manual cycling of the pistol.[555]

Apart from dimensional requirements, such as barrel bore dimensions and headspace, the functional requirements must allow the slide to operate smoothly without binding or sticking when operated by hand, or during firing tests; while the ejection mechanism shall eject cases without hang up and without hitting the shooter during the ejection test or the firing tests.[556]

Trigger pull is specified as follows: The single-action trigger pull force shall not be less than 13 N (3 lbs) nor more than 36 N (8 lbs). Double action trigger pull force must not be more than 80 N (18 lbs). For a pistol employing a striker-fire mechanism, the trigger pull force has to be not less than 22 N (5 lbs) nor more than 67 N (15 lbs).[557]

If the pistol is hammer-fired, the hammer shall operate smoothly without binding.[558] Moreover, the pistol shall have one or more design features to prevent inadvertent firing. Active safety devices, if provided, shall be designed so that the pistol can be made fire-ready by releasing the attached safety(s) with the shooting hand. The magazine shall have a capacity of six rounds, minimum, and shall be capable of being released without removing the shooting hand from the pistol.[559]

Firing Requirement
NIJ Standard-0112.03 focusses on four calibers: 9 mm Luger, .357 SIG, .40 S&W, and .45 Auto.[560] After checking if the barrel bore diameter is in accordance with SAAMI Standards for the caliber for which the pistol is chambered, headspace will then be gauged. The headspace gauges to be used in this testing shall be standard commercial headspace gauges, except that they shall be modified to permit the installation of

a standard primer, as well as a vent of 2 mm (0.078 in) in diameter located on the longitudinal axis of each gauge.

Unfired primers will be in place in each headspace gauge during the testing, and will be used to verify whether the hammer struck the firing pin with sufficient force to cause the pistol to fire during the headspace testing. With the "No-Go" headspace gauge chambered, it will be verified that the slide did not reach its mechanically-locked position, and that the hammer either will not fall or is blocked from striking the firing pin with sufficient force to cause the primer to detonate when all safeties are disengaged and the trigger is pulled.[561]

The unloaded pistol will also be operated in all of its action modes, also the slide will be pulled fully to the rear and released to the battery position. In each case, any sticking, binding, grittiness, or hesitation will be noted, after which the pistol will be loaded with a full magazine and then fired until the magazine is empty. Failure to eject, or if any ejected cases hit the shooter, and whether the slide remains open after the magazine's last round has been fired will be noted. With the pistol empty, the trigger pull will be checked with the barrel vertical and muzzle up.[562]

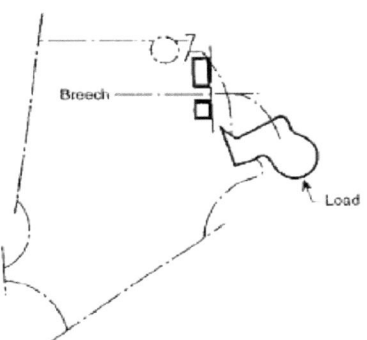

Fig. 4.16: Hammer "Push-Off" test setup.
(NIJ Standard-0112.03, 1999 p. 12)

If the pistol is hammer-fired, a hammer push-off test is to follow. With the pistol emptied and fully cocked, a load with a 46 N ± 1 N (10.25 lbs ± 0.25 lbs) force is applied to the rearmost part of the hammer spur and tangential to hammer's arc. The hammer shall not release under this load applied (Fig. 4.16).[563]

It will be verified that all of the safety parts are present, that they operate in the manufacturer's intended manner, and that the feature(s) perform their intended functions. Also, a primed case will be chambered and attempted to fire the pistol, with the safety device engaged to determine whether the round discharges. If a pistol has more than one safety device, all but one will be disengaged to conduct the test, and repeated using the next safety device, and so on, until all safety devices have been tested. Besides this, the ease of insertion and removal of each magazine provided with the pistol will be checked.[564]

For the actual firing test, a total of 600 rounds in increments of 100 rounds will be fired with no delays except to reload or to determine causes of malfunctions. After every 200 rounds, the pistol will be checked and cleaned according to the manufacturer's recommendations in the provided user information.

The first six rounds will be fired in five seconds. The firing rate for the remainder of the test must be at least one round every two seconds and no greater than two rounds per second.

For pistols with both a single-action and double strike capability, the first round of each magazine is to be fired after decocking the firing mechanism.

Should three or more misfires occur during the 600-round test sequence, the primers in the misfired cartridges will be examined. If it is obvious that the misfires were caused by faults in the pistol, e.g. very shallow or no indentation of the primer, the pistol has failed to meet the requirements. If it is not obvious that the misfires are a result of faults present in the pistol, the firing test will be repeated. If the pistol passes the second 600 round test, then it meets the requirements. If three or more misfires occur during the second 600 rounds, and again it is not clearly due to the pistol, the tester should consult the ammunition manufacturer to determine the condition of the misfired ammunition.[565]

Drop Safety Test

A primed casing is inserted into the chamber and the slide hold-open mechanism released, allowing the slide to move forward under the impetus of the recoil spring. Now, a magazine loaded to capacity with dummy rounds is inserted and the pistol placed in a drop fixture capable of dropping the pistol from a drop height of 1.22 m (4 ft) onto an 85 ± 5 Durometer (Shore A) rubber mat, 2.54 cm (1 in) thick, backed by concrete. The mat and concrete shall be large enough so that when the pistol is dropped, it will fall and come to rest without interference within the perimeter of the mat. The drop height shall be measured from the surface of the rubber mat to the lower most point of the firearm. The pistol shall be cycled and returned to the specified testing condition after each drop.

The pistol shall not be dropped from a hand; a fixture is required. However, the pistol shall be dropped in the condition that the pistol would be in if it were dropped from a hand, that is, cocked, no manual safety applied, etc. If the design of the pistol is such that upon leaving the hand a "safety" is automatically applied by the pistol, this feature shall not be defeated. NIJ recommends a fixture consisting of one short piece of string with the pistol attached at one end and the other end held in an air vise until the drop is initiated.

The following six cardinal orientations and one on the rearmost point are required for each of the two pistols:

1. Normal firing position; barrel horizontal;
2. Upside down; barrel horizontal;
3. On grip; barrel vertical;
4. On muzzle; barrel vertical;
5. On left side; barrel horizontal;
6. On right side; barrel horizontal; and
7. If there is an exposed hammer or striker, on the rearmost point of that device; otherwise on the rearmost point of the pistol. Alternately, a weight equivalent to that of the pistol may be dropped onto the rearmost point.

After each drop, the primer will be inspected. If indentations are present, a fresh primed case must be used for the next drop. Firing of the primer constitutes failure of the test.

After completing the drops, the pistols will be examined for damage. For those pistols that passed the drop safety test without structural damage or damage that will affect the safe and proper functioning of the pistol, a fully-loaded magazine will be fired, re-loaded and repeated until 20 rounds have been fired. If there are no more than three malfunctions, the pistol meets the requirements. If there are more than three malfunctions, the 20 round firing test will be repeated.[566]

NIJ-Test

- Magazine capacity: six rounds, minimum
- Trigger pull force
- Design features to prevent inadvertent firing
- Caliber
- Dimensional Requirements (gauging)
- Firing Requirement
- Drop Safety Test

Fig. 4.17: NIJ-test in a nutshell.

Once the pistol model has successfully completed testing at an NIJ-approved testing facility, and a letter of compliance for the model tested has been issued to the manufacturer by the *National Law Enforcement and Corrections Technology Center*, the manufacturer may place the following statement in the user information:

The manufacturer certifies that this model of autoloading pistol has been tested and found to comply with the requirements of NIJ Standard-0112.03 (Revision A), dated July, 1999.[567]

4.5 Evaluation of New Firearms, ANSI/SAAMI Z299.5

Since the Consumer Product Safety Commission (CPSC) was created by U.S. Congress in 1972 to set safety standards for most consumer products, there have been attempts to put firearms and ammunition under the Consumer Product Safety Act. Recognizing that firearms are not traditional "consumer products," Congress exempted the firearms and ammunition industries in 1976 saying "The Consumer Product Safety Commission shall make no ruling or order that restricts the manufacture or sale of firearms, firearm ammunition, or components of firearms ammunition, including black powder or gunpowder for firearms." [568] The Safety Act also excludes products of other industries that expressly lie in another federal agency's jurisdiction, for example food, motor vehicles, aircraft, and boats.

If firearms had been put under the Consumer Product Safety Act, according to 15 USC §§ 2051–2089[569] that would give the Safety Commission the power to develop safety standards and pursue recalls for products that present unreasonable or substantial risks of injury or death to consumers. It also would allow the CPSC to ban a product if there is no feasible alternative.

With the federal government boxed out, gun safety standards are set by the *Sporting Arms and Ammunition Manufacturers' Institute, Inc.* (SAAMI). Since 1926, SAAMI has been the principal organization in the United States actively engaged in the development and promulgation of product standards for firearms and ammunition, but compliance with its standards is strictly voluntary.[570]

The primary work of SAAMI is in the setting up of industry standards for firearms and ammunition. SAAMI is an accredited standards developer for the American National Standards Institute. As an accredited standards developer, SAAMI's standards for industry test methods, definitive proof loads, and ammunition performance specifications are subject to the American National Standards Institute's (ANSI) review and various ANSI criteria. Other key areas of SAAMI's technical expertise and standardization include pressure measurement, muzzle loading, and working towards universal, internationally recognized standards by working with the *Commission Internationale Permanente* (CIP). The CIP is an international association of proof houses. In Europe, proof houses or testing facilities for firearms and ammunition, have been the ones to set up European standards since the 1800s.[571]

In contrast to the mandatory CIP proof testing enforced by the firearms proof law, the use of American National Standards, including SAAMI, is voluntary. According to *ANSI/SAAMI Z299.5-2016*, the existence of American National Standards does not preclude anyone from manufacturing, marketing, purchasing, or using products, processes, or procedures not conforming to the standards.[572]

Among others, the following companies are SAAMI members: BROWNING ARMS COMPANY, FEDERAL CARTRIDGE COMPANY, OLIN CORPORATION/WINCHESTER DIVISION, REMINGTON ARMS CO., INC., SIG SAUER, INC., SMITH & WESSON HOLDING CORP., STURM, RUGER & CO., INC. and TAURUS HOLDINGS, INC.[573]

Companies such as COLT'S MANUFACTURING CO., SAVAGE ARMS, INC. and NORTH AMERICAN ARMS have shown interest in a voluntary industry performance standard to provide the firearm designer and manufacturer with recommendations for test procedures to evaluate new designs of firearms and initiated the introduction of ANSI/SAAMI Z299.5 in 1985.[574] This standard, *Criteria for Evaluation of New Firearms Designs Under Conditions of Abusive Mishandling for the Use of Commercial Manufacturers*,[575] provides recommendations for test procedures to evaluate new designs of centerfire and rimfire rifles, shotguns and handguns as defined under the Gun Control Act of 1968. Test parameters replicate conditions where abusive mishandling could possibly result in an accidental discharge.

The purpose of this standard is to provide test procedures for drop test, exposed hammer test, jar-off test, and rotation test, that will aid the design engineer and manufacturer in evaluating the ability of the firearm to withstand abusive mishandling without discharging. This standard does however not apply to muzzle loading and black powder firearms of any type, and the requirements of this standard are not appropriate for firearms primarily intended for formal target shooting, and therefore this standard does not apply to firearms whose trigger pull is designed to be less than 3 lbs or 1.36 kg [13.3 N; PD].[576]

Test Procedures, ANSI/SAAMI Z299.5-2016

The following applies to any of the four tests: The firearm shall not fire a chambered, empty primed case of its designated cartridge when tested. In a multi-chambered gun, the primed cases shall be inserted in the chambers directly in front of any firing pin.

The test shall be conducted with the magazine, clip or remaining revolver cylinder chambers fully loaded with SAAMI-compliant gun functioning dummy cartridges and locked in place. Any test shall be conducted with the trigger pull force set at the minimum force specified by the manufacturer, and shall be conducted with firearms of minimum and maximum weight configurations of a given model, including weight variations introduced by accessories catalogued by the manufacturer.

Drop Test

With the firearm in the "Safe Carrying" condition, the firearm shall be capable of passing the drop testing from a height of 1.22 m (4 ft) onto an 85 ± 5 Durometer (Shore A) rubber mat, 2.54 cm (1 in) thick, backed by concrete. The mat and concrete shall be large enough so that when the gun is dropped, it will fall and come to rest without interference within the perimeter of the mat. The drop height shall be measured from the surface of the rubber mat to the center of gravity of the firearm. The center of gravity shall be determined to an accuracy of ±2.54 cm (±1 in) by any recognized method for finding the center of gravity of an irregular shaped object. The firearm shall be re-cocked and reset in the "Safe Carrying" condition after each drop or a separate firearm may be used for each drop. As an alternative to free dropping, other methods may be substituted if they provide equivalent impact characteristics.

The firearm or firearms shall be dropped in such a way as to cause them to strike the rubber mat surface in each of the following attitudes:

1. Barrel vertical, muzzle down;
2. Barrel vertical, muzzle up;
3. Barrel horizontal, bottom up;
4. Barrel horizontal, bottom down;
5. Barrel horizontal, left side up; and
6. Barrel horizontal, right side up.

Parts breakage or other damage resulting from drop testing does not constitute failure as long as the empty primed case does not fire and the firearm can be unloaded safely after each drop.[577]

Exposed Hammer Test

This test applies to handguns with exposed hammers or strikers. Handguns with exposed hammers or strikers shall be capable of passing the exposed hammer test with the firearm in the "Safe Carrying" condition. The firearm shall be dropped from a height of 0.914 m (36 in), striking the rear of the hammer spur or exposed striker upon a mild steel block of at least 22.7 kg (50 lbs) weight with the barrel vertical, muzzle up, a total of six times. Another acceptable procedure is: Instead of dropping the firearm, a mild steel weight equal to the weight of a fully loaded firearm and accessories as catalogued by the manufacturer, may be dropped 0.914 m (36 in), striking the exposed hammer or striker with the firearm held with barrel vertical and muzzle down, its muzzle resting on a mild steel block of at least 22.7 kg (50 lbs) weight, a total of six times.

In any case, the same firearm shall be used throughout the test. The height shall be measured from the impact surface to the contact point on the exposed hammer of the firearm, and the test shall be conducted with the trigger pull force set at the minimum force specified by the manufacturer.

If at any time during the test, there is any observable damage to a part of the firearm without the firing of the primed case, the said part may be replaced and the test con-tinued, unless the damaged part bears the serial number of the firearm. Damage to the serial-numbered part without discharge of the primed case after all six drops, shall not constitute failure of this test, as long as the firearm can be unloaded safely after each drop.[578]

Jar-Off Test

This test simulates the bumping of the firearm against a hard surface with the firearm in a condition of maximum readiness. With the firearm cocked and in the ready-to-fire condition (safety "Off"), the firearm shall be capable of passing a jar-off shock equiva-lent to being dropped from a height of 0.305 m (12 in) onto an 85 ± 5 Durometer (Shore A) rubber mat, 2.54 cm (1 in) thick, backed by concrete. The mat and concrete shall be large enough, so that when the gun is dropped, it will fall completely within the perim-eter of the mat. The drop height shall be measured from the surface of the rubber mat to the lowest point on the firearm.

The firearm or firearms shall be dropped in such a way as to cause them to strike the rubber mat surface one time only in each of the following attitudes:

1. Barrel vertical, muzzle down;
2. Barrel vertical, muzzle up;
3. Barrel horizontal, bottom up;
4. Barrel horizontal, bottom down;
5. Barrel horizontal, left side up; and
6. Barrel horizontal, right side up.

The gun shall be caught after its first bounce from the mat so that it strikes the mat only once. The firearm shall be re-cocked and reset in the ready-to-fire condition after each drop or a separate firearm may be used for each drop. As an alternative to free dropping, other methods may be substituted, if they provide equivalent impact characteristics.

Parts breakage or other damage resulting from drop testing, does not constitute failure as long as the empty primed case does not fire and the firearm can be unloaded safely after each drop.[579]

Rotation Test

This test simulates the abusive fall of a long gun in the "Safe Carrying" condition from an upright position with its butt resting on a surface, when left leaning against a vertical surface. The test is to be performed with rifles and shotguns only, but not with handguns.[580]

ANSI/SAAMI Z299.5-2016 Evaluation

- Drop Test
- Exposed Hammer Test
- Jar-Off Test
- Rotation Test (rifles and shotguns)

Fig. 4.18: SAAMI Voluntary Industry Performance Standards in a nutshell.

4.6   IPSC Production Division

The International Practical Shooting Confederation (IPSC)[a] is a shooting sport associ-ation formed on the concept of practical shooting. Accuracy, power and speed are all required to achieve a maximum score. Competitors are divided into different divisions based on firearm and equipment features. Competitive divisions for handguns are Open Division, Standard Division, Classic Division, Production Division, and Revolver Division. The Production Division is of particular interest for manufacturers of hand-guns with standard triggers and open sights, because minimum trigger pull for the first shot is 2.27 kg (5 lbs) and handguns with optical or electronic sights are excluded from this division.

Handguns for Production Division must be chambered in caliber 9 mm or more and the cartridge case length starts from 19 mm. Handgun size is not limited, but the barrel length must not be over 127 mm [5 in; PD]. Magazine capacity is not limited but the first magazine must not contain more than 15 rounds at the Start Signal.[581]

Handguns deemed by IPSC to be single-action-only are expressly prohibited. Only handguns listed as approved on the IPSC website may be used in Production Division. The list of approved handguns is published annually in April.

Original parts and components offered by the original firearm manufacturer as stand-ard equipment, or as an option, for a specific model handgun on the IPSC approved handgun list are permitted, but modifications to them, other than minor detailing (the removal of burrs and/or adjustments unavoidably required in order to fit replacement original firearm manufacturer parts or components), are prohibited. Other prohibited modifications include those which facilitate faster reloading (e.g. flared, enlarged and/or add-on magazine wells, etc.), changing the original color or finish of a handgun, or adding stripes, stippling or other embellishments.

---

[a] In addition to IPSC handgun competition rules International Defensive Pistol Association (IDPA) and U.S. practical shooting Association (USPSA) have competition rules. In the USPSA Production Divi-sion the handgun with empty magazine installed must fit wholly within a box with internal dimensions of 8 $^{15}/_{16}$ × 6 × 1 $^{5}/_{8}$ inches, tolerance $+^{1}/_{16}$ inch [227.01 × 152.4 × 41.28 mm +1.59 mm; PD] (USPSA, 2014 p. 79).

Semi-automatic handguns permitted for IDPA matches have to be chambered in 9 mm Luger or must be for larger cartridges, the firearm with the largest magazine installed must fit in the IDPA gun test box measuring 8 $^{3}/_{4}$ × 6 × 1 $^{5}/_{8}$ inches [222.25 × 152.4 × 41.28 mm; PD], the unloaded firearm with the heaviest magazine must weigh 43 oz or less [1219 g; PD] (IDPA, 2019 pp. 26–29) There are no re-strictions in regards to the construction type of the handgun.

Aftermarket magazines and aftermarket open sights are permitted, provided their installation and adjustment requires no alteration to the handgun. Also, aftermarket grip panels which match the profile and contours of the original firearm manufacturer standard or optional grip panels for the approved handgun or the application of tape on grips are permitted. However, rubber sleeves are prohibited.[582] Handguns with shoulder stocks or fore grips of any kind are prohibited.[583]

IPSC Production Division Conditions

- Minimum projectile caliber: 9 mm (0.354 in)
- Minimum cartridge case length: 19 mm (0.748 in)
- Minimum trigger pull: 2.27 kg (5 lbs)
- Maximum barrel length: 127 mm
- Handguns deemed by IPSC to be Single-Action-Only are prohibited

---

Fig. 4.19: IPSC Production Division Conditions in a nutshell.

4.7   Summary "Requirements"

The graphical overview below summarizes the requirements for the importation and sales of pistols in the United States. The bundling of these requirements in the form of a quick-read graphic should help to facilitate the design of new products for the U.S. market. Fig. 4.20 provides a structured overview of the similarities and differences in the requirements for this purpose.

California compliance is a special case. No semi-automatic handgun may be sold in California unless it incorporates microstamping and it is certified as safe for sale in California. Some arms manufacturers might respond by offering Derringers[584] and re-volvers.

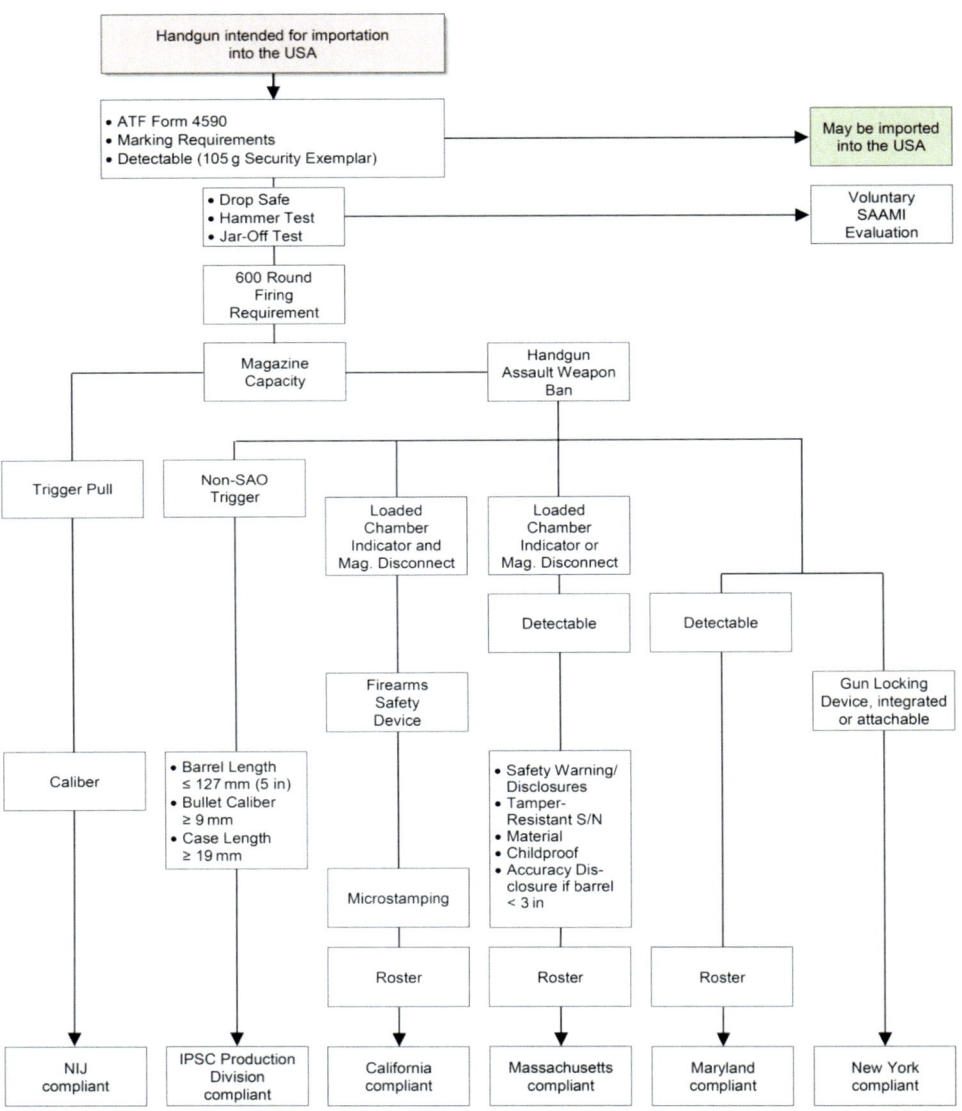

Fig. 4.20: Graphic summary for centerfire pistols with removable magazine.

Fig. 4.21: Cup holders are old-school.
We want gun compartments.

5   Survey of Various Pistols

Starting with an analysis showing the most popular caliber groups, and which U.S.-based manufacturers have the largest annual output per caliber group, we will then continue with a survey of nine popular pistols.

5.1   Analysis of the U.S. Market

ATF releases the Annual Firearms Manufacturing and Export Report (AFMER) which exhibits statistics on frames, receivers and guns manufactured in and exported from the United States. All federally licensed manufacturers of firearms and destructive devices are required to submit a production report of manufacturing and export activity to the ATF by April 1 of each year. The ATF compiles the submitted data and releases it on the ATF Web page each January with a one-year delay because the proprietary data furnished by filers is protected from immediate disclosure by the *Trade Secrets Act*. For example, data released in January 2020 was for calendar year 2018.

AFMER lists production quantities of pistols, revolvers, rifles, shotguns, and miscellaneous firearms. The latter are any firearms not specifically categorized in any of the firearms categories defined on the ATF Production Report Form 5300.11. Examples of miscellaneous firearms would include pistol grip firearms, starter guns, and firearm frames and receivers.[585] The AFMER report excludes production for the U.S. military but includes firearms purchased by domestic law enforcement agencies.

The annual production quantities of pistols and revolvers are divided into caliber groups. The report also lists the quantities of firearms exported by each manufacturer, but these numbers are not broken down into caliber groups. In 2018, more than 3.8 million pistols were manufactured in the United States, 333,266 of which were exported; i.e. about 8.6 percent of the pistols manufactured in the United States were exported (7.5 percent in 2017, 3.7 percent in 2016, 4.0 percent in 2015 and 3.5 percent in 2014).[586]

The production quantities mentioned below were taken from AFMER. The total numbers of handguns made in the United States are shown, exports were ignored because AFMER does not exhibit exports per caliber group. Production quantities by companies that produce in the United States at several locations were grouped together; for example, REMINGTON, RUGER and SIG SAUER.[587]

AFMER shows the numbers of units of each manufacturer, and breaks them down into the following six ATF-designated calibers:

1. *Pistols to .22,*
   (in the following referred to as 0 < Cal. ≤ .22");

2. *Pistols to .25,*
   (in the following referred to as .22" < Cal. ≤ .25");

3. *Pistols to .32,*
   (in the following referred to as .25" < Cal. ≤ .32");

4. *Pistols to .380,*
   (in the following referred to as .32" < Cal. ≤ .380");

5. *Pistols to 9MM,*
   (in the following referred to as .380" < Cal. ≤ 9 mm);

6. *Pistols to .50,*
   (in the following referred to as 9 mm < Cal. ≤ .50").

The evaluation of the figures published in AFMER for the year 2018 (current state of statistics published in 2020) translates to the shares in caliber categories shown in Fig. 5.1. In addition, the leading U.S. manufacturers are listed below and in the tables of Appendix 10. The manufacturer with the largest production numbers for each caliber group can be taken from there.

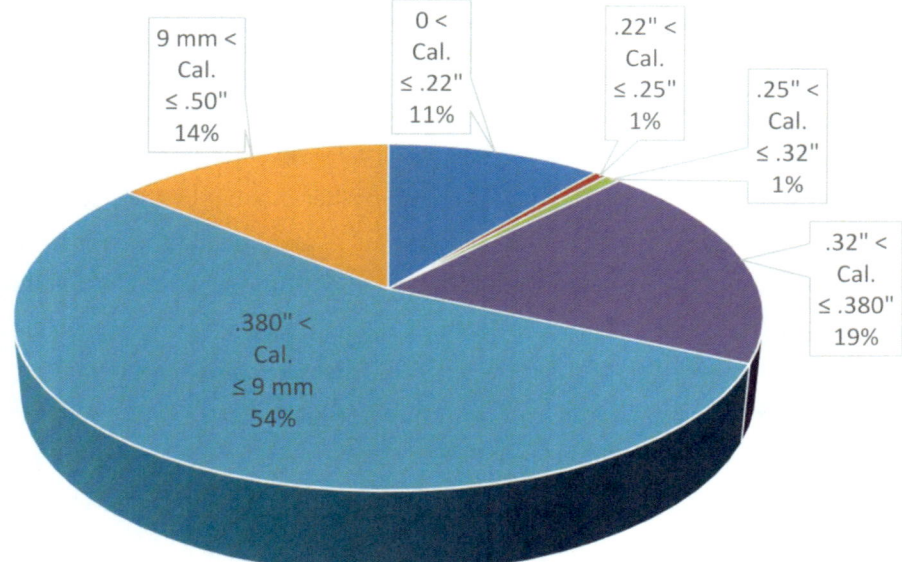

Fig. 5.1: Market segmentation by ATF-designated pistol calibers;
U.S. Production, Year 2018
(compiled from AFMER, Year 2018, see Appendix 10).

Pistols chambered for cartridges larger than .380 ACP and up to 9 mm Luger represent the largest share (54 percent) of manufactured pistols in the United States in 2018, followed by the caliber category greater than .32 ACP to .380 ACP (19.6 percent), pistols with calibers up to .50 (14.1 percent), and pistols chambered in calibers up to .22 (10.8 percent). As we have just seen, said four out of the six ATF-designated caliber categories represent the most popular calibers in the U.S.

In the year 2018, BROWNING and RUGER sold over 100,000 pistols each in the caliber category up to .22 cal. Also, quantities in the lower six-digit range were placed on the market by RUGER and SMITH & WESSON in the caliber category up to .380 ACP. GLOCK, RUGER, SCCY, SIG SAUER, and SMITH & WESSON produced considerable quantities in the category up to 9 mm, the output of SMITH & WESSON reached the six-digit range in the caliber category up to .50 cal.

Exhibited in Table 5.1 are the top five rankings of U.S. gun manufacturers in the four major caliber categories.

Table 5.1: Top 5 ranking of U.S. manufacturers per caliber category in 2018 (compiled from Appendix 10).

Manufacturer	Unit Production, 2018, 0 < Cal. ≤ .22"	Manufacturer	Unit Production, 2018, .32" < Cal. ≤ .380"
RUGER	125,636	RUGER	255,786
BROWNING	113,128	SMITH & WESSON	220,298
SMITH & WESSON	54,632	TAURUS	89,771
KEL-TEC	53,463	SIG SAUER	46,605
PHOENIX ARMS	16,390	SPRINGFIELD	44,730
Total, Top 5	363,249	Total, Top 5	657,190

Manufacturer	Unit Production, 2018, .380" < Cal. ≤ 9 mm	Manufacturer	Unit Production, 2018, 9 mm < Cal. ≤ .50"
SIG SAUER	505,749	SMITH & WESSON	154,741
SMITH & WESSON	457,160	SPRINGFIELD	77,107
RUGER	299,443	SIG SAUER	70,585
GLOCK	237,309	KIMBER	67,428
SCCY	169,818	COLT	35,399
Total, Top 5	1,669,480	Total, Top 5	405,260

5.2   Methodology

Nine Pistols were surveyed. Following a checklist, the pistols were inspected and then shot. The objective of this first analysis was to use given criteria to get a standardized basis for the comparison of different handguns.

In a second step, the samples were examined based on their respective technological level and innovative features. For this step, the guns were detail stripped to get full insight. Finally, a patent research should show if certain features were patented or rather if disclosures of new technologies were available. The patent search was carried out online at www.DEPATISnet.dpma.de provided by Deutsches Patent- und Markenamt and at www.uspto.gov provided by United States Patent and Trademark Office.

Samples

Nine pistols were closely examined:

1. CARACAL *F* (9 mm Luger)
2. GLOCK *G19 Gen4* (9 mm Luger)
3. HECKLER & KOCH *VP9* (9 mm Luger)
4. KEL-TEC *PF-9* (9 mm Luger)
5. RUGER *LC9* (9 mm Luger)
6. SMITH & WESSON *M&P9* (9 mm Luger)
7. SPRINGFIELD *XD^M* (9 mm Luger)
8. TAURUS *PT 24/7 PRO* (9 mm Luger)
9. WALTHER *PPX* (9 mm Luger)

These pistols are a representative sample of popular handguns with a variety of features and different sizes. Among the samples were two deep conceal pistols (KEL-TEC *PF-9* and RUGER *LC9*) of manufacturers that were part of the top 15 list of manufacturer's handguns in the 9 mm caliber group of 2014 (KEL-TEC, RUGER, SMITH & WESSON and SPRINGFIELD), and pistols produced by a young manufacturer, CARACAL. Aside these, the HECKLER & KOCH *VP9*, which marked the upper price limit of our sample in the year 2014 when the pistol was launched, was also examined. Research was also carried out on ARSENAL FIREARMS *Stryk B*, BERETTA *Nano* and *Pico*, BOND ARMS *Bull-Pup9*, HI-POINT *C9*, LAUGO ARMS *Alien*, SCCY *CPX*, SIG SAUER *P320* and *P365*, as well as TARA *TM-9* in order to gather information on these pistols and also to find additional innovative features.

Checklist

A template of the checklist is shown in Appendix 2. The checklist had four main parts, A) Pistol, B) Slide, C) Barrel and D) Frame. Essentially the evaluation by checklist focused on dimensions, weights and technical specifics. The validation of dimensions and weights of the samples and comparing them with the specs published by the manufacturers was interesting, because it would show if manufacturers publish information in their favor.

For instance, would the user manual of the CARACAL pistol just specify the length of the slide, and not include the total length of the gun. In several cases, the total width is another subject of consideration. Protruding levers on the sides of pistols were not considered by some manufacturers.

Fire-control mechanisms were categorized according to their technical design in three kinds. This approach set the results on a neutral basis for good comparison and ignored any of their manufacturer's marketing terms. Additionally, the trigger characteristic was checked. This included the slide racking force, and the copper crusher indent. These tests as well as the evaluation of the safeties was important from a technical point of view. The results in combination with design features provided might be in the interest of some readers, which highlight the innovative nature of the gun industry.

Measuring the projectile's muzzle velocity, uncoupling-travel, and slide travel was done to evaluate the safety inherent to the designs, and allowed a rough comparison of how much a gun is stressed as a shot is being fired. Other sources have published information on the same topics, however in this study it was deemed necessary to carry out our own testing rather than rely on these sources. With our own testing, we were able to yield reproducible findings under laboratory conditions.

Inspecting slide, barrel, and frame allowed us to evaluate how the samples were manufactured and the results might inspire some readers for future designs. The comparison of the manufacturer's suggested retail price (MSRP) with the actual street prices allow for some conclusions on how much a consumer may be willing to pay in reality for the products in our sample. The criteria listed on the checklist were established in the following manner:

A) Pistol

A.1 Serial number

The pistol's serial number is noted under this item in the checklist.

A.2 Dimensions L×W×H, with (and without) Magazine

The outside dimensions are to be measured with a coordinate measurement machine TESA *30S MS8101*, where the pistol's total length is parallel to the bore, the total height perpendicular to the bore, and the total width is measured at the widest point. The height is to be measured with and without the magazine in the pistol. Height without mag is shown in parentheses. Any dimension will be rounded to one decimal of a millimeter.

A.3 Weight, with (and without) Magazine

The weight of the gun with and without an empty magazine is measured with a digital scale SARTORIUS *PRO 28/38C*. The weight is rounded to the nearest integer in grams.

A.4 Fire-control mechanism

A classification is done between hammer-fired and striker-fired, and the firing mechanism is specified. Manufacturer's might use their own names for trigger actions, which can be narrowed down to standard specification. Known are restrike capability, single-action, and partially cocked (see p. 101). If a pistol offers a combination of trigger actions, each one is specified and noted. The first one on the list will be the pistols standard trigger, that is, the trigger action for the first shot when the gun would be carried on duty.

A.5 Tigger pull force, travel, and work

Trigger pull force, travel, and work are measured electronically and shown in a diagram. An electronic spring
scale ZWICK/ROELL, *Model Z0.5* is used for the measurement. The results are calculated with software
ZWICK/ROELL, *testXpert II V3.2* software (Fig. 5.2).

The test sample is lightly lubricated, and the magazine
loaded to full capacity with dummy rounds. The measurement is done with the frame fixed and muzzle pointing
vertically down.

The trigger is pulled via a polymer cylinder of 13 mm
outside diameter. The cylinder is guided on both sides
and the complete guide skeleton can pivot so that the cylinder can follow the trigger's trajectory (Fig. 5.3). Trigger
blade and cylinder are lined up perpendicular to each
other. The cylinder is pulled at a testing speed of
100 mm/min, and it is being pulled parallel to the bore
axis. The pre-load is 2 N.[588] The actual recording of the
measurement starts when the force measured equals the
pre-load. In case of a trigger safety, the force of the trigger
safety is recorded from the point where the pro-load is
met. At the sear break of the firing mechanism, the cylinder is pulling parallel to the bore axis and has to be at the
trigger blade's deepest face.

Fig. 5.2: Spring Scale
ZWICK/ROELL *Z0.5*

If a gun has more than one trigger action, each one will
be scaled.

Fig. 5.3: Guided cylinder
activates trigger blade.

As the trigger is being pulled the Software determines two values of interest: the
over-all maximum trigger pull force during trigger travel, and that of the pull force at the
sear break. The former may be greater than the pull force right at the point of release.
The greater magnitude of the two values will be recorded in the checklist, because a
user will have to overcome the maximum trigger pull force to discharge a round.

The trigger work correlates with the area underneath the curve in the force vs. trigger
travel diagram. The area and the value of the trigger work will be calculated by the
software.

The test charts are attached in Appendix 16. The values of trigger pull force, travel and work are to be rounded to the nearest integer and entered on the checklist.

We account for deviation, which can be seen when trigger pull forces are measured. Decimals indicate an accuracy, which does not correspond to reality. The result is greatly influenced by the exact positioning and adjustment of the cylinder, as well as the testing speed. The effect of the positioning of trigger blade and cylinder during measurement, and the manufacturing tolerances of the gun causes tolerances that are far beyond decimals. A trigger pull tolerance of ±4 N is absolutely common for serial production.

For validation of the numbers obtained, a control test is done with a traditional trigger weight. The results of the control test delivered trigger pull forces which were in a range of ±3 N from the results of the electronic spring scale. Results of the control test are not shown in the checklist.

A.6 Slide racking force

The evaluation of the slide racking force by pulling the slide to the rear is done with the same software parameters and orientation of the gun as implemented for the trigger pull measurement with the electronic scale ZWICK/ROELL, *Model Z0.5*.

The slide is moved from the forward position to the rear with an arrangement of levers, which are attached to the slide's front end via a cross-bar, between the barrel's muzzle and the recoil spring's guide rod (Fig. 5.4). The test sample is lightly lubricated, the magazine is filled with dummy rounds to its capacity and the firing mechanism is decocked. The maximum resistance is of interest, as the slide is being pulled to the rear most limit of the recoil travel. A slide racking force of not more than 100 N[589] is worthwhile.

Fig. 5.4: Slide racking force arrangement.

If the design of the slide's front end does not allow the attachment of the cross-bar between muzzle and recoil rod, a spot above the muzzle may be used, or rather the front sight.

The result of the measuring will be a force vs. slide travel curve (Appendix 17), from which the maximum racking force will be read. The value of the slide racking force will be rounded to the nearest integer and entered on the checklist.

Fig. 5.5: Box with Copper Crushers.

A.7 Copper crusher indent, maximum diameter

Copper crusher indentation is used as a reference for the transfer of energy from the striker nose to a primer. The copper crusher gauges used here and the ones used in measuring chamber pressure are of the same type.[590] They are manufactured by WILHELM HANDKE GMBH, Germany, type 5 × 7 mm.

A copper crusher is inserted into a modified headspace gauge for the measurement of the indent (Fig. 5.6). A Go-headspace gauge 19.15 mm[a] is used as an adapter. It will hold a copper crusher cylinder just where the primer cup on a live round is normally placed. The depth of the indent will be measured with a dial indicator (Fig. 5.7). The radius of the stylus' point is R 0.15 mm.

Fig. 5.6: Adapter and Copper Crusher with indent.

---

[a] Dimensions of the adapter cartridge is shown in Appendix 4 of ER (ER "Pistolen", 2008).

Without the magazine installed, the slide is pulled back and locked in the open position with the slide stop lever engaged. A copper crusher cylinder is put in the adapter, and the adapter then placed in the firing chamber. Care needs to be taken when the slide is released. The slide must be guided forward so as not to slam the adapter. If this is not done carefully, the extractor claw or other parts might get damaged.

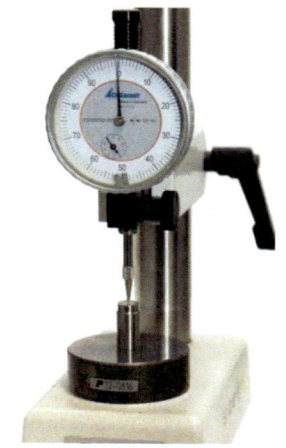

With the slide fully forward and the gun held with barrel horizontal, pointing in a safe direction, the trigger is pulled once, then the adapter removed from the chamber. The crusher remains in the gauge when it is placed on the measuring platform of the dial indicator with the indented end upwards. The gauge is moved until the stylus point has reached the indent. Rotating the gauge will allow the stylus point to bottom out. At this point, the dial is set to zero. Now the stylus point is lifted, the gauge moved far enough so the stylus point is contacting the rim of the copper crusher. Record the reading, which is to be rounded to two decimals of a millimeter.

Fig. 5.7: Dial Indicator for the measurement of copper crusher indent.

Note: The dial is set to zero as the stylus point is bottomed out. The zeroed dial must have a pre-load, e.g. 1 mm. Zeroing is done after the indentation. This will eliminate set and other influences.

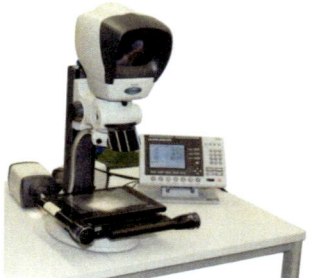

Next: The maximum diameter of the indent is measured with a measuring microscope *Kestrel Mono Dynascope* by VISION ENGINEERING (Fig. 5.8). The diameter will be rounded to within one decimal of a millimeter.

Fig. 5.8: Measuring Microscope, *Kestrel Mono Dynascope*

One copper crusher gauge will be used per indent test. Under normal conditions the test result will be clear enough (see also p. 106) for proper viewing. If a gun has more than one trigger action, one indent test will be done per trigger action.

A.8 Safeties

The survey distinguishes between manual safeties, automatic safeties, passive safeties and other safety features.

A.8.1 Manual Safety Features

A.8.1.1 Manual Safety

Manual safeties are devices that must be deliberately applied by the user. Most of them offer a choice between "Safe" and "Fire." Typically, they can be designed as external pivoting levers, which engage in "on-safe" and "off-safe" position.

A.8.1.2 Locking Device

Locking devices in this survey are defined as internal locks which are supposed to prevent unauthorized use of the firearm. They can be locked and unlocked with a proprietary appliance, which is not part of the gun.

Attachable gun locking devices are not to be considered under this category.

A.8.2 Automatic Safeties

The default mode of an automatic safety is "on-safe." Automatic safeties will automatically become disengaged prior to discharge. Examples are hammer block, striker safety, magazine disconnect, grip safety and trigger safety.

A.8.2.1 Hammer block

A hammer block is a safety that automatically engages the hammer. For example, the WALTHER *PP* pistol uses a safety which would block the hammer in the instance where the user's thumb might slip off the hammer.[591] Other designs might use the trigger bar to engage the hammer and block it.

Some hammer-fired pistols use a rebounding hammer where the hammer is automatically moved slightly rearward by a spring or cam after striking the firing pin. For example, hammer and hammer strut may be designed to be in a stable position to each other when at rest.[592] As a result, the hammer is kept at a distance from the firing pin, but the hammer is not blocked.

A.8.2.2 Striker Safety

An internal safety that blocks the striker or firing pin[a]. As the trigger is being moved backwards, the striker safety is forced to become disengaged and the striker can hit the cartridge's primer.

---

[a] To improve readability the term striker is used below as a proxy for both, striker and firing pin. Usually, the term striker is used for a spring driven pin, whereas a hammer would transfer kinetic energy on a firing pin. (Wille, 1902 p. 23)

A.8.2.3 Magazine Disconnect

A pistol without magazine will not discharge a round. Two basic principles are common, which depend on the mechanical design of the magazine disconnect: with magazine removed, the trigger can either be blocked or it may still be moveable but without causing a discharge.

A.8.2.4 Grip Safety

A safety device integrated in the grip handle of the pistol that prevents a discharge if the gun is not gripped correctly.

A.8.2.5 Trigger Safety

A trigger safety is supposed to prevent unintended movement of the trigger. The design of this safety depends on the specific gun. What they got in common is, that a safety feature is somehow integrated into the trigger mechanism. When the trigger is being used in intended manner, i.e. at the instance where the shooter places a finger on the trigger blade and starts pulling, the safety will get disengaged.

A.8.3 Passive Safeties

In contrast to automatic safeties, passive safeties include all those safeties that are permanently in disengaged position and are not operated to discharge a round, but they will automatically engage under certain extreme circumstances. Examples include an out-of-battery disconnect or a drop safety.

A.8.3.1 Out-of-battery disconnect

An out-of-battery disconnect prevents a discharge if the breech face is not in a position, which allows firing a shot without risk or as long as the component parts of the gun are not properly configured for firing.[593] For example, if the case is not fully chambered, that is, its thin wall does not receive sufficient radial supported by the chamber, the case might swell or burst during discharge.

The test is very theoretical. It is to be tested if the firing mechanism can already be released again, right at the moment when the coupling of barrel and slide starts. A magazine loaded with a dummy round is placed in the gun. The slide will be fully retracted and then guided forward slowly. At the moment where the coupling starts,[a] the trigger will be pulled. If the firing mechanism is not released, the test is considered

---

[a] On vertically tilting barrel actions, this will be the point where the barrel starts to be cammed up.

"passed" and there is an entry "out-of-battery disconnect" made in the checklist under passive safeties.

If the firing mechanism releases at the beginning of the coupling, the gun has failed this test. Testing will then carry on with the checklist's next criteria without further analysis regarding other possible means that might have prevented a discharge of an unlocked breech at this point, that is, the combination of disconnector function[a] and maybe other safeties or circumstances.[b] As a consequence, no further comments will be added to the checklist.

A.8.3.2 Other Passive Safeties

Drop safeties (see Chapter 3.8) or other safeties that work independently of the user in order to keep him safe from accidental discharge, are noted under this item on the checklist.

A.8.4 Other Features

Other features extend beyond the features already mentioned and contribute to the safety of the user, such as a loaded chamber indicator or any visual characteristics.

A.8.4.1 Loaded Chamber Indicator

A loaded chamber indicator (LCI) allows the user to identify the presence of a cartridge in the firing chamber, without requiring the user to perform any manipulation on the handgun. The embodiment can be an indicating element, which is displaceable in response to contact by the cartridge, for example an indicator pin, a lever or a part of the extractor, or it may be by a viewport in the chamber area, which is typically close to the slide's breech face area, which facilitates a visual determination of the chamber status.

A.8.4.2 Disassembly and assembly with loaded magazine

It is checked whether the handgun can be field-stripped while a loaded magazine is in it. And also, if the pistol can be assembled with a loaded magazine inserted and locked in place in the frame's magazine well.

The magazine is filled with three dummy rounds, locked in place, and then the slide is cycled to chamber a dummy round. The sample gun is then field-stripped, following

---

[a] Note: The disconnect function avoids that a gun will shoot full-auto, whereas an out-of-battery disconnect contributes to a risk-free discharge of a round. On some semi-auto pistols, both objectives are accomplished by the disconnect function.
[b] For example, hammer-fired pistols might not discharge if the firing pin can't be hit by the hammer when the slide is not fully closed, as the lower rear end of the slide is blocking the hammer.

the instructions of the manual that comes with the handgun, but without prior safety check. If the firearm can be field-stripped with magazine loaded, a "yes" will be noted in the checklist under "Strips w/Mag."

Second test: Make sure the chamber is clear. Magazine is then loaded with three dummy rounds, inserted into the magazine well and locked in place. Now it is tried to assemble the gun as instructed in the manual; any warnings in the manual are ignored when doing so. If the test sample can be assembled with magazine loaded, a "yes" will be noted in the checklist under "Assembles w/Mag."

A.8.4.3 Others

At this point, features can be listed, that do not fall under any of the above categories.

A.9 Number of Magazines

The number of the magazines included in the box is noted here.

A.10 Magazine capacity

The magazine capacity as specified by the manufacturer is noted here. Also, noted are the number of cartridges that can actually be loaded when testing the magazine.

A.11 Uncoupling start, end, and total slide recoil travel

The values are determined without recoil spring in the weapon and by manually retracting the slide starting from the closed position to the respective position.

The displacement of the slide relative to the frame is measured with a Vernier caliper. Three values are taken into consideration: the start of the uncoupling of the barrel (uncoupling start x_{u1}) or ending (uncoupling end x_{u2}), and how far the slide can be retracted to the rear (total slide recoil travel x_d). Any dimension obtained will be rounded to one decimal of a millimeter.

A.12 Muzzle velocity and muzzle energy

The muzzle velocity is determined from the average of five shots. Initially, ten shots are to be fired to shoot the barrel free of oil and to bring it to a certain "operating temperature."[594] This is followed by the measurement series with five shots. The rate of fire is approximately one round per second.

An electronic ballistic chronograph is used to measure the velocity of each projectile. It consists of DRELLO BALLISTICS *BAL 4043-09* chronograph, chronograph screen *LS 23*, and software *INLOC 2040*. The chronograph screens are located three meters ahead of the muzzle. Following the common parlance, the values in the tables are called v_0 and E_0, although the measurement of the velocity is done at a distance of three meters from the barrel's muzzle.[595]

The mean values of v_0 and E_0 are to be rounded to nearest integer and entered in the checklist. The individual test protocols are listed in Appendix 18.

The ammunition used is RUAG *Sintox* 9 mm Luger, 8.0 g FMJ, Lot: DAG13E0710. In the following, this ammunition is referred to as "reference ammunition."

A.13 Mean slide recoil velocity

Determined are theoretical and empirical mean slide recoil velocities, the former is to be established by rough calculation, while the latter is by shooting and high-speed videoing. The values are used as an indicator of how intense the load on the gun is when a shot is fired; velocities from 6 to 8 m/s are considered to be worthwhile (see Appendix 19).

The recoil velocity values are mean values, because the recoil velocity of the slide decreases during recoil. The decrease is caused by the influence of the recoil spring, friction and mechanical operations in the gun.

Mean slide recoil velocity, theoretical

The *mean slide recoil velocity, theoretical* $(\overline{v}_{St})$ is determined by the average of the speeds at the beginning and end of the recoiling slide. The calculation of the initial slide velocity v_{S1} is based on the principle of linear momentum[596]:

$$\boldsymbol{m_P v_0 = m_r v_{S1}}$$

5.1[597]

where: m_P Mass, projectile [kg]
            m_r Mass, recoiling [kg]
            v_0 Muzzle velocity of projectile [m/s]
            v_{S1} Velocity of recoiling mass [m/s]

The recoiling mass m_r of a straight blowback type pistol would be the mass of its slide, whereas in the case of a semi-automatic with coupled barrel and slide the recoiling mass m_r would be the combination of the mass of the barrel m_B and slide m_S. For the sake of simplification, the mass of the recoil spring and of the case is neglected.

Assuming the effect of the recoil spring's mass may be neglected, and the decoupled slide will continue to recoil with the same velocity v_{S1} as the coupled barrel and slide initially had, the following formula was thus derived in Appendix 19 for the mean slide recoil velocity, theoretical ($\overline{v}_{Vt}$, see Formula A19.9, p. 416):

$$\overline{v}_{St} = \frac{1}{2}\left(v_0 \frac{m_P}{m_r} + \sqrt{\left(v_0 \frac{m_P}{m_r} \right)^2 - \frac{2}{m_S} W_{fl}} \right)$$

5.2

where: $\quad \overline{v}_{St}$ Mean slide recoil velocity, theoretical [m/s]
$\quad\quad\quad\quad m_P$ Mass, projectile [kg]
$\quad\quad\quad\quad m_r$ Mass, recoiling, e.g. barrel and slide [kg]
$\quad\quad\quad\quad m_S$ Mass, slide [kg]
$\quad\quad\quad\quad v_0$ Muzzle velocity of projectile [m/s]
$\quad\quad\quad\quad W_{fl}$ Work, recoil [J], 3.5 J (cf. A19.9, S. 416)

The value for the muzzle velocity v_0 is taken from Appendix 18. Appendix 18 lists the obtained muzzle velocity for each gun as per Chapter 5.2 A.12. The recoiling mass m_r is also derived from the technical data in said appendix. It is calculated by adding the mass of the barrel m_B and the mass of the slide m_S. The value for the *mean slide recoil velocity, theoretical* $\overline{v}_{St}$ will be rounded to one decimal and is given in m/s (see Appendix 19, Table A19.1, p. 417).

The influences, caused by the mass of the casing and the mass of the powder charge, and the impulse force caused by the gases are ignored to simplify the calculation (the influence of the powder gases might be considered by including half the powder charge's mass). Furthermore, the momentum transmitted by the recoil spring or a hammer onto the slide and then onto the frame is to be ignored.[598]

Whether components in the considered guns will actually fail can only be determined by endurance testing of the guns (see also Appendix 13). Still, Formula 5.2 provides a theoretical comparison value that can be used as an indicator for a possibly existing intense load which is transferred to the weapon during firing.

Mean slide recoil velocity, empirical

To provide certainty about the validity of the calculated value, that is, *mean slide recoil velocity, empirical* $\overline{v}_{Se}$ is determined through evaluation of a high-speed video for each sample pistol. The high-speed camera used is the model NAC *MEMRECAM fx K5* in conjunction with the software *MEMRECAM fxLink* Version 2.36 by NAC IMAGE TECHNOLOGY.

"Laboratory conditions," as in optimal conditions for the experimental arrangement, shall provide the unhindered recoil of the slide and thus a maximum slide recoil velocity relative to the frame. For this purpose, the gun is to be shot without a magazine in the pistol to avoid the upper cartridge in the magazine rubbing against the slide's pickup rail. In case a test sample is equipped with a magazine disconnect, an empty magazine is to be put in the magazine well only as far as necessary, or the magazine is disassembled and just the magazine body put into the mag well.

The grip of the pistol is held securely in a vice at the pistol's trigger guard area. Fixing the sample on the trigger guard prevents deforming the frame and an impairment of the slide's recoil movement, e.g. caused by deformed slide rails. By means of the vice, the frame is rigidly mounted; it will not move as the slide cycles, and the camera records the maximum recoil velocity of the slide relative to the rigidly clamped frame.

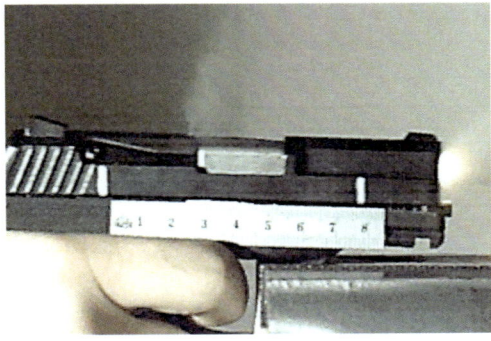

Fig. 5.9: Picture details from high-speed video of a KEL-TEC *PF-9*; projectile has just cleared the muzzle. NB: Small plume from the rear of the chamber.

The camera uses a frame rate of 20,000 frames per second. The number of images that were recorded during the slide's recoil equates to the duration of the slide's recoil travel (Formula 5.3), and taking into account the total slide recoil travel (see Chapter 5.2 A.11, p. 230), the mean slide recoil velocity, empirical $\bar{v}_{Se}$ may be obtained from Formula 5.4 or 5.5.

$$t_S = \frac{n_F}{f_F} \qquad\qquad \textbf{5.3}$$

where: t_S Duration of the slide's recoil travel [s]
               f_F Frame rate [s^{-1}]
               n_F Number of frames

With the values obtained above for the total slide recoil travel, the mean slide recoil velocity is given by:

$$\bar{v}_{Se} = \frac{x_d}{t_S} \qquad\qquad \textbf{5.4}$$

where: $\bar{v}_{Se}$ Mean slide recoil velocity, empirical [m/s]
               t_S Duration of the slide's recoil travel [s]
               x_d Total slide recoil travel [m]

or by combining both Formulae 5.3 and 5.4, the resulting formula is:

$$\overline{v}_{Se} = \frac{x_d\, f_F}{n_F}$$

5.5

The value for the *mean slide recoil velocity, empirical* $\overline{v}_{Se}$ will be rounded to one decimal and is given in m/s (Appendix 19).

A.14 Magazine Release

Design specifics of the magazine release are to be noted under this category. Typical characteristics might be:

- unilateral magazine release;
- reversible magazine release (i.e. end-users can reverse the magazine release to either side of the frame depending on preference);
- ambidextrous magazine release;
- push-button magazine release; or
- paddle magazine release

A.15 Slide Stop

Design specifics of the slide stop or slide lock are to be noted under this category. Typical characteristics might be:

- internal slide lock
- unilateral slide stop
- ambidextrous slide stop

A.16 MSRP and Street Price

In the United States, the recommended retail price or list price of most consumer products is normally referred to as the manufacturer's suggested retail price (MSRP). It is the net price the buyer has to bear before adding taxes, purchase-related fees etc.

In addition to the suggested price, a *street price* is to be noted. Street prices are determined by checking offers of online retailers (see Appendix 20). The offer with the lowest price is to be included in the checklist. All prices are rounded to nearest integer in U.S. Dollars.

Please note: Purchasers need to have a Federal Firearms License (FFL) to purchase firearms and have them shipped directly, or they need to have the firearm delivered to an FFL holder such as a local shop. When shipping to a shop the buyer will be required to undergo the usual NICS background check as if the firearm were purchased from said shop.

B) Slide

B.1 Dimensions L×W×H, with (and without) Recoil Lug

Length, width, and height of the slide are precisely measured with a Vernier caliper to an accuracy of a 10th of a millimeter, the alignment is parallel or perpendicular to the slide's rails. Only the slide is to be measured, while removable attachments are ignored. The height is measured with and without the recoil lug (front end of the slide on which the recoil spring rests upon, Fig. 5.10). If sights are removable, they will be ignored when measuring the height. The height without recoil lug will be noted in parentheses.

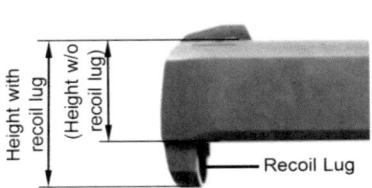

Fig. 5.10: Slide height definitions (schematic).

B.2 Slide Gross Weight

The recoiling mass of the slide is to be noted. This is the slide assemblies' weight, including all components that are typically attached to it when the gun is ready to use (i.e. gross), but without barrel and recoil spring. The weight is determined using a scale, model SARTORIUS *PRO 28/38C*. The weight is rounded to nearest integer in grams.

B.3 Manufacturing Technology, Slide

Findings with regard to the manufacturing method of the slide. Often manufacturing of slides is done by means of a material-removing process, stamping technology, MIM technology or a combination of several manufacturing techniques.

B.4 Retracted Striker

Specifies whether the striker nose will be withdrawn behind the breech face after firing a round. This is a feature which may contribute to reset a striker safety or to prevent an accidental discharge during feeding, such as when a cartridge is stripped from the magazine and pushed into the chamber.[599]

C) Barrel

C.1 Barrel Weight

The barrel's weight is determined using a scale, model SARTORIUS *PRO 28/38C*. The weight is rounded to nearest integer in grams.

C.2 Barrel Length

Barrel length is defined as the length of the barrel from its muzzle to the slide's breech face, with the slide closed. On barrels of the type shown in Fig. 5.11 the barrel length can be measured on the outside from the muzzle to the barrel's hood extension. The barrel length is measured with a Vernier caliper and rounded to one decimal of a millimeter.

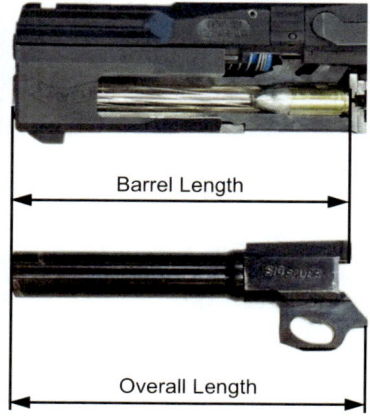

Fig. 5.11: Definition of *Barrel Length*.

In the case of a recessed breech face, the barrel length is measured from the muzzle to the breech face by pushing the depth rod of the Vernier caliper from the muzzle down to the breech face. Before doing so, the pistol must be cleared and slide closed.

C.3 Barrel Profile

Specifics of the barrel profile, such as lands and grooves, or polygonal shaping.

Interior dimensions (lands and groove diameter, or bore and groove area of polygonal barrels, twist rate etc.) are not to be measured. This information was considered irrelevant for commercially useful innovative features.

C.4 Manufacturing Technology, Barrel

Findings with regard to the production type of the barrel. For example, the production by means of a material-removing process, MIM technology, the use of a barrel sleeve, or a combination of different techniques.

D) Frame

D.1 Frame Gross Weight

Mass of the frame including all components (gross). If it is necessary to remove parts of the frame for field-stripping, then these parts will be taken into account for scaling. For instance, the weight of a removeable slide stop lever will be added to the weight of the frame.

The frame's weight is determined using a scale, model SARTORIUS *PRO 28/38C*. The weight is rounded to nearest integer in grams.

D.2 Manufacturing Technology, Frame

Often frames are manufactured by polymer injection molding. Sheet metal fabrication techniques, MIM technology, or a combination of several techniques may be found for instance, for components. The latter will be mentioned only, if the employed manufacturing technology or specifics on the components appears to be of particular interest for the survey.

5.3 CARACAL *F* (9 mm Luger)

The measurements and technical data according to
the checklist are summarized in Table A21.1,
p. 463. The gun was available in two sizes, namely
CARACAL *C* (compact) and CARACAL *F* (full size) and
the latter was surveyed. The manufacturer listed
193 × 28 × 135 mm (*L×W×H*) as outside dimensions. The measurement, however, showed
192 × 32 × 138 mm, which implies that the sample

Fig. 5.12: CARACAL *F*

was 1 mm shorter, 3 mm higher and at its widest
point, 4 mm wider than specification given by the
manufacturer.

When the trigger characteristic was tested, a short trigger
travel of only 5 mm was noticed. Trigger pull was 24 N. The
slide racking force amounted to 92 N, where racking of the slide
by hand proved to be of little comfort, since the height of the
slide serrations was limited.

Fig. 5.13: Striker in-
dent with drag.

The primer cup from fired casings of reference ammunition
did not exhibit severe markings (Fig. 5.13). The indentation
caused by the striker nose onto the copper crusher cylinder
was 0.24 millimeters. This figure was smaller than the desirable 0.3 millimeters (see
Chapter 3.6, Primer Initiation: Kinetic Energy, p. 106). Whether the 1.26 mm diameter
could compensate for the low depth of the indentation for the benefit of reliable ignition,
remained questionable and might be further examined by endurance tests.

The gun could only be field-stripped if the magazine was removed and the firing
mechanism had been decocked by pulling the trigger. The pistol had two safeties: a
striker safety and a trigger safety. In addition, a viewport served as a visual loaded
chamber indicator. A cocking indicator would signal the condition of the weapon. The
function of an out-of-battery disconnect could not be verified. The firing mechanism
could be released by pulling the trigger even when the barrel was not fully coupled to
the slide. Assembly of the CARACAL *F* was not possible with a loaded magazine in the
frame. Only when the user pressed the magazine's upper cartridge downwards, the
handgun could "accidentally" be assembled with a magazine locked in place.

The gun was equipped with a single-sided slide stop lever, and an ambidextrous magazine release. It came with two 18-round magazines, as well as a separate locking device. MSRP was $625 U.S.; while the handgun was available for $460 U.S. online as of September 2014.

The obtained values for the beginning and end of the uncoupling, and total slide recoil travel were 3.2 mm, 5.4 mm, and 50.1 mm respectively. Shooting the pistol gave a muzzle velocity of 351 m/s and a mean slide recoil velocity of 4.8 m/s. Mean slide recoil velocity calculated was 5.3 m/s.

The top of the slide (Fig. 5.14) seemed to be lower than on other pistols with low bore axis height. In shooting tests, the high tang grip seemed to have a positive effect on perceived recoil: muzzle jump was moderate and allowed the shooter to track the sights more accurately.

Fig. 5.14: Slide of CARACAL F

For the width of the slide, 28 mm was measured, while the height at the recoil lug was 31 mm. The weight of the slide assembly amounted to 337 g, and that of the barrel, 110 g. Barrel length was 104 mm.

Examination Regarding Innovative Features

The slide of the striker-fired pistol held an MIM'ed insert, which carried parts of the firing mechanism (Fig. 5.15). The front end of this MIM'ed insert formed the breech face, and its rear end's design resembled a fixed rear sight (Fig. 5.16). According to page 3 of a disclosure published under EP 1,490,643 B1, this preferred one-piece design would allow for a reduction in the height of the slide, since no additional flesh is needed for a dovetail-cut to attach a rear sight.

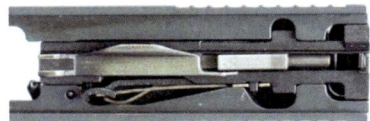

Fig. 5.15: MIM'ed insert with members of the firing mechanism.

Also, the inventor, Wilhelm Bubits named in patent EP 1,601,923 B1 dated February 9, 2004, discloses for the patent holder CARACAL, Abu Dhabi, the specific embodiment of a leaf spring. According to the disclosure, this spring is also attached to the MIM'ed insert and is used to mount a load on the striker safety as well as the extractor.

Fig. 5.16: Rear Sight (left) and Breech Face were part of the insert.

The spring disclosed in the patent was located on the right side of the MIM'ed insert (Fig. 5.17). It was affixed by two studs on the insert. The lower leg of the leaf spring would put a load on the striker safety while the upper leg would push against the extractor.

Fig. 5.17: Leaf spring with two legs, on the insert's right side.

The striker safety was designed as a stamped and bent part. The safety was on the right side of the striker and was laterally deflected by a leg of the trigger bar (Fig. 5.18).

Fig. 5.18: Striker safety and striker shown from above.

Basically, all small parts of the gun, such as striker and insert, were designed for MIM production technology, and thinner components designed to be manufactured as sheet metal stampings (Fig. 5.19). In contrast to other striker-fired pistols, the striker spring was not concentrically mounted on the striker, but above it, parallel to the striker on a separate guide rod. According to EP 1,490,643 B1, page 2, this design is advantageous to use. The constructive possibilities of the slide insert are described to work more efficiently, and this design would allow more space for the mainspring.

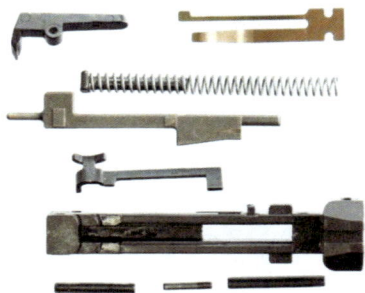

Fig. 5.19: Insert and components of the firing mechanism.

Barrel and slide were machined finished parts. Using the slide insert with integral breech face, the manufacturer avoided a separate machining operation of the breech face, which is usually required in the production of conventional pistol slides (Fig. 5.20). However, this approach shifted tolerances and the headspace dimension in a complex shaped MIM'ed part which can be subjected to tolerances in production accordingly. On the studied gun, the insert had nearly no clearance on the inside of the slide. Based on this observation, it was assumed that the adjustment of the tolerances and the reliable production of functional surfaces that are part of different members would be a challenge for CARACAL.

Fig. 5.20: Slide with insert (bottom) and without.

The frame of the CARACAL F was injection molded without molded-in rail inserts. The slide rails were part of a removeable chassis (Fig. 5.21), which carried a locking block, ejector and trigger assembly as well as other small parts. Two pins secured the chassis to the frame.

Fig. 5.21: Chassis with rails, ejector and components.

The separate locking device (Fig. 5.22) was shaped similarly to a magazine and could be inserted into the magwell until it engaged the magazine release latch. As a result of a large engagement edge on the locking device, the device could not be removed anymore from the gun by pressing the magazine release. A key was needed whenever the device was to be removed. The key was of the same kind used on regular door locks. According to the manufacturer, a replacement key can be obtained if a security card is presented. This card came with the gun at the time of purchase.

Fig. 5.22: Locking device and keys.

The internal space of the magazine was optimized to provide enough room to accommodate 18 cartridges. To accomplish this, the designers eliminated the locking plate and instead, the magazine spring engaged the magazine base plate to keep it in position (Fig. 5.23). This design reduced the usual number of parts of the magazine from five to four.

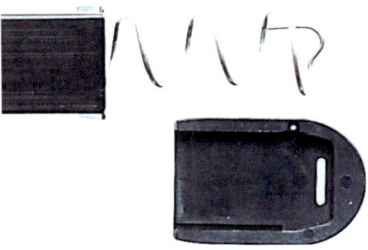

Fig. 5.23: Magazine spring and base plate.

The disconnector of the CARACAL F would move transversely to the direction of the pistol (Fig. 5.24). The firing mechanism could be released when the breech was not locked; but at the same time, the striker safety was still engaged and prevented the gun from firing.

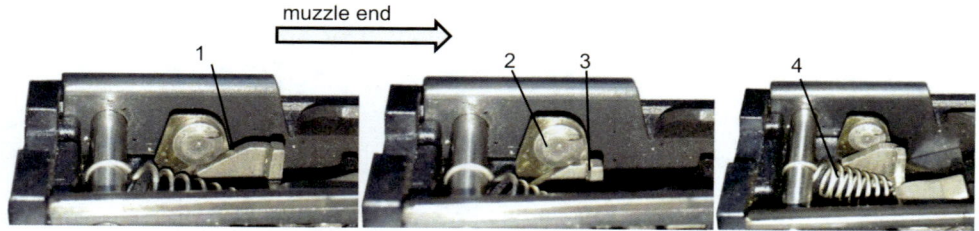

Fig. 5.24: Firing a shot and disconnect function
The ramped end (1) of the trigger bar is cammed downward as it hits a pin (2). This lowers the sear (3), which is part of the trigger bar, until the striker (not shown) is finally getting released. While the slide recoils, it pushes the spring-loaded pin to the left, and the traction spring (4) pulls the trigger bar far enough up to engage the striker as the slide closes.

Patent research CARACAL

The research focused on patent applications of the company CARACAL and registrations under Wilhelm Bubits, who worked for CARACAL as a product engineer.

According to a patent description, the optimal utilization of the internal volume of a magazine body should be achieved through optimization of the magazine spring and how the magazine base plate is kept in place. The magazine capacity of 18 cartridges should be realized by a magazine spring which occupies less space using coils, which are tapered tighter, and by eliminating the locking plate. Patent US 7,797,871 B2 reveals both ideas; claim 1 (c) in particular refers to the reduction of the magazine spring's solid length using multiple smaller spring coils which fit into the rest of the spring's larger diameter coils (see item 22 in Fig. 5.25) when the spring is compressed.

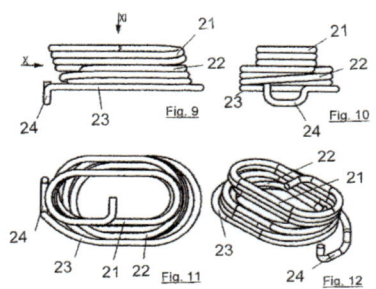

Fig. 5.25: US 7,797,871 B2, sheet 3

International application WO 2007/090,219 A1, dated February 6, 2007, by assignee CARACAL, describes a feature where a vertically pivotable lever puts a load on the top of the barrel's chamber end (Fig. 5.26) to eliminate any vertical play of a moveable barrel.

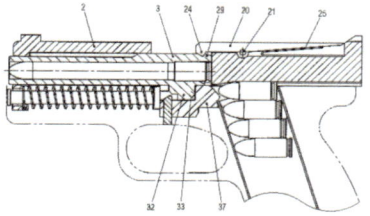

Fig. 5.26: WO 2007/090,219 A1, sheet 4

In WO 2010/065,911 A1, dated December 7, 2009, Bubits discloses for CARACAL a slide for a striker-fired pistol similar to the CARACAL F, but in contrast to the CARACAL F, the breech face is part of the slide and not part of a removeable insert (item 10 in Fig. 5.27). The application also discloses an insert, which could preferably be made of polymer.

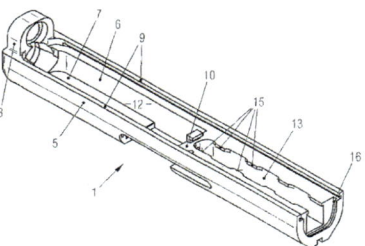

Fig. 5.27: WO 2010/065,911 A1, page 15

With patent AT 511,266 B1 2012-10-15, dated April 4, 2011, Bubits discloses a magazine release for assignee MERKEL JAGD- UND SPORTWAFFEN GMBH. The ambidextrous magazine release of the CARACAL F (WO 2005/057,121 A1) comprises a push bar and two bending rods (Fig. 5.28). As in other pistols, the magazine rests on the magazine release's push bar. Bubits changed this principle: he uses only one bending rod instead of two; the single rod has a bent upper end. This bent upper end then acts as an engagement for the magazine. The magazine release latch serves solely as a push-button, which is operated from either side.

Fig. 5.28: CARACAL F magazine release with two bending rods.

On April 4, 2011, Bubits discloses a slide made of profiled material. The specific feature of the disclosure is the focus on minimal costs of materials for the slide, as there is no material for the slide's recoil lug involved. Patent specification AT 510,291 B1 2012-03-15 explains that the U-shaped material can be produced sufficiently by extrusion or pultrusion with precise detail. After cutting bars of suitable length from the extruded material, recesses are milled on the inside of the blanks (Fig. 5.29). The front recess holds a slide recoil lug insert, which may include the barrel bushing, a recess in the middle takes a breech face insert, and the rear recess holds a polymer insert for the firing mechanism.

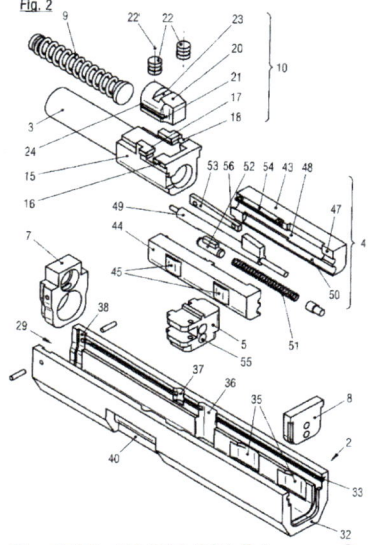

Fig. 5.29: AT 510,291 B1, page 8

According to the disclosure, the front insert can be fixed to the slide with two pins. Bubits proposes a roller lock slide and emphasizes the adaptability to different cartridge types. The adaptability is supposed to be made possible by interchangeable components inserted into the U-shaped slide profile. However, Bubits does not discuss the effects of tolerance stack-ups in the explanations of the patent.

In patent AT 511,979 B1 2013-08-15, Bubits reveals a cartridge in the caliber range from 7.5 to 8 mm (Fig. 5.30) with a bullet weight of 4 to 5 grams. Filing date is October 4, 2011, and assignee is the company MERKEL JAGD- UND SPORTWAFFEN GMBH. It's a bottle-necked cartridge. The overall length and gas pressure is suggested to be within the specifications of the 9 mm Luger. Objective of the invention is to achieve higher performance than can normally be achieved with a 9 mm Luger cartridge despite its similar overall length.

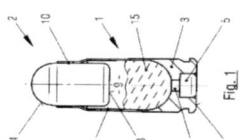

Fig. 5.30:
AT 511,979 B1,
page 5

Summary CARACAL *F*

The shape of the grip, which allowed the hand to be positioned more in line with the slide as it reciprocates and the low bore axis height seemed to reduce perceived recoil. The capacity of the magazine proved to be remarkable; in practice, however, loading the last two cartridges could only be achieved with considerable force.

The CARACAL *F* seems to be designed for low-cost manufacturing. A remarkably large scale of polymer injection molded parts was used, stampings, and MIM'ed parts. Whether this concept proves successful, and can be handled reliably in full production was doubted. The CARACAL *F* avoids a demanding machining of the breech face, but important dimensions for headspace are defined by a MIM'ed part, which is inserted into the slide with applicable dimensional tolerances. The control of the tolerances and the process stability in the production process of functional surfaces arranged in various components are likely challenges for CARACAL.

The patent research resulted in patents by Wilhelm Bubits, which revealed opportunities for cost-effective manufacturability of weapon parts. Whether the ideas of Bubits will take a full production of parts into account, could not be concluded ultimately by studying the patents.

5.4   GLOCK *G19 Gen4* (9 mm Luger)

The measurements and technical data according to the checklist are summarized in Table A21.2, p. 464. An actual width of almost 32 millimeters was measured; this value is two millimeters above the manufacturer's specification. A similar discrepancy was observed with respect to trigger pull. According to the manufacturer's specifications, the trigger pull is 25 N, and this value could not be verified. Instead, 41 N was measured in conjunction with 8 mm trigger travel. 115 N were required to manually move the slide. This was the highest slide racking force observed for the sample guns in this survey. Nevertheless, operating the slide could be done comfortably. Maybe, the generous dimensions and straight sides of the slide have a beneficial effect on grip.

Fig. 5.31: Striker indent with bulge and striker drag.

The GLOCK's lance-shaped striker nose delivered a 0.34 mm deep, rectangular indentation in the copper crusher cylinder. The overall dimensions of the indent were 0.60 × 0.96 millimeters. Fired casings' primer cups of reference ammunition exhibited a rectangular bulge around the striker nose's oval indentation (Fig. 5.31).

GLOCK pistols typically sport three safety mechanisms, so also here: a guided trigger bar in combination with a striker safety and a bladed trigger safety. The design of the weapon allowed for the release of the firing mechanism only, if the breech was locked, this was rated as an out-of-battery disconnect safety. In addition, the extractor served as a loaded chamber indicator; a cocking indicator did not exist. Field-stripping, disassembly and assembly was only possible with magazine removed from the pistol.

The gun has had a unilateral slide stop lever, a reversible magazine release, and included two 15-round magazines. The manufacturer's suggested retail price was $649 U.S.; as of September 2014, this model was available online for a price of $538 U.S.

The obtained values for the beginning and end of the uncoupling, and total slide recoil travel of the gun at hand amounted to 2.5 mm, 5.0 mm and 46.9 mm respectively. During shooting of the pistol, a muzzle velocity of 358 m/s and a mean slide recoil velocity of 4.7 m/s were observed. The theoretical mean slide recoil velocity resulted in a value of 5.4 m/s.

In respect to the slide, a width of 25.5 mm and a height of 32.9 mm was measured at the recoil lug. The weight of the slide assembly was 347 g, the barrel 104 g, with a barrel length of 102 mm.

Examination Regarding Innovative Features

In the 1980s, Gaston Glock introduced a pistol into the market that delivered a scope of features that was unique within the industry as well as from a weapon technology standpoint. Maybe his design was ahead of its time, or why else would competitors underestimate and smile at the gun, while at the same time consumers were reluctant to accept it?

The latter would hold true, however, only in the first few years after launch. Meanwhile, the products of this company have written a success story that is second to none. Competitors learned a lot from the market success of Gaston Glock's pistols, and caught up in the meantime. As a result, the actual difference in performance of products vanished over the years. Nevertheless, the reputation of GLOCK pistols is still enormous. Thanks to good positioning of the brand, a real difference in the performance of competitor's pistols is less important than the perceived differ-

Fig. 5.32: GLOCK *G19 Gen4*

ence of consumers.[600]

When Gaston Glock introduced his pistol, he differentiated it from offerings from established brands by combining injection molded polymer parts and ease of use. In doing so, he created a new market that today represents a worldwide standard: polymer pistols without a manual safety but automatically working safeties instead, in addition to consistent trigger pull for each shot. In the product information for the U.S. market, the company con-

Fig. 5.33: GLOCK Safe Action Fire-control System, shown during disassembly.

tinually repeats its mantra: "polymer receiver, independent automatic safeties and GLOCK safe action fire-control system."[601] The fire-control mechanism of the safe action system consists of five parts: the striker, the trigger assembly with trigger spring, a connector (i.e. disconnector) and the trigger mechanism assembly (Fig. 5.33). The compact design of the trigger mechanism assembly with its limited number of parts and its uncomplicated form in combination with the efficient manufacturability of the gun has been a role model for other manufacturers since its introduction along with the idea of having the "connector," i.e. a member of the firing mechanism, move crossways to the orientation of the pistol during the firing sequence.

Since 1986, GLOCK pistols have been imported from Austria into the United States.[602] GLOCK pistols hold a market share of 65 percent[603] of the law enforcement market, and in 2013 the record of 10 million[604] manufactured pistols was exceeded. All these years, GLOCK held on to the original concept of the gun's design and revised the product family only moderately through sustained engineering. There is one exception: To satisfy the distinctive demands of the German authorities, GLOCK developed a model *G46* duty pistol in 2017 which differs from the rest of the platform.

In August of 2017 the company headquartered in Deutsch-Wagram, Austria, announced a Generation 5. Its predecessor, the GLOCK Generation 4 (Gen4), was introduced in January of 2010 [605]. GLOCK created the Gen4 platform based on the .40 S&W round and models chambered for less powerful cartridges followed, derived from this basic design. While some manufacturers base the initial design of a new pistol model on a popular cartridge like the 9 mm Luger, GLOCK's approach is said to avoid the risk of durability problems when a model is chambered for a more powerful ammunition type later in the process.[606]

The model surveyed was a GLOCK *G19 Gen4*, that is, a GLOCK pistol chambered in 9 mm Luger in a size specified by GLOCK as "compact." The obvious advantages of the Gen 4 according to the manufacturer was: a smaller grip handle,[a] a recoil spring unit (Fig. 5.34), which reduces the recoil transmitted to the shooter, a larger magazine release button,[607] which is now reversible from left-side operation to right-side operation, the grip texture, and an adjustable grip size.

Fig. 5.34: *G17 Gen3* recoil spring assembly vs. *G19 Gen4* dual recoil spring assembly (below).

Included in the box of the *G19 Gen4* pistol were additional backstraps in sizes medium and large. A full pistol grip formed the basis of the adjustable grip handle, and

---

[a] So-called "Short Frame" (SF), with 70 mm trigger reach. Other pistol model's trigger reach is typically in the range from 72 to 75 mm.

was slightly smaller than the grip of previous generations. The separate backstraps were designed as additional attachments on this fully functional grip.[608]

The gun came with a total of two pairs of backstraps. The first pair served exclusively to make the grip handle larger increasing trigger reach at the same time. The second pair were of the same size, but additionally, each one incorporated a beavertail at the upper end (Fig. 5.35), which could be used to extend the frame's tang rearwards.

Fig. 5.35: Large Backstrap with Beavertail.

The combination of the two, the idea of an adjustable grip size in combination with a complete grip handle and backstraps with beavertail, could be seen as a "small innovation" on a GLOCK pistol. GLOCK technology has hardly changed during the last 30 years and grip ergonomics has only recently been raised to the level of products of other manufacturers with the introduction of the third generation in late 1998 by introducing finger grooves on the grip handle. GLOCK calls the changes from 1982 until 2012 "the evolution of a revolution" (Fig. 5.36).

Fig. 5.36: G17 Gen4 Anniversary Model "The Evolution of a Revolution."

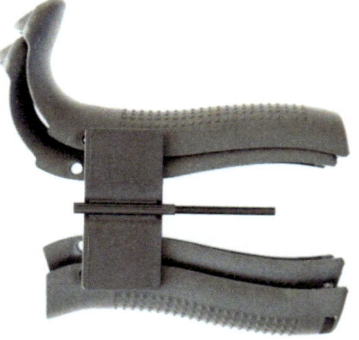

The various backstraps were not included separately in the box, but snapped on to a backstrap carrier (Fig. 5.37). The carrier is described by GLOCK as a "multifunction clip."[609] It could be used as a tool for replacing the backstrap and also provides an additional firing group housing pin for the attachment of the backstraps on the frame.

Fig. 5.37: Multifunction Clip with polymer pin punch, backstraps and additional firing group housing pin.

Putting a backstrap on the frame proved to be self-explanatory: the backstrap had locking lugs at its lower end for fastening it at the heel of the gun's grip, and the backstrap's top end had to be fixed with the additional firing group housing pin (Fig. 5.38).

Fig. 5.38: Locking lugs on the lower end of the backstrap.

The rear sight of the GLOCK *G19 Gen4* provided both elevation and windage adjustments.[610] A miniature screwdriver for the adjustment of the rear sight was included in the scope of delivery. This type of rear sight has existed since the 1980s, and was made of polymer with a compact design (Fig. 5.39).

According to the Austrian gun writers Mötz & Schuy, the barrel and slide of GLOCK *Gen4* pistols are gas nitrided. Previous generations have been tenifer treated since 1984.[611] There was a polymer liner found to be inserted inside the slide's striker channel. As described in the gun literature, this liner should provide a defined smooth surface inside the striker channel, a surface finish that is not possible in this quality just by drilling.[612]

Fig. 5.39: Adjustable rear sight, shown with mini screwdriver attached.

Since the magazine release of the GLOCK *G19* Gen4 was reversible from left side to right side operation, the magazine body had two notches accordingly, for the engagement of either orientation of the magazine release. In the area of the cut-outs for the engagement, a sheet metal part inside the polymer magazine body was clearly visible (Fig. 5.40). There's possibility of variation in the production of magazines with different interior volume, but keep-

Fig. 5.40: Metal insert in magazine body (full metal lined mag).

ing the same external dimensions appeared remarkable, that is, the magazine well dimensions inside the pistol grip were independent of the cartridge size. Noteworthy too, is the fact that they used polymer surfaces where the magazine release engaged.

According to *Sweeney*, the polymer engagement surfaces have the reputation of a lower load carrying capacity in contrast to components made of metal, and the polymer material of the mag body would require a counterpart made of similar soft material. In particular, when a fully loaded magazine is in the gun, and, for example, when the user takes it out by pressing the mag release, it can cause a maximum load of the affected planes,[a] right at the moment, where the resting surface of the magazine release moves far enough until the planes are just in contact at their edges. In this position, the edges bear an extreme load and wear eventually results.[613]

The engagement plane on conventional magazines would typically be part of the sheet metal body, which would be manufactured by stamping a cut-out. This opening however, might allow dirt to enter into the magazine body, and may result to malfunctioning. The front of the *G19* magazine was indeed closed, but the back wall had openings that were witness holes (Fig. 5.41).

Fig. 5.41: Rear wall with viewing ports.

In order to reverse the magazine release, the magazine release spring had to be pried out from the magazine release (Fig. 5.42). Then, the magazine release could be removed from the frame, inserted from the opposite side, and the spring could be placed back on.[614] This principle was already used on the first generation of GLOCK pistols, however, the magazine release was not reversible at that time.

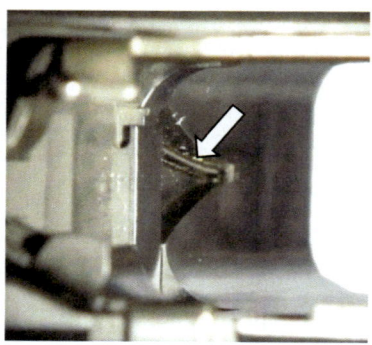

Fig. 5.42: Magazine release spring.

The barrel had a polygonal profile with six "lands" and "grooves" (Fig. 5.43),[615] and had a two-dimensional code at the bottom of the feed ramp. A similar code was located inside the frame under the firing group housing. On other components of the weapon, markings existed in the form of numbers (Fig. 5.44). No other sample in this survey had more parts with identification marks than found on this GLOCK pistol. Maybe these markings serve for the traceability of parts when constructive changes are introduced. For both, users and manufacturers, it

Fig. 5.43: Barrel profile of a *G19 Gen4*.
Note: apparent residue on both sides of "lands."

---

[a] When a magazine is in a pistol with slide closed, the magazine's top cartridge is pressed against the slide, and the magazine release is the counter bearing for this load. The load is caused by the magazine spring, and increases with the number of rounds in the magazine.

represents a real improvement if there are clearly visible marks to identify the construction status of components. Manufacturers, who have already had to carry out a recall, might appreciate this advantage.

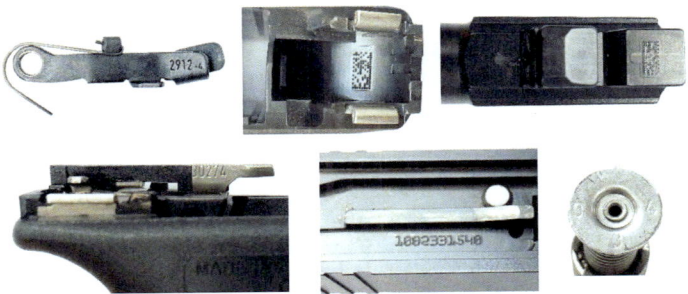

Fig. 5.44: Markings on components:
  Slide Stop Lever, Frame, Barrel, Recoil Spring Assembly, Slide and Ejector of a *G19 Gen4* (clockwise).

Is the GLOCK *G19 Gen4* a full size or a compact size pistol? NIJ tries to submit a definition for pistol sizes. According to the organization, a full size pistol is defined as a pistol with a barrel length from 4.26 in to 5.2 in [108 to 132 mm; PD], whereas compact pistols shall have a shorter barrel with a minimum length of 3.75 in [95.3 mm; PD], and minimum magazine capacity would be 16, or 14 cartridges respectively.[616] The answer to the initially asked question is: compact.

Apparently, the *G19's* 102 mm barrel fits smoothly into NIJ's definition, but this is just a coincidence. There is no general industry standard either for pistol size categories or for their appropriate designations or barrel length. Sometimes, this causes confusion within the gun community. This usually occurs when a manufacturer offers pistols in only two sizes: often the normal size is then called standard size (full size) and the smaller version of this model is referred to as a compact, whereas GLOCK might list the same sizes under compact and subcompact. GLOCK's product range, with its five size categories, can be considered to be the template for the identification of size classes. GLOCK is sizing its pistols as shown in Table 5.2.

Table 5.2: Size Categories of GLOCK pistols.

| | Caliber | | | | | |
| | .380 ACP, 9 mm Luger, .40 S&W, .357 SIG, .45 GAP | | | 10 mm Auto, .45 Auto | | |
	Length [a] [mm]	Height [b] [mm]	Width [mm]	Length [a] [mm]	Height [b] [mm]	Width [mm]
Subcompact	163 – 167	106	30	175 – 179	113 – 122	32.5
Compact	185 – 189	127	30	N/A	N/A	N/A
Standard	202 – 206	138 – 140	30	204 – 208	139	32.5
Slimline	159	108	26	177	122	28
Competition	222 – 226	138	30	226 – 230	139	32.5
Long Slide	N/A	138	30	241 – 245	139	32.5

[a] Variation in length is caused by backstraps in increments of 2 mm
[b] Height with magazine

Patent research GLOCK

Patent application EP 1,327,849 A1, dated November 28, 2002, relates to a method for producing at least one marking on the inner side of a barrel. If a projectile was to be recovered by forensics, such markings could help to identify the barrel it was fired from. The objective of the application is to transfer traits to the projectile as it runs down the bore. The invention relates to three areas: A method, a device to apply the method and barrels manufactured by this method. According to the pa-

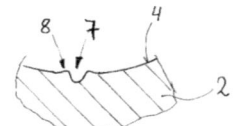

Fig. 5.45:
EP 1,327,849 A1,
p. 5: Groove (7)

tent, some of the hitherto known methods are not suitable for practical applications, since they affect the dimensional accuracy of the bore or that the production costs are rather too high. The method disclosed describes that at least one marking that follows the rifling is produced in the barrel during manufacture by at least a short section from the muzzle end, but not over the entire length of the barrel. The marking is groove shaped and a code is generated by correspondingly arranging the grooves within pre-determined circumferential distances (Fig. 5.45).

A disclosure DE 10,2012,017,637 A1, dated September 6, 2012, also addresses markings inside barrels. The disclosure describes bars of the same caliber diameters inside of polygonal barrels. According to the disclosure, these bars are shaped similarly to lands of a conventional rifled barrel.

The EP 2,660,551 A2, dated March 25, 2013, relates to pistols for smaller calibers or less powerful ammunition, where a part of the slide is made of polymer (see p. 73), and GLOCK's patent AT 512,771 B1, dated July 9, 2012, discloses a certain method to attach the backstrap to the frame by means of an undercut at the top of the backstrap, which will attach to the upper end of the frame (Fig. 5.46).

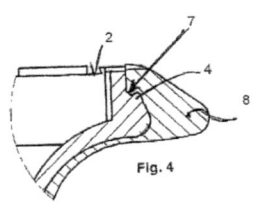

Fig. 5.46:
AT 512,771 B1,
page 6

Patent application US 2017/0,198,993 A1, filed December 9, 2016, by GLOCK TECHNOLOGY GMBH, discloses a pistol with rotary barrel. Model *G46* uses a rotary barrel.[617]

Summary GLOCK *G19 Gen4*

Depending on their design, pistols with an adjustable grip size offered by some manufacturers cannot be gripped comfortably without a backstrap attached, which makes the use of the firearm difficult in the event where a backstrap got damaged or is lost. The idea of GLOCK, to build a grip of the smallest size to be a complete grip handle, can be a buying feature for some prospective users. The option of having multiple backstraps which include an easy to install beavertail might be a selling feature as well.

The identification of individual gun parts can be considered exemplary within the industry. The field-strip procedure, however, was less exemplary: like with any GLOCK pistol, the trigger had to be pulled to de-cock the firing mechanism prior to disassembly.

The fourth generation of pistols was upgraded externally, apart from the adjustable grip size, by means of a modern texture on the grip handle and a reversible magazine release, but genuine innovations could not be found on the GLOCK *G19*. Nevertheless, the success of this brand with more than 10 million pistols sold is a fact that cannot be ignored.

5.5 HECKLER & KOCH *VP9* (9 mm Luger)

The measurements and technical data according to the checklist are summarized in Table A21.3, p. 465. The manufacturer's technical data[618] stated 186.5 × 33.5 × 137.5 millimeters. Actual measured data differed in the range of tenths of millimeters; rounded to nearest integer (187 × 34 × 138 mm), the values were then identical. The same was true for the data of the trigger action. The values measured, 6 mm and 24 N, confirmed the values given by the manufacturer for trigger travel and trigger pull respectively.

Models of *VP9* are offered in Europe under the model name *SFP9*. Because the *SFP9* can be regarded as a service pistol, trigger work was also considered. A minimum of 150 Nmm[619] is required by the German TR "Pistolen," but the evaluation revealed 67 Nmm. Thus, the sample *VP9* of this survey would not qualify as a duty pistol in Germany. The slide racking force was at 91 N, which was within the acceptable range of not more than 100 N required by LE in Germany.

Fig. 5.47: Striker indent with drag.

The striker nose of the *VP9* produced a 0.31 mm deep marking with a diameter of 1.23 mm in the copper crusher cylinder. The fired primers of reference ammunition exhibited a small drag at 12 o'clock of the indent (Fig. 5.47). The barrel had a stepped chamber, which left a ring-shaped mark around the case's neck (Fig. 5.48).

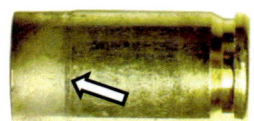

Fig. 5.48: Stepped chamber ring-mark.

The *VP9* had a trigger safety and a striker safety, as well as a loaded chamber indicator and a cocking indicator. The constructive design of the trigger mechanism allowed the release of the firing mechanism only when the breech was locked. This was assessed to be an out-of-battery disconnect. Field-stripping was only possible with magazine out of the pistol and with slide back. Assembly could only be done if no magazine was in the gun. The gun came with an ambidextrous slide stop lever and an ambidextrous magazine release; two 15-shot magazines were included.

The *VP9* pistol started selling in the United States in July 2014. The MSRP was $719 U.S., while online the gun retailed for $575 U.S. as of September 2014.

The obtained values for the beginning and end of the uncoupling and total slide recoil travel were 4.6 mm, 7.7 mm and 47.1 mm respectively. When shooting the pistol, a muzzle velocity of 365 m/s and a mean slide recoil velocity of 5.0 m/s were recorded. Mean slide recoil velocity calculated was 5.6 m/s.

The width of the slide measured 28.8 mm while the height at the recoil lug was 33.5 mm. The total weight of the slide assembly was 352 g while the weight of the barrel was 98 g, with a barrel length of 104 mm.

Examination Regarding Innovative Features

The pistol fired each shot in single-action. In other words, HECKLER & KOCH offered a pistol with a consistent trigger pull for every shot. The trigger did not have a restrike capability. HK claimed the trigger take-up weight to be less than those found on other pistols, the point of break could be clearly felt, and the reset travel to be short.[620] On the gun on hand, the trigger take-up weight was approximately 10 N (see Appendix 16). According to the user manual, trigger reset distance should be 3 mm[621]; reset was not measured.

Fig. 5.49: HECKLER & KOCH *VP9*

The user manual explained the gun would only disassemble with magazine removed and slide locked back. This was called a fourth safety feature, in addition to striker safety, trigger safety, and disconnect function.[622] The decocking of the firing mechanism was automatically done during disassembly practice; however, this function would cause the slide to halt after about two millimeters, when the automatic decocking feature began to function. At this point the user had to pull harder to get the slide off the frame.

Fig. 5.50: Main assembly groups.

In contrast to other pistols was the presence of a recoil spring assembly (Fig. 5.51) fully made of metal. It consisted of three parts: a flat wire recoil spring, a washer, and a guide rod. The front end of the guide rod had a circumferential undercut and was slotted in longitudinal direction. This formed a flexible abutment for the washer, and it seemed it could be mounted in the manufacturer's assembly line with little effort.

Fig. 5.51: Snap-fit on recoil spring assembly (shown with spring slightly retracted for clarity).

Generally, the components of the weapon consisted of sheet metal, MIM'ed, and polymer parts. The largest polymer part of the pistol was the frame.

The grip handle was adjustable in both hand fit and trigger reach (Fig. 5.52). The adjustment in width was provided by grip panels and trigger reach through palmswells. The inserts were attached by a tongue-and-groove joint type of connection, and the entire group was secured with a pin at the heel of the grip handle. Each insert came in three sizes.

Fig. 5.52: Interchangeable grip panels and palmswells.

The polymer frame carried a front assembly, which contained trigger components, and a rear assembly that held parts of the firing mechanism. The housing of these assemblies consisted of U-shaped sheet metal stampings, which simultaneously served as slide guides. Part of the front assembly was a block, which controlled the coupling/uncoupling of the barrel (Fig. 5.53). This locking block was welded to the U-shaped stamping; the welding

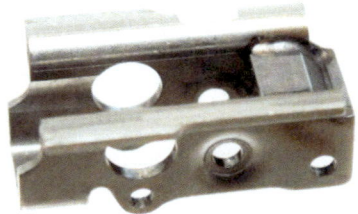

Fig. 5.53: Front rail insert with punched holes, throat and welded-in locking block.

seams were clearly visible on both sides of the stamping. The side walls of this housing had holes as well as a throat on each side which served as a bearing for the shaft of the right side slide stop lever. A recess was milled in the right side to carry a torsion spring. Milling toolmarks on the slide guides suggested a secondary operation was used to manufacture an exact width of the rails.

The sear assembly was self-contained in a polymer housing and this was then inserted as a sub-assembly into the rear insert (Fig. 5.54). A pin kept the sub-assembly in place, and the complete rear insert was mounted into the frame. The entire rear insert was secured by a second pin. The latter was visible on the pistol from the outside.

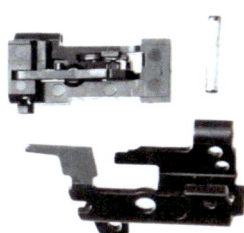

Fig. 5.54: Pin, firing mechanism in polymer block, rear insert with ejector (below).

The ambidextrous slide stop lever arrangement consisted of four components: right side and left side slide stop lever, a polymer retainer clip, and a compression spring. The right side slide stop lever was a sheet metal stamping of long and slender appearance, with a welded joint shaft on its front end. The form-fitting connection to the left slide stop lever was carried out by a cross-shaped fit (Fig. 5.55).

Fig. 5.55: Clip, right and left slide stop lever (clockwise).

The shaft of the slide stop also served as bearing for the trigger and carried the trigger's torsion spring. The previously mentioned retainer clip secured the shaft against lateral movement, and was installed from the top of the frame into a recess of the shaft.

The slide stop spring (Fig. 5.56) was located on top of the left slide stop lever and was covered by the front insert's slide rail.

Fig. 5.56: Slide stop spring

The VP9's trigger safety was integrated into the trigger blade in a similar manner to the triggers of GLOCK pistols (Fig. 5.57). The lower end of the trigger blade reached into a long recess inside the trigger guard. It was assumed that this construction detail should prevent snagging a user's finger and jamming the trigger as a consequence when shooting with gloves.

Fig. 5.57: Trigger safety and paddle magazine release.

At the top of the trigger blade, where it bridged into the frame, distinct openings were observed (Fig. 5.58). If these openings can be compromised by lint, sand and dust, it might lead to malfunctions; however, this theory was not further investigated.

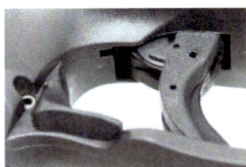

Fig. 5.58: Openings in the trigger area.

The magazine release paddles were on the left and right side of the trigger guard (Fig. 5.57 and Fig. 5.58). The levers were short and had to be pushed downward to release the magazine. This design is known with the HECKLER & KOCH P7. The paddle-design may prevent an unintentional release of the magazine when the pistol is carried in a Holster, because the paddle does not respond to lateral pressure; a condition which could occur when carrying inside the waistband or wearing a seatbelt while driving.

Fig. 5.59: Loading indicator holes and front side with engagement notch (below).

The magazine (Fig. 5.59) could be disassembled without any special tools. A 9 mm cartridge was enough to push the locking plate up into the magazine in order to remove the base plate (Fig. 5.60). The magazine base plate had a hole large enough to do this field-strip using a 9 mm round.

The magazine body's front (Fig. 5.59) was marked "Made in Germany," and unlike various other mags it had no openings for the mag release. This was achieved by an outward curved beading which was machined to form a notch at the bottom for the engagement of the magazine release.

Fig. 5.60: Magazine base plate with amply locking detent for the locking plate.

The rear wall of the magazine body provided viewing ports. They were commonly known holes in one line on top of each other on the rear wall's left side Fig. 5.59.

On the slide, two interesting features were identified: loading aids that extended up on the slide, and a swivel-mounted striker safety. What the manufacturer calls "charging supports," were polymer ridges at the rear end of the slide serrations on both sides of the slide. According to HK, these charging supports enable better purchase on the slide to ease the

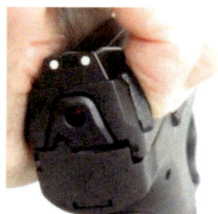

Fig. 5.61: Racking the slide. Shown here: overhand rack.

effort needed to retract the slide. This would be beneficial, especially for people with reduced hand strength.[623] The described benefit was clearly noticeable. The ribs provided a good abutment when working the slide slingshot style or by overhand rack (Fig. 5.61 and Fig. 5.62).

The charging supports were replaceable, described HECKLER & KOCH. The rear sight must be removed from the slide to install flush-fitting inserts. Flush inserts were not included and the rear sight could not be removed without tools.[624] Front sight and rear sight were drift adjustable. The sights had three phosphorescent dots.

The extractor doubled as a loaded chamber indicator (Fig. 5.63). For this function, it had a red marking on the top of its front end. If the extractor claw held a cartridge, the extractor's front end slightly protruded outwards and a red bar indicated a round was in the chamber.

On the rear of the slide was a cocking indicator. The back of the striker was marked in red and the tip was visible through a hole in the slide plate when the striker was cocked. (Fig. 5.64).

The technical implementation of the striker safety appeared to be an interesting solution. In contrast to vertically moveable safety plungers, the *VP9* used a horizontally swivel-mounted striker safety. In doing so, the striker on the *VP9* was blocked by a pivot-mounted lever. As the trigger bar moved rearward, it swept the safety lever to the side, disengaging the safety (Fig. 5.65).

Fig. 5.62: Rear sight covers charging support inserts.

Fig. 5.63: Extractor also serves as a loaded chamber indicator.

Fig. 5.64: Cocking indicator.

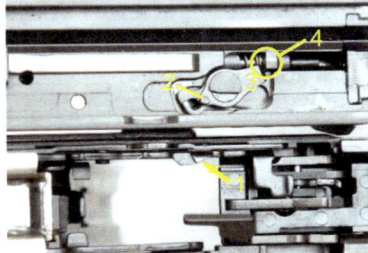

Fig. 5.65: Ramp on trigger bar (1), its counterpart (2) on the striker safety (3), which blocks the striker (4).

The firing pin assembly comprised of six parts: striker, mainspring, retract spring, a polymer sleeve and two polymer keepers (Fig. 5.66). The keepers were fluted and the striker was made of solid metal.

Fig. 5.66: Striker assembly components: Keepers (1), striker (2), mainspring (3), retract spring (4), sleeve (5).

Patent research HECKLER & KOCH

Patent DE 10,2013,022,080 B3, dated December 23, 2013, reveals the break down function of the *VP9* pistol. The invention relates to a disassembly without additional steps to decock the striker-fired action. This is obtained by a trigger blade that pivots forward as the take-down lever is swiveled downwards for disassembly (Fig. 5.67 and item 5 in Fig. 5.68), which simultaneously brings the trigger bar to the front as well. A leg (24) on the trigger bar rides below a disruptor lever (21) and is cammed down as soon as the slide is being pulled off the frame. A cam (33) on the disruptor lever introduces this downward movement, which in turn is controlled by the slide's forward movement.

Fig. 5.67: Position of the trigger blade during disassembly (standard position shown transparent).

According to the invention, the take-down lever can only be swiveled downward for the disassembly of the pistol, if the slide is locked back and no magazine is present in the pistol; for the latter, see also WO 2015/000,773 A1.

The horizontal pivot striker safety or drop safety is described in HK's disclosure WO 2015/096,891 A1.

When HECKLER & KOCH announced its *VP9* pistol, a patent for the charging handles was mentioned. The first claim in their publication WO 2014/180,574 A1 is: "[...] characterized in that a loading aid [..] underneath a sighting element [..] is secured against falling out by the base [..] thereof."

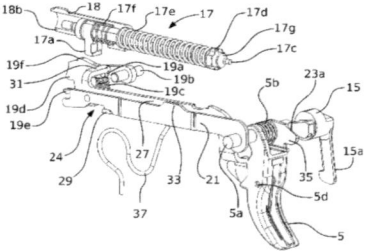

Fig. 5.68: HK *VP9* take-down function in DE 10 2013 022 080 B3, p. 28, Fig. 17.

Summary HECKLER & KOCH *VP9*

With its *VP9* pistol, HECKLER & KOCH follows other manufacturers who already offer striker-fired pistols with comfortable trigger action, but the unique selling proposition of the *VP9* is its user friendliness: customizable grip size, which includes width and trigger reach, ambidextrous slide stop switch and magazine release, consistent trigger pull for each shot, and a take-down procedure which eliminates negligent discharges. Despite the MSRP of $719 U.S., HK seemed to have made concessions for the benefit of the user.

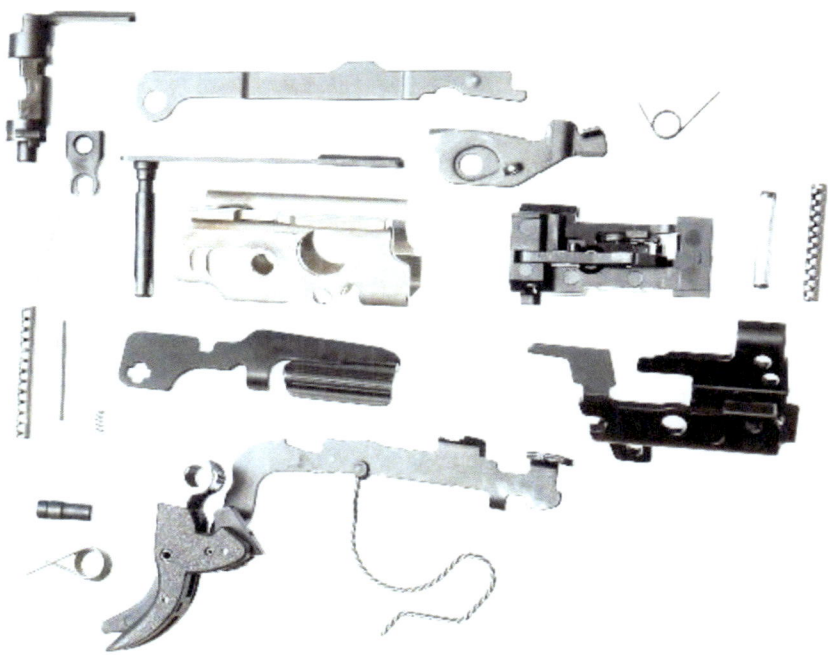

Fig. 5.69: HK *VP9* frame components.

5.6   KEL-TEC *PF-9* (9 mm Luger)

The measurements and technical data according to the checklist are summarized in Table A21.4, p. 466. KEL-TEC[625] claimed in the user manual of its *PF-9* pistol that it has the lightest and flattest pistol ever chambered in 9 mm Luger;[626] – perceived recoil was similarly tough. The dimensions specified in the technical data of the manufacturer could not be confirmed. In place of the stated 149 × 22 × 109 mm, a slightly larger dimension of 151 × 25 × 112 mm was measured.

Trigger pull on the *PF-9* felt long and hard. The measurement delivered a surprisingly low trigger pull force of 25 N, associated with a trigger travel of 16 mm. The slide racking force amounted to 89 N. The depth of indentation was only 0.23 mm in the copper crusher cylinder, and the diameter of the indent was 1.1 mm. Primers from fired cartridges of reference ammunition exhibited clearly a bulge on an off-center striker indent (Fig. 5.70).

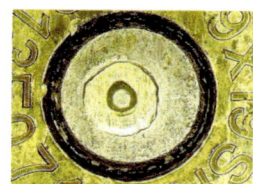

Fig. 5.70: Striker indent with bulge and little drag.

An automatic hammer "lock" safety, a safety notch that would keep the hammer in a partially cocked condition, was the only safety device on the sample gun. According to the description in the manufacturer's user manual, this was supposed to be sufficient. Disassembly and assembly was possible even if a magazine was in the gun.

Both slide stop and magazine release were on the left side of the pistol. Included in the box was a seven-round magazine. The manufacturer suggested a retail price of $333 U.S., while the gun was available online for $228 U.S. as of September 2014.

KEL-TEC was the only manufacturer in this survey that made a statement regarding the locking principle in his user manual. According to the manufacturer, the coupled rearward travel of slide and barrel would be 6 mm.[627] This value could not be verified. The uncoupling of the barrel started at about 2.9 mm and at about 5.2 mm, the barrel was completely decoupled from the slide. The total slide recoil travel was 41.3 mm.

Shooting the gun gave a muzzle velocity of 343 m/s, and a mean slide recoil velocity of 8.8 m/s was observed. The value for the theoretical mean slide recoil velocity was 11.0 m/s.

The slide width was measured to be 22.2 mm, and the height was 28.3 mm at the recoil lug. The weight of the slide assembly was 171 g, while the weight of the barrel was 58 g, with a barrel length of 79 mm.

Examination Regarding Innovative Features

The sample was a short-recoil-operated pistol with vertically tilting barrel and partially cocked hammer. The firing mechanism was inside of the frame in a chassis made of aluminum. This chassis also provided the slide rails. According to KEL-TEC, the accuracy of the gun would be enough for its intended use and both practical accuracy and perceived recoil would be comparable to those of much larger handguns. This was said to be possible due to the pistol's locking dynamics in combination with its superior ergonomics. The manufacturer markets its pistol to be particularly suitable for plainclothes police officers or as a secondary weapon for military personnel. In addition, the handgun would be suitable for female shooters owing to the small grip size and light trigger pull.[628]

Fig. 5.71: KEL-TEC *PF-9*

A separate extended magazine base plate was included, in addition to a single-stack seven-round magazine with flush fit base plate, and a trigger-lock (Fig. 5.72). If the trigger-lock was attached, part of it would sit behind the trigger blade and its outside would also cover the trigger blade, making it impossible to pull the trigger fully to the rear. The trigger lock was made of a left and right side, which had to get connected by a lock screw. The screw had a proprietary shaped screw head with cut-outs and could be used only with a designated screwdriver, which came with the lock.

Fig. 5.72: *PF-9* with trigger lock, tool key, magazine, and additional mag base plate.

According to the manufacturer, the slide and barrel were made of SAE 4140 ordnance steel. KEL-TEC mentioned the barrel was hardened and tempered to 48 HRC. The trigger mechanism was part of an aluminum alloy chassis, which should be made of solid 7075-T6 aluminum billet. The manual also pointed out the pistol's frame was made of Dupont ST-8018 polymer.[629]

The pistol had a bobbed hammer that protruded only slightly as the trigger was being pulled (Fig. 5.73). A tension spring was used to drive the hammer (Fig. 5.74). The pistol had a partially-cocked firing mechanism, that is, it did not offer a restrike capability. The manufacturer referred to the trigger action as "double action only" in the manual and mentioned further that this would contribute to the safety of the user and the user friendliness of the weapon.

Fig. 5.73: Shrouded hammer shown in partially cocked condition.

The end of the mainspring was fixed by a pin, which in turn was inside a cover plate on the heel of the pistol grip, next to the magwell. With this arrangement, the cover plate was kept in place (Fig. 5.74).

The gun did not have a manual safety. KEL-TEC's statement in the manual regarding the safety of the pistol pointed out that the hammer would never be fully cocked. And further: "the hammer is of a novel design," which would be useful, because the gun operates at high velocities as a result of the light weight

Fig. 5.74: Mainspring and cover plate close beside the magwell.

of the pistol. The trick would be to concentrate most of the hammer's mass close to its pivot point, giving "an inertia in the critical direction of the firing pin close to zero." Also, the partially cocked hammer would be secured by a hammer block, and only by pulling the trigger, this hammer block would become disengaged.[630] The manufacturer also explained in the manual that there could be no ignition when the breech was not fully locked, because the firing pin could not hit the primer as long as the barrel's chamber end is tilted downward.

There were no mechanical safeties found on the sample, with the exception of a hammer lock safety and a disconnector. The disconnector's function could easily be tested and allowed the release of the firing mechanism prior to coupling.

The user manual had a warning to not load a cartridge directly into the chamber.[631] A more detailed explanation on why KEL-TEC gave out the warning could not be found. One likely reason might have been to protect the extractor claw from damage.

The disassembly of the gun was done by locking the slide back, and then pushing out a takedown pin from right to left. Removing this pin required a tool. KEL-TEC's user manual recommended to use the rim of a cartridge case.

For basic cleaning, the user had six major components in front of him: barrel, recoil spring assembly, slide, magazine, frame, and take-down pin (Fig. 5.75). Some practice was necessary to do the assembly. Positioning the barrel properly demanded distinct attention. It sometimes needed to be pushed downward into the frame, as the slide was placed back on the frame.

Fig. 5.75: *PF-9*, main assembly groups.

The slide was milled from solid steel and seemed to be deliberately light: externally, on its muzzle end on the left and right side of the slide were serrations, and inside were recesses (Fig. 5.76), which had no function and could be interpreted only as a reduction of the slide's weight. Maybe this explains the snappy cycle time.

Fig. 5.76: Recesses in the slide.

The Extractor was located on the right side of the slide. It was spring loaded by an outer leaf spring. The leaf spring was in turn attached to the slide by an external screw. This seemed unusual to some viewers (Fig. 5.77). Normally, design engineers try to avoid screws on semi-automatics as screw heads tend to break, as a consequence of enormously intense accelerations and high loads during firing.

Fig. 5.77: Extractor with spring.

The recoil spring was not captured. It consisted of a guide rod, and an inner and outer recoil spring (Fig. 5.78).

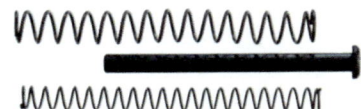

Fig. 5.78: Recoil Spring Components.

The pistol came in a surprisingly small sized box. This black plastic box had almost the same outer dimensions as the handgun (Fig. 5.79).

Patent research KEL-TEC

Several patents from KEL-TEC could be found, but only a few were related to handguns. KEL-TEC seems to focus on deep conceal carry pistols, as the publications refer to very specific solutions related to such models. Typically, these guns are too small and could not get the approval for import into the U.S., if they were manufactured outside of its shores. Therefore, the contents of these patents are not discussed here.

Fig. 5.79: *PF-9* shown in box.

Summary KEL-TEC *PF-9*

The company KEL-TEC has taken up the case of "innovative products," "polymer technology," and "modern production methods."[632] These key notes in combination with the impressions found during the survey when inspecting the *PF-9* lead to the conclusion that the company seeks to be the price leader on the U.S. market and wanted to create with the *PF-9* a handgun that is reduced to the essentials and chambered for the powerful 9 mm cartridge. Trying to find special features was to no avail, but the pistol's selling price was low. The *PF-9* was available online for $228 U.S. It was the second-lowest-priced handgun, after the HI-POINT *C9,* which retailed online for as low as $122 U.S.

5.7   RUGER *LC9* (9 mm Luger)

The measurements and technical data according to the checklist are summarized in Table A21.5, p. 467. RUGER quoted 152 × 23 × 114 mm for the outer dimensions of the pistol. Overall length and height could be confirmed, but 27 mm was measured at the widest point of the pistol.

Measuring the trigger gave a trigger pull of 35 N, at a trigger travel of 17 mm. The penetration depth in the copper crusher cylinder amounted to 0.26 mm with a diameter of 1.13 mm. On the primer of the reference ammunition, a bulge was clearly noticed (Fig. 5.80), which was concentric with the striker indent. The slide racking force was measured to be 89 N, which was at the same level as the one found on the KEL-TEC.

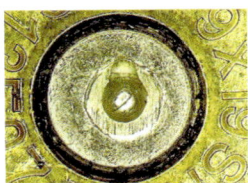

Fig. 5.80: Striker indent with bulge and drag.

Compared to the relatively small size of the pistol, it was equipped with a relatively large package of safeties: firing pin safety, thumb safety, locking device and magazine disconnect. Instead of a simple viewport, RUGER's *LC9* had a protruding loaded chamber indicator. Disassembly and assembly could be done with a loaded magazine in the gun.

The handgun was equipped with left side slide stop and magazine release. The scope of supply included a seven-round magazine. MSRP was $449 U.S., while online the pistol was available for $321 U.S. as of September 2014.

A long trigger travel was noticed during shooting, and an extremely rough texture on the grip's front could cause pain on the shooting hand. When shooting the gun with both hands and thumbs forward, the magazine release would hit the ball of the left hand's thumb if the weapon was not held firmly. The magazine release was sharp-edged and gave a clear sense of this. This effect did not occur when the pistol was gripped properly and firmly enough.

The obtained values for the beginning and end of the uncoupling and total slide recoil travel of the gun at hand amounted to 3.5 mm, 5.2 mm and 42.6 mm respectively. When shooting, a muzzle velocity of 339 m/s and a mean slide recoil velocity of 8.0 m/s were observed. The calculated mean slide recoil velocity was 8.7 m/s.

On the slide, its width gave a value of 22.8 mm and a height of 28.5 mm at the recoil lug. The weight of the slide assembly was 227 g, barrel weight 57 g and the barrel length was 80 mm.

Examination Regarding Innovative Features

Available since 2011, RUGER's *LC9* pistol seemed to look back toward the KEL-TEC *PF-9* launched in 2006. Both, the *LC9* and the *PF-9*, are small pistols for deep con-

cealment. They were in line with the trend towards very small handguns chambered in 9 mm, a trend that could be noticed on the U.S. market at the first half of this decade. The RUGER *LC9* seemed to be the more sophisticated equipped of both models, because it had a built-in locking device, a magazine disconnect and a thumb safety.

Fig. 5.81: RUGER *LC9* with flat fit magazine base plate.

The *LC9* was a short recoil-operated pistol with vertically tilting barrel chambered in 9 mm Luger. Slide stop lever, manual safety and magazine release were left side only.

A separate extended magazine base plate came with the gun, which enabled the replacement of the magazine's flat fit base plate.

Additionally, a padlock-style gun lock with keys and an additional key (Fig. 5.82) were included. The latter was needed to activate an internal lock and could also be used as a disassembly tool (Fig. 5.83). This "RUGER key" was labeled with "STORE SAFELY AND SECURELY." If this tool is lost or missing, the pistol can not be locked, unlocked or comfortably disassembled.

Fig. 5.82: Internal lock with "RUGER key" inserted into internal lock keyway.

A takedown pin located between frame and slide had to be removed for field-stripping. The pin was secured by a takedown plate on the left side of the frame. After pushing the plate down, the pin could be pushed out using the "RUGER key" from right to left. At the same time, the slide had to be pushed back a little. The other steps to strip down the gun were not in any way unusual.

Fig. 5.83: Disassembly and loaded chamber indicator; shown with empty chamber.

The *LC9* was hammer-fired and had a partially cocked firing mechanism. A magazine disconnect feature blocked the trigger (Fig. 5.84) when no magazine was in the pistol. A horizontally slidable plate was the main member of this magazine disconnect feature. A spring urged the slidable plate toward the magazine well to engage it with the trigger. When the magazine was removed from the pistol, the rear end of the slider would reach into the magazine well. In response to inserting a magazine, the slidable plate was moved to a non-blocking position and the trigger could then be pulled.

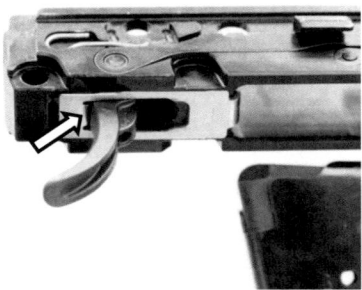

Fig. 5.84: Slidably mounted mag disconnect engages the trigger.

The loaded chamber indicator was carried out in a California-compliant manner. Its top was marked "LOADED WHEN UP" (Fig. 5.83, also Fig. 4.8, p. 186). The embodiment of the indicator resembled the extractor on Luger pistols, which served as both extractor and loaded chamber indicator.

The firing pin safety was in a conventional manner: a vertically moveable spring-loaded plunger, which blocked the firing pin when the plunger was in its lower position. The firing pin had a reset function, that is, it was a retracted firing pin. A spring would push the firing pin back to its original position behind the breech face. With this feature, the firing pin safety would engage again and keep the firing pin's nose behind the breech face as the next round was fed into the chamber.

The barrel had a traditional lands and grooves rifling, and according to RUGER, the twist rate was one revolution to 10 in [254 mm; PD], right hand twist. Fine scratches were visible. These toolmarks followed the twist of the grooves in longitudinal direction. RUGER mentioned an alloy steel is used for the barrels.[633]

Typically, the pistol's individual parts were either machined or MIM'ed. The slide was made from one solid piece and showed normal machining marks. According to the manufacturer, the slide was made of through-hardened alloy steel, and was blued.[634]

The frame, an injection molded polymer part, contained a chassis made of aluminum, which carried the components of the fire-control mechanism. The chassis was fixed inside the frame with one polymer pin per each of its ends. The user manual referred to the polymer used for the frame module as fiberglass-reinforced nylon.[635] At the lower end of the grip handle was a cover, which was just snapped on to the heel of the grip behind the magwell (Fig. 5.85).

The descriptions in the user manual seemed to discuss the gun quite in detail.[636] The instructions mentioned, for instance, the specific warnings needed for a variety of states.[a] It further discussed the controls and the benefits of the *LC9*. The no-menclature section explained that the slide's open top design would minimize the possibility of jam-ming, and would enable the shooter to clear any

Fig. 5.85: Snap-on cover plate, in-ternal side shown.

malfunction easily by hand. The manual even made a statement on loading individual cartridges. The necessary procedure was not described in detail, but it was expressly stated "Cartridges can be loaded singly if desired," which allowed the conclusion that the extractor claw would not be compromised if cartridges were loaded individually. On some other pistol models, the user might risk damage to the extractor claw if a cartridge is not stripped from a magazine into the chamber. If the slide is released from the slide stop lever on an already chambered cartridge and with mag well empty, chances are that the extractor claw could become damaged as it hits the head of the cartridge.

The manual explained how the slide stop was sup-posed to be used. The preferred operation would be to disengage the slide-lock by pulling back the slide slightly. The reason for this explanation might just have been the description of a standard tactical pro-cedure. But from a plain technical standpoint, the latch surface of the slide stop catch as well as the corresponding notch where it engages the slide, might accelerate wear over time, if the slide stop lever was just to be pressed down to release the slide, instead of operating the slide.

Fig. 5.86: Safety lever notches and engaging pawl.

The manual safety (Fig. 5.86) could only be moved upward to the "safe" position if the firing mechanism was cocked. Like on a Model 1911 pistol, the safety lever blocked

---

[a] California, Connecticut, Florida, Maine, Maryland, Massachusetts, New Jersey, New York City, North Carolina, Texas and Wisconsin.

the slide. Hence, a safetied weapon could not be cleared. Also, the safety blocked the trigger, thus it could not be pulled unless the safety was disengaged.

Engaging the thumb safety could barely be done without twisting the gun in the hand. Firstly, the safety was at a location where it could hardly be reached with the shooting hand's thumb, and secondly, quite some efforts were necessary to move this thumb safety upward. The thumb safety positively engaged in upper and lower position, which could protect the user from the possible consequences of an undeliberate manipulation of the safety.

The gun was equipped with an internal lock (Fig. 5.87), which could be activated with an internal lock key. This was the same key that would be used to break down the pistol. To activate the internal lock, the manual safety had to be engaged and then the key was rotated about one quarter clockwise until it stopped. With the internal lock engaged, the trigger would move freely without releasing the firing mechanism, because the trigger bar was tilted downward and was thus, no longer in service.

Fig. 5.87: Internal lock and trigger bar.

When releasing the trigger to go forward, two "clicks" were noticeable. The first click could have been misinterpreted as the trigger's reset, but pulling the trigger at this point proved to be useless. Only after the trigger was allowed to move fully forward, the trigger mechanism resetted and another shot could then be fired.

The frame could be stripped down beyond the extent normally needed for field-stripping, by pushing out aforementioned plastic pins. As it turned out, a third pin, which was a 3 mm metal pin, should not be removed, because this pin holds the spring-loaded sear; an act not required for disassembly and the sear will be difficult to install.

A padlock with a particularly long shackle came with the pistol. With slide in the retracted position, the shackle could be applied to block magwell and ejection port, in order to secure the weapon.

Patent research RUGER

Two RUGER patents matched the individual features of the *LC9*. One refers to the magazine disconnect and the other one, to the internal lock. The patent *Magazine Disconnect Mechanism for Firearms*, US 8,438,768 B2, dated May 14, 2013, refers to a hammer-fired pistol, and discloses a flat longitudinally slidable plate that engages the trigger blade. Under "Summary of the Invention," it is explained that a pistol would be provided that disables the fire-control mechanism under "normal trigger pull pressure."

The preferred embodiment of RUGER's patent *Lockable Safety for Firearm*, US 8,464,455 B2, dated June 18, 2013, focuses on a hammer-fired firearm, just like the patent on the mag disconnect feature. Claim #1 describes a trigger bar, which is engageable with the hammer, wherein, an eccentric camming member engages and displaces the trigger bar downwards away from the hammer into standby position.

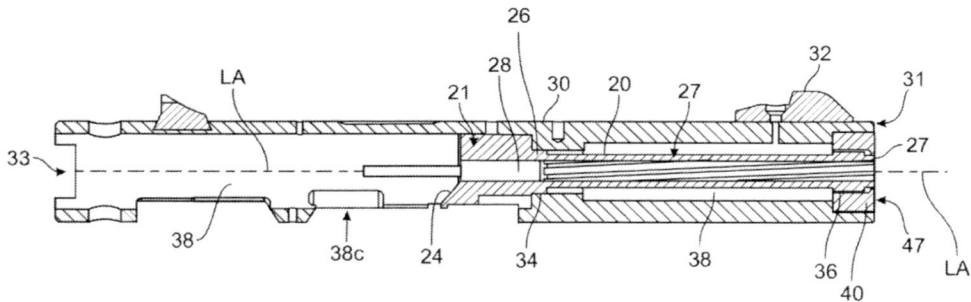

Fig. 5.88: Releasably secured barrel insert, US 8,701,326 B2, Fig. 5.

The search yielded no further patents with respect to the RUGER *LC9*, but patents from the year 2013 describe a pistol with fixed barrel and a specific embodiment of a follower.

US 8,701,326 B2, *Pistol Barrel System and Method*, dated April 22, 2014, discloses an interchangeable barrel system for a pistol, wherein the barrel insert includes a threaded locking ring on the barrel's muzzle end to releasably secure the barrel in the receiver (Fig. 5.88). It describes

Fig. 5.89: Pistol shown in US 8,701,326 B2, Fig. 1.

an inner durable steel barrel insert with integral feed ramp, which is an embodiment already known from the WALTHER *P22* pistol. As a special advantage of the described embodiment, the patent describes the flexibility in the selection of the caliber. In contrast, the pistol drawings in the patent (Fig. 5.89) remind the reader a lot about a pistol chambered in .22 Long Rifle.

Magazine for Firearm, patent US 8,752,318 B2, dated June 17, 2014, discloses an embodiment that is supposed to improve ejection of a last expended case, that is, of a gun which is fired with an empty magazine. According to the introduction to the invention, autoloaders sometimes "eject spent casings by impacting the next round in the magazine [...]." When the magazine is empty, the casing "may slip off the extractor and does not reach the ejector [...]." According to the patent, this could be overcome by a two-piece follower with a spring-loaded ridge-shaped insert in longitudinal direction in the center of the follower (see item 74 in Fig. 5.90).

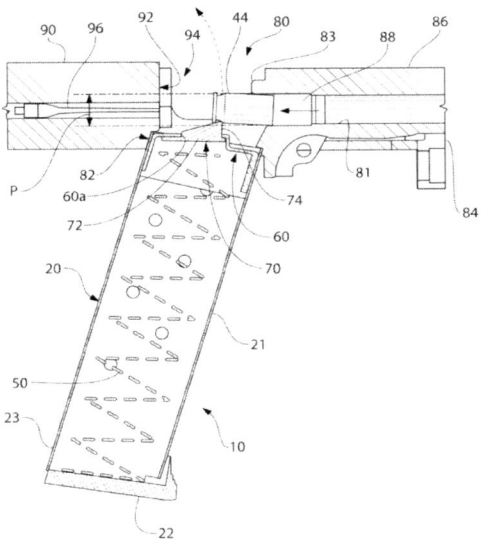

Fig. 5.90: Follower assembly providing consistent feed and ejection of last round; US 8,752,318 B2, Fig. 9.

Summary RUGER *LC9*

Despite being chambered in 9 mm, the fairly small RUGER *LC9* pistol combined polymer-, MIM-, and milling technology in an attractive design and with plenty of features: internal lock, magazine disconnect, thumb safety and loaded chamber indicator.

It was interesting to see the heel side cover behind the magwell. This snap-on cover plate concealed the mounting for the mainspring. Covers attached with snap-on connection on the outside of weapons are rare since they might eventually come off and be lost.

The embodiment of the internal lock seemed to be carried out nicely. The rear end of the trigger bar would become displaced through a simple key-operated rotary eccentric camming member, and when locked, the trigger would move freely without releasing a shot. Not blocking the trigger mechanism saved the associated trigger components from becoming damaged by a user who might use excessive trigger pull force under stressed conditions.

In contrast, the magazine disconnect would block the gun's trigger blade. If a user were to apply excessive force he might do damage to the pistol's mechanics. Apparently, RUGER was aware about this condition and addressed "normal trigger pull" in the summary of the patent.

RUGER's mag disconnect patent refers explicitly to hammer-fired weapons; that is, a workaround might be possible, if a pistol is striker-fired. The successor to the LC9 is the striker-fired LC9s which is also equipped with a magazine disconnect;[637] however, no patent publications existed at the time of the research carried out here.

The use of an aluminum alloy chassis made a simple but effective design for the mag disconnect possible. If RUGER had designed the polymer frame in the regular integral design for an immediate installation of the functional components, it certainly would have resulted in higher costs for the injection mold as the functional integration typically demands moveable tool-slides in the injection mold to form cavities for the magazine disconnect members as well as other members of the frame module's integral design. Also, the width of the frame might have been bigger.

5.8 SMITH & WESSON *M&P9* (9 mm Luger)

The measurements and technical data according to the checklist are summarized in Table A21.6, p. 468. SMITH & WESSON stated the dimensions for its *M&P9* as 194 × 30 × 139 mm. Measuring the outside dimensions showed the sample was both, three millimeters wider (33 mm) and higher (142 mm).

Fig. 5.91: SMITH & WESSON *M&P9*

Testing the trigger gave a trigger pull force of 25 N at 7 mm trigger travel. The slide racking force was just 84 N. The penetration depth of the striker nose was on a level that is normal for duty pistols. The indent was 0.32 mm, and had a diameter of 1.1 mm. On the primer of the reference ammunition was a bulge, and drag was also noticed (Fig. 5.92).

Fig. 5.92: Striker indent with bulge and drag.

Features of the sample were a manual safety, also a striker safety, and a trigger safety. A loaded chamber viewport was on the end of the barrel hood (Fig. 5.93). The port was generously funneled to allow better recognition of a chambered round. If the trigger was being pulled as the barrel was still decoupled from the slide, the firing mechanism would not release. The latter was considered an out-of-battery disconnect. Loading the first round into the magazine was a difficult task. Field-stripping and assembly was only possible, if no magazine was in the gun.

Fig. 5.93: Loaded chamber viewport.

The pistol had an ambidextrous slide stop lever, a reversible magazine release, and it came with two 17-round magazines. MSRP was $569 U.S.; online, the *M&P9* was available for $442 U.S. as of September 2014.

Uncoupling start and end, and total slide recoil travel of the sample amounted to 2.0 mm, 4.8 mm and 47.4 mm respectively. A muzzle velocity of 361 m/s was measured and a mean slide recoil velocity of 4.4 m/s was determined. The theoretical mean slide recoil velocity was 5.2 m/s.

The slide had a width of 27.4 mm, and at the recoil lug it had a height of 32.8 mm. The weight of the slide assembly was 368 grams, the barrel's weight was 103 g, and a barrel length of 108 mm was measured.

Examination Regarding Innovative Features

After the *SIGMA* and *SW99* pistols, the *M&P* line of pistols was added to SMITH & WESSON's product range of polymer pistols. With this pistol generation, the company re-activated its old product name Military & Police (M&P), which it first used for a revolver in 1899.[638]

Fig. 5.94: Magwell, mag latch with spring rod.

The embodiment of the magazine release and its spring reminded one of the GLOCK design. It was a spring-loaded latch, and the spring was a metal rod (Fig. 5.94).

M&P9 was available either with or without an ambidextrous thumb safety. Apparently, SMITH & WESSON took both types into account right from the beginning when designing this model, which probably had beneficial effects for the manufacturer regarding the variety of components and the respective interchangeability. The sample gun was equipped with an ambidextrous thumb safety (Fig. 5.95). When the lever was pushed up, the gun was on-safe, and the lever had to be pushed down to disengage the safety. During shooting, the safety lever was very close to the trigger hand's thumb, and depending on the hand placement, chances were that the lever could get pushed upward unintentionally to "on-safe."

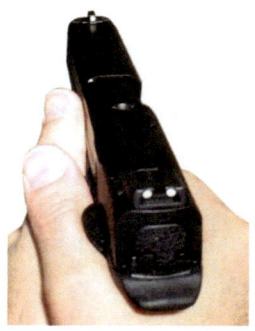

Fig. 5.95: Hand placement and position of the thumb safety.

A tool was required to disassemble the gun for cleaning: the user had to remove the magazine, then the frame tool had to be rotated one quarter turn to disengage it from the heel of the grip (Fig. 5.96). With its bar-shaped front end, the sear deactivation lever had to be lowered into the magazine well. This lever (Fig. 5.97) was painted yellow for better recognition.

Fig. 5.96: Removing the frame tool.

Lowering the sear deactivation lever renders the sear inactive and is compulsory prior to field-stripping. It required the removal of the magazine from the pistol and to lock the slide back. As a consequence, this procedure made sure the pistol was cleared prior to field-stripping. However, this technical solution required an additional tool for the break down procedure, which the user better not lose.

SMITH & WESSON called the firing mechanism "Striker Fired" and argued it was a double-action-only trigger. The survey of the weapon however concluded it was a fully cocked firing mechanism, hence, it was listed in the checklist as a single-action trigger. *Sweeney* mentions the trigger on the *M&P* pistol was crisper than the trigger of a GLOCK pistol and would come close to a COLT *Model 1911*.[639]

The sear housing (Fig. 5.98) was a modular assembly, which carried parts of the firing mechanism and the manual safety. The upper part of the MIM'ed housing formed the rear slide rails. The front slide rails were part of the locking insert, also an MIM'ed part.

SMITH & WESSON used primarily stamped, MIM'ed, injection molded and a few machined parts for the *M&P9* (Fig. 5.99). The slide stop lever component was made of a left and right side sheet metal stamping, connected in the center by a spot weld.

The hinged trigger was made of two polymer parts (see Fig. 3.80, p. 131), the bottom half worked as a trigger safety.

The trigger reset was driven by a tension spring. Tension springs usually break at the transition from the working coils to the loop ends. SMITH & WESSON attempted to increase the service life of this spring by the addition of a foam insert, which should reduce the vibrations of the spring body (Fig. 5.100). This application has been known since the *SIGMA* pistol was introduced.

Fig. 5.97: Lowering the sear deactivation lever (shown underneath the ejector).

Fig. 5.98: MIM'ed sear housing with slide rails and components.

Fig. 5.99: Frame components.

Fig. 5.100: Trigger spring with shock absorber (inside of the spring).

The interchangeable saddle-style backstrap covered parts of the side panels (Fig. 5.101). Thus, by
selecting the appropriate backstrap, the user could
at the same time adjust the width of the pistol's grip.
Three sizes of backstraps were included.

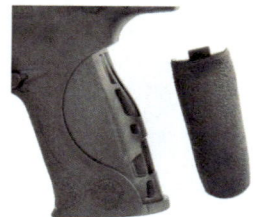

Fig. 5.101: Inter-
changeable back-
strap.

The slide had a black finish, even though it was
marked "STAINLESS." SMITH & WESSON supposedly
nitrated the slide despite the presence of the stainless steel material. According to *Sweeney*, SMITH &
WESSON heat treats slides and barrels, before they
melonite them.[640] In the striker channel was a polymer liner, similar to those present in the slides of
GLOCK pistols (Fig. 5.102).

Fig. 5.102: Polymer sleeve (white,
see arrow) inside striker channel.

Patent research SMITH & WESSON

Patent U.S. 8,695,262 B2, dated November 11,
2011, discloses a sear housing assembly configured
for "drop-in" insertion into a compartment inside the
frame. The sear housing assembly is secured in
place, and is provided with a set of slide rails. The
slide rails partially or wholly replace the slide rails
integrally formed on other frames. Images shown in
the patent remind one of the *SIGMA Series SW380*
pistol, a model launched in 1996 (Fig. 5.103).

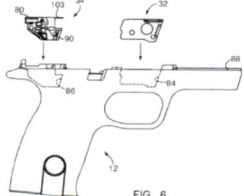

Fig. 5.103:
US 8,695,262 B2,
sheet 4

Patent US 8,296,990 B2, dated December 30,
2009, shows an interchangeable front sight that is
capable of quick and easy installation by hand without the use of tools. The front sight is attached to the
slide in the traditional way by a dovetail joint, but the
base of the front sight post snap-fits into a recess in
the slide (item 304 in Fig. 5.104); it is a snap-on
dovetail front sight.

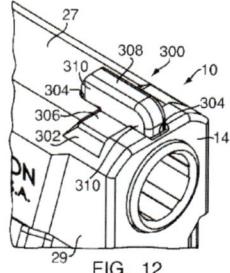

Fig. 5.104: Snap-on
dovetail front sight,
US 8,296,990 B2,
sheet 8.

Interestingly, the said patent US 8,296,990 B2 depicts other images showing a hammer-fired pistol of compact size (Fig. 5.105). These pictures can also be found in US 8,276,302 B2 and US 8,132,496 B2, of the same filing date. Said patents disclose a manual slide and hammer lock safety, and an automatic firing pin block safety.

With US 7,810,269 B2, *Frame-Mounted Trigger Safety and Well Extension*, filing date January 10, 2008, SMITH & WESSON discloses a manual safety. The title is somehow misleading as the embodiment of the patent does not refer to a trigger safety per se, but to a frame mounted ambidextrous thumb safety, and it also relates to a magazine well extension for a compact frame pistol facilitating the use of a full size magazine, in other words, a "grip extension" (Fig. 5.106).

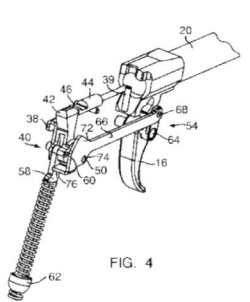

FIG. 4

Fig. 5.105: Hammer-fired pistol, US 8,296,990 B2, sheet 2.

The magazine well extension allows for a full size magazine with a higher capacity to be used with a compact frame pistol. The patent describes the extension may also have ergonomic benefits as it increases the size of the grip of the pistol. The well extension has several protrusions; it is latched into the bottom of the frame, and is also secured directly to the pistol via S&W's frame tool. SMITH & WESSON explains that the invention would overcome drawbacks with known extensions that are mountable directly to a magazine as opposed to a pistol frame. SMITH & WESSON highlights this functionality would allow a user to discharge multiple full size magazines without having to remove a magazine collar. This overcomes a significant disadvantage of prior art magazine collars, which attach directly to the magazine and are thereby difficult to remove.[641]

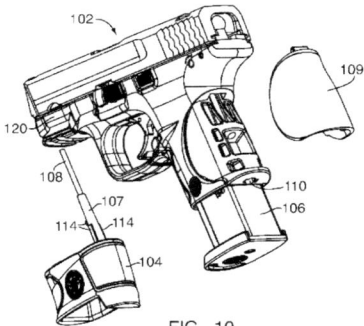

FIG. 10

Fig. 5.106: Well extension, US 7,810,269 B2, sheet 6.

Interestingly, the patent claims do not mention the well extension. It may be concluded that this grip extension is not protected by this patent. Other manufacturers might copy the idea, but can not file a patent anymore, because the idea is already disclosed; hence the idea is no longer a novelty. No further research has been carried out on this behalf; it may be possible other patents exist.

SMITH & WESSON'S patent is reminiscent of WALTHER'S application, WO 2007/128 252 A1, dated May 4, 2006, which discloses a customizable grip handle (Fig. 5.107). WALTHER'S idea, however, goes further than the one of SMITH & WESSON. WALTHER included saddle-style covers for the grip panel section. Thus, a modular design for grip handles to allow for customization of the in-size category for compact or full size pistol frames.[642]

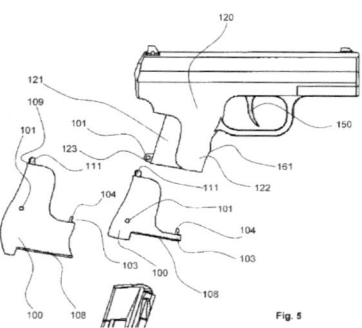

Fig. 5.107: Adjustable grip handle. WALTHER, WO 2007/128,252 A1, sheet 5.

Two patents appeared interesting from the perspective of patent eligibility: S&W's patents US 6,161,322, filing date May 15, 1998, and US 6,493,977 B1, filing date September 1, 2000. The former describes a loaded chamber viewport carried out as a notch extending through the barrel hood extension at the top of the firing chamber. The latter, discloses a further improvement of the port by an observation aperture. It suggests to include a second notch formed in the slide, opening forwardly through the breech face and upwardly in registry with the barrel's viewport.

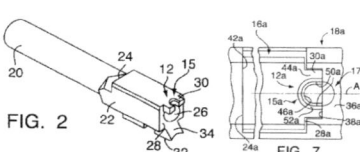

Fig. 5.108: Loaded chamber viewport as in US 6,493,977 B1, on barrel hood and on top of breech face (Fig. 7).

This second notch is described to be defined by a partial-conical surface (Fig. 5.108). Other gun manufacturers should not feel constrained by these patents. The novelty of the first patent's inspection port is omitted, since publications with photos of the AMT *Hardballer* pistol already appeared in 1978, which showed this kind of a loaded chamber viewport. If the second patent were still valid, it could be circumvented by selecting a design for the notch which differs from the specific shape described in the patent.

Summary SMITH & WESSON *M&P9*

With its *M&P9*, SMITH & WESSON delivered a package consisting of adjustable grip size, ambidextrous thumb safety, ambidextrous slide stop and reversible magazine release. The field-strip function protected the user from negligent discharge, but required the handling of a tool and to perform additional manipulations. Take down was found to be less demanding on other guns. Also uncomfortable were loading the first cartridge in the magazine as well as the manual safety. Controls should be designed as ergonomically as possible, easy to use, and with the intention to reduce the risk of user mistakes.

5.9 SPRINGFIELD *XD*ᴹ (9 mm Luger)

The measurements and technical data according to the checklist are summarized in Table A21.7, p. 469. The information specified by the manufacturer for length and width could not be confirmed. Instead of the 193 × 30 × 146 mm specified by the manufacturer, 201 × 33 × 146 mm were obtained for the gun's dimensions.

The measurement of the rigger provided 24 N for trigger pull force, and 7 mm trigger travel. The slide racking force amounted to only 84 N. With a depth of 0.32 mm and a diameter of 1.17 mm, the striker nose provided a distinct indentation in the copper crusher cylinder. The primers from spent casings of reference ammunition exhibited a bulge around the striker nose's indentation.

Fig. 5.109: Striker indent with bulge and drag.

The *XD*ᴹ was the only pistol in this survey equipped with a grip safety. Additionally, it had both trigger safety and a striker safety, also a cocking indicator and a loaded chamber indicator. The firing mechanism would not release as long as the barrel was decoupled from the slide. This was counted as an out-of-battery disconnect. Field-stripping was possible even when a magazine was in the pistol. Assembly with a loaded magazine in the gun was possible, if the top cartridge was pressed down by hand during assembly.

The gun had a left side slide stop lever and an ambidextrous magazine release. MSRP was $700 U.S.; online, the pistol was available for $537 U.S. as of September 2014.

Beginning and ending of uncoupling, and total slide recoil travel of the gun were measured 3.1 mm, 6.4 mm and 46.8 mm respectively. During shooting, a muzzle velocity of 371 m/s was recorded. The value of the theoretical mean slide recoil velocity was 5.6 m/s. Of all pistols considered in this survey, the SPRINGFIELD *XD*ᴹ had the highest muzzle velocity, and a moderate empirical mean slide recoil velocity of 5.2 m/s.

The slide was 26.3 mm wide and the height at the recoil lug was measured as 36.1 mm. The weight of the slide assembly amounted to 349 g, the weight of the barrel was 107 g, and barrel length was 117 mm.

Examination Regarding Innovative Features

The *XD* series of pistols (e**X**treme **D**uty) is manufactured in Croatia by HS PRODUKT and imported into the USA by SPRINGFIELD ARMORY, INC.[643] According to *Sweeney*, the components of the *XDM* are not interchangeable with those of its predecessor *XD*.[644] A generously dimensioned polymer case (Fig. 5.110) was included in the scope of delivery. The case housed the *XDM* pistol with a total of three 19-round magazines, two extra backstraps, a holster, a double magazine pouch, a cable lock, a magazine loader and a user manual. With this scope of delivery, the *XDM* was the gun with the most extensive accessory package of this survey.

Fig. 5.110: SPRINGFIELD *XDM* with accessories.

The pistol was in the same size category with the GLOCK *G17*. The *XDM* was relatively high, but apparently a long grip handle was necessary to cover the long 19-round magazine.

A special feature of the gun was its grip safety (Fig. 5.111). If the pistol's grip was not held appropriately with the grip safety adequately depressed, the gun was still safetied and the slide could not be racked because the grip safety blocked the sear, and the rear end of the slide's stripper rail would collide with the blocked sear (Fig. 5.112).

Fig. 5.111: Grip safety and cocking indicator.

When the trigger was squeezed while the grip safety was fully depressed, the sear would drop and release the striker. Slide, along with the striker are then reciprocated rearward while the pickup rail keeps the sear down.

This interaction between the cartridge rail and sear was particularly important for the field-strip function, which would require no additional manipulation by the user in decocking the firing mechanism prior to taking the gun down. The *XDM* could only be disassembled if the slide was locked to the rear. This

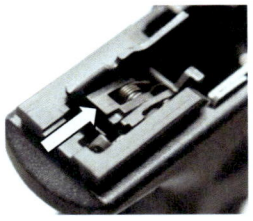

Fig. 5.112: Sear's tear-off edge which holds and releases the striker.

first step forced the user to depress the grip safety properly and demanded a proper pistol grip. With the slide locked back, the takedown lever could be rotated clockwise.

The rotation of the lever would activate a leverage, which engaged the sear to hold it down (Fig. 5.113). A break down function without having to pull the trigger, had not been available on the previous *XD* pistol generation.

The magazine consisted of four components. A locking plate was not included; instead, the end of the magazine spring engaged directly on the magazine base plate (Fig. 5.114). The mag body was made of stainless steel.

Patent research SPRINGFIELD

The search showed no results for the assignee "Springfield," but was available for "HS Produkt." Patent EP 2,313,734 B1, dated July 22, 2009, discloses the *XD^M's* take-down mechanism. The mechanism prevents the disassembly of the weapon when the magazine is in the gun, and is combined with the aforementioned mechanism offering field-stripping without the need for the user to decock the firing mechanism (see US 8,371,058 B2).

Patent US 8,646,200 B2, dated November 28, 2011, discloses a magazine where the base is locked by the magazine spring. As it is explained, advantages would be an improved safety of magazine base locking to the magazine body and a simplified construction. Interestingly, the actual embodiment of the sample pistol's magazine base (Fig. 5.114) was without the patent's claimed embodiment of a raised lock (item 1a in Fig. 5.115).

Fig. 5.113: Sear up in standard position, and sear blocked in downward position for take down (below).

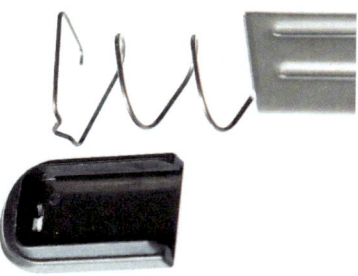

Fig. 5.114: Magazine base plate and mag spring.

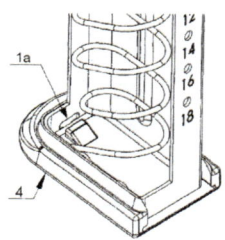

Fig. 5.115: US 8 646 200 B2, sheet 5

Summary SPRINGFIELD *XD*[M]

The scope of delivery of the SPRINGFIELD *XD*[M] pistol comprised a wide range of accessories, including three 19-round magazines. A potential buyer would obtain a package that he could just take out to a day at the range immediately after purchase, so to speak.

The pistol's grip handle was adjustable in size. Not all the pistol's controls were ambidextrous, only the magazine release was. However, the field-stripping procedure on this gun proved to be safe for the user. Even though it was striker-fired, it would not require pulling the trigger prior to disassembly, as long as the user noticed the grip safety. The latter was something the shooter will need to get used to. If it was not properly pressed, the operator wouldn't even be able to draw the slide back.

5.10   TAURUS *PT 24/7 PRO* (9 mm Luger)

The measurements and technical data according to the checklist are summarized in Table A21.8, p. 470. The dimensions of length and height were found to be different when compared to the values specified by the manufacturer. TAURUS had specified 181 × 32 × 140 mm, but the measurement showed 185 × 32 × 146 mm.

When measuring the trigger of the *PT 24/7 PRO*, two measurements were carried out, because this pistol had two trigger actions. The single-action trigger produced a trigger pull force of 21 N at 11.1 millimeters trigger travel, double-action offered 27 N and 10.8 mm. It is normal to have different trigger travels on guns with two trigger actions. The reason is the mechanical separation of the break points (see Chapter 3.6, Combinations of Trigger Actions, p. 102). The slide racking force amounted to 88 N.

On the primer from cartridge casings of reference ammunition, a bulge could be found around the striker indent (Fig. 5.116). The indentation depth in the copper crusher cylinder could be determined only for the SA trigger. It delivered an indent of 0.31 mm in combination with a diameter of 1.1 mm. A measurement of the indentation for repeat strike capability was not possible because after loading the pistol, the firing mechanism would be fully cocked and the weapon would not offer any ability to decock the firing mechanism other than pulling the trigger.

Fig. 5.116: Striker indent with bulge and drag.

Besides an internal lock, the gun was equipped with a manual safety, a striker safety, an internal trigger safety, and a loaded chamber indicator. The disassembly and assembly was not possible with a loaded magazine in the gun.

The gun had a left side slide stop lever, a reversible magazine release, and two 17-round magazines accompanied it. MSRP was $492 U.S.; online, the handgun was available for $395 U.S. in September 2014.

The obtained values for the beginning and end of the uncoupling, and total slide travel were 3.4 mm, 5.2 mm and 41.8 mm respectively. During shooting, a muzzle velocity of 356 m/s was recorded, and a mean slide recoil velocity of 4.4 m/s was observed. Calculating the value for the empirical mean slide recoil velocity gave 4.9 m/s.

The width of the slide was 25.6 mm, and its height was 35.0 mm at the recoil lug. The weight of the slide assembly amounted to 383 g, and the weight of the barrel was 103 g. Barrel length was 108 mm.

Examination Regarding Innovative Features

The striker-fired *PT 24/7 PRO* sample was a pistol with SA/repeat strike trigger action. The pistol's standard trigger was single-action. In the case of a failure to fire, the trigger would allow to be repeatedly pulled and it would cock and release the firing mechanism for every pull. The pistol did not have any decocking feature that would allow to decock the firing mechanism in a safe way.

Fig. 5.117: TAURUS *PT 24/7 PRO.*

On the pistol's left side were the slide stop lever, a thumb safety, a take-down lever, and a magazine release. The magazine release was reversible. The front strap "checkering" of the rubberized grip area looked like fins, which allowed the user to hold the pistol comfortably. TAURUS called this grip handle design "Ribber Grip." What TAURUS called "Index Memory Pads," was promoted to always allow positive and intuitive positioning of the trigger finger outside and above the trigger guard and always at the exact same spot (Fig. 5.118).

Fig. 5.118: Index Memory Pad.

Breaking down the TAURUS *PT 24/7 PRO* (Fig. 5.119) was different from other pistols. It involved the removal of the take-down lever as well as decocking by pulling the trigger as a second step. The full procedure was this: first, the slide had to be locked back. Next, the take-down lever had to be rotated down and then pulled out from the left side of the pistol. At this point, the slide could not just be simply removed from the frame, because closing the slide cocked the SA trigger and caused the slide to stop approximately in battery position. At this point the trigger then had to be pulled, and while keeping it to the rear, the slide could be pulled off the frame. If one would not keep the trigger fully to the rear,

Fig. 5.119: Main assembly groups of TAURUS *PT 24/7 PRO.*

chances were that the slide would stop again; now caused by the trigger bar's repeat strike firing function.

An internal lock was on the right side of the slide (Fig. 5.120). It reminded of a screw head with integral keyway. After a 180 degrees clockwise turn, it blocked the striker and also the slide, that is, the slide could not be retracted if the lock was engaged. In the unlocked state, the head was flush with the slide's contour; when the lock was engaged, the head protruded over the contours of the slide. With the head being flush or protruding from the slide, the user was able to tell if the gun could be fired or if it was locked. The key for the internal lock was not a traditional key, but a hexagonal key, with a hole in the front face (Fig. 5.120).

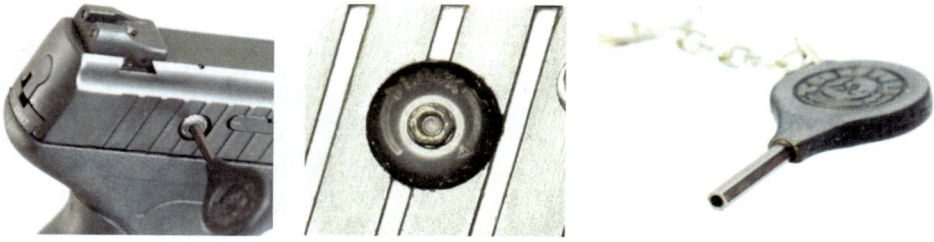

Fig. 5.120: Internal lock, keyway and key (from left).

Although the gun had two trigger actions, the internal lock could only be applied if the firing mechanism was uncocked. When the striker was "down," a recess in the striker-shaft would be in the correct position for the locking mechanism to engage and block it (Fig. 5.121). If a user wanted to use the locking function, he was required to clear the gun and

Fig. 5.121: Recess (arrow) on striker for engagement of internal locking mechanism.

then decock the firing mechanism by pulling the trigger, because the standard trigger mode of the *PT 24/7 PRO* was single-action, whereby the striker is fully cocked.

The striker assembly consisted of a generously-sized, hollow striker. It contained a captured mainspring in a rearward compartment. The design included a retract spring which would push the striker back to a resting position in case of a failure to fire or when dry firing the gun. A separate guide rod was provided for the retract spring, which was parallel to the striker.

The complete striker assembly (Fig. 5.122) was secured in the slide by a retainer plate. This rectangular plate was located in vertical recesses inside the slide. Installing the striker assembly required the user to keep pressing the captured mainspring's guide rod from the rear into the striker, while sliding the retainer plate back in position. If everything was correct, a pin of the mainspring's guide rod would engage the retainer plate to secure it in place. Additionally, the firing pin channel was then covered up with a polymer cover at the rear of the slide. This cover would snap on to the retainer plate.

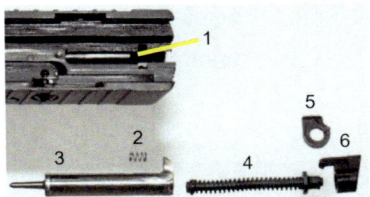

Fig. 5.122: Striker components: Guide rod (1) for retract spring (2), striker (3), mainspring (4), retainer (5), cover (6).

The extractor was a combination of two parts (Fig. 5.123). On the top was a thin sheet metal part, which served as a loaded chamber indicator. The tip of this sheet metal part pointed towards the breech face, and it would be pushed back by the casing of a chambered round. When a cartridge was in the chamber, the loaded chamber indicator protruded on the right side of the frame and a red marking could be seen on its top.

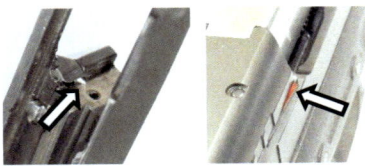

Fig. 5.123: Loaded chamber indicator: feeler and red marking (right).

An aluminum alloy chassis (Fig. 5.124) was secured by two pins and an undercut (Fig. 5.125) inside the frame. The slide rails were part of the chassis, which was also marked with the pistol's serial number at the rear end.

The chassis carried the trigger blade, parts of the firing mechanism, the slide stop lever, and the thumb safety. The latter would put the gun "on-safe" when the safety lever was rotated upward. A positive engagement of the manual safety was achieved by a ball. A spring-loaded plunger would press a steel ball against the safety lever. During assembly, the lever had to be inserted on the left side of the chassis. In this condition, it was not further secured against falling out. Only when the chassis was installed in the frame the side wall of the frame would keep the safety lever in place.

Fig. 5.124: Chassis with components and slide rails.

Fig. 5.125: Recess to keep rear end of the chassis in place.

When "on-safe," the thumb safety would engage the trigger bar, the curved lever of the safety mechanism would block the rear end of the trigger bar (Fig. 5.126).

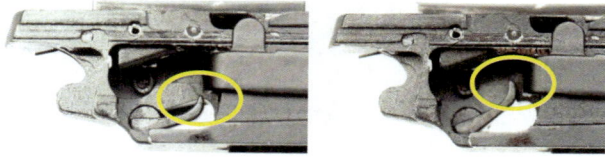

Fig. 5.126: Function of the manual safety: Hook-shaped device behind trigger bar, disengaged (left), engaged (right).

The *PT 24/7 PRO* had a trigger safety, in addition to the manual safety and the striker safety. The part of the trigger blade, which was visible from the outside, seemed to be made of one piece. However, the trigger consisted of a trigger blade, a hook and several other parts. The trigger safety was a hook-shaped part, coming from the trigger blade and reached rearward (see item 6 in Fig. 5.127 and Fig. 5.128). This hook would stay engaged with the trigger bar until the trigger was being pulled. The rotation of the trigger blade would lift the hook and free the trigger bar. TAURUS discloses this embodiment in patent U.S. 7,690,144 B2.

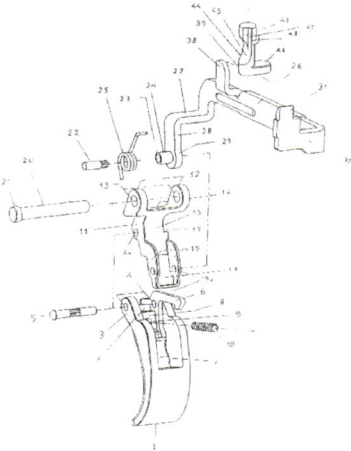

Fig. 5.127: US 7 690 144 B2, sheet 5

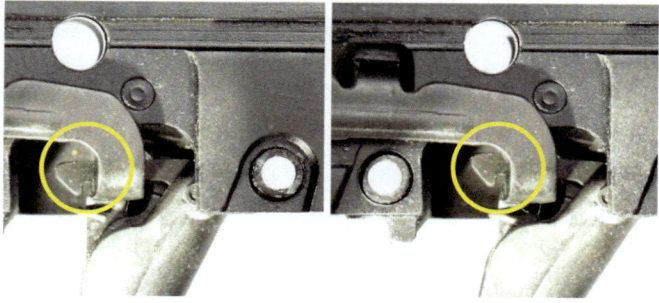

Fig. 5.128: Trigger safety:
Engaged: Hook couples trigger blade and trigger bar (left);
Disengaged: Hook is lifted as trigger is being pulled.

The patent however does not consider how reliably this type of a trigger safety works. If a trigger blade is made of metal, which can be heavy, it should be noted that in some cases where the gun is dropped, the trigger blade might move by inertia at the moment the weapon is grounded.

Patent research TAURUS

Assignees' search terms were "Forjas Taurus" and "Taurus International." At the time the survey was carried out, the most recent TAURUS patents were US 9,127,903 B2, US 8,752,322 B2 and US D687,505 S, filing date April 21, 2014, January 11, 2013, and January 13, 2012, respectively. They described a gun with a curved outline, which is referred to in the disclosures as "body contoured handgun." The curved outline is said to be better contoured to a person's body than regular straight handguns, thus, it would enable the user to carry the gun comfortably for extended periods of time. The patents were followed by the launch of the TAURUS *180CRVL* chambered in .380 ACP in 2015. This pis-

Fig. 5.129: TAURUS *180CRVL*, Body Contoured Handgun.
(© TAURUS)

tol also known as "Curve," has a built-in laser aiming device and two white light LEDs (Fig. 5.129). An integrated laser is a feature already known from the SMITH & WESSON *Bodyguard 380* and other manufacturers.

With its disclosure EP 2,525,185 A1, dated January 15, 2010, TAURUS publishes details of a "pistol with a firing mechanism that can easily be adapted to various modes of operation." Background of the invention is to reduce cost in production and engineering by a firing mechanism designed with interchangeable components. This firing mechanism may be converted to operate in double-action, single-action and alternative double-action/single-action mode.

Summary TAURUS *PT 24/7 PRO*

The "Ribber Grip" of the *PT 24/7 PRO* was perceived as a unique feature. So far, no other manufacturer has used two different polymers on the grip handle. How long-lasting and durable this embodiment is, is not known at this time; but TAURUS does not use this feature on the second generation of its *24/7*.

The constructive implementation of the trigger safety proved to be user-friendly because it used internal components; it would not affect the form of the trigger blade and the feel when the trigger was being pulled. Whether this trigger safety would reliably protect the weapon against accidental discharge when the pistol was dropped, was not tested.

The field-strip procedure of this model can be considered as the negative feature of the guns in this survey. Disassembly required more steps than for other pistols. Taking the gun down was not possible without reading the user manual, and the trigger had to be pulled to decock the gun. The trigger also had to be pulled prior to securing the pistol with the internal lock.

Fig. 5.130: TAURUS *180CRVL* with detachable Belt Loop Clip.
(© TAURUS)

5.11   WALTHER *PPX* (9 mm Luger)

The measurements and technical data according to the checklist are summarized in Table A21.9, p. 471. A check on the overall dimensions almost confirmed the specifications stated by the manufacturer. 186 × 34 × 143 mm was specified by WALTHER. Checking the dimensions revealed a different value for the width. The pistol measured 33 millimeters.

Measuring the trigger values observed a trigger pull force of 22 N at 7 mm trigger travel. 111 N were required to rack the slide. Manually moving the slide to the rear could be done comfortably. Much like it had been already seen with the GLOCK sample, the generously-sized surfaces on the sides of the slide seem to provide great traction when racking the slide.

Fig. 5.131: Striker indent with minimal drag.

The indent in the copper crusher was 0.33 mm with a diameter of 1.26 mm. On the primers from fired casings of reference ammunition there were no flashy signatures visible (Fig. 5.131).

The PPX offered a loaded chamber viewport at the rear end of the barrel hood, which permitted a visual control of the chamber. Included in the design of the slide were a firing pin safety and two drop safeties. Disassembly could be done by locking the slide back, and then rotating the take-down lever clockwise. Field-stripping did not require any manual decocking manipulation; and was possible to do even with a magazine locked in place. Assembly was possible with a magazine installed, if the user levered the rear end of the slide over the first cartridge of the magazine.

The PPX had a left side slide stop lever, a reversible magazine release, and came with two 16-round magazines. MSRP was $449 U.S.; online, the gun was available for $355 U.S. in September 2014.

The obtained values for the beginning and end of the uncoupling, and total slide recoil travel of the sample gun were 3.1 mm, 8.1 mm and 49.1 mm respectively. During shooting, a muzzle velocity of 358 m/s was measured, and an average slide recoil velocity of 5.4 m/s was recorded. The calculated value for the theoretical mean slide recoil velocity was 5.4 m/s.

The slide had a width of 29 mm, and its height measured at the recoil lug was 37 mm. The weight of the slide assembly was 345 g, and the weight of the barrel was 110 g. Barrel length was 102 mm.

Examination Regarding Innovative Features

The WALTHER *PPX* pistol's price was at the lower end, compared to the rest of WALTHER's law enforcement and home defense product line up. WALTHER promoted this hammer-fired pistol as its entry-level model that wouldn't sacrifice on quality and technology while selling for an attractive price. Three features appeared interesting from a technical standpoint: (1) two passive safeties, (2) use of standard mechanical components and (3) two major components were optimized for economical production; barrel and slide.

Fig. 5.132: WALTHER *PPX*

The slide was milled from solid steel and its geometry would allow efficient machining because its design involved barely any functional dimension. Any functional dimensions that would involve high demand of conventional machining, and thus, cause expenditure in time and result in higher costs of manufacturing of the slide, were integrated into a separate assembly made of polymer. The polymer held small parts and was inserted during assembly into the rear of the slide, behind the breech face (Fig. 5.133). Thus, the manufacturer could mill out steel generously in front and behind the breech face, by simple and time-saving roughening.

Fig. 5.133: Slide insert (1), guide block (2), extractor (3) and slide (4).

The earlier mentioned polymer insert carried most of the slide's functional parts (Fig. 5.134). An interesting embodiment was the design of the left side breech face guide block. This guide block was a simple in design MIM'ed part (pos. 2 Fig. 5.133), which could be used to adjust the left side locating face for the base of the cartridge, according to the caliber the pistol was chambered in. The actual part was shaped like a lever and had a location peg in the middle. With this peg, the lever-shaped guide block was held in place, as the peg would sit in a matching hole inside the slide. Eliminating the need to machine rectangular-shaped breech face guide blocks out of a solid bar of raw material seemed to allow a more efficient machining of

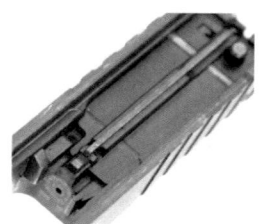

Fig. 5.134: Assembled slide with insert.

the slide. From a logistic standpoint, it even adds flexibility to production planning, because this approach would enable the manufacturer to decide at the end of the assembly line which caliber to build. Apparently, this eliminates inventory of machined slides of dedicated calibers.

The extractor was held in place in the same manner as the lever-shaped guide block insert, but on the other side of the breech face. The location peg served as a pivoting axis, which allowed the extractor claw to engage the extractor groove of the cartridge. A compression spring in the slide insert would place a load on the rear end of the extractor.

The polymer of the slide insert was protected at its center base line in longitudinal direction by a long thin metal plate. It actually formed the pickup rail. If it wasn't made of metal, this rail might be subject to wear as it rides on top of the uppermost cartridge of the magazine when the slide reciprocates.

The function of the firing pin safety was further supported by two passive safeties, which were supposed to work separately from each other in various drop positions. The first passive safety was a slide bar, designed to block the safety plunger, i.e. the actual firing pin safety, the function of the second safety was to engage the firing pin directly.

If the weapon was being dropped with muzzle pointing upward, the momentum of the slide bar would cause it to block the safety plunger as the rear end of the gun would hit the ground (pos. 2 in Fig. 5.135). This passive safety device might be necessary in the previously described drop position, where the momentum of the trigger would be enough to release the fire-control mechanism and theoretically disengage the firing pin safety. As the trigger and trigger bar are forced to the rear in this event, they would do what they are supposed to do when they are pushed backward.

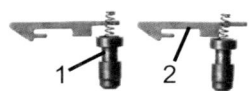

Fig. 5.135: Safety plunger (1) blocked by slide bar (2) (right).

The second passive safety was in the form of a rather simple shaped piece of mass made of MIM'ed material. It appeared to be designed to work whenever the firing pin safety would become accidentally forced out of its default position, i.e. its engagement with the firing pin. The piece of mass was moveable in the same orientation as the firing pin safety. Both

Fig. 5.136: Firing pin shown disengaged (top), and engaged by drop safety (bottom, schematic).

components would be set in motion when this gun was grounded upside down. The piece of mass would block the firing pin while the safety was disengaged (Fig. 5.136).

Two assemblies (Fig. 5.137) were installed in the frame. The front assembly carried the slide stop lever as well as trigger parts, while the rear group consisted of the firing mechanism. Other pistols in this survey used chassis, which carried the trigger mechanism and most part of the firing mechanism. These chassis were milled out of solid aluminum alloy. The PPX inserts seemed to be comparatively simple sheet metal stampings, easy to manufacture, and made of hefty material.

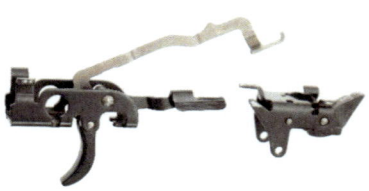

Fig. 5.137: Stampings: front assembly and rear assembly (right).

The hammer-fired action was a fully cocked SA firing mechanism. Although the hammer would pivot rearward, the cocking condition of the firing mechanism would remain the same. In carry condition, the hammer was flush with the slide. When the firing mechanism was decocked, the bobbed hammer was down, slightly below the slide's rear end (Fig. 5.138). In the case of a misfire, the trigger could not just be pulled again, but the slide had to be racked before the trigger could be pulled again to discharge a new round.

Fig. 5.138: Hammer in carry condition, and uncocked (right).

The barrel was a three-piece design: barrel tube, barrel block, and feed ramp (see Fig. 3.19, p. 70, Fig. 5.139 and Fig. 5.145); barrel block and feed ramp were MIM'ed parts. The barrel tube was screwed into the barrel block, and the feed ramp was attached to a rim at the barrel's chamber end. This combination was perceived to be unique in the field of pistols, however, the basic idea of screwing a barrel into a receiver has been commonly known with rifles for centuries.

Fig. 5.139: PPX barrel

The control for the coupling and uncoupling of the barrel was done by a downward facing cam. This design is common practice, but the PPX used a simple, standard cylindrical pin as a counterpart, which engaged the barrel block's cam to control the decoupling and also safeguard the coupling. The bearing for this pin was in the front insert, above the trigger, where it was mounted in the side panels' holes (Fig. 5.140). Once the assembly had been removed from the frame, this cylindrical pin could simply be pulled out of the assembly.

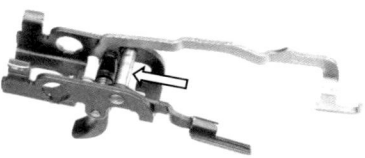

Fig. 5.140: A cylindrical pin (see arrow) is used to control coupling and uncoupling.

Patent research WALTHER

At the time of this survey, WALTHER's latest patents were related to the PPX, which were the slide insert, the recoil spring assembly, and a manufacturing method for the barrel.

The disclosure DE 10,2010,035,962 A1, dated August 31, 2010, focuses on a slide, which is customizable with simple means and with relation to the barrel as well as different calibers. Essentially, this should be done with a replaceable breech face guide block next to the breech face. This would allow a cost-effective manufacturing of slides, because the functional surfaces of slides can be manufactured with less effort (Fig. 5.141).

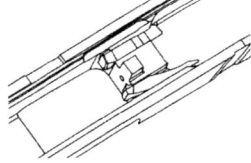

Fig. 5.141: DE 10,2010,035,962 A1, page 7

Patent DE 10,2010,047,500 B4, dated October 5, 2010, shows the polymer slide insert (Fig. 5.142). It describes a slide, that's production costs are lower than the costs of conventional slides, but without affecting quality and reliability of the pistol. As it is explained, conventional slides would have the disadvantage of combining several functional features, which are coupled to functional dimensions, and the production costs of these functional dimensions would be high. This could be remedied, by replacing the functional space behind the breech face with a compact insert. This insert can be an injection-molded component and can be connected detachably to the slide.

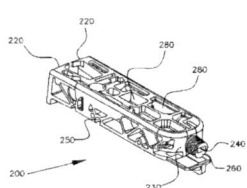

Fig. 5.142: DE 10,2010,047,500 A1, page 6

Patent DE 10,2011,115,771 B4 dated October 12, 2011, de-scribes a recoil spring assembly that is supposedly easy to as-semble by hand, and does not rely on a snap-fit connection between the guide rod and a retaining member to capture the spring. According to the described embodiment, two keepers are used, which were inserted into the last coils of the recoil spring during assembly of the captured recoil spring (Fig. 5.143).

Fig. 5.143: PPX recoil spring assembly: Components and assembled (below).

Patent DE 10,2011,114,038 B4, dated September 8, 2011, discloses a barrel block and a feed ramp (pos. 150 in Fig. 5.144) on a pistol barrel. According to the embodiment of the invention, this barrel is no longer made of one piece, as this is supposed to increase cost of material and it would be labor intensive in production. Also, the modular design allows a higher adaptability to different calibers, since the components which depend on the size of the cartridge can be manufactured independently from each other. In addition, the connection be-tween barrel tube and barrel block by a threading is described. The barrel has a rim at its chamber end, which engages a cut out on the feed ramp to permanently embrace the feed ramp.

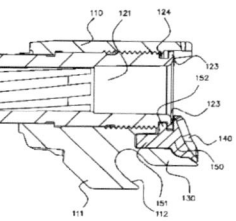

Fig. 5.144: DE 10,2011,114,038 A1, page 5

Summary WALTHER *PPX*

While other manufacturers are changing over from hammer-fired pistols to striker-fired guns, e.g. HK *VP9* or SIG SAUER *P320*, WALTHER was going the opposite direction, coming from striker-fired pistols, the company introduced the hammer-fired *PPX*. Not only because of this, but *PPX* took a prominent position in WALTHER's product portfolio. Especially from a production perspective, it seemed to have been the most interesting among the samples in this survey.

Judging from the machine marks, the breech face was economically milled with an end mill cutter. The breech face was part of the slide, milled out from solid steel, hence, it could be assumed that it was produced with reliably controlled manufacturing toler-ances. A polymer insert carried most functional parts of the slide. It contained two drop safeties, an ordinary safety plunger and other parts. The insert carried the functional parts reliably.

By a simple change of the breech face guide block, extractor, barrel, and magazine, the PPX could theoretically be converted to a different caliber, if WALTHER were to offer this option.

The barrel consisted of a novel combination of different production techniques: a barrel tube made of solid steel was connected with the barrel block by a threading and secured the feed ramp at the same time. When compared to other pistols chambered in the same caliber, the slightly larger dimensions of the PPX barrel block were obvious.

The frame carried two assemblies. The load-bearing components of the two inserts were rather simple appearing stamped and bent sheet metal components. The controls on the PPX's frame were unilateral, but the magazine release was reversible.

Recently the company started selling a facelifted model of the *PPX*, the WALTHER *CREED*.

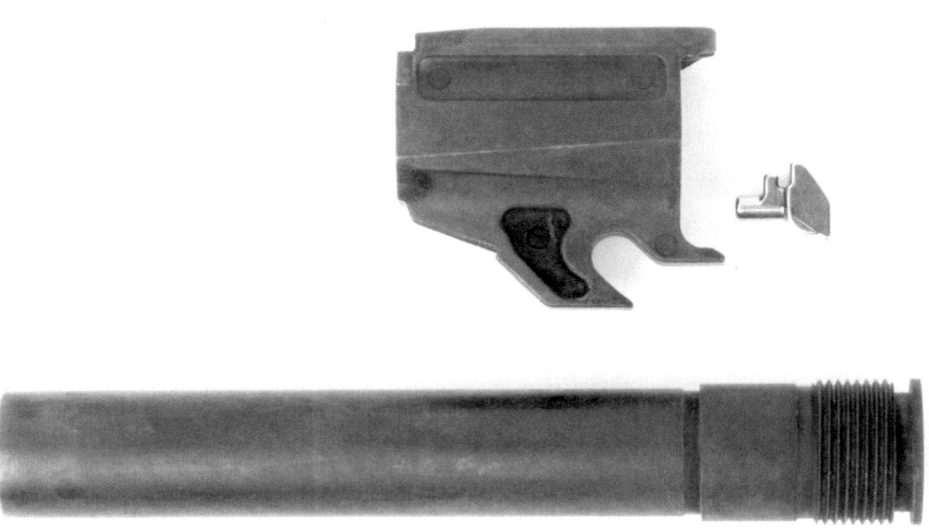

Fig. 5.145: PPX barrel components: Barrel Block, Feed Ramp, Barrel Tube (clockwise).
(© WALTHER)

5.12   Brief Discussion of Random Pistol Models

ARSENAL FIREARMS *Stryk B (9 mm Luger)*
The *Stryk B* is the successor of the *Strike One* made by ARSENAL FIREARMS. In the U.S. a more compact version of this gun is offered as ARCHON FIREARMS *Type B*. Proprietary to both versions is the low bore axis of a barrel, which moves along a straight line parallel to the slide's feedway and neither tilts nor rotates. The link between barrel and slide is carried out by a vertically moving, locking block. The coupling is released after a certain recoil travel by guiding the connecting element transversally to the barrel downwards.[645]

Disclosure   WO 2013/014656 A1   describes this design. This was followed by an   application   WO 2017/182843 A1, which relates to an improvement of the

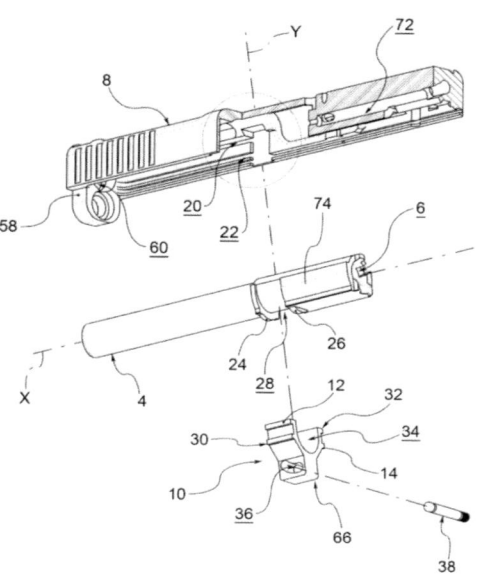

Fig. 5.146: *Strike One* components, patent application WO 2013/014656 A1.

locking block. The tab-like locking projections of the block's bifurcated arms are described to be heavily stressed, and the modification proposed in the 2017 application sets out to reduce the wear of the locking element.

The locking block appears to be an additional individual part, which has to be remembered during take down for cleaning. Usually one expects to deal with five components: frame, barrel, slide, recoil spring assembly and magazine. When reassembling the *Stryk B*, one should remember to install the locking block.

Other features: The striker will be cocked as the slide travels rearward. The shape of the breech face allows, in principle, the installation of barrels chambered in different calibers and does not require replacement of the slide when changing to a new caliber. The polymer shell is a non-serialized interchangeable module. The magazine release can be installed on either side. The slide's sight cuts will accept any accessory sights available for GLOCK pistols. In combination with a simple caliber change, its noted features include a certain degree of individualization.

BERETTA *Nano (9 mm Luger),* BERETTA *Pico (.380 ACP)*

BERETTA *Nano* chambered in 9 mm Luger and BERETTA *Pico* in .380 ACP are examples of low-profile concealed carry pistols that cannot be imported into the USA because of the pistols' small overall dimensions, but can be produced and sold in the United States. The *Nano* was 143 mm, 106 mm and 23 mm in length, height and width respectively, while the *Pico* was 130 mm long, 100 mm high and 18 mm wide. In 2016, BERETTA calls the *Pico* the "thinnest .380 ever made for super concealment."

Fig. 5.147: BERETTA *Pico*

The one-piece slide is made of steel. Both models have a vertical tilt-barrel design, short-recoil-action and according to information from the manufacturer, theoretically the *Pico* could also be modified to chamber the .32 ACP cartridge. No safety was found in the slide (Fig. 5.148).

Fig. 5.148: Slide (*Pico*)

The somewhat smaller of the two, the *Pico*, was hammer-fired. The larger *Nano* was striker-fired and could be decocked for disassembly without requiring the user to pull the trigger; the striker would become deactivated by depressing a lateral plunger with a pointed object. Firing mechanism and trigger mechanism were in a steel chassis (Fig. 5.149). This chassis was of similar construction as previously described with other sample pistols of this survey. The chassis could be removed from the frame and slide rails were part of the chassis' design.

Fig. 5.149: Frame with chassis installed (*Pico*).

The slide stop lever of the *Pico* was nearly flush with the slide and rested in a recess molded into the polymer frame (Fig. 5.150). According to the manufacturer, the purpose of the slide-lock is limited to just locking back the slide on an empty magazine.

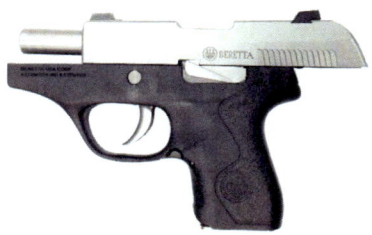

Fig. 5.150: Slide locked back (*Pico*).

The slide stop lever was almost too flat to lay the thumb on it and release the slide comfortably.

Pico offered an ambidextrous paddle magazine release, which was incorporated into the rear of the trigger guard (s. Fig. 3.46, p. 91), the *Nano* was equipped with a reversible push-button mag release.

Patent research BERETTA

Patent US 8,151,511 B2 dated April 8, 2010, claims a mounting mechanism for gun sights, which provides simple and easy replacement. The invention avoids the disadvantage of a fastening by means of a dovetail, which usually requires special tools to cut the dovetail, and are expensive to manufacture because of the required close tolerances. The proposed embodiment uses detent balls, which register with locking sockets (pos. 31–35, 41, and 42 in

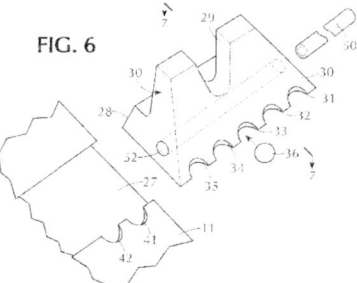

FIG. 6

Fig. 5.151: US 8,151,511 B2, sheet 3

Fig. 5.151), formed in the slide (item 11) and sight when engaged by a sliding lock pin (50). Detachment is achieved by removal of the lock pin. The patent gave no information on how much the point of impact would be affected per increment when a pistol's sights need adjustment for windage and if the lock pin has the tendency to start drifting.

In patent application US 2011/0,119,979 A1, "Electronic Device for a Firearm," dated April 20, 2009, BERETTA discloses general thoughts about its i-PROTECT Black Box (see p. 145f). According to the objective of the invention, an electronic device for a firearm suitable to supply information on the armed staff actively in a reliable manner, and which is practical to use and easy to install in a gun. The electronic device transmits information to a "Central Office," is designed to be attached to a firearm's Picatinny rail. It is also designed to be removable. Moreover, it has no external electrical interfaces to the weapon; it is pointed out that the data would be only transmitted when a shot was fired.[646] Starting from claim number 44, a device for locating a firearm is described. Claims 1 to 43 have been deleted, as these were not potentially patentable. The remaining notes are very generic and are concerned with execution details such as the arrangement of control elements. A patent, granted to these descriptions is US 8,720,092 B2, dated May 13, 2014.

In patent US 8,572,878 B2, filed May 28, 2010, BERETTA describes a decocking mechanism of a striker-fired pistol, which can be field-stripped without pulling the trigger. Also, the embodiment would protect the handgun from any unwanted discharge of the striker caused by accidental drops or rough handling. According to the patent, the mechanism has a minimum of parts and lends itself to manufacture using inexpensive, high volume techniques such as stamping and casting, rather than comparatively expensive machining processes; and utilizing off-the-shelf elements as well. The decocking does not work automatically. Described is a push-pin decocker, which requires a simple tool, such as a pointed object to operate it. The push-pin (22) (see Fig. 5.152) moves the sear transverse to the direction of the gun and releases the striker. The striker falls and its forward travel is blocked and stopped by the striker safety (47). This description reminds one of the mechanism found on the *Nano* pistol.

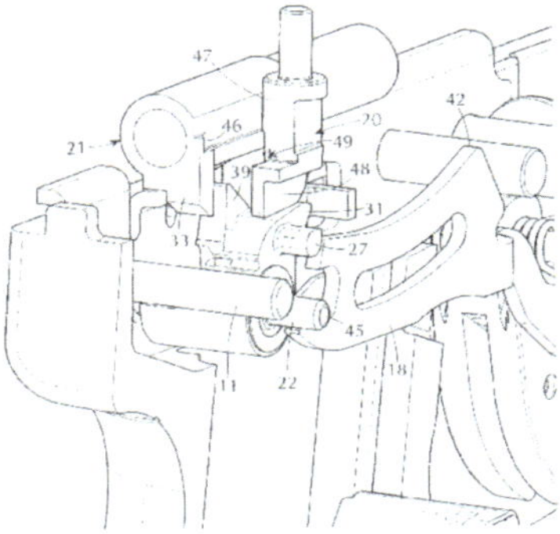

Fig. 5.152: US 8,572,878 B2, sheet 9, Fig. 9

BOND ARMS *BullPup9 (9 mm Luger)*

BOND ARMS INC. manufactures the *BullPup9* pistol chambered in 9 mm Luger. Its design is based on the *XR Series* by BOBERG ARMS. These pistols do not strip the cartridges toward the front of the magazine for feeding, but to the rear (Fig. 5.153).[a] What BOBERG originally called "reverse feed technology" allows the use of space above the magazine, whereby the barrel extends over the top of the magazine, which means the pistol's barrel can be longer

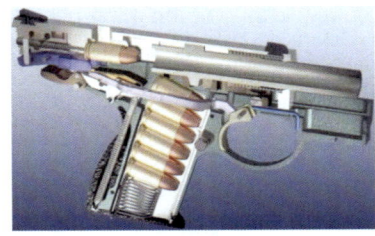

Fig. 5.153: BOBERG ARMS *XR Series* with chamber above the magazine.

(© BOBERG)

without adding to the total length of the pistol. A longer barrel ultimately may contribute to an increase in muzzle energy.[647]

Patent research BOND ARMS

In 2016, BOND ARMS INC. acquired the patent and rights for the *XR9*. Patent US 8,061,255 B1 discloses the cartridge pick-and-place mechanism.

---

[a] The functional principle of the BOBERG pistol is similar to the MARS model 1899 pistol. (Sterett, 1961 p. 90)

HI-POINT *C9 (9 mm Luger)*

The HI-POINT *C9* by STRASSELL'S MACHINE INC. (formerly BEEMILLER, INC.) was available for $150 at retail. When observing the striker-fired pistol, its simple look and a rough finish were apprehensible. According to HI-POINTFirearms.com, its pistols come with a lifetime, "no questions asked" warranty. For cost reasons, the company builds all parts of the gun. The slide consists of a zinc aluminum alloy, which is powder-coated. Despite the fact that HI-POINT pistol models are chambered in powerful calibers, all models are blowback designs.

Fig. 5.154: STRASSELL'S MACHINE, INC. HI-POINT *C9*

An external slide stop switch did not exist on the sample. Instead, the pistol had an internal slide-lock, similar to a WALTHER *PP*. Aside from this, the gun had a magazine disconnect.

Pulling the sample's slide fully to the rear worked only after several attempts. This made it initially impossible to lock the slide back in the open position. Eventually, when the slide was locked back, the striker nose would protrude approximately 7 millimeters over the breech face (Fig. 5.156). With reference to the other pistols of this survey, this was found to be somewhat disturbing, even though it might not affect the reliability of the pistol.

Fig. 5.155: HI-POINT *C9*, front view.

Field-stripping the gun required a tool. Take-down procedure was to lock the slide in open position with the safety switch, and to push a roll-type cross-pin out of the frame with a punch.

Patent research STRASSELL'S MACHINE, INC.

The patent research with regard to publications by Strassell, Beemiller and Hi-Point was inconclusive.

Fig. 5.156: Striker nose while slide is retracted to the rear.

LAUGO ARMS *Alien (9 mm Luger)*

Ján Lučanský started to work on a radically new pistol in 2013. He carried forward a previous concept, which he had begun working on as early as in 2004. Later on, in 2017, LAUGO ARMS CZECHOSLOVAKIA S.R.O. was established by former CZ employees and an investor. In September 2018 their handgun was unveiled to the press.

Fig. 5.157: LAUGO ARMS *Alien*.
(© LAUGO ARMS)

This first model of the *Alien* series is optimized for competition shooting. The pistol is equipped with a left side slide release, an interchangeable backstrap, a reversible magazine release and SA trigger. The latter is named H.I.H.S, which is short for Hybrid Inverted Hammerless System.

The gun operates on the principle of a gas retarded blowback action. The barrel doesn't move during firing and the sights remain motionless as well. The latter are part of an upper carrier (see pos. 56 in Fig. 5.158). Thus, attaching accessories to the muzzle or any aiming devices on the upper does not affect the firing sequence and feeding. In addition, shooters can switch between various aiming devices by exchanging the upper carrier. This can be done without any tools.

The design objective was to move the bore axis closer to the user's hand and to achieve a low center of gravity. For space reasons the recoil spring was moved upwards. Its orientation is parallel to the fixed barrel, but, in contrast to the FN BROWNING *Model 1900*, slightly to the left of the gun's vertical center plane.

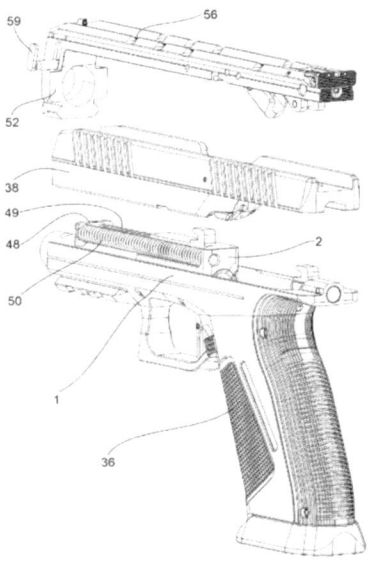

Fig. 5.158: Components,
US 2019/0310036 A1, sheet 9.

With this architecture it would be difficult to fit a hammer in the frame. The internal hammer was relocated to the top. Not just the sights but also the sear, mainspring, and hammer are part of the upper carrier. The hammer uses an up-side-down design similar to the IZHEVSK *IJ35* .22 cal. competition pistol.[648]

The slide and frame are not connected by rails in the conventional sense. The frame is a fusion of several components. The frame's top is made from steel. An aluminum grip handle is bolted on to the frame. The trigger guard is part of this substructure, which also carries interchangeable grip panels and the backstrap.

Patent search LAUGO ARMS CZECHOSLOVAKIA S.R.O.

Patent application WO 2017/217939 A1 lists Ján Lučanský as the inventor. This document explains in the *Prior Art* section that aiming devices attached to rapidly moving gun parts can be a disadvantage of current gun designs, as well as the requirement for a precise frame to slide and barrel fit to achieve utmost accuracy of a pistol. The patent application centers around a pistol design with a super-low bore axis, and the unique layout is disclosed in detail in the application's claims.

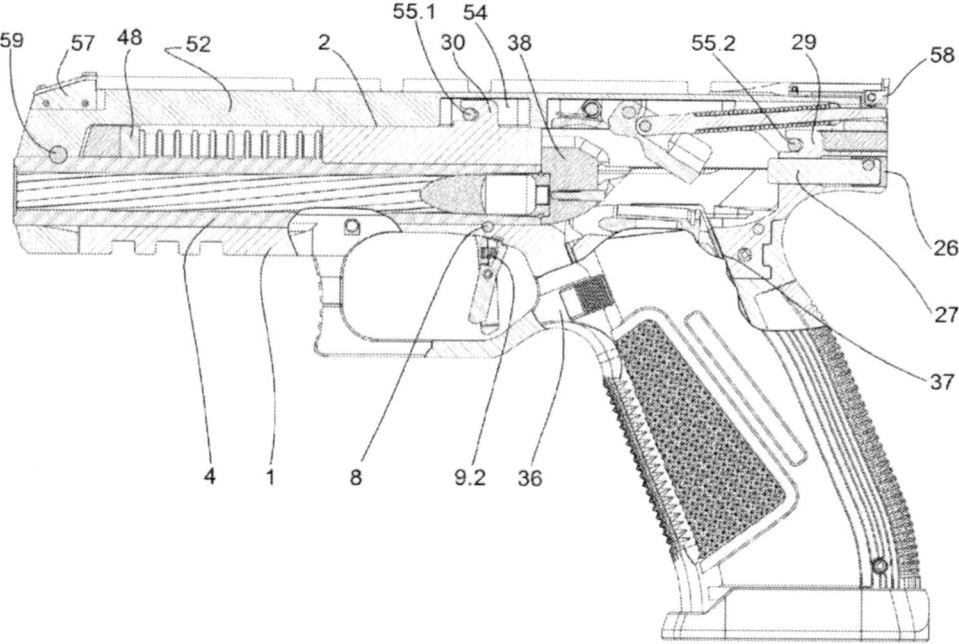

Fig. 5.159: US 2019/0310036 A1, sheet 9, Fig. 11.

SCCY CPX-1 and CPX-2 (9 mm Luger)

Joe Roebuck established SCCY INDUSTRIES, LLC in Florida in 2003. The product range of its pistols is built around two base models, *CPX-1* and *CPX-2*, chambered in 9 mm. The *CPX-1* has a thumb safety, while the *CPX-2* doesn't (Fig. 5.160). All SCCY 9 mm variants are based on the two and are available in a variety of colors for slide and frame, as well as chambered in .380 ACP.[649] Similar to KEL-TEC *PF-9* and RUGER *LC9*, the SCCY pistols are hammer-fired, with a bobbed hammer.[650] The street price was $270 U.S. for the *CPX-1* and $247 U.S. for its sibling without thumb safety – as of September 2014.

Fig. 5.160: SCCY *CPX-2*
(© SCCY)

Barrels and slides are made of 416 stainless steel profile material, and the chassis inside the frame is made from a 7075-T6 aluminum alloy.[651] The barrels have a peculiarity: the rifling has seven lands and grooves and a 16" rifling pitch (406 mm). According to Roebuck, this leads to an improvement in the accuracy of the *CPX* pistols.

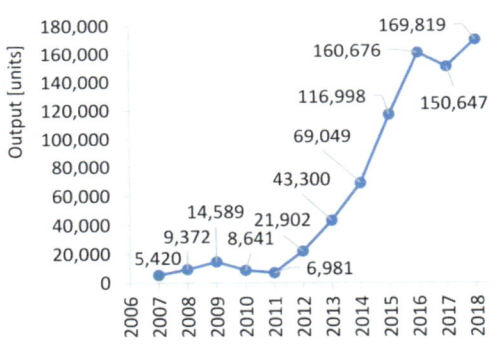

Fig. 5.161: SCCY, Annual Pistol Production, 9 mm and .380 cal. pistols.
(Compiled from ATF AFMER)

Patent research SCCY

The patent research did not show any publications. Key words were SCCY and Joe Roebuck.

SIG SAUER P320 (9 mm Luger)

With the *P320* pistol, SIG SAUER added a striker-
fired pistol to its product range previously dominated
by hammer-fired handguns. A special feature of the
gun was its short trigger reset.[652] The gun was available in full size (203 mm overall length) and carry
(183 mm), chambered in either 9 mm Luger,
.357 SIG or .40 S&W. The magazine release was
reversible, and the slide stop was ambidextrous.
The grip size could be modified by interchanging the
polymer grip module. Inside was a metal chassis, a

Fig. 5.162: SIG SAUER *P320 Carry*

serialized component that is technically the firearm. The pistol had a striker safety and
an out-of-battery disconnect, and could only be disassembled if the magazine was removed from the pistol and the slide locked back. SIG SaUER calls this a "3-Point Take
Down Safety System." With this system, the magazine must be removed, the slide
must be locked back, and the user does not need to pull the trigger to decock the
weapon before field-stripping,[653] while the stainless steel slide of this striker-fired pistol
comes down. Four other safeties are available upon request: trigger safety, frame
mounted ambidextrous manual safety, loaded chamber indicator, and magazine disconnect.[654]

The modular design goes beyond the *P320* series of pistols; *P250* and *P320* share
the same barrels, magazines, and frame modules. Should a user require a grip module
to adjust the grip size, he can get them for $46 U.S. each.[655] There is a choice of three
sizes: small, medium and large.

Patent research SIG SAUER

SIG SaUER's patent US 8,561,334 B2, dated December 15, 2010, discloses an ambidextrous magazine release, with a spring that is integrated into the
latch. The spring has two tasks, it holds the magazine and the latch (Fig. 5.163) at the same time.

The described embodiment appeared interesting
and of a certain quality. Which is in contrast to the
commonly encountered embodiments of recent
years which employ simple looking metal rods as

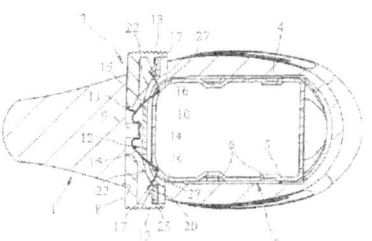

Fig. 5.163: US 8,561,334 B2,
sheet 2, Fig. 2

mag catch springs. Depending on an individual manufacturer's brand positioning, a cheap looking metal rod, which is visible in the magazine well, might be perceived as low quality by consumers.

SIG SAUER P365 (9 mm Luger)

Explicitly slim 9 mm pistols are regarded to be ideal concealed carry guns. To make a carry pistol less prone to imprint its shape on garments concealing it, such handguns often use single-stack magazines with consequently limited magazine capacity. The P365 micro-compact uses double-stack magazines, which can hold 10 rounds or more. For larger pistols in the next size class, double-stack magazines are common, but their pistol grips are bulky. Not so with the SIG's. The main part of the P365 is a fire-control chassis, which sits in a particularly thin-walled polymer module.

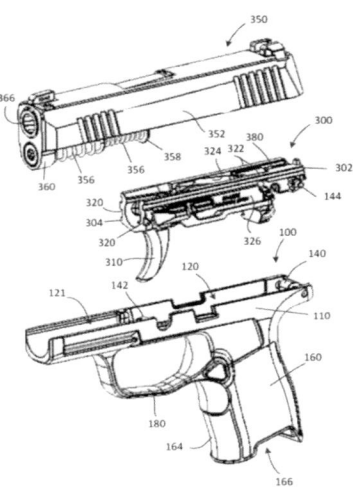

Fig. 5.164: US 2019/0195596 A1, sheet 7, Fig. 13.

Similar to the magazine of the HK P7M13 pistol, the magazine of the SIG SAUER has a stepped shape between the double-stack portion and the upper single-stack end, just where the chassis occupies space in the frame. Patent applications US 2019/0195584 A1 and WO 2019/126699 A1 attempt to derive a patent claim from this distinctive feature.

The chassis carries the fire-control group and other functional components. The majority of these elements are metal injection molded parts.

Application US 2019/ 0107353 A1 reveals a specific design of the sear, US 2019/0195596 A1 discloses a reinforcing metal bracket molded into the polymer module for better distribution of recoil forces every time the slide hits the back stop.

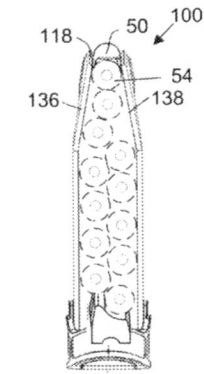

Fig. 5.165: US 2019/0195584 A1, sheet 10, Fig. 19B.

Muzzle velocity was a mere 305 m/s. Consequently, the slide velocity v_{S1} was 7.7 m/s (applying Formula A19.7, p. 416).

TARA *TM-9 (9 mm Luger)*

TARA[656] described its *TM-9* pistol as a weapon that can be disassembled without having to pull the trigger to drop the striker. The manufacturer referred to the *TM-9* pistol's trigger action as double action rapid engagement. It was a trigger with 2.8 kg trigger pull weight and three millimeters reset, where the firing mechanism is in a partially cocked condition after these initial three millimeters. Only when the shooter released the trigger fully, can the firing mechanism be completely decocked.[657] From this forward position, the trigger offered restrike capability and trigger travel was 10 mm.[658]

Fig. 5.166: TARA *TM-9*

The pistol allowed for high firing-hand placement and its bore axis was just 22 mm above the firing hand. It was equipped with an ambidextrous push-button magazine release. The magazine capacity was 17 rounds. The grip size could be adjusted by interchangeable backstraps, and under the backstrap was a tool that could be used as both punch and attaching a lanyard.[659]

Patent research TARA

No publications were found under the key word "Tara."

5.13   Results and Discussions

As for the results, a range of criteria from the checklist was selected. These criteria appeared to be of particular interest for a final comparison. The results were combined in groups for the following discussion; an overview in table format is in Appendix 22.

Trigger Characteristics, Copper Crusher Indent and Slide Racking Force

Table 5.3:   Trigger characteristics, copper crusher indent and slide racking force.

	CARACAL F	GLOCK G19 Gen4	HECKLER & KOCH VP9	KEL-TEC PF-9	RUGER LC9	SMITH & WESSON M&P9	SPRINGFIELD XDᴹ	TAURUS PT 24/7 PRO	WALTHER PPX	Average
Trigger Pull [N]	24	41	24	25	35	25	24	21	22	27
Trigger Travel [mm]	5	8	6	16	17	7	7	11	7	9
Trigger Work [Nmm]	69	122	67	280	380	77	68	63	84	134
Copper Crusher Indent [mm]	0.24	0.34	0.31	0.23	0.26	0.32	0.32	0.31	0.33	0.30
Diameter of Crusher Indent [mm]	1.26	0.6 × 0.96	1.23	1.1	1.13	1.05	1.17	1.1	1.26	1.16*
Slide Racking Force [N]	92	115	91	89	89	84	84	88	111	94

* w/o GLOCK

The average trigger pull was 27 N. The TAURUS *PT 24/7 PRO* had the lowest trigger pull (21 N), while RUGER *LC9* and GLOCK *G19 Gen4* were at 35 and 41 N respectively at the upper end of the pistols measured. For a pistol to qualify for IPSC production division, a minimum trigger pull of 22.27 N (2.27 kg or 5 lbs) is required for the first shot. TAURUS *PT 24/7 PRO* and WALTHER *PPX* would not have met this requirement.

No clear trend could be seen with regard to trigger characteristic. Trigger actions were noticed that combined a light trigger force with a short travel (such as CARACAL *F*: 24 N and 5 mm), as well as combinations with long travel (KEL-TEC *PF-9*: 25 N at 16 mm). RUGER's *LC9* had the longest at 17 mm.

Perhaps the decisive factor when designing the trigger mechanism was the manufacturer's target group as well as the intended use: the CARACAL F might be carried safely as a duty pistol in a holster; on the other hand, a KEL-TEC PF-9 might be carried loosely in a jacket pocket and may therefore require a heavy trigger pull, which is intended to prevent unintentional discharge.

Note: We are not condoning any of the guns were safe to be carried without a carry holster.

The trigger work is tested only as a criterion for police duty weapons in Germany while most part of the world might not have any interest in this evaluation requirement. KEL-TEC PF-9 and RUGER LC9 met and surpassed the minimum value of 150 Nmm demanded by the German police. The average value of the test samples was 134 Nmm.

The average indentation of the striker nose in copper crusher cylinder was 0.30 mm deep, that is, some of the samples gave an indentation which was less and thus did not fulfil the requirement of minimum depth of 0.3 millimeters as a theoretical characteristic of a sufficient impact. Especially the pocket pistols KEL-TEC PF-9 and RUGER LC9, but also the CARACAL F with 0.23 to 0.26 mm were 13 to 23 percent below the set value. The diameter of the impressions which the samples left in the copper crusher cylinder were in a close range from 1.05 to 1. 26 mm.

The GLOCK G19 Gen4 with its lance-shaped striker nose created a unique striker mark different from other pistols. It produced an oval instead of a spherical indent on the primers, and the marking was surrounded by a raised rectangle (see Fig. 5.31, p. 245); the latter was a consequence of the rectangular striker hole in the breech face.

GLOCK G19 Gen4 and WALTHER PPX had slide racking forces of more than 100 N. SMITH & WESSON M&P9 and SPRINGFIELD XD^M had the lowest measured value at 84 N. The mean value of all tested guns was 94 N.

Barrel and Slide Data, Uncoupling, and Muzzle Velocity

Table 5.4: Travels, masses and velocities.

	CARACAL F	GLOCK G19 Gen4	HECKLER & KOCH VP9	KEL-TEC PF-9	RUGER LC9	SMITH & WESSON M&P9	SPRINGFIELD XD^M	TAURUS PT 24/7 PRO	WALTHER PPX	Average
x_ℓ [mm]	1.6	1.6	1.6	2.3	1.9	1.6	1.8	1.6	1.5	1.7
Uncoupling, start x_{u1} [mm]	3.2	2.5	4.6	2.9	3.5	2.0	3.1	3.4	3.1	3.1
Uncoupling, end x_{u2} [mm]	5.4	5.0	7.7	5.2	5.2	4.8	6.4	5.2	8.1	5.9
Slide travel x_d [mm]	50.1	46.9	47.1	41.3	42.6	47.4	46.8	41.8	49.1	45.9
Slide gross weight (m_S) [g]	337	347	352	171	227	368	349	383	345	320
Mass, barrel (m_B) [g]	110	104	98	58	57	103	107	103	110	94
Barrel length [mm]	104.3	102.0	104.3	78.9	79.7	108.0	116.7	108.4	101.5	100.4
Lands/grooves, polygonal rifling	L/G	P	P	L/G	L/G	L/G	L/G	L/G	L/G	N/A
v_0 [m/s]	351	358	365	343	339	361	371	356	358	356
E_0 [J]	492	513	533	470	459	520	550	508	513	507
Recoiling Mass ($m_S + m_B$) [g]	447	451	450	229	284	471	456	486	455	414
$\bar{v}_{St}$ [m/s]	5.3	5.4	5.6	11.0	8.7	5.2	5.6	4.9	5.4	6.3
$\bar{v}_{Se}$ [m/s]	4.8	4.7	5.0	8.8	8.0	4.4	5.2	4.4	5.4	5.6

On average, the recoil mass had retreated by about 1.7 mm (x_ℓ) when the projectile cleared the bore. The average coupled rearward travel of barrel and slide was 3.1 mm. After this distance, the uncoupling of the barrel from the slide would begin. The fully coupled rearward travel on the SMITH & WESSON M&P was particularly short at 2.0 mm and the longest travel was found on the HECKLER & KOCH VP9 at 4.6 mm. After an average of 5.9 mm, uncoupling of the barrel from the slide was completed. The total slide recoil travel was averaging 45.9 mm, the longest was 50.1 mm (CARACAL F), and the shortest being 41.3 mm (KEL-TEC PF-9).

The average muzzle velocity of the projectiles was 356 m/s. The weapon having the longest barrel resulted in the maximum measured muzzle velocity: the SPRINGFIELD XD^M with the 116.7 mm barrel delivered a muzzle velocity of 371 m/s, which translated to a muzzle energy of 550 joules. The RUGER and KEL-TEC, even with the rather short barrel of the pocket pistols category, measured 339 and 343 m/s respectively, which equals to muzzle energy of 459 and 470 Joule. Interestingly, the projectiles fired from the KEL-TEC *PF-9* had a muzzle velocity at four meters per second faster and its muzzle energy was 11 Joule more than the RUGER *LC9*, despite the fact that the KEL-TEC *PF-9's* barrel was one millimeter shorter. A similar observation was noticed when CARACAL and GLOCK were compared: despite having a two millimeters shorter barrel, the polygonal barrel of the GLOCK pistol delivered a higher muzzle velocity and energy.

A general explanation for higher values from a shorter barrel could be in the distribution of measured values, in the shape of the barrel profile, that is, lands and grooves vs. polygonal rifling, or narrower dimensions of the bore and groove area of barrels with same barrel profile. The explanation of this phenomenon was not the subject of this survey.

If the SPRINGFIELD *XD^M* with its particularly long barrel was taken out from the evaluation, barrel lengths from 78.9 to 108.4 mm would remain, and despite the significant differences in barrel length, the determined values for v_0 and E_0 were within a narrow range from 339 to 365 m/s or 459 to 533 Joule. In this consideration, the HECKLER & KOCH *VP9* would lead the field. Fig. 5.167 shows values for v_0 and E_0 in the form of a bar chart, and the consistent increase in level becomes more obvious.

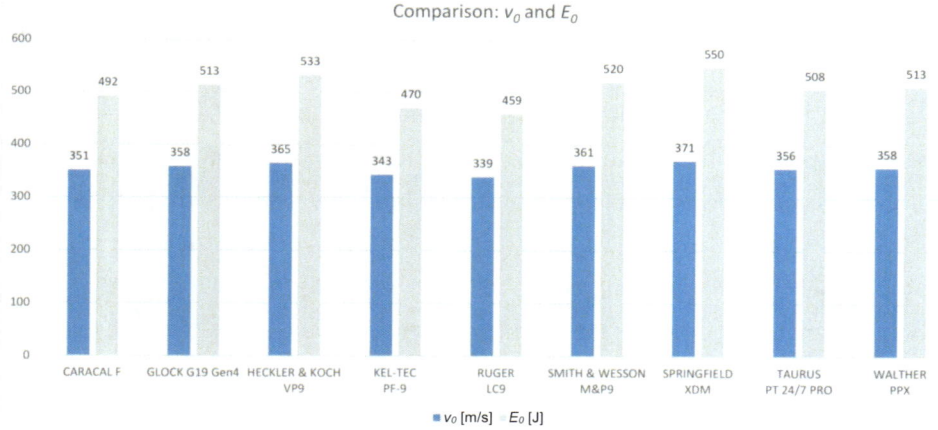

Fig. 5.167: Diagram with v_0 and E_0 values of the sample guns.

Mean Slide Recoil Velocity

Table 5.5:   Recoiling mass and mean slide recoil velocity.

	CARACAL F	GLOCK G19 Gen4	HECKLER & KOCH VP9	KEL-TEC PF-9	RUGER LC9	SMITH & WESSON M&P9	SPRINGFIELD XD^M	TAURUS PT 24/7 PRO	WALTHER PPX	Average
v_0 [m/s]	351	358	365	343	339	361	371	356	358	356
Recoiling mass $(m_S + m_B)$ [g]	447	451	450	229	284	471	456	486	455	414
$\bar{v}_{St}$ [m/s]	5.3	5.4	5.6	11.0	8.7	5.2	5.6	4.9	5.4	6.3
$\bar{v}_{Se}$ [m/s]	4.8	4.7	5.0	8.8	8.0	4.4	5.2	4.4	5.4	5.6
$\bar{v}_{Se} / \bar{v}_{St}$ [%]	91	87	89	80	92	85	93	90	100	90
$\Phi_{\ell u1}$	1:1.9	1:1.6	1:2.8	1:1.2	1:1.8	1:1.2	1:1.7	1:2.1	1:2.0	1:1.8
ζ_{SB}	1:3.0	1:3.3	1:3.5	1:2.9	1:3.9	1:3.5	1:3.2	1:3.7	1:3.1	1:3.3

The mass ratio ζ_{SB} for pistols from CARACAL, KEL-TEC, SPRINGFIELD and WALTHER was in the lower range of the test field (see Formula A13.5, p. 377). The values for $\Phi_{\ell u1}$ provide information about the safety taken into account in the design of the unlocking travel (see Formula A19.62, p. 434). KEL-TEC and SMITH & WESSON marked the lower range of the measuring field.

We observed that pistols with light barrels and slides resulted in high values for mean slide recoil velocity. The KEL-TEC *PF-9* had the fastest slide velocity in this survey; empirically, 8.8 m/s were measured, and a value of 11 m/s was calculated. Faster cycling slides are considered to be critical if they exceed 6 to 8 m/s, and KEL-TEC explains on its website in the FAQ section that, "All Kel-Tec firearms have an expected life of 6,000 rounds or more with proper care and maintenance."[660] Within the industry, a service life of 10,000 rounds is considered the usual minimum requirement, which meant KEL-TEC's was 40 percent lower than the industry standard.

Almost all empirical mean slide recoil velocities $\bar{v}_{Se}$ were lower than the calculated, theoretical values $\bar{v}_{St}$., with one outlier: WALTHER *PPX*. According to the empirical measurements, all slides reached as little as 80 to 100 percent of calculated velocities (see Table 5.5). The deviation was uniformly in the same direction, but with the exception of *PPX* the calculated values were higher than the measured ones (cf. Appendix 19).

Maybe the discrepancy was caused by friction and mechanical processes, which took place during recoil, and which would not allow a loss-free movement[a] of the slide. Also, the work needed to compress the recoil spring might have helped to reduce the slide's velocity. This work was included in the calculation by an estimated value.

Slide Outside Dimensions

Table 5.6: Slide outside dimensions.

	CARACAL F	GLOCK G19 Gen4	HECKLER & KOCH VP9	KEL-TEC PF-9	RUGER LC9	SMITH & WESSON M&P9	SPRINGFIELD XD^M	TAURUS PT 24/7 PRO	WALTHER PPX	Average
Width W, Slide [mm]	28.0	25.5	28.8	22.2	22.8	27.4	26.3	25.6	29.1	26.2
Height H, Slide [mm]	31.0	32.9	33.5	28.3	28.5	32.8	36.1	35.0	37.0	32.8
Cross-sectional area ($B \cdot H$) [mm²]	868	839	965	628	650	899	949	896	1077	863

The height of the CARACAL F slide appeared to be shorter, however the actual height (31.0 mm, see Table 5.6) was just 1.9 mm less than the value of the GLOCK G19 Gen4 slide (32.9 mm), if we want to use GLOCK as a benchmark. Since the slide of the GLOCK G19 was slimmer, the theoretical cross-sectional area for the raw material of the GLOCK G19 Gen4 amounted to 839 mm² and that of the CARACAL F was 868 mm². Theoretically, this could have an impact on procurement costs of raw material for the slide. In reality, the cost difference will be negligible, if the external dimensions of the slides do not differ dramatically, as shown here.

---

[a] Pistols were shot without magazine, i.e. any mag-induced influence was eliminated in the test.

Features

Table 5.7:   Features

	CARACAL F	GLOCK G19 Gen4	HECKLER & KOCH VP9	KEL-TEC PF-9	RUGER LC9	SMITH & WESSON M&P9	SPRINGFIELD XD^M	TAURUS PT 24/7 PRO	WALTHER PPX	Average
Magazine capacity [rounds]	18	15	15	7	7	17	19	17	16	15
Magazine release *	a	r	r	l	l	r	a	r	r	N/A
Slide stop *	l	l	a	l	l	a	l	l	l	N/A
Consistent trigger pull	yes	yes	yes	yes	yes	yes	yes	yes	yes	N/A
Hammer- (H) or Striker-fired (S)	S	S	S	H	H	S	S	S	H	N/A
Take down w/o pulling trigger	no	no	yes	yes	yes	yes	yes	no	yes	N/A
Strips w/mag	no	no	no	yes	yes	no	yes	no	yes	N/A
Assembles w/mag	no	no	no	yes	yes	no	no	no	yes	N/A

* ambidextrous (a), left side only (l), reversible (r)

The single-stack deep conceal pistols had a magazine capacity of seven shots. Full size pistols had a minimum magazine capacity of 15 rounds, and could go as high as 18 (CARACAL *F*) or 19 rounds (SPRINGFIELD *XD^M*).

Reversible magazine releases (5 out of 9 pistols, i.e. 55 percent) or ambidextrous magazine releases (on 2 of 9 pistols, 22 percent) were overrepresented in comparison to unilateral magazine releases; only two deep conceal carry pistols were equipped with fixed, one-side-only magazine releases. A different issue was the slide stops: ambidextrous slide stops were offered only by *VP9* and *M&P9*. The remaining guns, that is, 78 percent of the samples, had its slide stop lever on the left side.

All guns had a consistent trigger pull from the first to the last round. The TAURUS *PT 24/7 PRO* offered a second trigger action on top of its standard trigger. The firing mechanism of six out of nine guns (67 percent) was striker-fired, while three were hammer-fired (33 percent).

Particular attention was paid to the take-down function. It was checked whether it was possible to field-strip the pistols without the risk of an unintentional discharge. If the trigger had to be pulled prior to disassembly, it was regarded as a potential risk for the user. Six out of the nine sample pistols (67 percent) protected the user from an unintentional discharge. On CARACAL, GLOCK and TAURUS, the user had to pull the trigger.

Pistols from KEL-TEC, RUGER, SPRINGFIELD and WALTHER would allow to be disassembled when a loaded magazine was locked in place. KEL-TEC, RUGER and WALTHER could be assembled with a loaded magazine locked in place.

MSRP and Street Price

Table 5.8:   MSRP and street price (as of September 2014).

	CARACAL F	GLOCK G19 Gen4	HECKLER & KOCH VP9	KEL-TEC PF-9	RUGER LC9	SMITH & WESSON M&P9	SPRINGFIELD XD^M	TAURUS PT 24/7 PRO	WALTHER PPX	Average
MSRP [USD]	625	649	719	333	449	569	700	492	449	554
Street Price [USD]	460	538	575	228	321	442	537	395	355	428
Street/MSRP [%]	74	83	80	69	72	78	77	80	79	77

Street price picked for the comparison was the cheapest price from an online research carried out in September 2014. Street prices were on average 23 percent lower than the manufacturer's suggested retail prices. Fig. 5.168 shows a comparison based on MSRP (i.e. 100 percent) versus lowest retail prices.

Deep conceal pistols from KEL-TEC and RUGER had a street price of just 69 and 72 percent respectively of the MSRP. The then new *VP9* by HECKLER & KOCH reached 80 percent, GLOCK *G19 Gen4* would even reach 83 percent of the MSRP.

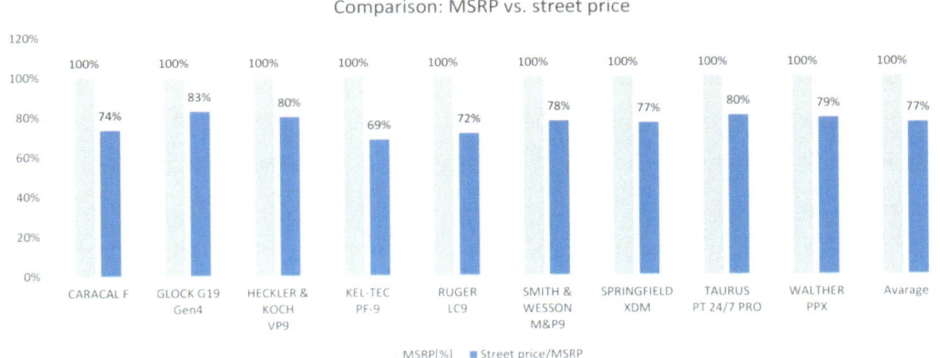

Fig. 5.168: Comparison of MSRP vs. street price in percent (as of September 2014).

Conclusions

Reference made earlier to computer and telecommunications products is to enable us to better comprehend how the price spiral keeps spinning downward, increasing the product variety and accelerating the pace of product life cycles. This worsens the manufacturer's situation by steadily decreasing innovation cycles: while the execution of marketing of an increasing number of product variants is required, ever shorter innovation cycles cause the sales numbers per specific pistol model to drop. These changes are pervasive, and apply to gun manufactures in the same way as they would for any company operating in other parts of the consumer goods industry.

Up to now the evolution as seen in the firearms industry has progressed by repeatedly recombining existing product features to improve the performance of established products. However, a feature that initially may be used as a company's unique selling proposition, sometimes quickly becomes a new industry standard. Adjustable grip size and trigger reach, introduced in the 1990s, is now a feature found on a plethora of pistols. GLOCK, HECKLER & KOCH, SMITH & WESSON, SPRINGFIELD, WALTHER and other manufacturers offer similarly-equipped handguns. In general, the features related to operator safety of pistols seem to strive toward a level which can be regarded as a common standard, and the convenience of guns is being improved continuously, e.g. ambidextrous controls.

Whether gun enthusiasts actually perceive certain features as a benefit or simply just accept these because the manufacturers wouldn't provide other options, could not be determined. Similarly, it can only be guessed what motivates firms to opt for ambidextrous slide-lock levers, magazine releases etc. (aka feature creep). A possible explanation can be found in Clayton M. Christensen's *The Innovator's Dilemma*. The best-selling author's point of view is that companies are obsessed with the idea of increasing the quality of their products and its performance, to exceed the products of competitors, and to penetrate into higher-margin markets.[661] The author refers to the upward mobility of the value system into more attractive markets, and sees a downward market immobility. The influential business theorist compares markets with hierarchies.

At the low-end, price is paramount and margins are low. High-end markets promise usually higher margins but a high quality is required. Companies would no longer develop down, once they have positioned themselves in a market at a higher level. Christensen explains, it is incomparably more difficult for a company whose cost structure is tailored to compete in high-end markets to penetrate and be profitable in low-end markets. An upward migration to premium segments will appear much more attractive and easier. This pattern of asymmetric mobility creates a vacuum in lower markets that draws in manufacturers with technologies and cost structures better-suited to competition. Often the upward drift will not be consciously perceived by the company, because companies together with the other participants of the market, e.g. competitors, consumers, and suppliers are migrating upmarket as well. Thus, companies are interested in offering products, whose scope of equipment is superior and could be interesting for consumers of previously neglected market segments, but the company would demand at the same time prices which a neglected consumer group is not willing to pay.[662]

Before we focus too much on product-specific details in this section, we should stop at this point and look at the big picture. For example, Christensen argues that only a few established companies would succeed to penetrate into the lower market segments. In his opinion, three barriers prevent downward mobility: the promise of higher margins in the upper market segments, the expectations of the company's customers, which will increase over the years, and the difficulty to reduce costs.[663] Thus, companies would neglect the lower market segments, and as a consequence, inadvertently make room for more simple, cheaper products, and allowing new entrants to have access to the market; or possibly access to markets, which were initially wrongly identified as niche markets.

Schumpeter coined the term "creative destruction" to describe the impetus for economic progress, while "disruptive innovation" denoted by Christensen, is believed to be the cause behind the "creative destruction" by reason of a new product or service. A radically new product might initially be demanded by just a small number or a new group of consumers[664] but later however, would become a serious threat, if it displaces established products because the new product is just cheaper or more convenient to use[665] than traditional products from established manufacturers.[666]

Ideas and their rapid implementation are now more valuable than a company's history with decades of tradition. In today's world, a simple smartphone-app might cause a technological change which has the potential to eliminate a complete field of an industry overnight.

Christensen distinguishes between evolutionary innovation, revolutionary innovation and disruptive innovation. A revolutionary innovation will hit the consumer by surprise when it is introduced on the market, whereas an evolutionary innovation is expected to take place before the actual launch of a new product generation is being announced. Both foster the improvement of product performance and they do not impose any danger for established companies; quite in contrast to disruptive innovation: the success of disruptive innovations hit companies and consumers by surprise and may threaten the existence of established structures. History provides numerous examples: Until the invention of the printing press in the 15th century, almost all books were copied by hand and this even limited who had access to literature. That changed with the advent of the printing press. Scribes were no longer in demand while literature became more readily available. The combustion engine created an automobile industry that literally displaced horse-drawn vehicles, which had been used for centuries, in a matter of decades. Today mobile phones are commonplace worldwide while landlines are disappearing, and streaming media is displacing cable TV and audio CDs alike.

Christensen thinks that most companies have established a trajectory of performance improvement over time, where they are focusing on the improvement of products or services along with their customer's requirements in existing markets. Such companies can gradually reach innovation leadership in established markets but, long-lasting success can spoil businesses and reduce a company's sensitivity to change.

The essential characteristic of evolutionary innovation is that it will not cause the demise of leading companies. On the contrary, *The Innovator's Dilemma* explains that disruptive innovations are inherently unpredictable and can cause established companies to lose market dominance. Leading companies could fail if they are not able to use other business models than they are used to, for the introduction of new products.

The advantage of less established companies lies in their openness to new approaches: to lower and offer cheaper technology without greater profit, they start in markets which leading firms experience as insignificant markets because mainstream consumers would not want a new product which offers lower performance initially; and at a certain point, it might be just too late for established companies to follow. If established companies despite many years of leadership did not see it coming and fail to catch up with a new technology on the market, Christensen calls this a disruptive innovation.[667]

Henry Ford is reported to have said, "If I had asked people what they wanted, they would have said faster horses." We all know examples of products that initially offered a lower performance than established options, e.g. early cars vs. trains or horses, early cell phone cameras vs. digital cameras etc. These low performers and supposedly "inferior" products had other attributes that were valued by a different kind of consumer, and which created a new customer segment for the manufacturers of these products.[668]

Today we live in an era of complex interdependence and there is a variety of interactions between the participants in the market. The manufacturers are in a dilemma: too often they develop products using the same methods, e.g. competitive analysis, focusing on market share, market research, innovation management, customer orientation, as well as the question of whether a certain novelty fits into the company's value system[669]. And because the decisions of competitors are based on similar practices and information, their products tend to mimic each other. The environment in which a company operates, that is, external influences they are subject to, as well as internal factors, have a decisive influence on how flexible a company is to react.

For instance, when a premium brand has to make a decision if it would be compliant to its value system when it includes a simpler "entry-level" product in its portfolio, and the company might worry the new product will take sales away from its high-end products with higher margins. According to Christensen, a cannibalization by a new product

in the portfolio can be healthy when the new product has other features that are perceived by consumers to create a very different value proposition.[670]

At the same time, the cost-consciousness of consumers is more pronounced than
ever. The competition does not take place based on differentiation, but by using special
promotions, discounts and the sales price; and while manufacturers continue to push
their products in markets, although, the supply already exceeds the demand by far,
these manufacturers are trying to differentiate their products from those of the other
competitors by adding little details in the product features. Is there any way out of this
situation?

With regard to the five competitive forces previously mentioned in the Introduction,
Porter addresses the question of how these five forces determine the intensity of competition and profitability within industries. Instead of understanding these five forces as
an unalterable "paradigm of competition," Porter's concept vice versa may also be used
as a tool to analyze competitive structures. Companies can gain an overview of the
conditions of competition existing on the market in this way and ultimately influence it.
W. Chan Kim has been engaged in this approach and outlines how companies can
enter new markets to break free from the situation of tough competition.

The situation in which a battle for market share in shrinking markets takes place, is
what Kim call's "red oceans," it is so to speak, a shark tank where bloody competition
is happening. In his book *Blue Ocean Strategy,* the author recommends companies to
break out of the competition and to create areas that previously did not exist, and are
thus not covered by competitors; he refers to these currently not existing fields as "blue
oceans." Companies can succeed in creating blue oceans if they stop focusing on the
competition and trying to beat their competitors on the market, but rather to follow a
strategy which Kim calls "value innovation." Value innovation is the cornerstone of the
blue ocean strategy; it can make the competition irrelevant, but it is only feasible if
innovation is on par with customer value, price and costs. Companies that fail to merge
innovation and customer value and to implement the resulting advantage soon, risk to
offer their ideas to their competitors.

Blue Ocean Strategy mentions the performance art troupe CIRQUE DU SOLEIL as a
prime example. It exceeded the limits of art and theatre, enhanced customer value by
a new kind of performance, which no other circus would offer, and in addition, costs
have been reduced by doing without animals. At the same time, new customers were
won, which had previously shown no interest in circus acts.

Kim is of the opinion that, the usual strategy of companies is to focus on better solutions for all kinds of problems; which are considered to be issues from the perspective of the industry. By changing the point of view towards alternatives and new customers, they could identify what the industry focused incorrectly on and create new customer value, which could go beyond the limits seen so far by the industry. Only through the creation of true innovations, coupled with customer value, a new demand could be created. The author explains that the first step would require to be able to detach from a self-absorbed industry and to find out what factors within this industry apply naturally but also wrongly as important and are just dragged along, although they do not contribute any more to satisfy the current needs of customers.

In a second step, it must be considered whether products or services were designed far beyond the requirements of the customers in this race against other market participants, thereby making the customer to pay for services that are not required. In the third step, it is key to recognize what compromises customers are currently living with and what factors about the currently available industry-standard could be improved beyond. In the fourth and final step, not-yet-known customer value and new demand should be created, and a higher price achieved as it would be common so far within the market.

Steps one and two would enable manufacturers to gain insight on how to improve their own cost structure and bring down their costs to a level which is below that of competitors. As a result of the remaining considerations, customer value can be increased and additional demand created. As a result, newly created customer value in new areas will emerge which would offer customers a completely new experience and at the same time the company's own costs were kept low. As Kim points out, industries are required to have the courage to reinvent themselves. Structures and boundaries within the industries exist only in the minds of the managers. Real achievers, however, are always looking for new customers and additional demand. The crux for a company is to create this additional demand. To get this done, a shift from the attention of the companies self-created range of services to the requirements of the consumers must take place. New demands could be created only by creating innovations in conjunction with customer value.[671]

In conclusion, it can be said: innovations – in the strict sense of the definition of this term – can rarely be found in the design of pistols. As we could see, sweeping changes are often initiated by new entrants from outside the industry. In the early 21st century, the smartphone was introduced by a computer manufacturer, and well-established automakers received a highly impressive demonstration of the automotive future that was virtually put overnight to reality through a market newcomer in the field of electric vehicles.

Within the wide range of the firearms industry, military applications might represent the far most lucrative market for outsiders and a disruptive innovation will probably start from there. This new technology is going to have spinoffs that will trickle down to the consumer market and will be sewn into the fabric of everyday life.

Some firms try to integrate electronics, others remain to manufacture purely mechanical products, and we see a trend to improve convenience of striker-fired pistols. Originally, pre-cocked striker-fired pistols required some kind of a manual decocking procedure; pistols such as HECKLER & KOCH *VP9*, RUGER *American Pistol*, SIG SAUER *P320*, and SPRINGFIELD *XD^M* do not require a manual decocking, thus, they offer more customer value.

Regardless of the technical potential in the field of handguns, the question must be asked, whether the prerequisite regarding a minimum weapon size in ATF Form 4590 is still useful today. In all 50 United States, it is legal – in some states a permit is required – to carry handguns either openly or concealed, and U.S. consumers de facto can buy extremely small pistols, even though this segment is exclusively covered by domestically manufactured products. It should be in the interest of American consumers, to promote the supply, and the competition on that domestic market.

The evaluation of handguns for the importation into the United States imposes a barrier to arms manufacturers producing outside of the United States. Whether the original objective to keep cheap weapons of low quality off the streets is achieved through an assessment on the suitability of imported guns for sporting purposes may be doubted. In reality, this requirement can be understood as a protection for domestically made products, as U.S. made handguns are not subject to the ATF evaluation by form 4590.

As long as the approval for the importation of handguns depends on a suitability for sporting purposes, non-U.S. manufacturers will have to make a compromise between new technologies, achieving enough points for importation and adding consumer value.

An important lesson might be to recognize that it is not the technically feasible that should be produced, but rather a bundle of properties that achieves a high degree of user benefit for potential customers. It will be impossible to implement all features in one product and to meet with this single product the expectations of more than 300 million Americans. Consumer preferences are too diverse, and how *Lugs* said, emphasized originality and originality at any cost is not purposeful; and of course, it should not get too technical, in today's user-friendly times of smartphones and tablets in conjunction with a focus on lifestyle and brand awareness. However, the analysis showed that there are opportunities for innovation in production technology and the integration of electronics into firearms.

Nevertheless, it will not matter for consumers such as our "*ToddG*," who is cited at the beginning of this book, whether ultimately "the little elves inside the gun that make it work have blue hats or green hats." The manufacturers are required to cater to shooters, listen to potential customers, invest in new ideas which give rise to new markets, and to join the journey toward new benefits together with their customers.

Types of Innovation
- *Sustaining*
  An innovation that does not affect existing markets. May be either:
  - *Evolutionary*
    An innovation that improves a product in an existing market in ways that customers are expecting.
  - *Revolutionary*
    An innovation that is unexpected, but nevertheless does not affect existing markets.
- *Disruptive*
  An innovation that creates a new market by applying a different set of values, which ultimately (and unexpectedly) overtakes an existing market.

(Clayton M. Christensen, 1997)

Appendices

[…] it is certainly no mere coincidence that usually such weapons, combining features which contribute to a reliably working mechanism and high-performance of a good self-loader, will at the same time rarely disappoint with regard to aesthetics. Whereas others, which can't deny their crude, bumbling design and implementation by the hands of an inexperienced beginner, will usually turn out to be a Knight of the Sad Countenance. This could easily be proved by a number of striking examples, however – nomina sunt odiosa.

(Wille, 1897 p. 66) [translated by PD]

Appendix 1   Principle of Linear Momentum

Newton's first law of motion tells us that, an object at rest stays at rest and an object in motion stays in motion with the same velocity and in the same direction unless acted upon by an external cause. External causes which change the state of motion of bodies are called *forces*. Forces are vector quantities and are described by magnitude, direction and point of application.

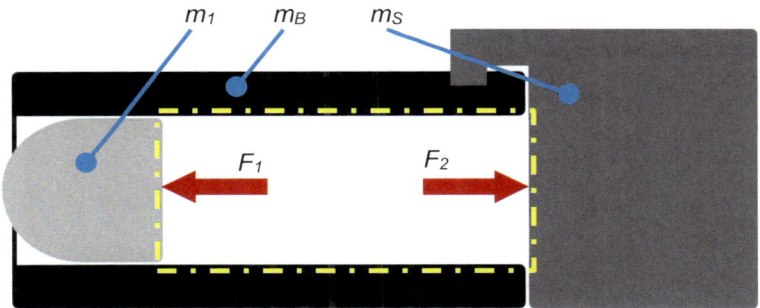

Fig. A1.1: System boundaries (dot and dash line) and forces
     by the example of a short-recoil-operated pistol.

For the calculation of forces, it is necessary to capture the observed processes in theory and to draw boundaries as narrow as possible, which just include the respective forces acting within the system under evaluation. These designated system boundaries define an isolated system, which is in theory so far removed from its surroundings, that it does not interact with them, i.e. it is not acted on by any external forces. In an isolated system matter and energy are conserved, the total energy stays constant even though it might be converted in form or transferred to objects within the boundaries, and consequently, the total momentum is constant.

The fact, that the total momentum is constant in an isolated system, is implied by Newton's laws of motion and is known as the law of conservation of momentum. The calculation of forces is based on the assumption that the sum of all force vector quantities, that is, the resultant force or net force, within an isolated system at any given time, is constant:

$$\vec{F}_{res} = \sum_{n=1}^{N} \vec{F}_n = const.$$

A1.1

where: $\quad \vec{F}_{res}$ Resultant force [N]
$\qquad \quad \vec{F}_n$ Vector quantity of a force n, with n = 1 to N [N]
$\qquad \quad N$ Total number of forces in isolated system

According to Newton's second law of motion, every individual force $\vec{F}_n$, is calculated by multiplying the mass m_n of a respective body n by its acceleration $\vec{a}_n$, which acts in a given direction on this body n (the acceleration vector size):

$$\vec{F}_n = m_n \, \vec{a}_n$$

A1.2

where: $\quad \vec{F}_n$ Force acting in the direction of acceleration $\vec{a}_n$ [N]
$\qquad \quad \vec{a}_n$ Acceleration vector size [m/s²]
$\qquad \quad m_n$ Mass [kg]

The masses m_n contained in Formula A1.2 remain constant during our consideration, however every acceleration $\vec{a}_n$ corresponds to a velocity change $d\vec{v}_n$ for an infinitesimally small period dt:

$$\vec{a}_n = \frac{d\vec{v}_n}{dt}$$

A1.3

where: $\quad dt$ Infinitesimally small period of time [s]
$\qquad \quad d\vec{v}_n$ Velocity change within dt,
$\qquad \qquad \quad$ in direction of acceleration $\vec{a}_n$ [m/s]

Substituting Formula A1.3 into Formula A1.2 gives

$$\vec{F}_n = m_n \frac{d\vec{v}_n}{dt}$$

A1.4

With Formula A1.4 the resultant force of Formula A1.1 is

$$\vec{F}_{res} = \sum_{n=1}^{N} m_n \frac{d\vec{v}_n}{dt} = const.$$

A1.5

The movement processes in a gun barrel take place in the barrel's longitudinal direction exclusively, hence the commonly used vector notation of the equations can be conveniently substituted by component expressions, which further simplify considerations. Only the amounts of forces and accelerations are used in place of the vector quantities and depending on their specific direction positive or negative signs are inserted:[672]

$$F_{res} = \sum_{n=1}^{N} m_n \frac{dv_n}{dt} = const.$$
A1.6

The masses within the system were at rest at the beginning of the discharge, hence F_{res} initially was zero. The latter applies also at any time after the beginning (law of conservation of momentum, see "= const." in A1.1). Under these conditions it follows that

$$F_{res} = \sum_{n=1}^{N} m_n \frac{dv_n}{dt} = 0$$
A1.7

For our application of a discharge from a pistol with coupled barrel and slide, we define system boundaries as shown in Fig. A1.1 and consider force F_1, which acts upon the projectile's base, and the resulting counterforce F_2 acting on the breech face (Newton's third law, action = reaction).

Transferring this into Formula A1.1 we obtain for the forces acting in longitudinal direction of the barrel:

$$F_{res} = \sum_{n=1}^{2} F_n = F_1 - F_2 = 0$$
A1.8

which can be written as

$$F_{res} = m_1 \frac{dv_1}{dt} - m_2 \frac{dv_2}{dt} = 0$$
A1.9

where: m_1 Projectile's mass [kg]

m_2 Combined mass of barrel m_B and slide m_S [kg]

and accordingly

$$m_1 \Delta v_1 - m_2 \Delta v_2 = 0$$
A1.10

In our application, the change of velocity Δv_n is equal to the top velocity achieved because the velocity of each observed mass was zero before the discharge; thus, we get:

$$m_1\, v_1 - m_2\, v_2 = 0 \qquad\qquad \textbf{A1.11}$$

Nothing but products of mass and velocity $(m_n v_n)$ remain in Formula A1.11, i.e. momentums or impulses. As noted in Chapter 3.1, the term impulse would describe the effect of an unbalanced force acting instantaneously on an object. Impulse force applied to an object changes its momentum; it is a concept frequently used in the analysis of collisions and impacts.[673] Based on our situation, the example of a discharge, the impulse force causing the projectile to travel down the bore is also applied to the breech face instantaneously in the opposite direction with the same magnitude, i.e. ejecta momentum = recoiling momentum. – To each force there is an equal and opposite force. – As per definition, *recoil* is the observed phenomenon of an opposing reaction of force that is utilized in the forward acceleration of an object. In the case of an auto-loading pistol, recoiling momentum is employed to accelerate the gun's breech. Hence, blow-back-operated pistols as well as pistols utilizing the Browning type action with coupled barrel and slide, are both grouped under the same classification "recoil-operated" in this book.

In fact, each momentum p_n is a vector and has the same direction as the velocity v_n at which a body m_n travels. As initially mentioned, for our consideration, only the effect in the longitudinal direction of the barrel is of importance, and the sum of all momenta remains constant at any time, that is, in our case, the sum of all momenta at any time must be zero, because the masses were at rest at the beginning of our consideration. Under these conditions, the momentum can be calculated using the following formula:

$$p_{res} = \sum_{n=1}^{N} m_n\, v_n = 0 \qquad\qquad \textbf{A1.12}[674]$$

where: p_{res} Resultant momentum [N s]
         m_n Mass of a body n [kg]
         v_n Velocity of a body n [m/s]

Appendix 2 Checklist, Specs Sheet (Template)

Table A2.1 Blank checklist template

	Criteria	Manufacturer's Specifications	Survey
A) Pistol	Serial Number		
	Dimensions $L \times W \times H$, with (w/o) Magazine [mm]		
	Weight, with (w/o) Magazine [g]		
	Fire-control Mechanism		
	Trigger Pull / Trigger Travel / Trigger Work		
	Slide Racking Force / Work		
	Copper Crusher Indent / max. Diameter [mm]		
	Manual Safeties: Manual Safety Locking Device Automatic Safeties: Hammer block Striker Safety Mag Disconnect Grip Safety Trigger Safety Passive Safeties: Out-of-battery disconnect Other passive safeties Other Features: Loaded Chamber Indicator Strips w/Mag. Assembles w/Mag. Others		
	Number of Magazines [pcs]		
	Magazine capacity [rounds]		
	Uncoupling start / -end, Total Slide Recoil Travel [mm]		
	v_0 / E_0 (RUAG *Sintox* 8.0 g FMJ)		
	$\overline{v}_{St} / \overline{v}_{Se}$		
	Magazine Release		
	Slide Stop		
	MSRP Street Price [USD]		
B) Slide	Dimensions $L \times W \times H$, with (w/o) recoil lug [mm]		
	Gross Weight		
	Manufacturing Technology		
	Retracted Striker		
C) Barrel	Weight		
	Barrel Length		
	Barrel Profile		
	Manufacturing Technology		
D) Frame	Gross Weight		
	Manufacturing Technology		

Appendix 3   ATF Form 4590 (5330.5)

U.S. Department of Justice
Bureau of Alcohol, Tobacco, Firearms and Explosives

Factoring Criteria for Weapons

NOTE: The Bureau of Alcohol, Tobacco, Firearms and Explosives reserves the right to preclude importation of any revolver or pistol which achieves an apparent qualifying score but does not adhere to the provisions of Section 925(d)(3) of Amended Chapter 44, Title 18 U.S.C.

Pistol				Revolver			
Model:				**Model:**			
Prerequisites				**Prerequisites**			
1. The pistol must have a positive manually operated safety device. 2. The combined length and height must not be less than 10" with the height *(right angle measurement to barrel without magazine or extension)* being at least 4" and the length being at least 6".				1. Must pass safety test. 2. Must have overall frame *(with conventional grips)* length *(not diagonal)* of 4 1/2" minimum. 3. Must have a barrel length of at least 3".			
Individual Characteristics	**Point Value**	**Point Sub Total**		**Individual Characteristics**	**Point Value**	**Point Sub Total**	
Overall Length				Barrel Length (Muzzle to Cylinder Face)			
For Each 1/4" Over 6"	1			Less Than 4"	0		
Frame Construction				For Each 1/4" Over 4"	1/2		
Investment Cast or Forged Steel	15			Frame Construction			
Investment Cast or Forged Hts Alloy	20			Investment Cast or Forged Steel	15		
Weapon Weight W/Magazine *(unloaded)*				Investment Cast or Forged Hts Alloy	20		
Per Ounce	1			Weapon Weight (Unloaded)			
Caliber				Per Ounce	1		
.22 Short and .25 Auto	0			Caliber			
.22 LR and 7.65MM to .380 Auto	3			.22 Short to .25 ACP	0		
9mm Parabellum and Over	10			.22 LR and .30 to .38 S&W	3		
Safety Features				.38 Special	4		
Locked Breech Mechanism	5			.357 Mag and Over	5		
Loaded Chamber Indicator	5			Miscellaneous Equipment			
Grip Safety	3			Adjustable Target Sights (Drift or Click)	5		
Magazine Safety	5			Target Grips	5		
Firing Pin Block or Lock	10			Target Hammer and Target Trigger	5		
Miscellaneous Equipment				**Safety Test**			
External Hammer	2			A double action revolver must have a safety feature which automatically (or in a single action revolver by manual operation) causes the hammer to retract to a point where the firing pin does not rest upon the primer of the cartridge,. The safety device must withstand the impact of a weight equal to the weight of the revolver dropping from a distance of 36" in a line parallel to the barrel upon the rear of the hammer spur, a total of 5 times.			
Double Action	10						
Drift Adjustable Target Sight	5						
Click Adjustable Target Sight	10						
Target Grips	5						
Target Trigger	2						
Score Achieved **(Qualifying Score is 75 Points)**				**Score Achieved** **(Qualifying Score is 45 Points)**			

ATF Form 4590 (5330 5)
Revised March 2008

Appendix 4   Positive Manually Operated Safety Device, Definition

(Published by kind permission of ATF.)

---

eMail

Betreff:	RE: ATF Form 4590; Prerequisite #1, positive manually 17.04.2014 19:18:38 operated safety device

Hi Peter,

In general, a positive manual safety would be a safety device which requires manipulation by the shooter to disengage, however, below is the additional guidance we use in determining whether a particular firearm incorporates a "positive manual safety":

```
6. Factoring Criteria, Pistols (ATF Form 4590)

    a. Prerequisites:

        1. Must have positive manually operated safety.
           Automatic safety as described in firing pin
           lock or block section also qualifies.
```

```
Firing pin lock or block, locks or blocks firing
pin.  Can be automatic or manual in operation
(Automatic - SIG, Glock, etc., manual - French
Model 1935).  Any automatic operating safety
which only unlocks or brings the firing pin in
line with the hammer at the moment of hammer
fall qualifies as a firing pin block, i.e. SIG
P230.  Note:  Trigger block safety does not
count as a firing pin lock or block.
```

I hope this information is helpful.

Firearms Technology Branch

From: Peter Dallhammer
Sent: Tuesday, April 08, 2014 5:18 AM
To: ATF
Subject: ATF Form 4590; Prerequisite #1, positive manually operated safety device

Dear Sirs,

May I kindly ask the question what the term "positive" means in the context of "positive manually operated safety device"? Will a pistol submitted with a automatically working internal safety qualify?

Sorry, but my English is not good enough and I was not able to find a proper translation for the word "positive" as being used in the context of said prerequisite. Originally I was thinking the term calls out for a safety that needs to be deliberately operated by the user demanding an extra step of operation before using the pistol.
I now think it means any kind of mechanical safety, and any safety is acceptable, even if it works automatically and independently from the user. Thus, a positive manual safety could be something like a thumb operated safety of a "1911" type, a P08 Luger type safety lever, a push-button type on the early Beretta pistols, a Glock-style trigger safety or e.g. an automatically working firing pin block.

Does the term "positive" imply that a automatically working firing block must be always automatically in "SAFE" position - or would a passive drop safety be acceptable, e.g. a passive drop safety that is always in "FIRE"-position and automatically blocks the firing pin only when the pistol is being dropped?

Kind regards
Peter Dallhammer

338

Appendix 5   HTS Alloy, Definition

(Published with friendly approval of ATF.)

eMail

Betreff:	FW: ATF Form 4590; Hts Alloy	19.03.2014 15:07:45

Hello Peter,

For purposes of the ATF Form 4590, the term "alloy" is generally interpreted to mean non-ferrous metal.

Points would be awarded for High Tensile Strength (HTS) alloy if the frame is manufactured from an alloy having a tensile strength of 50,000 psi or greater.

When submitting an alloy framed handgun, please include information regarding the type of alloy used to manufacture the frame.

Firearms Technology Branch

From: Peter Dallhammer
Sent: Wednesday, March 19, 2014 8:14 AM
To: ATF
Subject: ATF Form 4590; Hts Alloy

Dear Sirs,

What does the abbreviation "Hts" stand for as mentioned in "Frame construction: Investment Cast or Forged Hts Alloy"?

What is your definition of "Alloy" Is it a non-ferrous metal?
What tensile strength is required to qualify?

Kind regards
Peter Dallhammer

Appendix 6   Certification of Microstamping Technology

Information Bulletin 2013-BOF-03

California Department of Justice **DIVISION OF LAW ENFORCEMENT** Larry J. Wallace, Director		*INFORMATION* *BULLETIN*	
Subject: **Certification of Microstamping Technology** **pursuant to Penal Code section 31910,** **subdivision (b)(7)(A)**	*No:* 2013-BOF-03 *Date:* May 17, 2013	*Contact for information:* **Bureau of Firearms**	

TO: California Licensed Firearms Dealers, California Department of Justice Certified Laboratories, Firearm Manufacturers with Firearms listed on the Roster of Handguns Certified for Sale in California, and all other interested persons/entities

The purpose of this bulletin is to inform California licensed firearms dealers, California Department of Justice certified laboratories, firearm manufacturers with firearms listed on the Roster of Handguns Certified for Sale in California, and all other interested persons/entities of the Department of Justice's certification on May 17, 2013 pursuant to Penal Code section 31910, subdivision (b)(7)(A) that the microstamping technology is available to more than one manufacturer unencumbered by any patent restrictions.

Background

In 2007, Assembly Bill 1471 was passed and signed into law, requiring all semiautomatic pistols to be equipped with microstamping technology—"a microscopic array of characters that identify the make, model, and serial number of the pistol, etched or otherwise imprinted in two or more places on the interior surface or internal working parts of the pistol, and that are transferred by imprinting on each cartridge case when the firearm is fired." (Pen. Code, § 31910, subd. (b)(7)(A).) The legislation further provided that this requirement becomes effective when the Department of Justice "certifies that the technology used to create the [microstamp] imprint is available to more than one manufacturer unencumbered by any patent restrictions." (*Ibid.*)

Certification of the Microstamping Technology

On May 17, 2013, the Department of Justice issued a certification that the microstamping technology is available to more than one manufacturer unencumbered by any patent restrictions. A copy of the certification is attached to this bulletin.

Effect of the Department's Certification

Following the issuance of the Department of Justice's certification, the provisions of Penal Code section 31910, subdivision (b)(7)(A) are in immediate effect. Therefore, to be listed on the Roster of Handguns Certified for Sale in California, a semiautomatic pistol must be equipped with microstamping technology—i.e., a microscopic array of characters that identify the make, model, and serial number of the pistol, etched or otherwise imprinted in two or more places on the interior surface or internal working parts of the pistol, and that are transferred by imprinting on each cartridge case when the firearm is fired. (Pen. Code, § 31910, subd. (b)(7)(A).) Semiautomatic pistols already listed on the Roster of Handguns Certified for Sale in California will remain on the roster and need not incorporate the microstamping technology provided that the firearms comply with Penal Code sections 32015, 32020, and 32030.

Also, please consult the Department's regulations for more information regarding how microstamping technology should be incorporated within all semi-automatic pistols and tested for compliance with Penal Code section 31910, subdivision (b)(7)(A). (See Cal. Code Regs., tit. 11, §§ 4046 et seq.) A link to a copy of the applicable regulations can be found at the following website: http://oag.ca.gov/firearms.

For any questions regarding the Roster of Handguns Certified for Sale in California and/or the certification of microstamping technology, please contact Leslie McGovern at (916) 227-4024 or leslie.mcgovern@doj.ca.gov.

Sincerely,

STEPHEN J. LINDLEY, Chief
Bureau of Firearms

For KAMALA D. HARRIS
Attorney General

STATE OF CALIFORNIA
OFFICE OF THE ATTORNEY GENERAL
ROCHELLE C. EAST
CHIEF DEPUTY ATTORNEY GENERAL, LEGAL AFFAIRS

CERTIFICATION UNDER
CALIFORNIA PENAL CODE § 31910, SUBDIVISION (b)(7)(A)

Under California Penal Code § 31910, subdivision (b)(7)(A), a semiautomatic pistol not already listed on the firearm roster pursuant to California Penal Code § 32015 is an "unsafe handgun" unless it is "designed and equipped with a microscopic array of characters that identify the make, model, and serial number of the pistol, etched or otherwise imprinted in two or more places on the interior surface or internal working parts of the pistol, and that are transferred by imprinting on each cartridge case when the firearm is fired, provided that the Department of Justice certifies that the technology used to create the imprint is available to more than one manufacturer unencumbered by any patent restrictions."

The California Department of Justice has conducted a review of the known and available patent restrictions applicable to the microscopic-imprinting technology described in § 31910, subdivision (b)(7)(A). Based on this review, the Department certifies that, as of May 17, 2013, this technology is available to more than one manufacturer unencumbered by any patent restrictions.

Rochelle C. East
Chief Deputy Attorney General

1300 I STREET • SUITE 1730 • SACRAMENTO, CALIFORNIA 95814 • PHONE (916) 324-5435 • FAX (916) 327-7154

NEWS RELEASE

Second Amendment Foundation

12500 NE Tenth Place • Bellevue, WA 98005
(425) 454-7012 • FAX (425) 451-3959 • www.saf.org

GLOCK FILES AMICUS BRIEF SUPPORTING SAF'S CALIFORNIA CASE

For Immediate Release: 11/6/2013

BELLEVUE, WA – Attorneys for Glock, Inc. have filed an amicus curiae brief supporting the Second Amendment Foundation's case in California, Pena v. Lindley, a lawsuit challenging the state handgun roster requirements that include microstamping and magazine disconnects.

Glock produces some of the most popular pistols in the world, and their guns are carried by law enforcement professionals and legally-armed private citizens across the United States.

"We are proud of Glock for stepping up to the plate," said SAF founder and Executive Vice President Alan M. Gottlieb. "Glock believes, as do we, that California's requirements place an undue burden on both consumers and manufacturers."

According to the brief filed by attorneys Erik S. Jaffe of Washington, D.C. and John C. Eastman of Orange, Calif., Glock pistols are like the majority of semi-auto pistols manufactured today, because they do not include the magazine disconnect. Indeed, the brief notes that "the overwhelming majority of law enforcement agencies require pistols that do not have a magazine disconnect mechanism."

Glock pistols, nor any other handgun in common use, can comply with California's "microstamping" mandate, the brief notes. As a result the newest generation of Glock pistols is not on the California roster, and therefore cannot be sold to private individuals in that state.

"Under the First Amendment," Gottlieb observed, "California is not allowed to compile a list of books you can read, and under the Second Amendment the state should not be allowed to compile a list of handguns you can own."

Both Jaffe and Eastman clerked for Supreme Court Justice Clarence
Thomas, Gottlieb noted. Mr. Eastman has considerable experience in civil
and constitutional litigation, and was a candidate for California attorney
general in 2010. He is a law professor at Chapman University. Mr. Jaffe also
clerked for Judge Douglas H. Ginsburg of the U.S. Court of appeals in the
District of Columbia. He has litigated in Washington, D.C. and has
considerable experience in constitutional challenges.

"Glock definitely has an interest in this case," Gottlieb said, "and their
expertise could be crucial at this point. We're glad they have chosen to weigh
in."

The Second Amendment Foundation (www.saf.org) is the nation's oldest and
largest tax-exempt education, research, publishing and legal action group
focusing on the Constitutional right and heritage to privately own and
possess firearms. Founded in 1974, The Foundation has grown to more than
650,000 members and supporters and conducts many programs designed to
better inform the public about the consequences of gun control. In addition
to the landmark *McDonald v. Chicago* Supreme Court Case, SAF has
previously funded successful firearms-related suits against the cities of Los
Angeles; New Haven, CT; New Orleans; Chicago and San Francisco on behalf
of American gun owners, a lawsuit against the cities suing gun makers and
numerous amicus briefs holding the Second Amendment as an individual
right.

-END-

Appendix 8   Maryland Roster

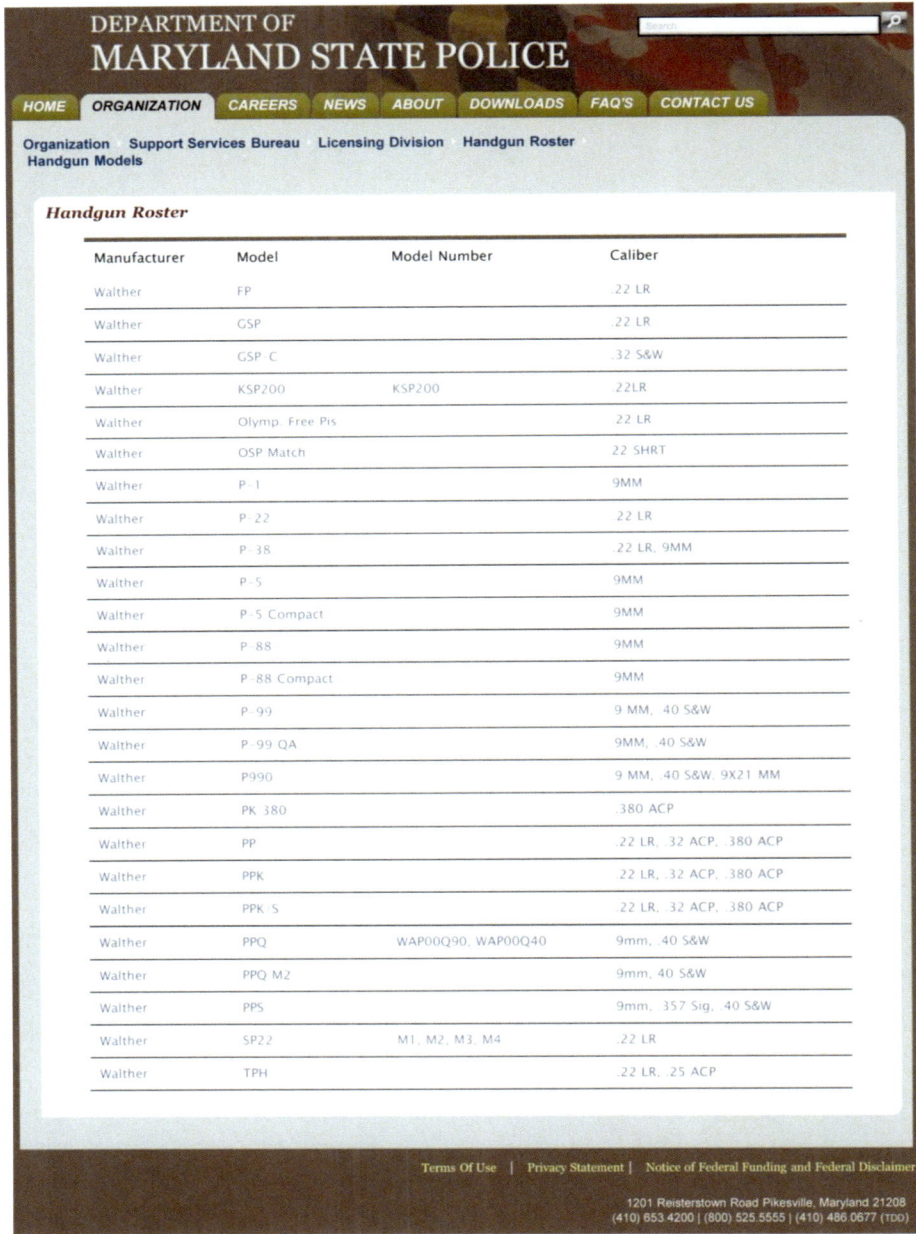

Fig. A8.1: WALTHER-Pistols listed on Maryland Roster.
(Maryland Handgun Roster, 2013)

Appendix 9   Compilation of U.S. Regulations

For the sake of clarity, a compilation of the requirements with cross references to the descriptions in Chapter 4 is given, and a tabular comparison as a summary.

Unintentional or accidental discharge
    ATF: 4.2.1 Factoring Criteria for Weapons
    NIJ: 4.4 NIJ test
    MD: 4.3.3 Maryland

Minimum length 152.4 mm
    ATF: 4.2.1 Factoring Criteria for Weapons

Height w/o Magazine, minimum 101.6 mm
    ATF: 4.2.1 Factoring Criteria for Weapons

Markings
    GCA: 4.2.2 Markings

Assault-Weapons
    CA: 4.3.1 California
    MA: 4.3.2 Massachusetts
    NY: 4.3.4 New York

Magazine capacity: 10 rounds (CA), (MA), (MD), (NY),
                    min. 6 rounds (NIJ),
                    max. 15 rounds (IPSC), (CO)
    CA: 4.3.1 California
    MA: 4.3.2 Massachusetts
    MD: 4.3.3 Maryland
    NY: 4.3.4 New York
    NIJ: 4.4 NIJ test
    IPSC: 4.6 IPSC Production Division
    CO: In Colorado, the Columbine High School shooting had happened in 1999, and a shooting during a film premiere in Aurora in July 2012. This was followed by a debate to limit the magazine capacity with the result, to limit them to 15 cartridges. (Ferner, 2013)

Reliability: 600 rounds
    CA: 4.3.1 California, Requirement Testing: "Firing Requirement for Handguns"
    MA: 4.3.2 Massachusetts
    NIJ: 4.4 NIJ test

Drop Safety Requirements:

- 1 meter @ concrete, gun in carry condition (CA)
- 1 meter @ concrete, gun in ready condition (MA)
- 1.22 meter @ rubber (NIJ)
- 0.305 meter @ rubber (SAAMI)
- 1.22 meter @ rubber, gun safetied (SAAMI)

CA: 4.3.1 California, Requirement Testing: "Drop Safety Requirement for Handguns"
MA: 4.3.2 Massachusetts, Accidental Discharge Test
NIJ: 4.4 NIJ test
SAAMI: 4.5 ANSI/SAAMI Z299.5

Loaded Chamber Indicator
CA: 4.3.1 California, Requirement Testing: Chamber Load Indicator
MA: 4.3.2 Massachusetts

Magazine Disconnect
CA: 4.3.1 California, Requirement Testing: Magazine Disconnect Mechanism
MA: 4.3.2 Massachusetts

Microstamping
CA: 4.3.1 California, Note on 4. Safety and Functionality Test, and Safety and Functionality Test Procedure

External Lock (CA), or internal locking device (NY)
CA: 4.3.1 California
NY: 4.3.4 New York

Roster
CA: 4.3.1 California, Note on 5. Certified for sale in CA by the Department of Justice
MA: 4.3.2 Massachusetts
MD: 4.3.3 Maryland

Detectability
MA: 4.3.2 Massachusetts
MD: 4.3.3 Maryland

Hidden Serial Number
MA: 4.3.2 Massachusetts

Material
MA: 4.3.2 Massachusetts

Childproofing
MA: 4.3.2 Massachusetts

Trigger Pull Force
MA: Childproofing
NIJ: 4.4 NIJ test

Accuracy, when barrel length < 76.2 mm
MA: 4.3.2 Massachusetts

Barrel Length max. 127 mm
IPSC: 4.6 IPSC Production Division

Owner's Manual: Safety Warning/Disclosures
  MA: 4.3.2 Massachusetts

Hammer Release: 46 N (NIJ), 0.914 m (SAAMI)
  NIJ: 4.4 NIJ test
  SAAMI: 4.5 ANSI/SAAMI Z299.5

A summary is given in Table A9.1, another graphic summary in Fig. 4.20 (Section 4.7, Summary "Requirements").

Table A9.1:   Comparison of U.S. regulations.

	GCA, ATF	CA	MA	MD	NY	NIJ	SAAMI	IPSC
Safety	x			x		x		
Length	≥ 152.4							
Height	≥ 101.6							
Barrel Length			(≥ 76.2)					≤ 127
Markings	x							
Assault Weapon		x	x	x	x			
Magazine Capacity		≤ 10	≤ 10	≤ 10	≤ 10	≥ 6		≤ 15
Reliability		600	600			600		
Drop Safety Reqmt.		1 m	1 m			1.22 m	0.305 m	
Loaded Ch. Indicator		x	(x)					
Mag Disconnect		x	(x)					
Microstamping		x						
External Locking Dev.		x						
Separate or int. Lock					x			
Roster		x	x	x				
Detectability			x	x				
Hidden S/N			x					
Material			x					
Childproofing			(x)					
Trigger Pull Force			(x)			x		
Accuracy			(x)					
Owner's Manual			x					
Hammer Release						46 N	0.914 m	

Legend: x Criteria must be met.
              ( ) Designations in parenthesis are bound to requirements or apply alternatively.

Appendix 10   U.S. Handgun Production and Net Supplies

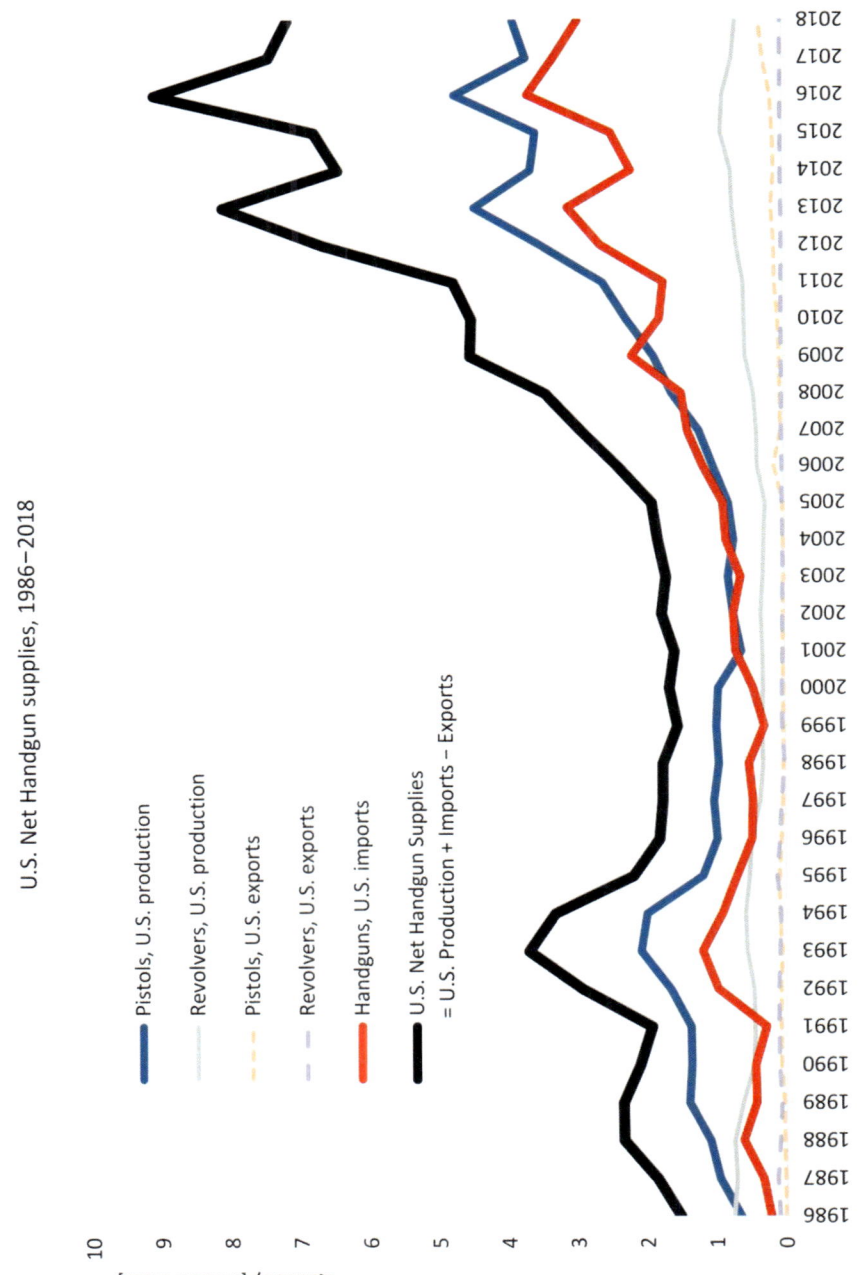

Fig. A10.1:   Net Handgun Supplies: exports and imports that complement U.S. based production (compiled from ATF AFMER and ATF Commerce Report 2019, see Endnote 587).

Table A10.1: Top 40 U.S. Manufacturers based on Units in Calendar Year 2018, all Caliber Categories, Calendar Year 2018 and 2017.

Rank 2018	Rank 2017	Company	0 < Cal. ≤ .22"	.22" < Cal. ≤ .25"	.25" < Cal. ≤ .32"	.32" < Cal. ≤ .380"	.380" < Cal. ≤ 9 mm	9 mm < Cal. ≤ .50"	Total Prod. 2018 Units	2018 % of Top 40	Total Prod. 2017 Units	2017 % of Top 40	Exports 2018 Units	Exports 2017 Units
1	1	SMITH & WESSON CORP	54,632	33	53	220,298	457,160	154,741	886,917	23.3%	1,032,450	28.8%	25,406	22,440
2	2	STURM, RUGER & COMPANY, INC	125,636	0	0	255,786	299,443	23,723	704,588	18.5%	781,623	21.8%	10,196	9,264
3	3	SIG SAUER INC	2,570	1,813	7,833	46,605	505,749	70,585	635,155	16.7%	536,774	15.0%	167,851	177,414
4	5	GLOCK INC	2,000	0	0	4,000	237,309	4,237	247,546	6.5%	175,696	4.9%	110,943	47,861
5	4	KIMBER MFG INC	8	0	0	20,478	113,224	67,428	201,138	5.3%	183,858	5.1%	2,225	2,605
6	6	SCCY INDUSTRIES LLC	0	0	0	0	169,819	0	169,819	4.5%	150,647	4.2%	270	270
7	7	SPRINGFIELD INC	0	0	0	44,730	18,200	77,107	140,037	3.7%	81,377	2.3%	693	152
8	13	BROWNING ARMS COMPANY	113,128	0	0	12,358	0	0	125,486	3.3%	50,331	1.4%	N/A	N/A
9	8	TAURUS INTERNATIONAL MANUFA	4,829	0	0	89,771	0	0	94,600	2.5%	69,123	1.9%	390	275
10	12	BERETTA USA CORP	1,529	0	4,053	19,224	53,739	887	79,432	2.1%	57,411	1.6%	5,145	4,750
11	11	KEL TEC CNC INDUSTRIES INC	53,463	0	5,260	1,896	6,532	0	67,151	1.8%	58,982	1.6%	213	351
12	9	FN AMERICA, LLC	0	0	0	0	39,602	12,241	51,843	1.4%	61,510	1.7%	2,377	636
13	16	COLT'S MANUFACTURING COMPAN	2	0	0	1,135	4,437	35,399	40,973	1.1%	31,987	0.9%	1,812	1,870
14	14	STRASSELLS MACHINE INC	0	0	0	10,300	26,600	0	36,900	1.0%	46,015	1.3%	N/A	N/A
15	19	DIAMONDBACK FIREARMS LLC	12,720	0	2,339	6,183	15,349	0	36,591	1.0%	24,270	0.7%	N/A	N/A
16	10	REMINGTON ARMS COMPANY LLC	0	0	0	6,049	13,141	14,624	33,814	0.9%	59,581	1.7%	827	3,022
17	17	COBRA ENTERPRISES OF UTAH, IN	4,180	26	13	7,038	19,066	7	30,330	0.8%	27,905	0.8%	N/A	N/A
18	N/A	EPP TEAM INC	0	0	0	732	28,610	0	29,342	0.8%	N/A	N/A	N/A	N/A
19	21	PHOENIX ARMS	16,390	3,610	0	0	0	0	20,000	0.5%	20,650	0.6%	N/A	N/A
20	18	JIMENEZ ARMS INC	4,031	62	197	8,669	6,968	0	19,927	0.5%	26,499	0.7%	N/A	N/A
21	22	BOND ARMS, INC	158	0	4	1,484	2,081	12,127	15,854	0.4%	13,333	0.4%	N/A	N/A
22	26	CZ-USA INC	0	0	0	0	9,489	5,719	15,208	0.4%	8,261	0.2%	N/A	N/A
23	31	AMERICAN TACTICAL INC	0	0	0	35	106	14,840	14,946	0.4%	5,005	0.1%	121	271
24	20	SAEILO, INC	0	0	0	2,245	5,241	5,963	13,449	0.4%	23,646	0.7%	N/A	N/A
25	23	HASKELL MANUFACTURING INC	0	0	0	0	0	12,800	12,800	0.3%	10,900	0.3%	N/A	N/A
26	35	PALMETTO STATE ARMORY, LLC	0	4,452	523	0	4,631	7	9,613	0.3%	3,123	0.1%	165	214
27	28	FMK FIREARMS INCORPORATED	0	0	0	0	8,359	0	8,359	0.2%	7,000	0.2%	N/A	N/A
28	58	DANIEL DEFENSE INC	0	5,259	2,305	0	0	1	7,565	0.2%	456	0.0%	24	N/A
29	27	IBERIA FIREARMS INC	0	0	0	0	0	7,400	7,400	0.2%	7,800	0.2%	N/A	N/A
30	39	FREEDOM ORDNANCE MANUFACT	0	0	0	0	6,229	0	6,229	0.2%	2,552	0.1%	N/A	N/A
31	30	WILSONS GUN SHOP INC	0	0	0	0	4,343	1,416	5,759	0.2%	5,203	0.1%	N/A	N/A
32	51	CMMG INC	0	1,655	369	0	2,602	1,104	5,730	0.2%	844	0.0%	103	111
33	38	TRAILBLAZER FIREARMS LLC	5,337	0	0	0	0	0	5,337	0.1%	2,723	0.1%	N/A	N/A
34	29	STI FIREARMS LLC	0	0	0	0	4,436	733	5,204	0.1%	6,416	0.2%	1,048	N/A
35	33	ALPHATECH INC	4,775	0	0	0	0	0	4,775	0.1%	3,450	0.1%	N/A	N/A
36	34	KRISS USA, INC	0	0	0	0	2,235	2,143	4,378	0.1%	3,254	0.1%	197	119
37	N/A	HENRY RAC HOLDING CORP	1,726	0	0	0	0	2,600	4,326	0.1%	N/A	N/A	720	N/A
38	95	HECKLER & KOCH, INC	0	0	0	0	0	4,308	4,308	0.1%	140	0.0%	N/A	N/A
39	N/A	PAUWAY CORP	0	0	0	0	4,250	0	4,250	0.1%	N/A	N/A	N/A	N/A
40	37	RADICAL FIREARMS LLC	2,517	0	1,351	0	29	10	3,907	0.1%	2,775	0.1%	N/A	N/A
		Total, Top 40	409,631	16,910	24,300	759,016	2,068,979	532,150	3,810,986		3,583,570		330,726	271,625
		% per Caliber Category, Top 40	10.7%	0.4%	0.6%	19.9%	54.3%	14.0%						
		Total, all U.S. Manufacturers	417,805	25,370	30,306	760,776	2,094,887	546,809	3,875,936		3,691,010		332,218	274,283
		% per Cal. Category, Total, all U.S. Manufacts.	10.8%	0.7%	0.8%	19.6%	54.0%	14.1%						
		ditto (2017)	(11.1%)	(0.3%)	(0.2%)	(23.0%)	(47.6%)	(17.8%)						

Source: Compiled from ATF AFMER, Year 2018 p. 1–34 and Year 2017 p. 1–27, see Endnote 587.

Table A10.2:　Top 40 U.S. Manufacturers based on Units in Calendar Year 2016, all Caliber Categories, Calendar Year 2016 and 2015.

Rank 2016	Rank 2015	Company	0 < Cal. ≤ .22"	.22" < Cal. ≤ .25"	.25" < Cal. ≤ .32"	.32" < Cal. ≤ .380"	.380" < Cal. ≤ 9 mm	9 mm < Cal. ≤ .50"	Total Prod. 2016 [units]	% of Top 40 2016	Total Prod. 2015 [units]	% of Top 40 2015	Exports 2016 [units]	Exports 2015 [units]
1	1	SMITH & WESSON CORP	84,529	0	0	229,835	823,260	291,827	1,429,451	30.5%	989,853	28.0%	17,542	12,084
2	2	STURM, RUGER & COMPANY INC	196,006	0	1	332,666	423,162	79,032	1,030,867	22.0%	748,364	21.2%	1,608	1,895
3	3	SIG SAUER INC	3,081	826	143	128,561	346,347	102,849	581,807	12.4%	459,655	13.0%	74,256	31,486
4	4	GLOCK INC	0	0	0	136,993	213,751	17,396	368,140	7.8%	216,616	6.1%	59,420	35,529
5	5	KIMBER MFG INC	155	0	0	61,089	65,165	94,395	220,804	4.7%	131,924	3.7%	1,649	1,562
6	6	SCCY INDUSTRIES LLC	0	0	0	0	160,676	0	160,676	3.4%	116,998	3.3%	50	N/A
7	8	TAURUS INTERNATIONAL	10,194	730	0	114,493	1	0	125,418	2.7%	91,386	2.6%	451	343
8	9	STRASSELLS MACHINE INC	0	0	0	21,000	69,900	0	90,900	1.9%	86,800	2.5%	N/A	N/A
9	16	REMINGTON ARMS COMPANY	0	0	0	29,206	36,414	19,232	84,852	1.8%	36,049	1.0%	N/A	N/A
10	13	SPRINGFIELD INC	0	0	0	0	21,935	50,078	72,013	1.5%	57,579	1.6%	102	697
11	10	COLT'S MANUFACTURING	20	0	0	7,380	17,129	46,705	71,234	1.5%	65,417	1.9%	1,140	761
12	11	KEL TEC CNC INDUSTRIES INC	44,607	0	3,233	4,480	5,482	0	57,802	1.2%	63,000	1.8%	721	N/A
13	12	ARMS TECHNOLOGY INC	39,113	0	0	17,672	0	0	56,785	1.2%	59,320	1.7%	N/A	N/A
14	7	BERETTA USA CORPORATION	13,303	0	2,353	12,379	14,967	0	43,002	0.9%	94,183	2.7%	4,430	3,636
15	20	PHOENIX ARMS	34,736	6,264	0	0	0	0	41,000	0.9%	23,850	0.7%	82	N/A
16	19	SAEILO, INC	0	0	0	12,280	17,714	10,280	40,274	0.9%	25,444	0.7%	82	76
17	14	JIMENEZ ARMS INC	6,617	449	11	15,208	14,709	0	36,994	0.8%	45,328	1.3%	N/A	N/A
18	18	HASKELL MANUFACTURING INC	0	0	0	0	0	32,400	32,400	0.7%	27,300	0.8%	N/A	N/A
19	21	IBERIA FIREARMS INC	0	0	0	0	0	24,100	24,100	0.5%	21,500	0.6%	N/A	N/A
20	22	BOND ARMS, INC	151	0	21	2,439	743	16,758	20,112	0.4%	15,697	0.4%	359	350
21	28	MAGNUM RESEARCH INC	0	0	0	655	2,885	12,815	16,355	0.3%	5,339	0.2%	14	77
22	25	FMK FIREARMS	0	0	0	0	12,247	0	12,247	0.3%	8,138	0.2%	N/A	N/A
23	24	CZ-USA INC	0	0	0	0	3,503	6,284	9,787	0.2%	8,962	0.3%	1,664	2,101
24	26	STI INTERNATIONAL LLC	2	0	0	0	4,663	3,121	7,786	0.2%	6,908	0.2%	N/A	N/A
25	40	HI TECH PLASTICS INC	7,250	0	0	0	0	0	7,250	0.2%	1,798	0.1%	N/A	N/A
26	N/A	HONOR DEFENSE LLC	0	0	0	0	7,170	0	7,170	0.2%	N/A	N/A	N/A	N/A
27	N/A	PAUWAY CORP	0	0	0	0	5,774	0	5,774	0.1%	N/A	N/A	160	N/A
28	44	KRISS USA, INC	0	0	0	0	2,638	2,529	5,167	0.1%	1,374	0.0%	82	153
29	29	WILSONS GUN SHOP INC	0	0	0	0	2,225	2,429	4,654	0.1%	5,052	0.1%	N/A	N/A
30	31	AMERICAN TACTICAL IMPORTS INC	0	0	0	0	1,233	2,908	4,141	0.1%	3,543	0.1%	N/A	N/A
31	37	MASTERPIECE ARMS HOLDING	239	0	0	3,055	478	0	3,772	0.1%	2,192	0.1%	97	147
32	N/A	WM C ANDERSON INC	0	0	0	0	0	3,666	3,666	0.1%	N/A	N/A	N/A	N/A
33	32	NIGHTHAWK CUSTOM LLC	0	0	0	0	713	2,653	3,366	0.1%	3,022	0.1%	N/A	N/A
34	33	VLH INC	0	0	0	0	669	2,431	3,100	0.1%	2,960	0.1%	N/A	N/A
35	34	RADICAL FIREARMS LLC	1,891	0	578	0	0	19	2,488	0.1%	2,896	0.1%	N/A	N/A
36	88	WALTHER MANUFACTURING	0	0	0	2,119	0	0	2,119	0.0%	91	0.0%	N/A	N/A
37	42	WHALLEY PRECISION INC	0	119	1,269	606	0	0	1,994	0.0%	1,494	0.1%	N/A	N/A
38	41	COONAN INC	0	0	0	0	0	1,885	1,885	0.0%	1,694	0.0%	N/A	N/A
39	38	CENTURY ARMS INC	0	0	0	0	0	1,418	1,418	0.0%	2,093	0.1%	N/A	N/A
40	39	ED BROWN PRODUCTS, INC	0	0	0	0	147	1,259	1,406	0.0%	1,894	0.1%	35	68
		Total, Top 40	441,894	8,388	7,609	1,129,061	2,278,277	828,947	4,694,176		3,530,013		163,862	100,370
		% per Caliber Category, Top 40	9.4%	0.2%	0.2%	24.1%	48.5%	17.7%						
		Total, all U.S. Manufacturers	447,315	13,141	10,175	1,130,459	2,281,450	837,535	4,720,075		3,557,199		172,408	140,787
		% per Cal. Category, Total, all U.S. Manufacts. ditto (2015)	9.5% (11.6%)	0.3% (0.3%)	0.2% (0.4%)	24.0% (23.0%)	48.3% (43.0%)	17.7% (21.6%)						

Source: Compiled from ATF AFMER, Year 2016 p. 1–21 and Year 2015 p. 1–20, see Endnote 587.

Table A10.3: Top 15 U.S. Manufacturers per Caliber Category,
0 > Cal. ≤ .22", .22" < Cal. ≤ .25", .25" < Cal. ≤ .32",
Calendar Year 2018 and 2017.

Pistol Production 2018, Top 15 U.S. Manufacturers by Units, Caliber from 0 to .32"

0 < Cal. ≤ .22"

Rank	Company	0 < Cal. ≤ .22"
1	STURM, RUGER & COMPANY, INC	125,636
2	BROWNING ARMS COMPANY	113,128
3	SMITH & WESSON CORP	54,632
4	KEL TEC CNC INDUSTRIES INC	53,463
5	PHOENIX ARMS	16,390
6	DIAMONDBACK FIREARMS LLC	12,720
7	TRAILBLAZER FIREARMS LLC	5,337
8	TAURUS INTERNATIONAL MANUFACT	4,829
9	ALPHATECH INC	4,775
10	COBRA ENTERPRISES OF UTAH, INC	4,180
11	JIMENEZ ARMS INC	4,031
12	SIG SAUER INC	2,570
13	RADICAL FIREARMS LLC	2,517
14	GLOCK INC	2,000
15	HENRY RAC HOLDING CORP	1,726
	Total, Top 15	407,934
	Total, all U.S. Manufacturers, Year 2018	417,805

.22" < Cal. ≤ .25"

Rank	Company	.22" < Cal. ≤ .25"
1	DANIEL DEFENSE INC	5,259
2	PALMETTO STATE ARMORY, LLC	4,452
3	PHOENIX ARMS	3,610
4	DEL-TON, INC	2,750
5	SIG SAUER INC	1,813
6	CMMG INC	1,655
7	LWRC INTERNATIONAL	1,095
8	BRAVO COMPANY MFG INC	855
9	FEDERAL ARMAMENT LLC	672
10	LUXUS ARMS LLC	375
11	WINDHAM WEAPONRY INC	309
12	PODUNK INC	286
13	ROCK RIVER ARMS INC	247
14	BLACK RAIN ORDNANCE INC	224
15	HARDENED ARMS LLC	186
	Total, Top 15	23,788
	Total, all U.S. Manufacturers, Year 2018	25,365

.25" < Cal. ≤ .32"

Rank	Company	.25" < Cal. ≤ .32"
1	SIG SAUER INC	7,833
2	KEL TEC CNC INDUSTRIES INC	5,260
3	BERETTA USA CORP	4,053
4	DIAMONDBACK FIREARMS LLC	2,339
5	DANIEL DEFENSE INC	2,305
6	RADICAL FIREARMS LLC	1,351
7	WHALLEY PRECISION INC	833
8	PALMETTO STATE ARMORY, LLC	523
9	PTR INDUSTRIES INC	522
10	PATRIOT ORDNANCE FACTORY INC	496
11	ARES DEFENSE SYSTEMS INC	483
12	CMMG INC	369
13	FEDERAL ARMAMENT LLC	326
14	U S ARMAMENT CORP	326
15	NORTH AMERICAN ARMS INC	271
	Total, Top 15	27,290
	Total, all U.S. Manufacturers, Year 2018	30,304

Pistol Production 2017, Top 15 U.S. Manufacturers by Units, Caliber from 0 to .32"

0 < Cal. ≤ .22"

Rank	Company	0 < Cal. ≤ .22"
1	STURM, RUGER & COMPANY, INC	196,582
2	SMITH & WESSON CORP	61,997
3	KEL TEC CNC INDUSTRIES INC	46,696
4	BROWNING ARMS COMPANY	30,859
5	PHOENIX ARMS	16,762
6	WALTHER MANUFACTURING INC	10,856
7	TAURUS INTERNATIONAL MANUFACT	8,948
8	BERETTA USA CORP	5,432
9	JIMENEZ ARMS INC	5,063
10	DIAMONDBACK FIREARMS LLC	4,706
11	COBRA ENTERPRISES OF UTAH, INC	3,485
12	ALPHATECH INC	3,450
13	TRAILBLAZER FIREARMS LLC	2,723
14	SIG SAUER INC	2,229
15	OUTDOOR COLORS LLC	1,712
	Total, Top 15	401,500
	Total, all U.S. Manufacturers, Year 2017	408,705

.22" < Cal. ≤ .25"

Rank	Company	.22" < Cal. ≤ .25"
1	PHOENIX ARMS	3,888
2	PALMETTO STATE ARMORY, LLC	1,357
3	EXTAR LLC	950
4	BRAVO COMPANY MFG INC	593
5	LWRC INTERNATIONAL	471
6	JIMENEZ ARMS INC	435
7	DEL-TON, INC	433
8	SIG SAUER INC	396
9	DANIEL DEFENSE INC	328
10	PODUNK INC	245
11	ROCK RIVER ARMS INC	198
12	HARDENED ARMS LLC	141
13	WINDHAM WEAPONRY INC	106
14	BLACK RAIN ORDNANCE INC	101
15	SAEILO, INC	100
	Total, Top 15	9,742
	Total, all U.S. Manufacturers, Year 2017	11,135

.25" < Cal. ≤ .32"

Rank	Company	.25" < Cal. ≤ .32"
1	RADICAL FIREARMS LLC	1,326
2	WHALLEY PRECISION INC	1,035
3	DIAMONDBACK FIREARMS LLC	777
4	PALMETTO STATE ARMORY, LLC	608
5	SIG SAUER INC	495
6	U S ARMAMENT CORP	441
7	NORTH AMERICAN ARMS INC	440
8	MODERN OUTFITTERS LLC	351
9	KEL TEC CNC INDUSTRIES INC	315
10	JIMENEZ ARMS INC	299
11	LUXUS ARMS LLC	193
12	BERETTA USA CORP	184
13	PTR INDUSTRIES INC	164
14	WINDHAM WEAPONRY INC	162
15	DANIEL DEFENSE INC	128
	Total, Top 15	6,918
	Total, all U.S. Manufacturers, Year 2017	8,152

Source: Compiled from ATF AFMER, Year 2017 p. 1–27 and Year 2018 p. 1–34, see Endnote 587.

Table A10.4: Top 15 U.S. Manufacturers per Caliber Category,
0 > Cal. ≤ .22", .22" < Cal. ≤ .25", .25" < Cal. ≤ .32",
Calendar Year 2016 and 2015.

Pistol Production 2016, Top 15 U.S. Manufacturers by Units, Caliber from 0 to .32"

Rank	Company	0 < Cal. ≤ .22"	Rank	Company	.22" < Cal. ≤ .25"	Rank	Company	.25" < Cal. ≤ .32"
1	STURM, RUGER & COMPANY INC	196,006	1	PHOENIX ARMS	6,264	1	KEL TEC CNC INDUSTRIES INC	3,233
2	SMITH & WESSON CORP	84,529	2	EXTAR LLC	1,085	2	BERETTA USA CORPORATION	2,353
3	KEL TEC CNC INDUSTRIES INC	44,607	3	SIG SAUER INC	826	3	WHALLEY PRECISION INC	1,269
4	ARMS TECHNOLOGY INC	39,113	4	TAURUS INTERNATIONAL	730	4	US ARMAMENT CORP	615
5	PHOENIX ARMS	34,736	5	PALMETTO STATE ARMORY	485	5	CMMG INC	609
6	BERETTA USA CORPORATION	13,303	6	ROCK RIVER ARMS INC	472	6	RADICAL FIREARMS LLC	578
7	TAURUS INTERNATIONAL	10,194	7	JIMENEZ ARMS INC	449	7	NORTH AMERICAN ARMS INC	318
8	HI TECH PLASTICS INC	7,250	8	BRAVO COMPANY MFG INC	338	8	WINDHAM WEAPONRY INC	149
9	JIMENEZ ARMS INC	6,617	9	WINDHAM WEAPONRY INC	230	9	SIG SAUER INC	143
10	SIG SAUER INC	3,081	10	DANIEL DEFENSE INC	226	10	DESTRUCTIVE DEVICES	115
11	RADICAL FIREARMS LLC	1,891	11	LWRC INTERNATIONAL	129	11	ALEX PRO FIREARMS LLC	52
12	DANIEL DEFENSE INC	541	12	WHALLEY PRECISION INC	119	12	BARNES PRECISION MACHINE	43
13	TANURY INDUSTRIES, INC.	442	13	PODUNK INC	94	13	ASYLUM WEAPONRY LLC	32
14	EXCEL INDUSTRIES INC	433	14	PATRIOT ORDNANCE FACTORY	80	14	INLAND MANUFACTURING LLC	29
15	ADAMS ARMS LLC	394	15	DEL-TON, INC	62	15	PARTISAN ENTERPRISES LLC	25
	Total, Top 15	443,137		Total, Top 15	11,589		Total, Top 15	9,563
	Total, all U.S. Manufacturers, Year 2016	447,315		Total, all U.S. Manufacturers, Year 2016	13,141		Total, all U.S. Manufacturers, Year 2016	10,175

Pistol Production 2015, Top 15 U.S. Manufacturers by Units, Caliber from 0 to .32"

Rank	Company	0 < Cal. ≤ .22"	Rank	Company	.22" < Cal. ≤ .25"	Rank	Company	.25" < Cal. ≤ .32"
1	STURM, RUGER & COMPANY INC	216,206	1	PHOENIX ARMS	3,816	1	COLT'S MANUFACTURING	5,352
2	ARMS TECHNOLOGY INC	49,607	2	SIG SAUER INC	2,759	2	KEL TEC CNC INDUSTRIES INC	3,309
3	KEL TEC CNC INDUSTRIES INC	41,968	3	EXTAR LLC	1,235	3	RADICAL FIREARMS LLC	973
4	SMITH & WESSON CORP	23,299	4	JIMENEZ ARMS INC	779	4	JIMENEZ ARMS INC	972
5	PHOENIX ARMS	20,034	5	DANIEL DEFENSE INC	653	5	BERETTA USA CORPORATION	862
6	BERETTA USA CORPORATION	15,368	6	TAURUS INTERNATIONAL	452	6	WHALLEY PRECISION INC	814
7	TAURUS INTERNATIONAL	13,649	7	ROCK RIVER ARMS INC	387	7	PTR INDUSTRIES INC	471
8	COBRA ENTERPRISES OF	7,510	8	WINDHAM WEAPONRY INC	215	8	DANIEL DEFENSE INC	359
9	JIMENEZ ARMS INC	7,273	9	BRAVO COMPANY MFG INC	133	9	COBRA ENTERPRISES OF	335
10	SIG SAUER INC	6,386	10	PHASE 5 WEAPON SYSTEMS	112	10	CMMG INC	269
11	HI TECH PLASTICS INC	2,480	11	MILTAC INDUSTRIES LLC	83	11	US ARMAMENT CORP	198
12	KEYSTONE SPORTING ARMS LLC	1,917	12	COBRA ENTERPRISES OF	82	12	NORTH AMERICAN ARMS INC	108
13	ENGINEERING & CYCLE CO INC	1,898	13	PRECISION SMALL ARMS INC	78	13	WINDHAM WEAPONRY INC	105
14	CHIAPPA FIREARMS LTD	1,755	14	DEER COUNTRY ARCHERY INC	71	14	LUXUS ARMS LLC	66
15	CDQ SOLUTIONS, LLC	1,585	15	CMMG INC	64	15	PRIMARY WEAPONS SYSTEMS	59
	Total, Top 15	410,935		Total, Top 15	10,919		Total, Top 15	14,252
	Total, all U.S. Manufacturers, Year 2015	413,230		Total, all U.S. Manufacturers, Year 2015	11,567		Total, all U.S. Manufacturers, Year 2015	14,763

Source: Compiled from ATF AFMER, Year 2015 p. 1–20 and Year 2016 p. 1–21, see Endnote 587.

Table A10.5:   Top 15 U.S. Manufacturers per Caliber Category,
.32" > Cal. ≤ .380", .380" < Cal. ≤ 9mm, 9mm < Cal. ≤ .50",
Calendar Year 2018 and 2017.

Pistol Production 2018, Top 15 U.S. Manufacturers by Units, Caliber from > .32" to .50"

Rank	Company	.32" < Cal. ≤ .380"	Rank	Company	.380" < Cal. ≤ 9mm	Rank	Company	9mm < Cal. ≤ .50"
1	STURM, RUGER & COMPANY, INC	255,786	1	SIG SAUER INC	505,749	1	SMITH & WESSON CORP	154,741
2	SMITH & WESSON CORP	220,298	2	SMITH & WESSON CORP	457,160	2	SPRINGFIELD INC	77,107
3	TAURUS INTERNATIONAL MANUFACT	89,771	3	STURM, RUGER & COMPANY, INC	299,443	3	SIG SAUER INC	70,585
4	SIG SAUER INC	46,605	4	GLOCK INC	237,309	4	KIMBER MFG INC	67,428
5	SPRINGFIELD INC	44,730	5	SCCY INDUSTRIES LLC	169,819	5	COLT'S MANUFACTURING COMPANY	35,399
6	KIMBER MFG INC	20,478	6	KIMBER MFG INC	113,224	6	STURM, RUGER & COMPANY, INC	23,723
7	BERETTA USA CORP	19,224	7	BERETTA USA CORP	53,739	7	AMERICAN TACTICAL INC	14,840
8	BROWNING ARMS COMPANY	12,358	8	FN AMERICA, LLC	39,602	8	REMINGTON ARMS COMPANY LLC	14,624
9	STRASSELLS MACHINE INC	10,300	9	EPP TEAM INC	28,610	9	HASKELL MANUFACTURING INC	12,800
10	JIMENEZ ARMS INC	8,669	10	STRASSELLS MACHINE INC	26,600	10	FN AMERICA, LLC	12,241
11	COBRA ENTERPRISES OF UTAH, INC	7,038	11	COBRA ENTERPRISES OF UTAH, INC	19,066	11	BOND ARMS, INC	12,127
12	DIAMONDBACK FIREARMS LLC	6,183	12	SPRINGFIELD INC	18,200	12	IBERIA FIREARMS INC	7,400
13	REMINGTON ARMS COMPANY LLC	6,049	13	DIAMONDBACK FIREARMS LLC	15,349	13	SAEILO, INC	5,963
14	GLOCK INC	4,000	14	REMINGTON ARMS COMPANY LLC	13,141	14	CZ-USA INC	5,719
15	SAEILO, INC	2,245	15	CZ-USA INC	9,489	15	HECKLER & KOCH, INC	4,308
	Total, Top 15	753,734		Total, Top 15	2,006,500		Total, Top 15	519,005
	Total, all U.S. Manufacturers, Year 2018	760,776		Total, all U.S. Manufacturers, Year 2018	2,094,870		Total, all U.S. Manufacturers, Year 2018	546,809

Pistol Production 2017, Top 15 U.S. Manufacturers by Units, Caliber from > .32" to .50"

Rank	Company	.32" < Cal. ≤ .380"	Rank	Company	.380" < Cal. ≤ 9mm	Rank	Company	9mm < Cal. ≤ .50"
1	STURM, RUGER & COMPANY, INC	376,304	1	SMITH & WESSON CORP	606,732	1	SMITH & WESSON CORP	250,394
2	SMITH & WESSON CORP	113,246	2	SIG SAUER INC	368,386	2	SIG SAUER INC	98,680
3	GLOCK INC	73,646	3	STURM, RUGER & COMPANY, INC	163,887	3	KIMBER MFG INC	57,828
4	SIG SAUER INC	66,588	4	SCCY INDUSTRIES LLC	150,235	4	SPRINGFIELD INC	57,459
5	TAURUS INTERNATIONAL MANUFACT	60,172	5	KIMBER MFG INC	98,385	5	STURM, RUGER & COMPANY, INC	44,850
6	KIMBER MFG INC	27,608	6	GLOCK INC	94,665	6	COLT'S MANUFACTURING COMPANY	25,977
7	BERETTA USA CORP	20,810	7	FN AMERICA, LLC	45,384	7	REMINGTON ARMS COMPANY LLC	18,587
8	BROWNING ARMS COMPANY	19,472	8	STRASSELLS MACHINE INC	31,210	8	FN AMERICA, LLC	16,126
9	REMINGTON ARMS COMPANY LLC	15,178	9	BERETTA USA CORP	30,741	9	HASKELL MANUFACTURING INC	10,900
10	STRASSELLS MACHINE INC	14,805	10	REMINGTON ARMS COMPANY LLC	25,781	10	BOND ARMS, INC	9,629
11	WALTHER MANUFACTURING INC	13,604	11	SPRINGFIELD INC	23,918	11	CENTURY ARMS INC	9,321
12	JIMENEZ ARMS INC	11,586	12	WALTHER MANUFACTURING INC	17,688	12	SAEILO, INC	8,108
13	COBRA ENTERPRISES OF UTAH, INC	11,475	13	DIAMONDBACK FIREARMS LLC	14,672	13	IBERIA FIREARMS INC	7,800
14	SAEILO, INC	5,360	14	COBRA ENTERPRISES OF UTAH, INC	12,852	14	GLOCK INC	7,385
15	STI FIREARMS LLC	4,890	15	SAEILO, INC	10,075	15	MAGNUM RESEARCH INC	7,217
	Total, Top 15	834,744		Total, Top 15	1,694,611		Total, Top 15	630,261
	Total, all U.S. Manufacturers, Year 2018	848,425		Total, all U.S. Manufacturers, Year 2018	1,756,618		Total, all U.S. Manufacturers, Year 2018	657,971

Source: Compiled from ATF AFMER, Year 2017 p. 1–27 and Year 2018 p. 1–34, see Endnote 587.

Table A10.6: Top 15 U.S. Manufacturers per Caliber Category, .32" > Cal. ≤ .380", .380" < Cal. ≤ 9mm, 9mm < Cal. ≤ .50", Calendar Year 2016 and 2015.

Pistol Production 2016, Top 15 U.S. Manufacturers by Units, Caliber from > .32" to .50"

Rank	Company (.32" < Cal. ≤ .380")	.32" < Cal. ≤ .380"	Rank	Company (.380" < Cal. ≤ 9mm)	.380" < Cal. ≤ 9mm	Rank	Company (9mm < Cal. ≤ .50")	9mm < Cal. ≤ .50"
1	STURM, RUGER & COMPANY INC	332,666	1	SMITH & WESSON CORP	823,260	1	SMITH & WESSON CORP	291,827
2	SMITH & WESSON CORP	229,835	2	STURM, RUGER & COMPANY INC	423,162	2	SIG SAUER INC	102,849
3	GLOCK INC	136,993	3	SIG SAUER INC	346,347	3	KIMBER MFG INC	94,395
4	SIG SAUER INC	128,561	4	GLOCK INC	213,751	4	STURM, RUGER & COMPANY INC	79,032
5	TAURUS INTERNATIONAL	114,493	5	SCCY INDUSTRIES LLC	160,676	5	SPRINGFIELD INC	50,078
6	KIMBER MFG INC	61,089	6	STRASSELLS MACHINE INC	69,900	6	COLT'S MANUFACTURING	46,705
7	REMINGTON ARMS COMPANY	29,206	7	KIMBER MFG INC	65,165	7	HASKELL MANUFACTURING INC	32,400
8	STRASSELLS MACHINE INC	21,000	8	REMINGTON ARMS COMPANY	36,414	8	IBERIA FIREARMS INC	24,100
9	ARMS TECHNOLOGY INC	17,672	9	SPRINGFIELD INC	21,935	9	REMINGTON ARMS COMPANY	19,232
10	JIMENEZ ARMS INC	15,208	10	SAEILO, INC	17,714	10	GLOCK INC	17,396
11	BERETTA USA CORPORATION	12,379	11	COLT'S MANUFACTURING	17,129	11	BOND ARMS, INC	16,758
12	SAEILO, INC	12,280	12	BERETTA USA CORPORATION	14,967	12	MAGNUM RESEARCH INC	12,815
13	COLT'S MANUFACTURING	7,380	13	JIMENEZ ARMS INC	14,709	13	SAEILO, INC	10,280
14	KEL TEC CNC INDUSTRIES INC	4,480	14	FMK FIREARMS	12,247	14	CZ-USA INC	6,284
15	BOND ARMS, INC	2,439	15	PAUWAY CORP	5,774	15	WM C ANDERSON INC	3,666
	Total, Top 15	1,125,681		Total, Top 15	2,243,150		Total, Top 15	807,817
	Total, all U.S. Manufacturers, Year 2016	1,130,459		Total, all U.S. Manufacturers, Year 2016	2,281,450		Total, all U.S. Manufacturers, Year 2016	837,535

Pistol Production 2015, Top 15 U.S. Manufacturers by Units, Caliber from > .32" to .50"

Rank	Company (.32" < Cal. ≤ .380")	.32" < Cal. ≤ .380"	Rank	Company (.380" < Cal. ≤ 9mm)	.380" < Cal. ≤ 9mm	Rank	Company (9mm < Cal. ≤ .50")	9mm < Cal. ≤ .50"
1	STURM, RUGER & COMPANY INC	169,083	1	SMITH & WESSON CORP	552,482	1	SMITH & WESSON CORP	256,806
2	SMITH & WESSON CORP	157,266	2	STURM, RUGER & COMPANY INC	272,551	2	SIG SAUER INC	125,576
3	GLOCK INC	150,008	3	SIG SAUER INC	234,962	3	STURM, RUGER & COMPANY INC	90,524
4	SIG SAUER INC	89,972	4	SCCY INDUSTRIES LLC	116,998	4	KIMBER MFG INC	77,444
5	TAURUS INTERNATIONAL	77,285	5	STRASSELLS MACHINE INC	65,100	5	SPRINGFIELD INC	41,381
6	KIMBER MFG INC	37,179	6	BERETTA USA CORPORATION	58,775	6	HASKELL MANUFACTURING INC	27,300
7	STRASSELLS MACHINE INC	21,700	7	GLOCK INC	47,504	7	IBERIA FIREARMS INC	21,500
8	JIMENEZ ARMS INC	18,825	8	COLT'S MANUFACTURING	47,401	8	GLOCK INC	19,104
9	BERETTA USA CORPORATION	18,658	9	JIMENEZ ARMS INC	17,479	9	REMINGTON ARMS COMPANY	18,236
10	COBRA ENTERPRISES OF	18,270	10	KIMBER MFG INC	17,077	10	FN AMERICA, LLC	18,125
11	REMINGTON ARMS COMPANY	17,714	11	SPRINGFIELD INC	16,198	11	BOND ARMS, INC	13,067
12	COLT'S MANUFACTURING	12,662	12	FN AMERICA, LLC	14,624	12	CZ-USA INC	6,924
13	ARMS TECHNOLOGY INC	9,713	13	KEL TEC CNC INDUSTRIES INC	13,666	13	SAEILO, INC	6,530
14	SAEILO, INC	7,981	14	COBRA ENTERPRISES OF	13,085	14	HECKLER & KOCH INC	4,904
15	DIAMONDBACK FIREARMS LLC	4,678	15	SAEILO, INC	10,933	15	MAGNUM RESEARCH INC	4,857
	Total, Top 15	810,994		Total, Top 15	1,498,835		Total, Top 15	732,278
	Total, all U.S. Manufacturers, Year 2015	819,103		Total, all U.S. Manufacturers, Year 2015	1,531,065		Total, all U.S. Manufacturers, Year 2015	767,471

Source: Compiled from ATF AFMER, Year 2015 p. 1–20 and Year 2016 p. 1–21, see Endnote 587.

U.S. Imports, Handguns (Revolvers and Pistols)

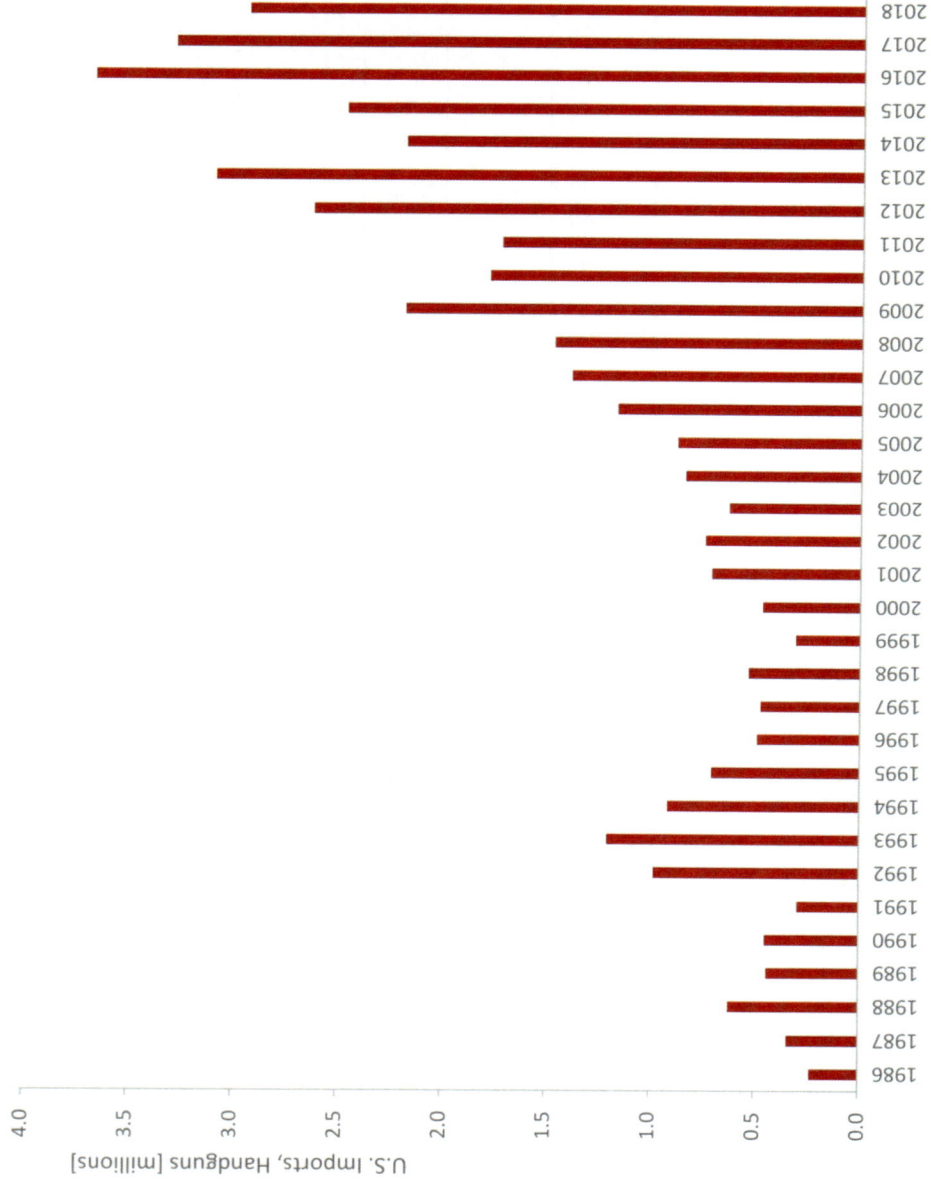

Fig. A10.2: U.S. Firearms Imports, Handguns, 1986–2018
(compiled from ATF Commerce Report 2019)

Handgun Demand, NICS

Mandated by the Brady Handgun Violence Prevention Act of 1993 and launched by the FBI on November 30, 1998, the National Instant Criminal Background Check System (NICS) is used by Federal Firearms Licensees to instantly determine whether a prospective buyer is eligible to buy firearms. The number of requests for review of reliability are more up to date (monthly basis) than AFMER and allows an assessment of the trend of the market volume broken down to the individual U.S. States and Territories. The FBI's statistics represent the number of firearm background checks initiated through the NICS but does not represent the number of firearms sold. Based on varying state laws and purchase scenarios, a one-to-one correlation cannot be made between a firearm background check and a firearm sale. Also included are applications for carry permissions and other applications in the field of firearms legislation. Occasionally, adjusted data is available, *adjusted NICS*, which exhibit the number of applications for the purchase of weapons.

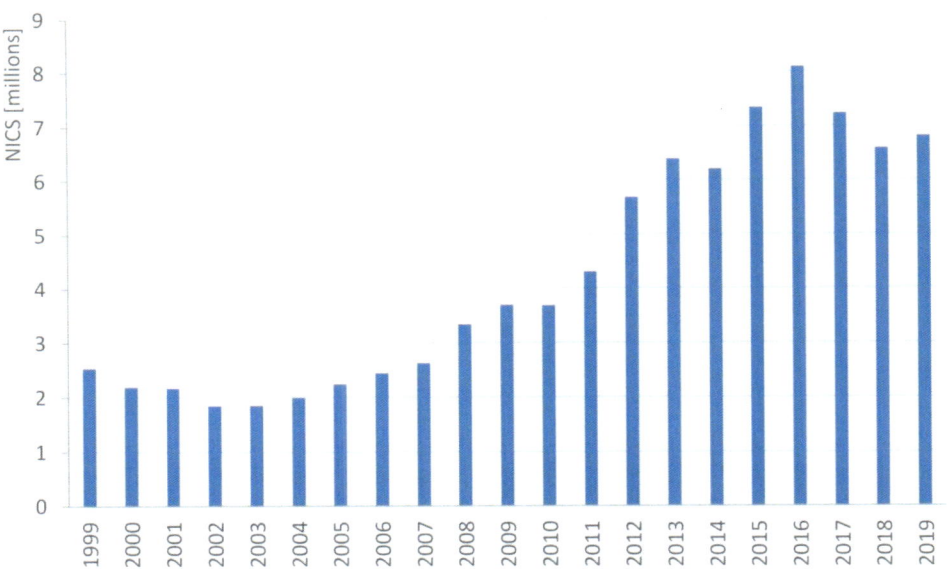

Fig. A10.3:   Yearly numbers of firearm background checks, 1999–2019.
(FBI NICS Firearm Background Checks: Year by State/Type, Handguns)

Table A10.7:   Total number of NICS checks, 2019, State by State NICS,
Pistols and Revolvers (CA, MA, MD and NY are highlighted).

Rank	State or Territory	NICS Handguns, 2019	Rank	State or Territory	NICS Handguns, 2019
1	Florida	608,924	29	Kansas	66,735
2	Texas	554,577	30	Connecticut	58,808
3	Pennsylvania	529,408	31	New Hampshire	58,376
4	California	432,239	32	Nevada	58,284
5	Tennessee	295,844	33	Idaho	50,565
6	Virginia	281,144	34	Massachusetts	50,515
7	Ohio	264,798	35	New Jersey	48,969
8	Illinois	248,616	36	Maryland	43,180
9	Indiana	228,087	37	Utah	41,881
10	Missouri	218,453	38	Maine	36,250
11	Colorado	210,909	39	Montana	34,421
12	Washington	206,107	40	Alaska	31,943
13	Oregon	166,667	41	South Dakota	29,458
14	Wisconsin	164,711	42	Puerto Rico	24,574
15	Georgia	162,135	43	Wyoming	23,090
16	Alabama	160,667	44	Delaware	20,228
17	Arizona	158,864	45	North Carolina	19,788
18	Oklahoma	145,542	46	North Dakota	19,291
19	Kentucky	126,820	47	Vermont	16,299
20	Michigan	126,582	48	Rhode Island	11,698
21	South Carolina	121,053	49	Iowa	2,098
22	Louisiana	119,091	50	Guam	1,446
23	Minnesota	105,932	51	District of Columbia	1,295
24	New York	103,076	52	Nebraska	1,269
25	Mississippi	97,345	53	Virgin Islands	370
26	New Mexico	74,072	54	Mariana Islands	88
27	West Virginia	71,036	55	Hawaii	0
28	Arkansas	68,549		TOTAL, NICS, 2019	6,802,167

Source: FBI NICS Firearm Background Checks: Year by State/Type, Handguns.

Table A10.8:   Total number of NICS checks, 2018, State by State NICS,
    Pistols and Revolvers (CA, MA, MD and NY are highlighted).

Rank	State or Territory	NICS Handguns, 2018	Rank	State or Territory	NICS Handguns, 2018
1	Florida	590,140	29	Minnesota	62,516
2	Pennsylvania	534,653	30	Connecticut	60,630
3	Texas	523,406	31	New Hampshire	59,794
4	California	433,106	32	Nevada	54,925
5	Tennessee	290,067	33	Massachusetts	53,039
6	Ohio	277,027	34	Maryland	52,671
7	Illinois	259,554	35	New Jersey	51,074
8	Virginia	253,744	36	Idaho	45,993
9	Missouri	224,305	37	Utah	42,347
10	Indiana	224,264	38	Maine	37,050
11	Colorado	216,617	39	Montana	32,778
12	Washington	202,826	40	Alaska	32,193
13	Oregon	166,211	41	South Dakota	27,838
14	Georgia	157,027	42	Puerto Rico	25,039
15	Wisconsin	155,441	43	Wyoming	21,617
16	Arizona	147,612	44	Delaware	18,953
17	Oklahoma	135,474	45	Vermont	18,532
18	Michigan	123,070	46	North Dakota	18,332
19	Louisiana	121,441	47	North Carolina	17,932
20	Kentucky	112,323	48	Rhode Island	12,252
21	South Carolina	111,804	49	Iowa	1,901
22	New York	108,936	50	Nebraska	1,461
23	Mississippi	95,091	51	Guam	1,145
24	Alabama	84,187	52	District of Columbia	975
25	West Virginia	75,539	53	Virgin Islands	437
26	New Mexico	68,549	54	Mariana Islands	82
27	Arkansas	66,321	55	Hawaii	0
28	Kansas	65,870		TOTAL, NICS, 2018	6,576,111

Source: FBI NICS Firearm Background Checks: Year by State/Type, Handguns.

Table A10.9:   Total number of NICS checks, 2017, State by State NICS,
Pistols and Revolvers (CA, MA, MD and NY are highlighted).

Rank	State or Territory	NICS Handguns, 2017	Rank	State or Territory	NICS Handguns, 2017
1	Florida	638,938	29	Connecticut	71,853
2	Texas	577,417	30	New Mexico	70,791
3	Pennsylvania	566,320	31	Massachusetts	66,966
4	California	512,465	32	New Hampshire	62,280
5	Ohio	332,879	33	New Jersey	58,989
6	Tennessee	303,513	34	Nevada	57,524
7	Virginia	292,199	35	Maryland	51,873
8	Illinois	282,267	36	Utah	45,085
9	Missouri	252,709	37	Idaho	44,939
10	Indiana	249,127	38	Maine	40,485
11	Colorado	229,708	39	Alaska	34,556
12	Washington	198,632	40	Montana	34,444
13	Georgia	172,447	41	South Dakota	31,493
14	Wisconsin	171,610	42	Delaware	22,308
15	Oregon	168,938	43	Wyoming	20,155
16	Arizona	153,522	44	North Dakota	19,693
17	Oklahoma	144,736	45	North Carolina	18,131
18	Michigan	141,062	46	Vermont	17,599
19	Louisiana	138,975	47	Puerto Rico	15,298
20	Kentucky	127,240	48	Rhode Island	12,959
21	New York	124,253	49	Iowa	2,376
22	South Carolina	123,035	50	Nebraska	1,669
23	Mississippi	103,312	51	Guam	1,041
24	Alabama	97,751	52	District of Columbia	756
25	Minnesota	94,383	53	Virgin Islands	298
26	West Virginia	79,813	54	Mariana Islands	108
27	Kansas	73,959	55	Hawaii	0
28	Arkansas	72,100		TOTAL, NICS, 2017	7,226,979

Source: FBI NICS Firearm Background Checks: Year by State/Type, Handguns.

Table A10.10: Total number of NICS checks, 2016, State by State NICS,
Pistols and Revolvers (CA, MA, MD and NY are highlighted).

Rank	State or Territory	NICS Handguns, 2016	Rank	State or Territory	NICS Handguns, 2016
1	Florida	662,308	29	Arkansas	80,244
2	Pennsylvania	642,232	30	Massachusetts	75,977
3	Texas	637,476	31	New Jersey	70,249
4	California	560,355	32	New Mexico	69,434
5	Ohio	394,570	33	New Hampshire	68,529
6	Illinois	352,411	34	Nevada	66,158
7	Tennessee	319,169	35	Maryland	52,305
8	Virginia	295,963	36	Maine	48,845
9	Indiana	278,182	37	Utah	47,401
10	Missouri	274,754	38	Idaho	46,523
11	Colorado	242,502	39	Montana	37,547
12	Washington	214,106	40	Alaska	37,491
13	Georgia	194,800	41	South Dakota	36,520
14	Wisconsin	186,300	42	Delaware	25,574
15	Oregon	178,716	43	Wyoming	22,569
16	Louisiana	171,588	44	North Dakota	19,975
17	Arizona	166,784	45	Vermont	18,320
18	Oklahoma	166,181	46	North Carolina	16,595
19	Michigan	164,878	47	Rhode Island	14,262
20	Alabama	153,123	48	Puerto Rico	12,779
21	Kentucky	140,721	49	Iowa	2,790
22	Minnesota	133,962	50	Nebraska	1,717
23	New York	133,285	51	Guam	1,249
24	South Carolina	132,473	52	District of Columbia	675
25	Connecticut	122,375	53	Virgin Islands	266
26	Mississippi	119,050	54	Hawaii	0
27	West Virginia	92,124	55	Mariana Islands	0
28	Kansas	81,116		TOTAL, NICS, 2016	8,085,498

Source: FBI NICS Firearm Background Checks: Year by State/Type, Handguns. (NICS, 2016)

Appendix 11   Comparison: Flat Wire vs. Round Wire Spring

Table A11.1:   Recoil spring data comparison: Flat wire vs. round wire.

	Flat-wire Spring		Round-wire Spring		Ratio: round-wire vs. flat-wire
Free Length	Lo	155,40 mm	Lo	91,05 mm	
Wire Size	b	0,61 mm	d	1,30 mm	
	h	2,40 mm			
Outer Diameter	De	12,45 mm	De	12,45 mm	
Total Coils	nt	30,3	nt	18,4	
Deflection 1	s1	82,10 mm	s1	17,75 mm	22%
Spring Length 1	L1	73,30 mm	L1	73,30 mm	
Load F1	F1	50,00 N	F1	22,00 N	44%
Deflection 2	s2	128,10 mm	s2	63,75 mm	50%
Spring Length 2	L2	27,30 mm	L2	27,30 mm	
Load F2	F2	78,00 N	F2	79,00 N	101%
Deflection to Solid	sc	132,93 mm	sc	67,10 mm	50%
Solid Length	Lc	22,47 mm	Lc	23,95 mm	
Force at Solid	Fc	80,91 N	Fc	83,15 N	103%
Working Stroke	sH	46,00 mm	sH	46,00 mm	
Material Tensile Strength	Rm	2050 MPa	Rm	2060 MPa	100%
Shear Stresses	$\tau 1$	905 MPa	$\tau k1$	330 MPa	36%
	$\tau 2$	1411 MPa	$\tau k2$	1184 MPa	84%
	τC	1464 MPa	τC	1075 MPa	73%
	$\tau C / Rm$	71% Rm	$\tau C / Rm$	52% Rm	73%
Stroke Stresses	τH	506 MPa	τH	854 MPa	**169%**
	$\tau H / Rm$	25% Rm	$\tau H / Rm$	41% Rm	
Spring Rate	R	0,609 N/mm	R	1,239 N/mm	**204%**
Work between L1 and L2	A12	2,944 J	A12	2,323 J	**79%**
Weight	G	11,68 g	G	6,79 g	58%

BAUMANN FEDERN AG

Phone:	0041 55 286 316	
E-Mail:	pius.grob@baumann-springs.com	
Name:	P. Grob	Date: Sept. 19, 2013

Appendix 12   Influence of the Frame Material on the Total Weight of Pistols

Comparison of total weight of pistols

It is often suggested over the past decades that pistols were getting lighter and lighter. This statement will be reviewed below. The objective is to determine the weight difference between handguns with frames made of steel, aluminum alloy or polymer.

For the comparison, each weight of three guns from each material category was taken. The pistols were weighed with empty magazine. Next, the average weight per material category was calculated. To exclude too much deviation in weight, care was taken to include only tilt-barrel design pistols in the survey, chambered in 9 mm Luger.

In addition to the comparison of the total weight, the weight of the frame assembly was measured. Considered was each frame's "gross" weight, that is, the weight of the frame including all frame components, which also included parts which had to be removed from the frame during field-stripping procedure.

Table A12.1:   Mass of steel frame pistols.

	CZ Model 75	FN Hi-Power	RADOM P35	Average
Mass, pistol w/empty magazine [g]	980	910	1016	969
Mass, frame, gross [g]	478	417	520	472

Table A12.2:   Mass of aluminum alloy frame pistols.

	SIG SAUER P220	S&W Mod. 39	WALTHER P88 Compact	Average
Mass, pistol w/empty magazine [g]	807	803	822	811
Mass, frame, gross [g]	314	293	329	312

Table A12.3: Mass of polymer frame pistols.

	GLOCK G19 Gen4	HK P2000 V2	WALTHER PPQ M2	Average
Mass, pistol w/empty magazine [g]	669	710	693	691
Mass, frame, gross [g]	130	200	175	168

Summary

The sample pistols chosen for the comparison represented a randomly picked selection of limited choice within a very extensive range of handguns and the result of this sample was strongly dependent on the selection made. In addition, further variations within the manufacturing tolerances of the firearms, different size classes, equipment features, etc. were not taken into account. As the tables also indicate, there was a large variation within each of the three categories. Steel frame pistol samples, for example, included a FN *Hi-Power* with a weight of 910 grams, but also a RADOM *P35* with 1016 grams.

Table A12.4: Comparison of average masses.

Frame material	Mass [g]		Frame material	Mass [g]	
	Pistol	Frame		Pistol	Frame
Steel	969	472	Al-Alloy	811	312
Al-Alloy	811	312	Polymer	691	168
delta	158	160	delta	120	144

Yet, a trend is visible: pistols with aluminum alloy frame instead of steel frames had a lower weight in the examined sample. On the basis of the semi-autos selected for the sample the average weight of steel frame pistols was 969 g, whereas pistols with aluminum alloy frames would have an average weight of 811 g, that is, the average weight of the samples with aluminum frame was 158 g less (see Table A12.4). The average weight of the considered polymer frame pistols was yet another 120 g lower than the average weight of the guns with aluminum alloy frame.

The weight reduction in the gross weight of the frames was particularly evident. The weight decreased here, starting from an average 472 g of steel frames, to 312 g of aluminum alloy frames, and 168 g of polymer frame assemblies. The weight reduction amounted to 160 g, and another 144 g, respectively.

The calculated average delta of 144 g between aluminum alloy frames and polymer frames was significantly greater than the average delta of 120 g of complete weapons (120 g) within the same categories. The conclusion is: when the manufacturers were going from pistols with aluminum alloy frames to "polymer pistols" the weight reduction of the frames did not lead to the same reduction of the handguns' total weight. Theoretically, the pistols could have been 144 g lighter, but they were actually only 120 g lighter on average. The difference of 24 g was obviously added to barrel, slide and maybe the magazine.

Appendix 13   Slide Mass, 9 mm Luger

Determination of approximate values for the mass of a slide depending on the desired mean slide recoil velocity from 6 m/s up to 8 m/s, for 9 mm Luger.

When designing a pistol, the engineer's challenge is to choose a slide's mass which allows to reliably shoot a wide variety of cartridge types without causing the slide to short stroke or to recoil too fast, that is, to over-ride, or to cause components to crack or fail.

The desirable range for the mean slide recoil velocity of 6 to 8 m/s referred to by *Maier* should be considered when determining the first approximations for the mass of the slide (cf. 3.7 Springs, p. 112ff). Based on this worthwhile velocity-range, a theoretical approach is shown below for 9 mm Luger, which may provide an initial approximation for the recoiling mass m_r, see Fig. A13.1, then an approximate value for the mass of the barrel m_B is provided based on barrel length, which ultimately takes us to an approximation of the slide's mass m_S.

Initially, it was considered which momentum can be expected when firing a cartridge 9 mm Luger, that is, the magnitude for minimum and maximum impulse force. Based on Formula 5.1, p. 231, the linear momentum was calculated using:

$$p_P = m_P \, v_0 \qquad\qquad \text{A13.1}$$

where: p_P Momentum of projectile [N s]
         m_P Mass of projectile [kg]
         v_0 Muzzle velocity of projectile [m/s]

For the calculation of the projectile's momentum p_P data for muzzle velocity and projectile mass was taken from professional literature. Frank C. Barnes, *Cartridges of the World*, shows data of current ammunition types. Details for those cartridges were pulled out from there, which resulted in minimum or maximum impulse force. The U.S. military load was on the upper level, with an 8 g projectile and a muzzle velocity of 396 m/s.

As a representative of the medium range, German police ammunition data were considered in addition, which has a projectile mass of 6.1 g, and delivers a muzzle energy E_0 of 500 J, according to TR "Pistole" of the German authorities.[675] With this data, the muzzle velocity for the 6.1 g projectile could be calculated, using the following formula:

$$E_0 = \frac{m_P}{2} v_0{}^2 \qquad\qquad \textbf{A13.2}[676]$$

where: E_0 Projectile's muzzle energy [J]
         m_P Mass of projectile [kg]
         v_0 Muzzle velocity of projectile [m/s]

This formula was dissolved after the muzzle velocity, v_0:

$$v_0 = \sqrt{\frac{2\,E_0}{m_P}} \qquad\qquad \textbf{A13.3}$$

Using Formula A13.3, 405 m/s was calculated for the muzzle velocity v_0 of the German police duty ammo.

Using Formula A13.1, the projectile's momentum p_P was calculated for cartridges with minimum or maximum momentum, as well as for ammunition specified by TR, and applying Formula A13.2, the muzzle energy E_0 of the projectile was calculated. The resulting performance data is listed in Table A13.1. The muzzle energy E_0 in this table is listed for comparison purposes only, because in gun literature muzzle energy some-times maybe of a certain interest to the reader.

Table A13.1: Performance data, cartridge 9 mm Luger.

	Min.	TR [a]	Max.
Mass of projectile [gr][677]	100	94	124
Mass of projectile m_P [g] [b]	6.5	6.1	8.0
Muzzle velocity [ft/s][678]	1150	1329	1300
Muzzle velocity v_0 [m/s] [c]	351	405	396
Momentum p_P [N s]	2.27	2.47	3.18
Muzzle energy E_0 [J]	398	500	631

After the momentum was identified, the resulting recoiling mass m_r could be calculated by assuming various slide recoil velocities v_{S1}. With the numbers from Table A13.1, this calculation was done for recoil velocities from 6 m/s to 8 m/s, in increments of 0.25 m/s, based on Formula 5.1, p. 231, which was transformed to:

---

[a] Values given are for a 6,1 g projectile with $E_3 = 500$ J as specified by *Technische Richtlinie*. (TR "Pistolen", 2008 p. 9)
[b] Conversion factor from grains to grams: 1/15,43 g/gr. (Meyer, 1986 p. 509)
[c] Conversion factor from ft/s to m/s: 0,3048 m/ft. (Meyer, 1986 p. 508)

$$m_r = \frac{p_P}{v_{S1}}$$

A13.4

where: m_r Recoiling mass [kg]
          p_P Momentum of projectile [N s]
          v_{S1} Slide recoil velocity, projectile leaves the muzzle [m/s]

An overview of the results is given in Table A13.2.

Table A13.2:   Recoiling mass as a function of mean slide recoil velocity.

Slide recoil velocity v_{S1} [m/s]	Recoiling Mass m_r [g]		
	Min. Momentum (m_P = 6.5 g, v_0 = 351 m/s)	TR Momentum (m_P = 6.1 g, v_0 = 405 m/s)	Max. Momentum (m_P = 8.0 g, v_0 = 396 m/s)
6	379	412	531
6.25	363	395	509
6.5	349	380	490
6.75	337	366	472
7 [a]	325	353	455
7.25	313	341	439
7.5	303	329	425
7.75	293	319	411
8	284	309	398

Pistols chambered in 9 mm Luger commonly use a modified Browning type locking mechanism, which consists of a coupled barrel and slide. Consequently, the recoiling mass m_r is the combination of the barrel's mass m_B, the mass of slide m_S, and a fraction of the recoil spring's mass. The mass of the recoil spring was ignored in our consideration. Using a ball-park figure for the barrel's mass m_B, made it easy to determine the mass of the slide m_S. Barrel masses exhibited in Table A13.3 can be used for orientation, if an estimation value is required for the mass of a barrel, intended for the use of the 9 mm Luger cartridge type. The data was taken from barrels of commercially available pistols, and was determined by weighing.

----

[a] Mean slide recoil velocity $\bar{v}_{Vt}$ = 7 m/s represents a preferred average value.

Table A13.3:   Mass of different barrels, 9 mm Luger; overview.

Brand	Model	Serial Number	Barrel Length L_B [mm]	Barrel Length [Inch]	Barrel Mass m_B [g]	Specific Mass [a] m_B/L_B [g/mm]
GLOCK	G19	GCN657	102	4.0	104	1.02
HK	P2000	116-000379	93	3.7	96	1.03
SIG SAUER	P250DCc	EA 003 591	98	3.9	104	1.06
WALTHER	PPS	AF1750	81	3.2	79	0.98
WALTHER	P99	FAE5879	101	4.0	93	0.92
Average					95	1.00

The numbers determined from Table A13.2 are depicted graphically below:

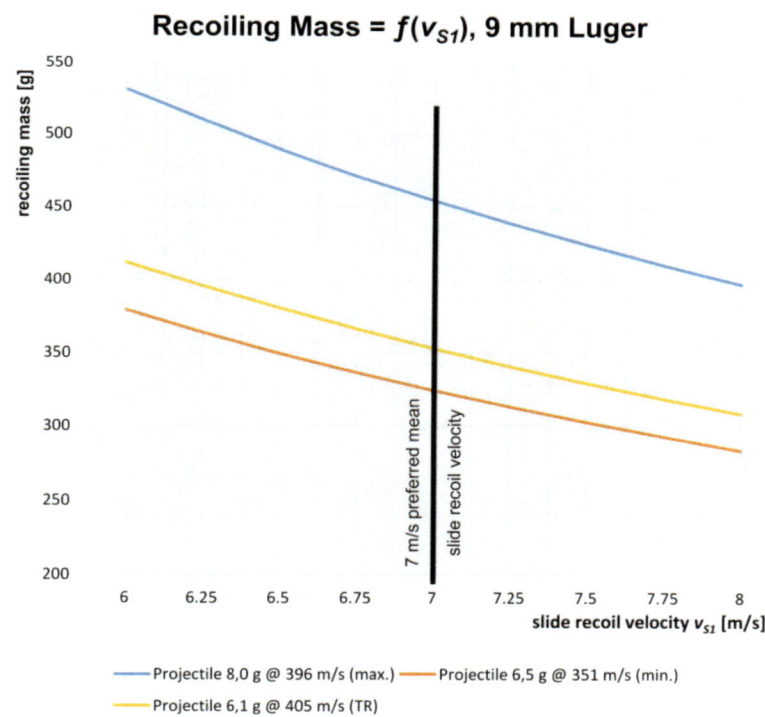

Fig. A13.1: Recoiling mass vs. slide recoil velocity.

---

[a] Specific mass is calculated by dividing mass m_B by length L_B of the barrel, i.e. the specific mass represents an average value of the thin walled tubular part and the barrel block.

A reading example: If a 9 mm pistol is to be engineered and a preferred initial slide recoil velocity of 7 m/s is the objective, a minimum recoiling mass of about 325 g and maximum 455 g can be read off from Fig. A13.1. Usually, a pistol chambered for this cartridge type will have a coupled barrel and slide, that is, the moving mass is the total of the mass of barrel m_B and the slide assembly's mass m_S. An approximation for the mass of a 100 mm barrel chambered in 9 mm is roughly 100 g (see specific mass in Table A13.3), thus, the mass of the slide assembly would be in the range from circa 225 g to 355 g.

A mass ratio of 1:3 to 1:4 is considered a balanced ratio between the moving masses of the barrel and the slide for designs with coupled barrel and slide:[679]

$$\zeta_{SB} = \frac{m_B}{m_S}$$

A13.5

where: ζ_{SB} Mass ratio [1] of moving masses, m_B vs. m_S
           m_B Mass [kg] of the barrel
           m_S Mass [kg] of the slide

Summary

The diagram can serve as a rough guidance for design engineers. A good rule of thumb for determining the recoil mass is: Use $v_{S1} = 7$ m/s in A13.4 to compute the recoiling mass. The mass of the barrel is dictated by its layout. Choose it as heavy as necessary and design the slide as heavy as possible.

The actual mean slide recoil velocity of a newly designed pistol model will have to be determined as part of shooting tests.

Specific design features of a pistol can have significant impact on the mean recoil velocity of the slide and may offer the engineer possibilities for fine-tuning. In this respect, reference is made to the considerations on slide dynamics in Appendix 19.

Appendix 14 Mean Magazines Between Failures

Description of reliability as a function of Mean Magazines Between Failures

The malfunction rate is considered to be the benchmark for the reliability of firearms.[680] The malfunction rate is determined by endurance tests. In Germany, a malfunction rate of maximum 0.2 percent would be acceptable for duty pistols.[681] To create a tangible indication of per mille values (‰, e.g. $^1/1000$) or percentage values (%, e.g. $^1/100$) to the less experienced reader, a new approach is suggested in the following. To do so, formulas were derived and a chart provided. This chart provides the average number of magazines which may be reliably emptied between failures, based on a gun's malfunction rate and its magazine capacity.

Generally, the malfunction rate P of firearms is:

$$P = \frac{n}{N} \hspace{4cm} \textbf{A14.1}[682]$$

where: P Malfunction rate
         n Number of malfunctions
         N Total number of shots

The mean number of shots between failures (Mean Rounds Between Stoppages (MRBS) or Mean Rounds Between Failures, *MRBF*) is equivalent to the inverse of the malfunction rate P:

$$MRBF = \frac{1}{P} \hspace{4cm} \textbf{A14.2}[683]$$

where: *MRBF*   Mean Rounds Between Failures

The new approach is to transform Formula A14.2 to a new formula which will indicate the number of magazines emptied without malfunction (Mean Magazines Between Failures, *MMBF*) by calculating the mean number of magazines in place of the average number of shots between failures. This is done by dividing MRBF by the magazine capacity for the test weapon:

$$MMBF = \frac{MRBF}{k_{Mag}} \hspace{4cm} \textbf{A14.3}$$

where: *MMBF*   Mean Magazines Between Failures
         k_{Mag} Magazine capacity

In Formula A14.3, *MMBF* can be replaced by *1/P* from Formula A14.2, if a malfunction rate is given and the *MMBF* is to be determined. Based on this, the mean number of magazines between failures can be calculated according to:

$$MMBF = \frac{1}{P\,k_{Mag}}$$
A14.4

where: *MMBF*   Mean Magazines Between Failures
        k_{Mag} Magazine capacity
        P   Malfunction rate

or, if the number of malfunctions *n* and the total number of shots *N* are known, the mean number of magazines between failures can be calculated as follows, when Formulae A14.1 and A14.2 are combined with A14.3:

$$MMBF = \frac{N}{n\,k_{Mag}}$$
A14.5

where: *n*   Quantity of malfunctions
        N   Total number of shots

The results for *MMBF* calculated with Formula A14.4 and under assumption of
- Magazine capacity from 6 to 17 rounds,
- Malfunction rate of 0.1, 0.2, 0.5, 1.0, 1.5, 2, 2.5, 5, 7.5 and 10 percent,

were put together in Table A14.1 and summarized in a graph.

Table A14.1: Mean Magazines Between Failures as a function of magazine capacity and malfunction rate.

Mag-Capacity	5	6	7	8	9	10	11
$MMBF_{0.1\%}$	200.0	166.7	142.9	125.0	111.1	100.0	90.9
$MMBF_{0.2\%}$	100.0	83.3	71.4	62.5	55.6	50.0	45.5
$MMBF_{0.5\%}$	40.0	33.3	28.6	25.0	22.2	20.0	18.2
$MMBF_{1.0\%}$	20.0	16.7	14.3	12.5	11.1	10.0	9.1
$MMBF_{1.5\%}$	13.3	11.1	9.5	8.3	7.4	6.7	6.1
$MMBF_{2.0\%}$	10.0	8.3	7.1	6.3	5.6	5.0	4.5
$MMBF_{2.5\%}$	8.0	6.7	5.7	5.0	4.4	4.0	3.6
$MMBF_{5.0\%}$	4.0	3.3	2.9	2.5	2.2	2.0	1.8
$MMBF_{7.5\%}$	2.7	2.2	1.9	1.7	1.5	1.3	1.2
$MMBF_{10\%}$	2.0	1.7	1.4	1.3	1.1	1.0	0.9

Mag-Capacity	12	13	14	15	16	17
$MMBF_{0.1\%}$	83.3	76.9	71.4	66.7	62.5	58.8
$MMBF_{0.2\%}$	41.7	38.5	35.7	33.3	31.3	29.4
$MMBF_{0.5\%}$	16.7	15.4	14.3	13.3	12.5	11.8
$MMBF_{1.0\%}$	8.3	7.7	7.1	6.7	6.3	5.9
$MMBF_{1.5\%}$	5.6	5.1	4.8	4.4	4.2	3.9
$MMBF_{2.0\%}$	4.2	3.8	3.6	3.3	3.1	2.9
$MMBF_{2.5\%}$	3.3	3.1	2.9	2.7	2.5	2.4
$MMBF_{5.0\%}$	1.7	1.5	1.4	1.3	1.3	1.2
$MMBF_{7.5\%}$	1.1	1.0	1.0	0.9	0.8	0.8
$MMBF_{10\%}$	0.8	0.8	0.7	0.7	0.6	0.6

Note: the ordinate in Fig. A14.1 has a logarithmic scale.

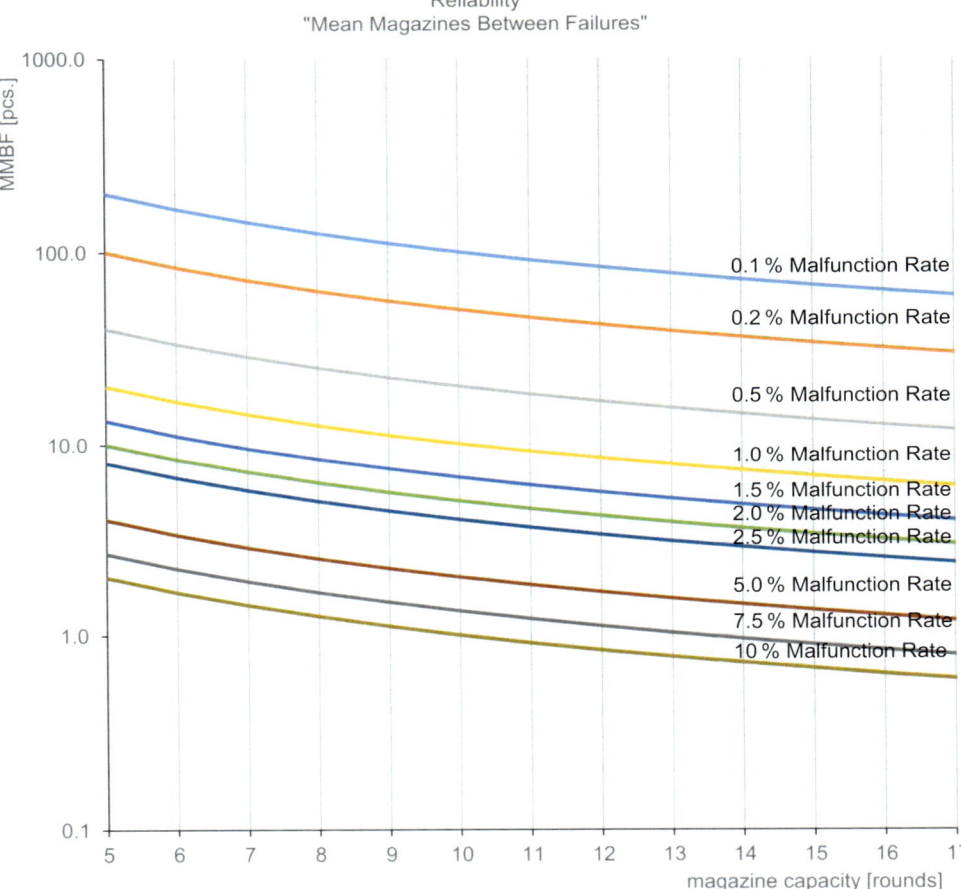

Fig. A14.1: *Mean Magazines Between Failures* as a function of Malfunction Rate and Magazine capacity.

Examples:

1. TR "Pistols" demands P = 0.2 percent (0.2 % is 0.2/100, i.e. 0.002). If a test gun would have a magazine capacity of 17 rounds, then according to Formula A14.4, the pistol must fire an average of 29.4 magazines ($\frac{1}{0.002 \cdot 17}$) between malfunctions. This number can also be obtained from Fig. A14.1 (intersection of 17 round line and 0.2 percent line).

2. Let us assume 3000 rounds were fired when testing a pistol with five-round magazine capacity and 75 malfunctions were counted. According to Formula

A14.5, this test weapon had one malfunction after every 8.0th magazine on average ($\frac{3000}{75 \cdot 5}$).

3. Let us assume the same conditions as in example #2, but with one exception: a magazine capacity of 15 rounds. *MMBF* is calculated according to Formula A14.5 and gives a malfunction after every 2.67th magazine on average ($\frac{3000}{75 \cdot 15}$).

By comparing the results for the *MMBF* of example #2 and #3, it can be seen how the magazine capacity affects the number of *MMBF* despite an identical malfunction rate. As the magazine capacity appears in the denominator of the formulae, the resulting *MMBF* in the example with a five-round mag capacity is triple the *MMBF* under same conditions but with a 15-round magazine. Hence, when comparing *MMBF* results, it might be advantageous to base the *MMBF* on a magazine size of an already known reference firearm. This can be simulated if the test weapon's magazine capacity k_{Mag} is replaced by the magazine capacity of the reference gun. Based on this, Formulae A14.4 and A14.5 will change as follows:

$$MMBF_{std} = \frac{1}{P\,k_{std}}$$
A14.6

where: $MMBF_{std}$ Standardized Mean Magazines Between Failures
k_{std} Standard mag capacity of the reference model
P Malfunction rate of test pistol

$$MMBF_{std} = \frac{N}{n\,k_{std}}$$
A14.7

where: n Quantity of malfunctions with test pistol
N Total number of shots fired with test pistol

If we assume the pistol model taken for reference has a magazine capacity of 17 rounds and the above-mentioned pistol in example #2 was tested in comparison, then the *MMBF* should be standardized to allow a comparison based on a magazine capacity of 17 rounds. In place of the $MMBF_5$ = 8 identified in example #2, an $MMBF_{17}$ can be calculated using Formula A14.7 giving us $MMBF_{17}$ = 2.35 (i.e. $\frac{3000}{75 \cdot 17}$). With this approach for a standardized *MMBF*, everyone involved receives a test result which can be compared with the failure rate of a reference pistol model they are familiar with and which gives them the chance to assess whether a received average failure rate is acceptable or not, that is, in our example, an *MMBF* of approximately three magazines.

382

Appendix 15   Barrel Heating

A WALTHER *PPQ* pistol, chambered in 9 mm Luger, barrel length 102 mm, lands and grooves rifling, is used for the determination of mean rise of external barrel temperature per shot. Up to 150 rounds are fired per sequence, and two test procedures will be used: approach i, where test firing will be interrupted after each 15 rounds to measure outside barrel temperature, and approach ii where the temperature is measured after shooting the intended number of cartridges without interrupting the sequence.

After the required number of shots, a non-contact temperature measurement is carried out similar to DIN EN ISO 13732-1. An infrared temperature measuring instrument TESTO *825-T2* is used; measuring temperature range −50 to +400 °C, measuring emissivity set to $\varepsilon = 0.7$, accuracy of the temperature measurements is ±2 °C or 2 percent of the measured value respectively; the larger value applies. The temperature is measured on the outside of the barrel, at two points: at a distance of approximately 30 mm ahead of the breech face (near the leade) and at 92 mm (in the following referred to as *muzzle end*). The distance between sensor and barrel surface is 25 mm. To better reach the reading points, the shooting is ceased after the required number of shots, the slide removed as quickly as possible, recoil spring assembly and barrel taken out of the slide, barrel temperature measured, and in the case of approach i, the gun then assembled to resume shooting with the next magazine already loaded with 15 rounds. The rate of fire is 30 shots per minute, that is, one shot every two seconds. For approach ii, barrel and slide will have to be cooled down to ambient temperature before a new test sequence is started.

The amount of time elapsed from the first shot is measured and time t_n will be noted at the same time when a temperature ϑ_n is measured. A spreadsheet is used to calculate the change in temperature $\Delta\vartheta_n$ for each 15 subsequent shots, also the average rate of temperature change based on the total rounds fired respectively.

The average rate of temperature change per shot $(\Delta\overline{\vartheta}_n)$ from the first shot to shot n is calculated with Formula A15.1:

$$\Delta\overline{\vartheta}_n = \frac{\vartheta_n - \vartheta_0}{n} \hspace{3cm} \textbf{A15.1}$$

where: $\Delta\overline{\vartheta}_n$ Average rate of temperature change after n shots [°C/shot]
n Number of shots fired
ϑ_0 Starting temperature [°C]
ϑ_n Temperature after n shots [°C]

Following approach i, a first test series (see 1st Series) was carried out, where temperature was measured at the barrel's leade area only. A second test series was done where both spots were checked, throat and muzzle end (2nd Series). For the 3rd Series, the shots were fired continuously, that is, following the idea of approach ii. Table A15.1 exhibits the summary of the survey.

Following approach i implied a certain time would be consumed for each sequence of safety check, disassembly, temperature measurements, assembly, and loading the pistol. As a consequence, the total time for one complete series was 710 or 730 seconds respectively. After 150 rounds fired, the measurements showed a final temperature ϑ_{150} at the leade of 120 °C or 126 °C respectively, and 130 °C at the muzzle end.[a] The barrel's temperature was almost evenly distributed in longitudinal direction, however, results of the 2nd Series indicated a slightly higher temperature at the muzzle end. According to the second law of thermodynamics, heat always flows from the hotter regions to colder regions.[684] If the discharge of a round would cause a temperature variation due to the longitudinal location of the barrel, then the duration of the shooting test would have a negative effect on the analysis, because the heat transferred to the barrel would flow from hotter spots to colder areas inside of the barrel material, and approximately 12 minutes of total time for the test apparently was enough time for an almost evenly heat distribution.

When the pistol was fired without ceasing fire, a lower temperature ϑ_{150} was measured at the throat area than on the muzzle end (96 °C vs. 129 °C). Following approach ii, the test would take 300 seconds, which reduced the time for a heat transfer from hotter to colder regions. Presumably, a lower temperature at the barrel's chamber end is a consequence of the cartridge. First, during chambering, heat will be transferred from the barrel block to the "fresh" cartridge. Then, during discharge, the casing will absorb some of the propellant gases' heat, which is going to be removed from the chamber together with the casing as the gun cycles. As a result, less heat can be transferred to the barrel.[685] Under the test conditions of approach i, this effect was not noticeable, since it allowed more time for the heat transfer within the barrel's material,

---

[a] The risk of burns on contact with the skin depends on the specific vulnerable part of the body, the temperature, the material, the surface and the size of the contact area, as well as the duration of the contact. Surface temperatures of up to 44 °C will not cause any tissue damage. (BAuA, 2014)

increasing the temperature at the chamber end, thus leading to an almost equally distributed heat from the barrel's front to its rear.[686]

After 150 shots, the average rate of change in temperature $\Delta\bar{\vartheta}_{150}$ at the leade reached from 0.5 to 0.7 °C/shot and 0.7 to 0.8 °C/shot at muzzle end.

For the graph of temperature ϑ_n vs. shots fired, the results of the 2nd and 3rd series for forcing cone area and muzzle end were put together in a diagram (top curves in Fig. A15.1), as well as the values of the change in temperature $\Delta\vartheta_n$ after each 15 rounds (bottom lines).

Table A15.1: External barrel temperature vs. shots fired.

i) 1st Series: @ Forcing Cone
WALTHER PPQ, 9 mm x 19, S/N: FAJ0639
Ambient Temperature: 21 °C
Ammunition: Sintox 8 g VM, Lot: DAG 14A0703

i) 2nd Series: @ Forcing Cone and Muzzle
WALTHER PPQ, 9 mm x 19, S/N: FAJ0473
Ambient Temperature: 20 °C
Ammunition: Sintox 8 g VM, Lot: DAG 14A0703

n	t_n [s]	ϑ_n @ forcing cone [°C]	$\Delta\vartheta_n$ @ forcing cone [°C]	$\Delta\bar{\vartheta}_n$ @ forcing cone [°C/shot]	t_n [s]	ϑ_n @ forcing cone [°C]	ϑ_n @ muzzle [°C]	$\Delta\vartheta_n$ @ forcing cone [°C]	$\Delta\vartheta_n$ @ muzzle [°C]	$\Delta\bar{\vartheta}_n$ @ forcing cone [°C/shot]	$\Delta\bar{\vartheta}_n$ @ muzzle [°C/shot]
0	0	18			0	17	17				
15	67	47	29	1.9	100	52	59	35	42	2.3	2.8
30	130	88	41	2.7	150	81	79	29	20	1.9	1.3
45	210	102	14	0.9	230	97	98	16	19	1.1	1.3
60	290	106	4	0.3	300	99	109	2	11	0.1	0.7
75	360	107	1	0.1	380	109	113	10	4	0.7	0.3
90	440	108	1	0.1	460	115	118	6	5	0.4	0.3
105	500	112	4	0.3	525	117	121	2	3	0.1	0.2
120	565	113	1	0.1	590	123	125	6	4	0.4	0.3
135	650	113	0	0.0	670	124	128	1	3	0.1	0.2
150	710	120	7	0.5	730	126	130	2	2	0.1	0.1

ii) 3rd Series: @ Forcing Cone and Muzzle, continuous fire, 30 shots per minute
WALTHER PPQ, 9 mm x 19, S/N: FAJ0473
Ambient Temperature: 20 °C
Ammunition: Sintox 8 g VM, Lot: DAG 14A0703

n	t_n [s]	ϑ_n @ forcing cone [°C]	ϑ_n @ muzzle [°C]	$\Delta\vartheta_n$ @ forcing cone [°C]	$\Delta\vartheta_n$ @ muzzle [°C]	$\Delta\bar{\vartheta}_n$ @ forcing cone [°C/shot]	$\Delta\bar{\vartheta}_n$ @ muzzle [°C/shot]
0	0	19	19				
15	30	46	51	27	32	1.8	2.1
30	60	62	68	16	17	1.4	1.6
45	90	71	87	9	19	1.2	1.5
60	120	76	96	5	9	1.0	1.3
75	150	88	116	12	20	0.9	1.3
90	180	90	120	2	4	0.8	1.1
105	210	92	125	2	5	0.7	1.0
120	240	94	126	2	1	0.6	0.9
135	270	95	128	1	2	0.6	0.8
150	300	96	129	1	1	0.5	0.7

After the first fifteen shots of the 2nd series the temperature change $\Delta\vartheta_{15}$ was 35 °C at the forcing cone, and 42 °C at the muzzle end. This translated to an average change in temperature of 2.3 and 2.8 °C/shot respectively. The average change in temperature of subsequent 15-round-series was following a decreasing trend. As it can be seen by the top temperature curves in the diagram, the temperature approaches a maximum, whereas the gradient of temperature change flattens out (bottom curves).

A comparison of the numbers reveals that the average increase in temperature $\Delta\bar{\vartheta}_{150}$ after 150 shots was 0.7 to 0.8 °C/shot, which is close to the value referred to by *Schmid*. It must be noted though, that the average rate of change in temperature of 1 °C/shot mentioned by Schmid was established for the *MK213C* machine canon, which is chambered in 20 mm and fires projectiles at a muzzle velocity of about 1050 m/s with a cyclic rate between 1200 to 1400 shots/min.[687] The technical data of this machine canon is significantly different than the technical data of a pistol chambered in 9 mm, and tells us, testing parameters such as firing mode, rate of fire, ammunition, or delays needed to take the temperature have an effect on exterior barrel temperature. But it may be concluded in general that the temperature of the test sample increases faster on the barrel's muzzle end than on the throat area. Furthermore, the temperature increases very rapidly at first, and finally approaches a maximum temperature. Up to this maximum, an average change in temperature just below 1 °C per shot can be expected.[688]

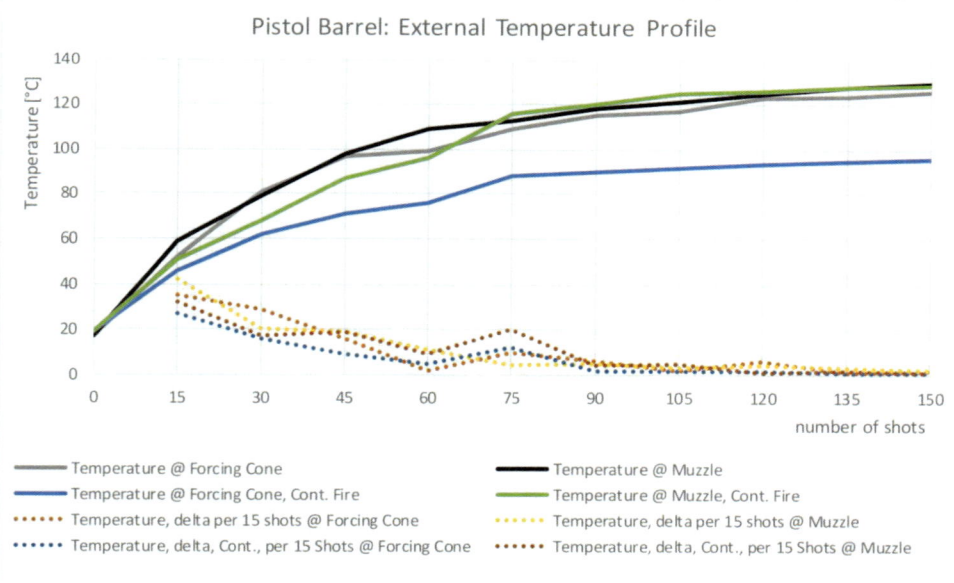

Fig. A15.1: Temperature curves of barrels chambered in 9 mm Luger.

Appendix 16   Trigger Pull Analysis

Legend for trigger pull test protocol

Make, model and serial number of each pistol tested is specified in the header of the test protocol. The nomenclature and abbreviations used in the protocols are:

(German)	(English)
Überschrift	Headline
Fabrikat	Brand
Modell	Model
Nummer	Serial Number
Prüfer	Examiner
Vorkraft	Pre-Load
Prüfgeschwindigkeit	Trigger Pull Velocity
F_{max}	Maximum force recorded during pull; maximum value of the trigger pull force between the beginning of trigger movement and trigger break.
F_B	Sear break force; trigger pull force at trigger break.
Abzugsweg	Trigger Travel
Abzugsarbeit	Trigger Work, i.e. the area underneath trigger force vs. trigger travel graph.

The interpretation of the readings is performed automatically by the software *testX-pert II V3.2* from ZWICK/ROELL. The reading will automatically start as soon as the trigger pull force reaches a magnitude of 2 N (as specified under "Vorkraft" (pre-load)). For example, if a sample pistol would have a trigger safety whose operation force would require more than 2 N, then the activation of the trigger safety would be included in the reading starting from 2 N. Hence the trigger safety would have an effect on the calculation of the trigger work.

Trigger Pull vs. Trigger Travel, CARACAL *F*

Prüfprotokoll, Abzugkraft

Überschrift : Prüfprotokoll, Abzugkraft
Fabrikat : Caracal
Modell : Caracal F
Nummer : LA516
Prüfer : Dallhammer

Vorkraft : 2 N
Prüfgeschwindigkeit : 100 mm/min

Prüfergebnisse:

Legende	Nr	F_{max} N	F_B N	Abzugsweg mm	Abzugsarbeit Nmm
(rot)	1	21,4	22,9	4,8	70,40
(grün)	2	22,8	23,9	4,7	68,28
(blau)	3	23,5	24,1	4,7	69,44

Trigger Pull vs. Trigger Travel

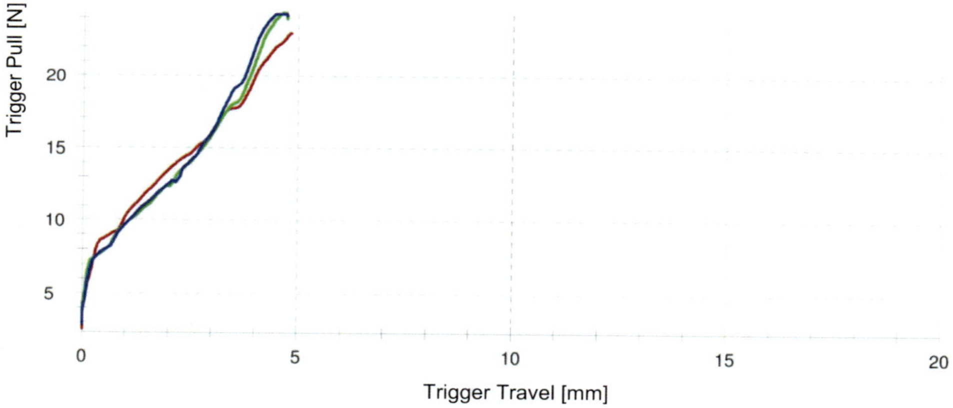

Statistik:

Serie n = 3	F_{max} N	F_B N	Abzugsweg mm	Abzugsarbeit Nmm
$\bar{x}$	22,6	23,7	4,8	69,37
s	1,04	0,647	0,0	1,06
ν	4,61	2,74	1,01	1,53

Trigger Pull vs. Trigger Travel, GLOCK *G19 Gen4*

Prüfprotokoll, Abzugkraft

Überschrift	: Prüfprotokoll, Abzugkraft	Nummer	: WSP758
Fabrikat	: Glock	Prüfer	: Dallhammer
Modell	: G19 Gen4		

Vorkraft : 2 N
Prüfgeschwindigkeit : 100 mm/min

Prüfergebnisse:

Legende	Nr	F_{max} N	F_B N	Abzugsweg mm	Abzugsarbeit Nmm
▬	1	40,5	36,1	8,0	117,33
▬	2	39,8	35,8	7,9	117,72
▬	3	41,5	37,8	8,1	129,81

Trigger Pull vs. Trigger Travel

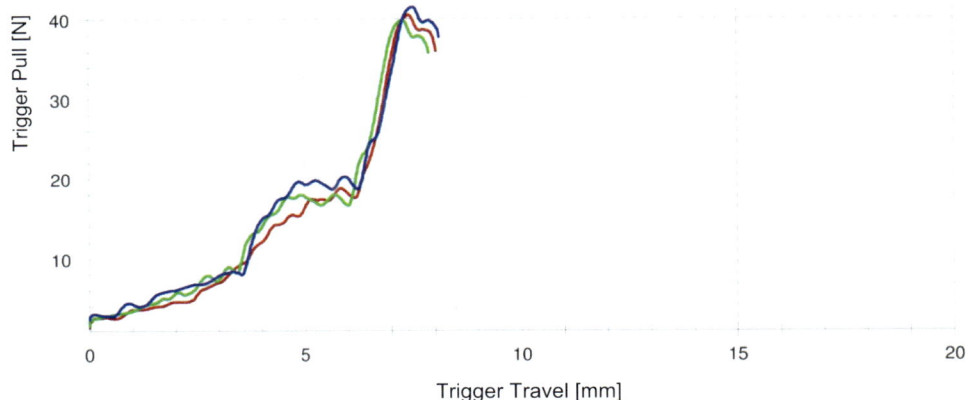

Statistik:

Serie n = 3	F_{max} N	F_B N	Abzugsweg mm	Abzugsarbeit Nmm
x̄	40,6	36,6	8,0	121,62
s	0,853	1,08	0,1	7,09
v	2,10	2,94	1,59	5,83

Trigger Pull vs. Trigger Travel, HECKLER & KOCH *VP9*

Prüfprotokoll, Abzugkraft

Überschrift : Prüfprotokoll, Abzugkraft
Fabrikat : Heckler & Koch
Modell : VP9

Nummer : 2204-001989
Prüfer : Dallhammer

Vorkraft : 2 N
Prüfgeschwindigkeit : 100 mm/min

Prüfergebnisse:

Legende	Nr	F_{max} N	F_B N	Abzugsweg mm	Abzugsarbeit Nmm
	1	23,8	25,2	5,8	67,24
	2	24,0	25,3	5,8	66,99
	3	23,9	25,0	5,8	66,85

Trigger Pull vs. Trigger Travel

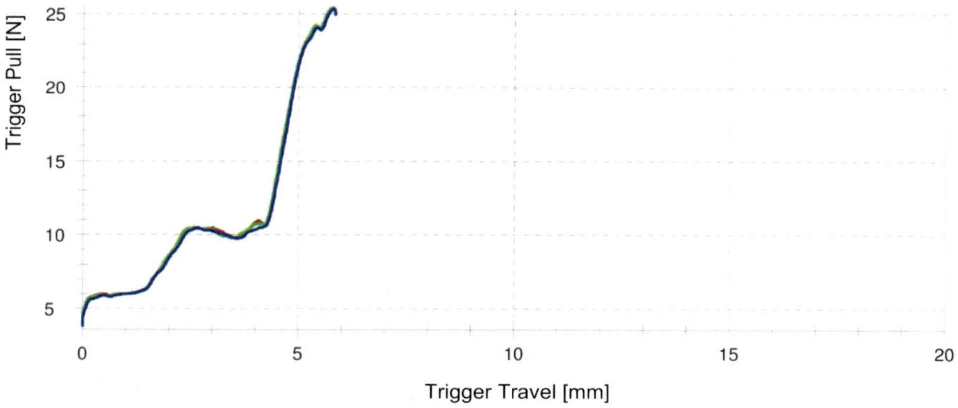

Statistik:

Serie n = 3	F_{max} N	F_B N	Abzugsweg mm	Abzugsarbeit Nmm
$\bar{x}$	23,9	25,2	5,8	67,03
s	0,0699	0,194	0,0	0,20
ν	0,29	0,77	0,28	0,30

Trigger Pull vs. Trigger Travel, KEL-TEC *PF-9*

Prüfprotokoll, Abzugkraft

Überschrift : Prüfprotokoll, Abzugkraft   Nummer : SOF52
Fabrikat : KelTec   Prüfer : Dallhammer
Modell : PF-9

Vorkraft :   2 N
Prüfgeschwindigkeit : 100 mm/min

Prüfergebnisse:

Legende	Nr	F_{max} N	F_B N	Abzugsweg mm	Abzugsarbeit Nmm
▬ (rot)	1	25,4	17,5	15,8	281,15
▬ (grün)	2	25,1	19,2	15,8	280,90
▬ (blau)	3	25,0	17,3	15,8	279,21

Trigger Pull vs. Trigger Travel

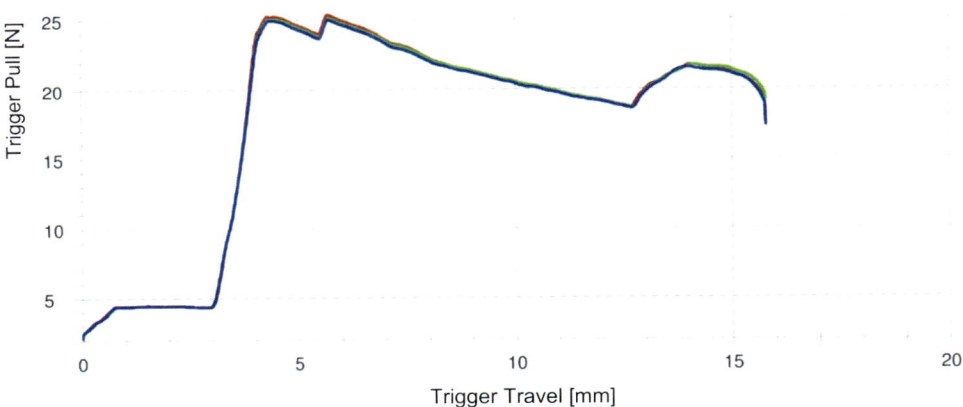

Statistik:

Serie n = 3	F_{max} N	F_B N	Abzugsweg mm	Abzugsarbeit Nmm
$\bar{\bar{x}}$	25,2	18,0	15,8	280,42
s	0,191	1,03	0,0	1,05
ν	0,76	5,74	0,06	0,38

Trigger Pull vs. Trigger Travel, RUGER *LC9*

Prüfprotokoll, Abzugkraft

Überschrift : Prüfprotokoll, Abzugkraft Nummer : 320-43011
Fabrikat   : Ruger   Prüfer : Dallhammer
Modell   : LC9

Vorkraft   :   2 N
Prüfgeschwindigkeit : 100 mm/min

Prüfergebnisse:

Legende	Nr	F_{max} N	F_B N	Abzugsweg mm	Abzugsarbeit Nmm
(rot)	1	35,4	24,2	17,3	382,19
(grün)	2	34,5	25,9	17,0	381,73
(blau)	3	33,9	23,3	17,2	375,15

Trigger Pull vs. Trigger Travel

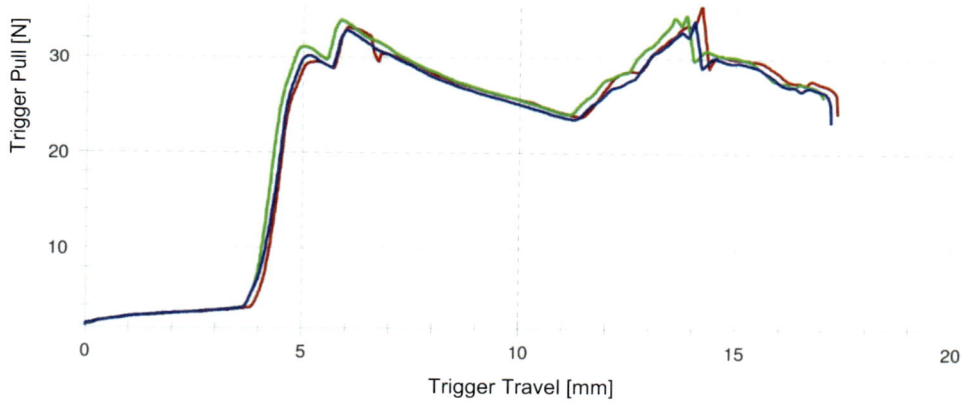

Statistik:

Serie n = 3	F_{max} N	F_B N	Abzugsweg mm	Abzugsarbeit Nmm
x̄	34,6	24,5	17,2	379,69
s	0,740	1,32	0,2	3,94
v	2,14	5,41	0,97	1,04

Trigger Pull vs. Trigger Travel, SMITH & WESSON *M&P9*

Prüfprotokoll, Abzugkraft

Überschrift : Prüfprotokoll, Abzugkraft   Nummer : DST7991
Fabrikat   : Smith & Wesson   Prüfer : Dallhammer
Modell   : M&P9

Vorkraft   :   2 N
Prüfgeschwindigkeit   : 100 mm/min

Prüfergebnisse:

Legende	Nr	F_{max} N	F_B N	Abzugsweg mm	Abzugsarbeit Nmm
(rot)	1	24,4	19,4	6,8	76,88
(grün)	2	24,3	19,1	6,8	76,82
(blau)	3	24,7	18,7	6,9	78,64

Trigger Pull vs. Trigger Travel

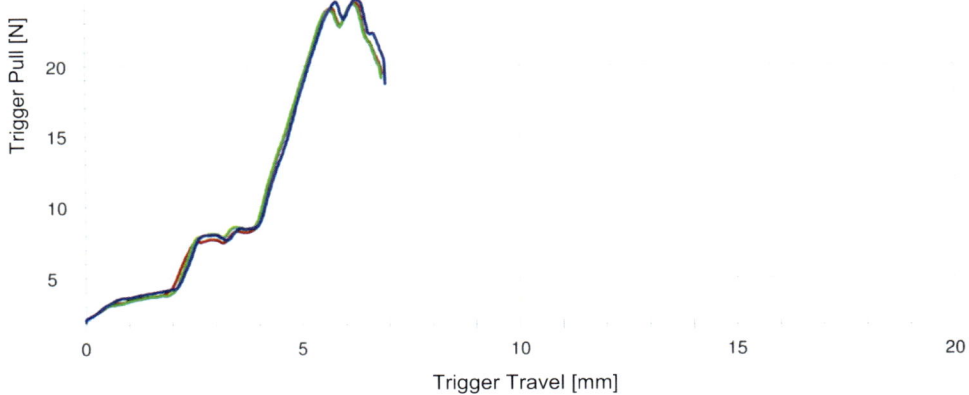

Statistik:

Serie n = 3	F_{max} N	F_B N	Abzugsweg mm	Abzugsarbeit Nmm
$\bar{x}$	24,5	19,1	6,9	77,44
s	0,171	0,373	0,1	1,03
ν	0,70	1,96	0,77	1,33

Trigger Pull vs. Trigger Travel, SPRINGFIELD *XD^M*

Prüfprotokoll, Abzugkraft

Überschrift : Prüfprotokoll, Abzugkraft   Nummer : MG957178
Fabrikat   : Springfield   Prüfer   : Dallhammer
Modell   : XD^M

Vorkraft   :   2 N
Prüfgeschwindigkeit   : 100 mm/min

Prüfergebnisse:

Legende	Nr	F_{max} N	F_B N	Abzugsweg mm	Abzugsarbeit Nmm
🟥	1	24,4	24,3	6,6	71,25
🟩	2	24,0	23,1	6,3	63,41
🟦	3	24,8	23,9	6,7	69,91

Trigger Pull vs. Trigger Travel

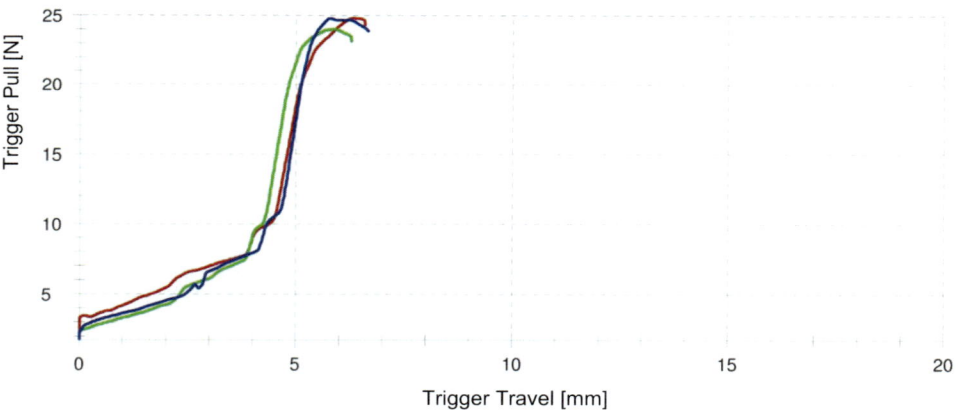

Statistik:

Serie n = 3	F_{max} N	F_B N	Abzugsweg mm	Abzugsarbeit Nmm
x̄	24,4	23,8	6,5	68,19
s	0,405	0,599	0,2	4,19
ν	1,66	2,52	3,15	6,15

Trigger Pull vs. Trigger Travel, TAURUS *PT 24/7 PRO*, SA

Prüfprotokoll, Abzugkraft

Überschrift	:	Prüfprotokoll, Abzugkraft	Nummer	:	TZJ22447
Fabrikat	:	Taurus	Abzug	:	SA
Modell	:	PT 24/7 PRO	Prüfer	:	Dallhammer

Vorkraft : 2 N
Prüfgeschwindigkeit : 100 mm/min

Prüfergebnisse:

Legende	Nr	F_{max} N	F_B N	Abzugsweg mm	Abzugsarbeit Nmm
🟥	1	20,0	20,9	11,1	63,85
🟩	2	19,7	20,8	11,1	62,96
🟦	3	19,5	20,0	11,0	62,58

Trigger Pull vs. Trigger Travel

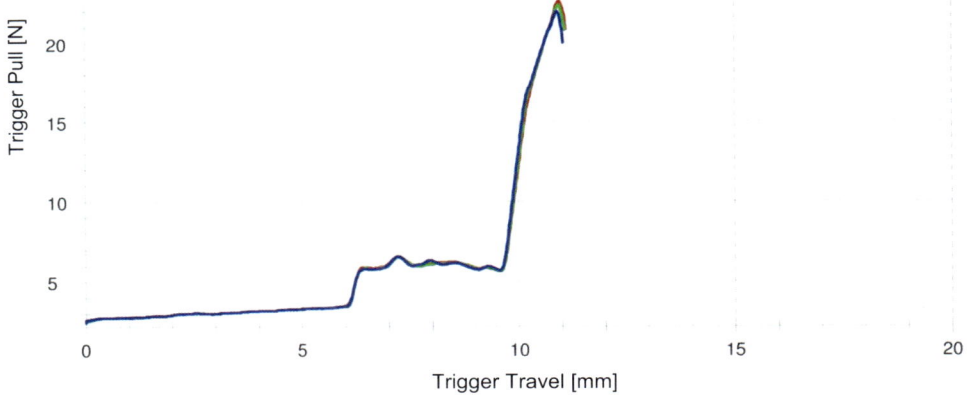

Statistik:

Serie n = 3	F_{max} N	F_B N	Abzugsweg mm	Abzugsarbeit Nmm
$\bar{x}$	19,7	20,6	11,1	63,13
s	0,239	0,487	0,0	0,65
v	1,21	2,37	0,30	1,03

Trigger Pull vs. Trigger Travel, TAURUS *PT 24/7 PRO*, Restrike Trigger Action

Prüfprotokoll, Abzugkraft

Überschrift	: Prüfprotokoll, Abzugkraft	Nummer	: TZJ22447
Fabrikat	: Taurus	Abzug	: Restrike Trigger
Modell	: PT 24/7 PRO	Prüfer	: Dallhammer

Vorkraft : 2 N
Prüfgeschwindigkeit : 100 mm/min

Prüfergebnisse:

Legende	Nr	F_{max} N	F_B N	Abzugsweg mm	Abzugsarbeit Nmm
▬	1	23,7	26,9	10,8	155,81
▬	2	23,6	26,7	10,8	154,42
▬	3	23,3	26,8	10,8	153,28

Trigger Pull vs. Trigger Travel

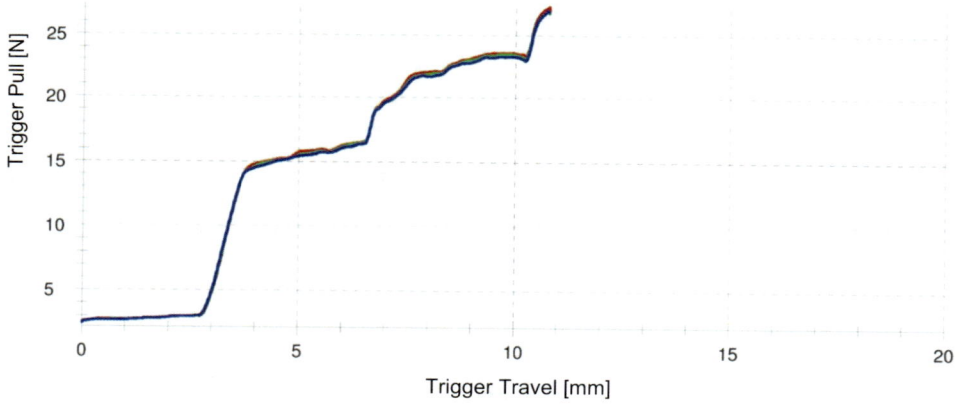

Statistik:

Serie n = 3	F_{max} N	F_B N	Abzugsweg mm	Abzugsarbeit Nmm
x̄	23,5	26,8	10,8	154,50
s	0,172	0,123	0,0	1,26
v	0,73	0,46	0,16	0,82

Trigger Pull vs. Trigger Travel, WALTHER *PPX*

Prüfprotokoll, Abzugkraft

Überschrift	:	Prüfprotokoll, Abzugkraft	Nummer	:	FAM7387
Fabrikat	:	Walther	Prüfer	:	Dallhammer
Modell	:	PPX			

Vorkraft : 2 N
Prüfgeschwindigkeit : 100 mm/min

Prüfergebnisse:

Legende	Nr	F_{max} N	F_B N	Abzugsweg mm	Abzugsarbeit Nmm
🟥	1	21,8	21,7	7,0	83,63
🟩	2	21,9	21,7	7,1	84,47
🟦	3	22,0	21,6	7,1	84,31

Trigger Pull vs. Trigger Travel

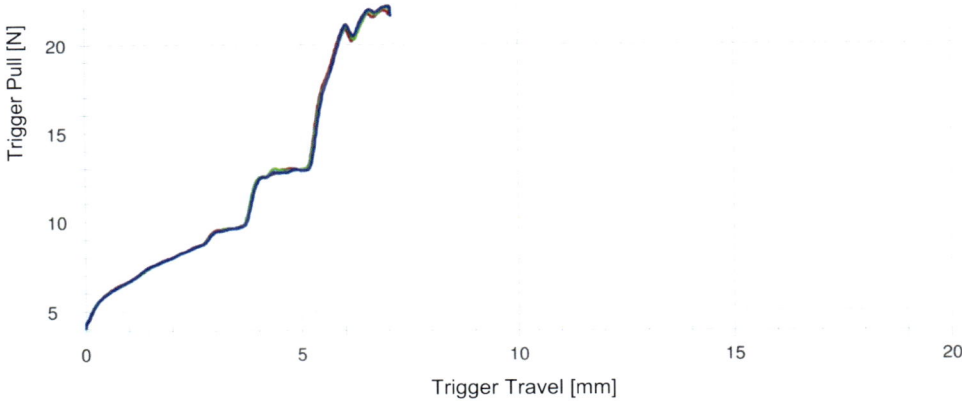

Statistik:

Serie n = 3	F_{max} N	F_B N	Abzugsweg mm	Abzugsarbeit Nmm
x̄	21,9	21,7	7,1	84,14
s	0,101	0,0463	0,0	0,45
ν	0,46	0,21	0,41	0,53

Appendix 17   Slide Racking Force Analysis

Legend for slide racking force test protocol

Make, model and serial number of each pistol tested is specified in the header of the test protocol. The nomenclature and abbreviations used in the protocols are:

(German)	(English)
Überschrift	Headline
Fabrikat	Brand
Modell	Model
Nummer	Serial Number
Prüfer	Examiner
Prüfgeschwindigkeit	Slide Racking Velocity

The interpretation of the readings was performed graphically by reading the maximum slide racking force from the graph in the force vs. slide travel diagram. The determined maximum slide racking force is marked up in each chart.

Slide Racking Force, CARACAL *F*

Durchladewiderstand

Überschrift : Durchladewiderstand   Nummer   : LA 516
Fabrikat   : Caracal   Prüfer   : Dallhammer
Modell   : Caracal F

Prüfgeschwindigkeit : 100 mm/min

Slide Retract Force vs. Slide Travel

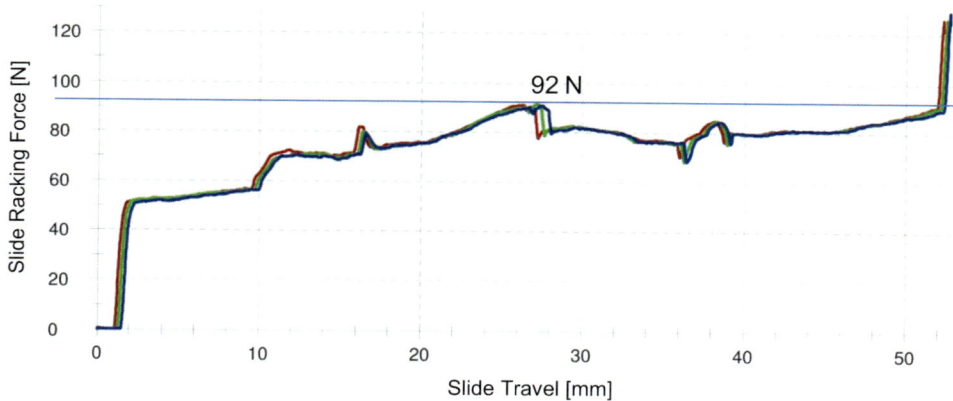

Slide Racking Force, GLOCK *G19 Gen4*

Durchladewiderstand

Überschrift : Durchladewiderstand   Nummer : WSB758
Fabrikat   : Glock   Prüfer : Dallhammer
Modell   : G19 Gen4

Prüfgeschwindigkeit : 100 mm/min

Slide Racking Force vs. Slide Travel

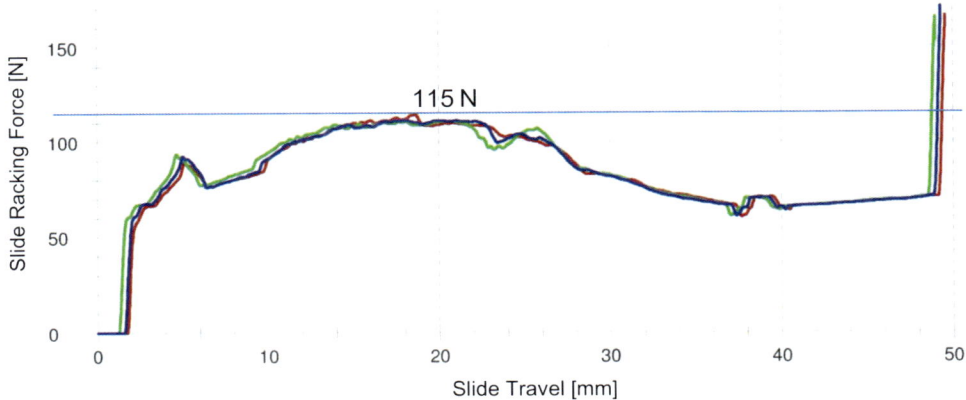

Slide Racking Force, HECKLER & KOCH *VP9*

Durchladewiderstand

Fabrikat : Heckler & Koch   Nummer : 224-001989
Modell   : VP9   Prüfer   : Dallhammer

Prüfgeschwindigkeit : 100 mm/min

Slide Racking Force vs. Slide Travel

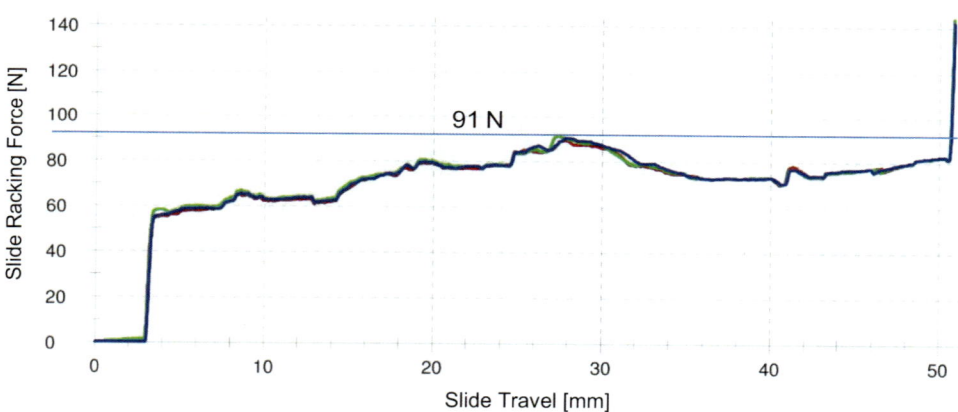

Slide Racking Force, KEL-TEC *PF-9*

Durchladewiderstand

Überschrift : Durchladewiderstand   Nummer   : SOF52
Fabrikat   : KelTec   Prüfer   : Dallhammer
Modell   : PF-9

Prüfgeschwindigkeit : 100 mm/min

Slide Racking Force vs. Slide Travel

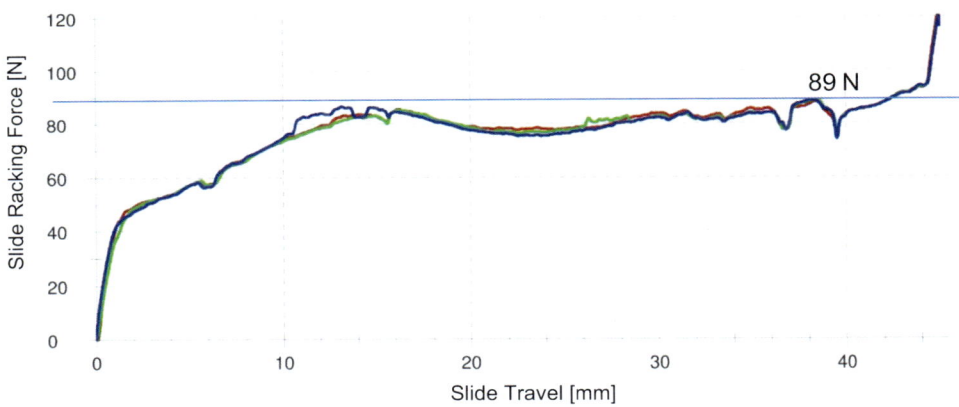

Slide Racking Force, RUGER *LC9*

Durchladewiderstand

Überschrift : Durchladewiderstand   Nummer : 320-43011
Fabrikat   : RUGER   Prüfer   : Riedel
Modell   : LC9

Vorkraft   :   2 N
Prüfgeschwindigkeit : 100 mm/min

Slide Racking Force vs. Slide Travel

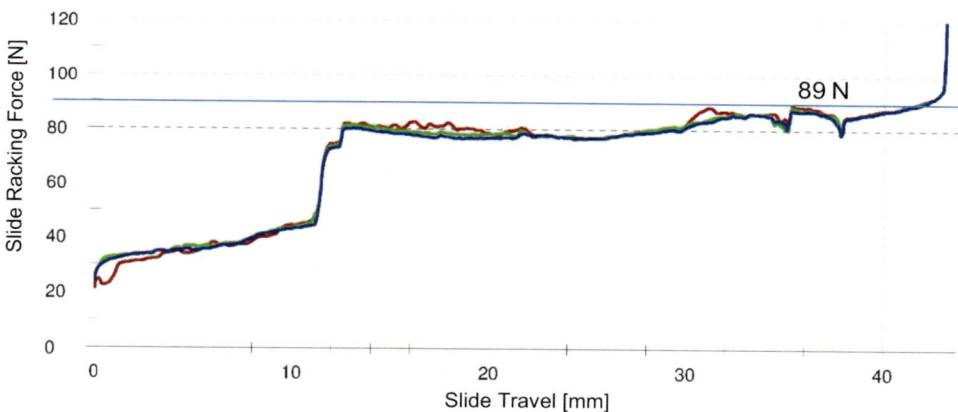

Slide Racking Force, SMITH & WESSON *M&P9*

Durchladewiderstand

Überschrift : Durchladewiderstand   Nummer : DST7991
Fabrikat : Smith & Wesson   Prüfer : Dallhammer
Modell : M&P9

Prüfgeschwindigkeit : 100 mm/min

Slide Racking Force vs. Slide Travel

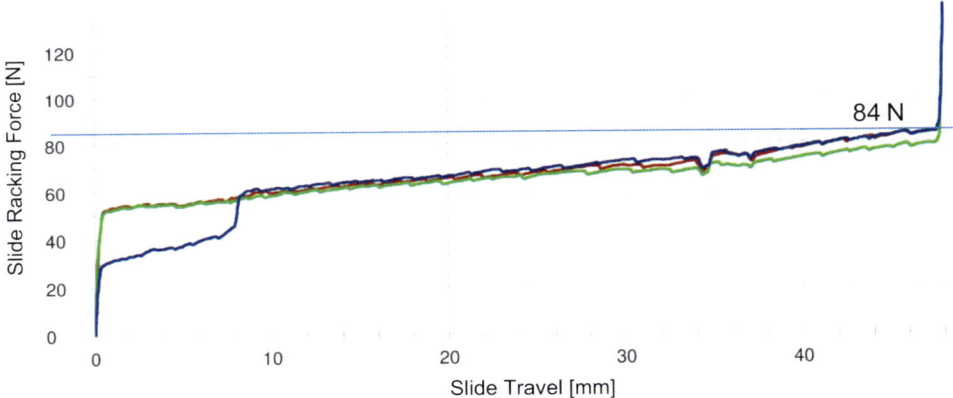

Slide Racking Force, SPRINGFIELD *XD^M*

Durchladewiderstand

Überschrift : Durchladewiderstand   Nummer   : MG957178
Fabrikat   : Springfield   Prüfer   : Dallhammer
Modell   : XD^M

Prüfgeschwindigkeit : 100 mm/min

Slide Racking Force vs. Slide Travel

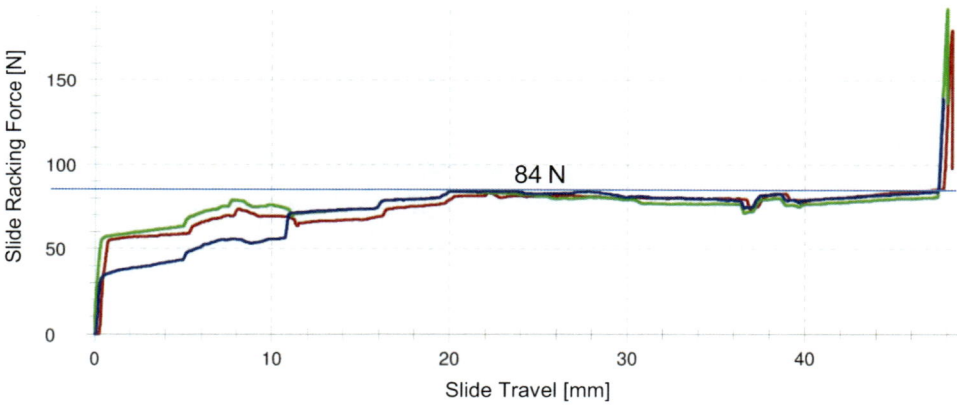

Slide Racking Force, TAURUS *PT 24/7 PRO*

Durchladewiderstand

Überschrift : Durchladewiderstand   Nummer : TZJ 22447
Fabrikat   : Taurus   Prüfer   : Dallhammer
Modell   : PT 24/7 PRO

Prüfgeschwindigkeit : 100 mm/min

Slide Racking Force vs. Slide Travel

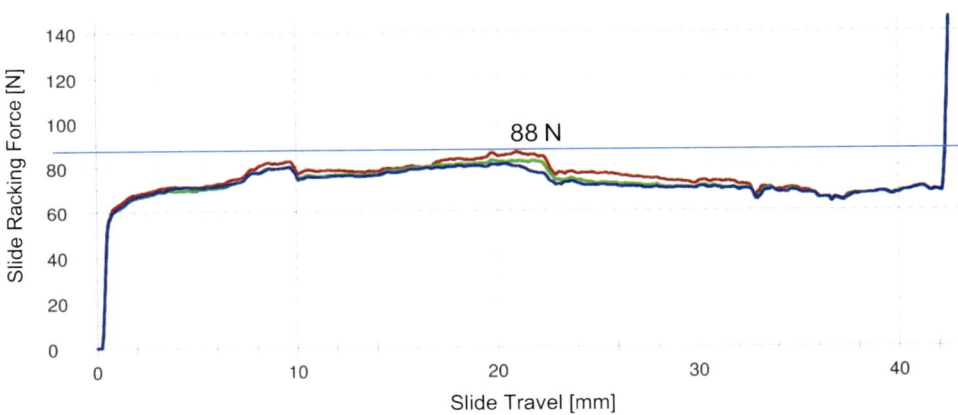

Slide Racking Force, WALTHER *PPX*

Durchladewiderstand

Überschrift : Durchladewiderstand   Nummer   : FAM7387
Fabrikat   : Walther   Prüfer   : Dallhammer
Modell   : PPX

Prüfgeschwindigkeit : 100 mm/min

Slide Racking Force vs. Slide Travel

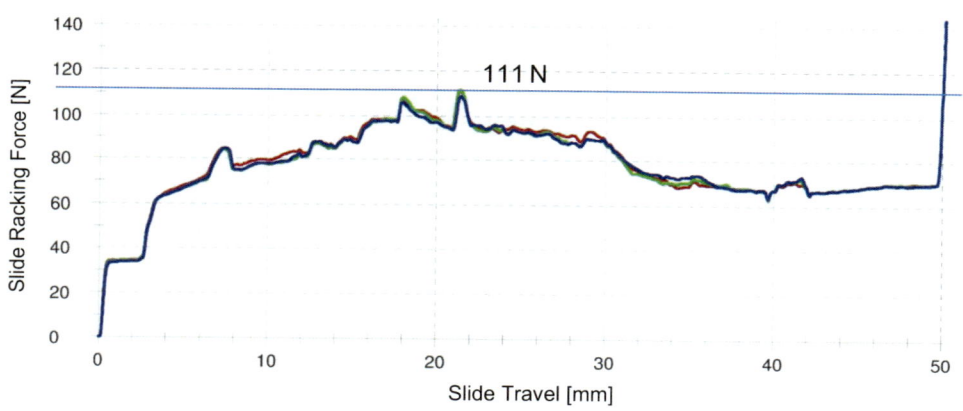

The area under the force-displacement graph corresponds to the mechanical work done to rack the slide. This work can be determined by planimetry of the force-travel graph. The following approximate values were determined:

Striker-fired:		Hammer-fired:	
CARACAL *F*	3.8 J	WALTHER *PPX*	3.8 J
GLOCK *G19 Gen4*	4.0 J		
HECKLER & KOCH *VP9*	3.5 J		
KEL-TEC *PF-9*	3.3 J		
RUGER *LC9*	3.0 J		
SMITH & WESSON *M&P9*	3.2 J		
SPRINGFIELD *XD^M*	3.6 J		
TAURUS *PT 24/7*	3.0 J		

Appendix 18   Muzzle Velocity and Muzzle Energy

Legend for muzzle velocity protocol

The header of the test protocol lists model, serial number, ammunition, ammunition brand, caliber, projectile's mass, lot number, average muzzle velocity (Ø Geschwindigkeit), average muzzle energy (Ø Energie), and comments, i.e. 10 shots to reach operating temperature. This is followed by five readings, including minimum, maximum and standard deviation.

Note: a summary of the test results can be found at the end of Appendix 18.

Waffe: **CARACAL *F***	Waffennummer: **LA516**		Datum: 16.07.2014 08:46
Munition: *Sintox Standard*	Hersteller: RUAG	Kaliber: 9mm x 19	Geschoßmasse: 8,0 g
Munitionslos: **DAG13E0710**	Ø Geschwindigkeit: **350,6 m/s**	Ø Energie: **491,7 J**	
Kommentar: Werte nach 10 Schuss			
Messwerte:			
1	348,6 m/s	486,0 J	
2	349,6 m/s	488,9 J	
3	350,2 m/s	490,5 J	
4	350,8 m/s	492,3 J	
5	353,9 m/s	500,9 J	
Anzahl Messwerte: 5	Min: 348,6 m/s	Max: 353,9 m/s	Standardabweichung: 2,0 m/s

Waffe: **GLOCK G19 Gen 4**	Waffennummer: **WSB758**		Datum: 16.07.2014 08:49
Munition: *Sintox Standard*	Hersteller: RUAG	Kaliber: 9mm x 19	Geschoßmasse: 8,0 g
Munitionslos: **DAG13E0710**	Ø Geschwindigkeit: **358,2 m/s**	Ø Energie: **513,1 J**	
Kommentar: Werte nach 10 Schuss			
Messwerte:			
1	357,4 m/s	510,9 J	
2	357,0 m/s	509,8 J	
3	356,6 m/s	508,7 J	
4	359,6 m/s	517,3 J	
5	360,2 m/s	519,1 J	
Anzahl Messwerte: 5	Min: 356,6 m/s	Max: 360,2 m/s	Standardabweichung: 1,6 m/s

Waffe: **HECKLER & KOCH** *VP9*	Waffennummer: **224-001989**		Datum: 02.10.2014 09:45

Munition: *Sintox Standard*   Hersteller: RUAG   Kaliber: 9mm x 19   Geschoßmasse: 8,0 g

Munitionslos: **DAG13E0710**   Ø Geschwindigkeit: **364,8 m/s**   Ø Energie: **532,5 J**

Kommentar: Werte nach 10 Schuss

Messwerte:

1	367,0 m/s	538,7 J
2	361,1 m/s	521,5 J
3	364,6 m/s	531,8 J
4	364,5 m/s	531,4 J
5	367,0 m/s	538,9 J

Anzahl Messwerte: 5   Min: 361,1 m/s   Max: 367,0 m/s   Standardabweichung: 2,4 m/s

---

Waffe: **KEL-TEC** *PF-9*	Waffennummer: **SOF52**		Datum: 16.07.2014 08:42

Munition: *Sintox Standard*   Hersteller: RUAG   Kaliber: 9mm x 19   Geschoßmasse: 8,0 g

Munitionslos: **DAG13E0710**   Ø Geschwindigkeit: **342,7 m/s**   Ø Energie: **469,8 J**

Kommentar: Werte nach 10 Schuss

Messwerte:

1	340,5 m/s	463,7 J
2	346,1 m/s	479,1 J
3	339,6 m/s	461,2 J
4	342,0 m/s	467,9 J
5	345,4 m/s	477,2 J

Anzahl Messwerte: 5   Min: 339,6 m/s   Max: 346,1 m/s   Standardabweichung: 2,9 m/s

---

Waffe: **RUGER** *LC9*	Waffennummer: **320-43011**		Datum: 16.07.2014 08:35

Munition: *Sintox Standard*   Hersteller: RUAG   Kaliber: 9mm x 19   Geschoßmasse: 8,0 g

Munitionslos: **DAG13E0710**   Ø Geschwindigkeit: **338,8 m/s**   Ø Energie: **459,1 J**

Kommentar: Werte nach 10 Schuss

Messwerte:

1	338,4 m/s	458,2 J
2	338,0 m/s	457,1 J
3	339,2 m/s	460,3 J
4	340,0 m/s	462,3 J
5	338,2 m/s	457,5 J

Anzahl Messwerte: 5   Min: 338,0 m/s   Max: 340,0 m/s   Standardabweichung: 0,8 m/s

Waffe: **SMITH & WESSON** **M&P9**	Waffennummer: **DST 7991**		Datum: 16.07.2014 08:51	
Munition: *Sintox Standard*	Hersteller: RUAG	Kaliber: 9mm x 19	Geschoßmasse: 8,0 g	
Munitionslos: **DAG13E0710**	Ø Geschwindigkeit: **360,7 m/s**	Ø Energie: **520,3 J**		
Kommentar: Werte nach 10 Schuss				
Messwerte:				
1	364,8 m/s	532,4 J		
2	361,9 m/s	524,0 J		
3	359,9 m/s	518,0 J		
4	359,8 m/s	517,9 J		
5	356,8 m/s	509,3 J		
Anzahl Messwerte: 5	Min: 356,8 m/s	Max: 364,8 m/s	Standardabweichung: 3,0 m/s	

Waffe: **SPRINGFIELD** **XD**[M]	Waffennummer: **MG957178**		Datum: 16.07.2014 08:38	
Munition: *Sintox Standard*	Hersteller: RUAG	Kaliber: 9mm x 19	Geschoßmasse: 8,0 g	
Munitionslos: **DAG13E0710**	Ø Geschwindigkeit: **370,6 m/s**	Ø Energie: **549,5 J**		
Kommentar: Werte nach 10 Schuss				
Messwerte:				
1	368,9 m/s	544,3 J		
2	372,2 m/s	554,0 J		
3	368,0 m/s	541,8 J		
4	371,1 m/s	550,7 J		
5	373,1 m/s	556,7 J		
Anzahl Messwerte: 5	Min: 368,0 m/s	Max: 373,1 m/s	Standardabweichung: 2,1 m/s	

Waffe: **TAURUS** **PT 24/7 PRO**	Waffennummer: **PZJ22447**		Datum: 02.10.2014 09:40	
Munition: *Sintox Standard*	Hersteller: RUAG	Kaliber: 9mm x 19	Geschoßmasse: 8,0 g	
Munitionslos: **DAG13E0710**	Ø Geschwindigkeit: **356,4 m/s**	Ø Energie: **508,2 J**		
Kommentar: Werte nach 10 Schuss				
Messwerte:				
1	352,6 m/s	497,2 J		
2	358,1 m/s	513,0 J		
3	358,7 m/s	514,6 J		
4	355,1 m/s	504,4 J		
5	357,7 m/s	511,7 J		
Anzahl Messwerte: 5	Min: 352,6 m/s	Max: 358,7 m/s	Standardabweichung: 2,6 m/s	

Waffe: **WALTHER PPX** Waffennummer: **FAM 7387**		Datum: 16.07.2014 08:57

Munition: *Sintox Standard*   Hersteller: RUAG   Kaliber: 9mm x 19   Geschoßmasse: 8,0 g

Munitionslos: **DAG13E0710**   Ø Geschwindigkeit: **358,2 m/s**   Ø Energie: **513,2 J**

Kommentar: Werte nach 10 Schuss

Messwerte:

1	358,0 m/s	512,6 J
2	359,9 m/s	518,1 J
3	357,3 m/s	510,8 J
4	359,9 m/s	518,2 J
5	355,8 m/s	506,4 J

Anzahl Messwerte: 5   Min: 355,8 m/s   Max: 359,9 m/s   Standardabweichung: 1,8 m/s

The summary of the test results is shown in Table A18.1, where the data is listed in descending order of E_0. Barrel length and barrel profile of the sample guns are shown for additional information.

Table A18.1:   Ballistics and Barrel Data.

Model	Caliber	Barrel Length [mm]	Barrel Profile	$\bar{v}_0$ [m/s]	$\bar{E}_0$ ** [J]
SPRINGFIELD *XD*^M	9 mm Luger	116.7	L/G	371	550
HECKLER & KOCH *VP9*	9 mm Luger	104.3	P	365	533
SMITH & WESSON *M&P9*	9 mm Luger	108.0	L/G	361	520
GLOCK *G19 Gen 4*	9 mm Luger	102.0	P	358	513
WALTHER *PPX*	9 mm Luger	101.5	L/G	358	513
TAURUS *PT 24/7 PRO*	9 mm Luger	108.4	L/G	356	508
CARACAL *F*	9 mm Luger	104.3	L/G	351	492
KEL-TEC *PF-9*	9 mm Luger	78.9	L/G	343	470
RUGER *LC9*	9 mm Luger	79.7	L/G	339	459

* L/G: Lands/Grooves; P: Polygonal Rifling
** Ammunition: RUAG *Sintox* 8.0 g Full Metal Jacket, Lot: DAG13E0710

Appendix 19   Slide Motion Analysis

Analysis: mean slide recoil velocity, empirical

Using the software PHOTRON *FASTCAM Viewer* Version 3.0, the high-speed video was analyzed frame by frame to determine the number of frames n_F, from the beginning of the discharge until the slide reaches the frame's rear stop. With the number of frames n_F, the camera's frame rate (20,000 frames per second), and Formula 5.3, p. 233, the duration t_S of the slide's recoil travel was calculated; which corresponds to the determined number of frames. Based on this and the values of the total slide recoil travel x_d shown in Appendix 21, the empirical mean slide recoil velocity was calculated.

Mean slide recoil velocity, empiric $\bar{v}_{Se}$

For the determination of the empirical value of the mean slide recoil velocity $\bar{v}_{Se}$, Formula 5.5, p. 234, was applied:

$$\bar{v}_{Se} = \frac{x_d f_F}{n_F}$$

Based on this formula, $\bar{v}_{Se}$ will reach a maximum (minimum) value, if total slide recoil travel x_d and frame rate f_F obtain maximum (minimum) values, and the number of frames n_F obtains a minimum (maximum) value. The number of frames was corrected with a value which reflected both, the inaccuracy in frame count at the beginning and the end of the reading, as well as the deviation caused by the variation in muzzle velocity, which was considered by increasing the number of frames arithmetically:

$$n_{F\ max} = n_F \frac{v_{0t\ max}}{v_0} + n_{F0} + n_{F1} \qquad \text{A19.1}$$

where: $n_{F\ max}$ Number of frames, maximum
  n_F Number of frames
  n_{F0} Number of frames, reading accuracy at the beginning
  n_{F1} Number of frames, reading accuracy at the end
  v_0 Muzzle velocity [m/s]
  $v_{0t\ max}$ Maximum muzzle velocity [m/s]

The following values were assumed for the determination of uncertainty of the measurement:

- Total slide recoil travel x_d: ±0.2 mm – caused by the Vernier caliper;
- Frame rate f_F: ±100 frames per second – according to the manufacturer of the camera, this would not be required because the camera's frame rate is supposed to be exact;

- Number of images n_{F0} and n_{F1}: ±1 frame – at the beginning and at the end of the reading respectively;
- Muzzle velocity v_0: min. and max. values ($v_{0\ min}$ and $v_{0\ max}$) were chosen from each respective test protocol, and on top was an uncertainty in measurement of 27.4 m/s[a] taken into account by increasing or decreasing the chosen value by 27.4 m/s respectively ($v_{0t\ min}$ or $v_{0t\ max}$, respectively).

The value for $n_{F\ min}$ was determined accordingly, but $v_{0t\ min}$ and the reading accuracy at the beginning and end of the reading were taken into account with negative sign. The amount of each maximum deviation was chosen for each tolerance shown (for example: CARACAL F: deviation, empirical: −0.4 m/s and +0.5 m/s, that is, a tolerance of 0.5 m/s was chosen). The results are summarized in Table A19.1, p. 417.

Analysis: mean slide recoil velocity, theoretical

The theoretical mean slide recoil velocity $\bar{v}_{St}$ is calculated as the average of the speeds at the beginning and end of the recoiling slide:

$$\bar{v}_{St} = \frac{v_{S1} + v_{S2}}{2} \qquad\qquad \textbf{A19.2}$$

where: $\bar{v}_{St}$ Mean slide recoil velocity, theoretical [m/s]
v_{S1} Slide velocity, initial [m/s]; projectile leaves the muzzle
v_{S2} Slide velocity, final [m/s]; slide impacts the backstop

The initial slide velocity v_{S1} is the maximum speed of the slide, and it occurs at the beginning of the firing cycle. The calculation of this velocity is based on the principle of linear momentum (Formula A19.7, p. 416).

In contrast to the initial velocity v_{S1}, an exact calculation of the final velocity v_{S2} is rather demanding. As a result of the impulse force transmission during the passage of the projectile through the bore, the recoiling mass has a certain initial kinetic energy $E_{Skin\ 1}$. As a consequence of the work W_{Rcomp} necessary to compress the recoil spring, frictional losses W_{fl} and possibly the effort W_k required to cock the firing mechanism, the slide has a lower kinetic energy when it strikes the backstop. Energy-consuming processes generally summarized under frictional losses are, for example, the disconnect mechanism, the uncoupling process, the disposal of the spent casing as well as other, gun-specific, individually different conditions. Not less important: The uppermost cartridge in the magazine, which is pressed against the slide by the magazine spring.

---

[a] "Ammunition tested subsequent to manufacture using equipment and procedures conforming to these guidelines [ANSI/SAAMI Z299.3-2017; PD] can be expected to produce velocities within a tolerance of ±90 fps of the tabulated values." (ANSI/SAAMI Z299.3-2015, p. 3) This tolerance translates to 27.4 m/s by applying a conversion factor of 0.3048 m/ft.

The slide's instantaneous velocity at the rear stop is hereinafter referred to as v_{S2}. We simplify its calculation by means of an approximate determination: the kinetic energy of the slide is directly proportional to the square of its speed v_{S2}, i.e. by making certain assumptions regarding losses, the velocity at the rear stop can be determined via the law of conservation of energy[689]. The kinetic energy of the slide at the end of the recoil travel x_d will be referred to as $E_{S\,kin\,2}$. Its amount is smaller than $E_{S\,kin\,1}$.

$$E_{S\,kin\,2} = E_{S\,kin\,1} - W_{R\,comp} - W_{fl} - W_k \qquad\qquad \textbf{A19.3}$$

where: $E_{S\,kin\,2}$ Kinetic energy of the slide [J], at end of the recoil travel
$E_{S\,kin\,1}$ Kinetic energy of the slide [J], after momentum transfer
W_{fl} Frictional losses [J] as the slide travels rearwards
W_k Work [J], e.g. to press down the hammer
$W_{R\,comp}$ Work [J], to compress the recoil spring

For simplification, the action of force of the recoil spring, frictional losses, the influence of the masses of the recoil spring and the fired case as well as the work needed to cock the mainspring will all be combined in the approximate value W_{fl}. We also assume that the instantaneous velocity of the slide during uncoupling is identical to that of the coupled barrel and slide. Equation A19.3 is simplified:

$$E_{S\,kin\,2} = E_{S\,kin1} - W_{fl} \qquad\qquad \textbf{A19.4}$$

where: W_{fl} Work [J] and losses during slide recoil travel

Independent of Formula A19.4, kinetic energy $E_{S\,kin\,2}$ is calculated in general:

$$E_{S\,kin\,2} = \frac{1}{2} m_S\, v_{S2}^2 \qquad\qquad \textbf{A19.5}$$

where: m_S Mass of the slide [kg]

When determining the kinetic energy $E_{S\,kin\,1}$ initially supplied to the system, the recoiling mass m_r must be taken into account:

$$E_{S\,kin\,1} = \frac{1}{2} m_r\, v_{S1}^2 \qquad\qquad \textbf{A19.6}$$

where: m_r Recoiling mass [kg]
e.g. if barrel and slide are coupled: $m_B + m_S$, where m_B is the barrel's mass

For this purpose, the initial slide velocity v_{S1} is calculated via the principle of linear momentum. Since the influence of gases, propellant weight and recoil spring mass proves to be small, it is ignored in the following. From the principle of linear momentum follows:

$$v_{S1} = v_0 \frac{m_P}{m_r}$$ **A19.7**

where: v_{S1} Slide velocity [m/s], when projectile's base leaves the muzzle
          m_P Mass, projectile [kg]
          v_0 Projectile's muzzle velocity [m/s]

Equating A19.4 with A19.5 and then resolving to final velocity v_{S2} results in

$$v_{S2} = \sqrt{v_{S1}^2 - \frac{2}{m_S} W_{fl}}$$ **A19.8**

By combining Formulas A19.8 and A19.2 one obtains:

$$\overline{v}_{St} = \frac{1}{2}\left(v_{S1} + \sqrt{v_{S1}^2 - \frac{2}{m_S} W_{fl}}\right)$$ **A19.9**

where: W_{fl} 3.5 J as an approximation[690]
          v_{S1} Slide velocity, calculated by applying Formula A19.7 [m/s]

When inspecting a pistol or conducting a feasibility study for a new design, parameters such as the data of the recoil spring or the total slide recoil travel x_d may not be known. Formula A19.9 provides useful results in these cases to determine whether $\overline{v}_{St}$ is approximately in the range of 6 to 8 m/s.

If the values for the spring rate R, pre-tension s_1 of the recoil spring and total slide recoil travel x_d are available, the tension energy of the recoil spring is calculated:

$$W_{R\,comp} = \frac{1}{2} R\big((s_1 + x_d)^2 - s_1^2\big)$$ **A19.10**[691]

where: R Spring rate, recoil spring [N/m]
          s_1 Deflection, [m]; pre-load of recoil spring; s_1 is the shortening
                 by which the recoil spring is compressed with slide closed
          x_d Total slide recoil travel [m]

One uses Formula A19.10 in A19.9 and gets for the theoretical mean slide recoil velocity:

$$\overline{v}_{St} = \frac{1}{2}\left(v_{S1} + \sqrt{v_{S1}^2 - \frac{R}{m_S}\big((s_1 + x_d)^2 - s_1^2\big)}\right)$$ **A19.11**

Even if the resulting formula gives the impression of precision, A19.11 is a rough calculation that neglects mechanical losses and any cocking of the firing mechanism in cases where the latter would be appropriate during recoil.

The recoiling mass m_r listed in Table A19.1 is the sum of the slide's gross weight m_S and the barrel's weight m_B. The data required for the calculation, as well as information for the muzzle velocity v_0 were taken from the checklists in Appendix 21. Next, the impulse force was calculated applying Formula A13.1, and by applying Formula A19.9, the mean slide recoil velocity $\bar{v}_{St}$ was calculated, based on a projectile's mass of 8 g (ammunition: RUAG *Sintox* 9 mm Luger, 8.0 g Full Metal Jacket, Lot: DAG13E0710; see Chapter 5.2 A.12, p. 230). The results are summarized in Table A19.1.

Table A19.1: Summary of Mean Slide Recoil Velocities;
empirical and theoretical values respectively.

					Analysis of Mean Slide Recoil Velocitiy							
	empirical				theoretical						cf. A19.9	
					applying A19.9 and using W_{fl} from Appendix 17							
	n_F	t_S	x_d	$\bar{v}_{Se}$	m_S	m_B	m_r	W_{fl}	v_0	$\bar{v}_{St}$	$\bar{v}_{St}$	W_{fl}
		[ms]	[mm]	[m/s]	[g]	[g]	[g]	[J]	[m/s]	[m/s]	[m/s]	[J]
CARACAL F	208	10,4	50,1	4.8 ±0.5	337	110	447	3,8	350,6	5,2	5,3	3,5
GLOCK G19 Gen4	199	10,0	46,9	4.7 ±0.5	347	104	451	4,0	358,2	5,3	5,4	3,5
HECKLER & KOCH VP9	189	9,5	47,1	5.0 ±0.6	352	98	450	3,5	364,8	5,6	5,6	3,5
KEL-TEC PF-9	94	4,7	41,3	8.8 ±1.1	171	58	229	3,3	342,7	11,1	11,0	3,5
RUGER LC9	107	5,4	42,6	8.0 ±1.0	227	57	284	3,0	338,8	8,8	8,7	3,5
S&W M&P9	214	10,7	47,4	4.4 ±0.6	368	103	471	3,2	360,7	5,3	5,2	3,5
SPRINGFIELD XD M	181	9,1	46,8	5.2 ±0.6	349	107	456	3,6	370,6	5,6	5,6	3,5
TAURUS PT 24/7 PRO	189	9,5	41,8	4.4 ±0.6	383	103	486	3,0	356,4	5,1	4,9	3,5
WALTHER PPX	182	9,1	49,1	5.4 ±0.6	345	110	455	3,8	358,2	5,2	5,4	3,5

Note:

The analysis showed that the mean slide recoil velocities of KEL-TEC *PF-9* and RUGER *LC9* were at 8.0 m/s and above. Generally, a mean slide recoil velocity of 6 to 8 m/s is considered desirable,[692] mean velocities exceeding this range may cause the mechanical functions of a handgun to override and will induce heavy impact loads, which in turn can damage a gun's components. Impact loads can induce vibrations, causing some of the pistol's components to vibrate heavily or move rapidly and cause these parts to break, e.g. slide stop levers.

Also, the durability of recoil springs is no longer satisfactory at impact velocities faster than 12 m/s; with stranded springs, the limit may be improved to approximately 14 m/s.[693] Slide recoil velocities slower than 6 to 8 m/s may cause the slide to short stroke, that is, the slide does not reach the rear stop with sufficient kinetic energy. Symptoms may be weak ejection, failures to feed, or the slide fails to stay open on an empty magazine.

Comments on dynamical systems

Classical mechanics distinguishes between three types of movement of masses: linear and circular, and periodically to and fro. The latter are also known as oscillations.[694] In the theoretical consideration of the slide's straight line displacement we find ourselves in the field of vibration theory of mechanical systems. Such systems are also briefly referred to as mechanical oscillators. Every mechanical system that has a stable equilibrium position is a system capable of oscillating. The equilibrium position of a mechanical system is called stable if, when this balance is agitated, restoring forces are created which bring the system back into the equilibrium position.

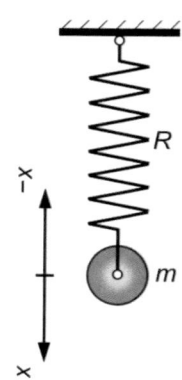

Fig. A19.1: Vertical spring-mass harmonic oscillator in equilibrium position; spring rate R, mass m, orientation of displacement x.

A well-known example of this is a tension spring suspended vertically from a rigid support. When a mass is attached to the free end of the spring, gravity stretches the spring beyond its natural length to an equilibrium position. At static equilibrium, the elastic force of the spring balances the weight of the suspended mass (Fig. A19.1). The only forces exerted on the mass are its weight and the force from the spring. If the spring is slightly disturbed from its stable equilibrium state, e.g. the mass is pulled downwards in the spring's direction of action by a small amount relative to the equilibrium position and released to make the spring and mass oscillate in the vertical plane, the system reaches its equilibrium position with a certain speed. Caused by inertia of the involved masses of oscillator and spring, the balanced position will not be taken back immediately, but the system will go on moving beyond. Restoring forces change their effective direction and the system comes to rest momentarily at a deflected position, and from this reversal point, the system will move again toward the zero position, which it will pass next time with a different velocity. This process is repeated periodically. If the spring mass system is assumed to be ideal, the mass would oscillate about the equilibrium position. Consequently, the spring stretches and compresses itself again and again. Such a motion is known as simple harmonic motion.[695]

Dilatation in the direction of action of the spring is normally associated with a positive sign, and displacement x is negative when the spring is compressed (see Fig. A19.1).

The origin for measuring the distance travelled is normally placed at the center of gravity of the oscillating mass, whenever the time course of the movement is in the focus of examinations. However, it is also possible to generate data at the point where the spring is attached to the mass. This proves to be useful if the center of gravity is not known. In both cases, the correct dimension for the displacement of mass and spring is obtained and the restoring force of the spring can thus be determined.

If a mechanical system swings about its equilibrium position, there will be a permanent exchange between kinetic energy E_{kin} of the system when passing through the middle position and potential energy E_{pot} at the upper and lower reversal point. In real systems, this energy exchange is not loss-free. Instead, a continuous part of oscillation-energy is transferred into other forms of energy, e.g. into heat energy caused by internal friction. Thus, the oscillation energy of the simple harmonic oscillator decreases continuously, because energy is only supplied during the initial disturbance.

A further variation for studying the oscillation behavior of another system with only one degree of freedom is the horizontal single-spring-mass oscillator. With this oscillator, we consider a mass m, which slides over a frictionless horizontal, flat, smooth surface. It is attached to a linear spring with spring constant R. The other end of the spring is anchored in an immovable wall. In this instance, the state of stable equilibrium is characterized by a tension-free

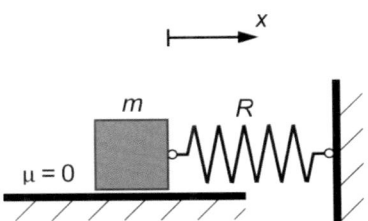

Fig. A19.2: Horizontal spring-mass oscillator in equilibrium position; friction is neglected.

spring. Any horizontal displacement of the oscillator from the equilibrium state leads to restoring forces of the spring. Thus, a system capable of oscillating is present. A one-dimensional coordinate system is introduced to describe the position of the mass, such that the x-axis is co-linear with the motion, and the origin is located at the body's center of gravity and where the spring is at rest. Normally, the positive direction corresponds to the spring being expanded. In the below we intentionally deviate from this, shifting the origin to where the spring is attached to the oscillating body and, contrary to common practice, the displacement x should assume positive values when the spring is being compressed (see Fig. A19.2). Applied to our specific application, this results in a positive sign as the slide recoils. A negative sign indicates counter-recoil.

In the case of oscillators with one degree of freedom, the displacement of the body can be unambiguously described by a single kinematic dependent variable at any given time. For a single-mass oscillator this is the horizontal displacement x. This only holds for the model of a rigid body, which is coupled to the environment in a correspondingly movable condition. Such a model concept proves to be permissible if the deformation of the body is very small compared to the dynamic displacements of the body.[696]

The oscillator equations of an undamped spring oscillator are derived by setting up the differential equation based on Newton's laws of motion and then solving it. One regularly obtains a sinusoidal course of the displacement as a function of time as shown in Fig. A19.3. The procedure is considered standard and has long been treated exhaustively in the relevant literature. For the sake of clarity, formulas will be developed here once again in great detail.

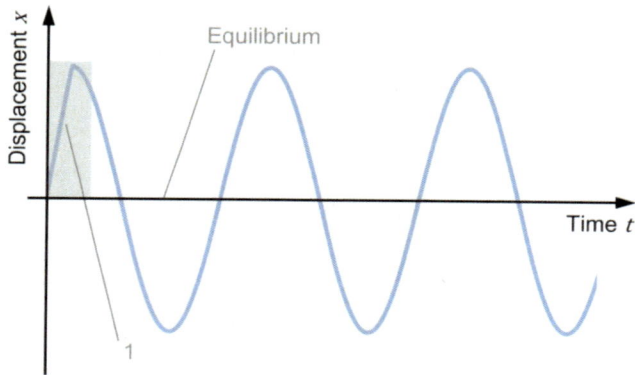

Fig. A19.3: Displacement-time diagram using the example of an undamped spring-mass oscillator: a disruption (1) leads to a displacement of the system, which then oscillates about the equilibrium position without coming to rest.

In the instance of the slide's motion, the displacement-time diagram corresponds to a sequence of sinusoidal sections. For example, the slide's recoil travel could be similar to the shaded area shown in Fig. A19.3. From a physical point of view, the recoiling components of a pistol can be reduced to an arbitrarily small body, which has a certain mass and is spring-loaded. The setup sufficiently known to us forms a system that can oscillate. Before the discharge, the compression spring and the oscillating body are at rest. At the time $t = 0$, a shock is introduced into the resting system, which leads to a deflection[697] of the spring-mass oscillator, and at the end of the cycle it returns to its initial position. With a sufficiently strong impact, the body, i.e. the slide, reaches the rear stop at x_d. The displacement x_d corresponds to the length of the slide's recoil travel. Our considerations focus on this displacement.

An impact caused by gases and particles following the projectile as it leaves the muzzle is not taken into account.[698] Furthermore, we are assuming that there are no retarding forces acting on the system. Also, losses caused by mechanical effects in the gun are neglected; for example, the uncoupling or coupling of barrel and slide, the friction between the slide and the magazine's top cartridge, resetting the firing mechanism,[699] the ejection of the fired case, chambering a round, the movement of the extractor claw or the control of safety devices, as well as other effects such as friction or damping and not least the acceleration of the frame during recoil.

The pistol is in normal shooting position with the barrel horizontal. The direction of action of the recoil spring is parallel to the orientation of the slide's feedway, the spring characteristic curve is linear. One end of the recoil spring is intended to close the slide, the other end is immovably mounted in the gun. The mass of this spring as well as the mass of the fired case and the propellant weight can be neglected in handguns. In a 9 mm pistol, there is a negligible amount of propellant of about 0.4 g in the cartridge, the case weighs in at about 3.6 grams, the recoil spring only 6 to 12 grams, of which an intrinsic mass of ⅓ would have to be considered.[700] To complete the computational method, these values are nevertheless included in the following. If you transfer the formulas into a spreadsheet calculation program, you get a complete calculation model and can simulate the influence of the variables. If this consideration is made on the basis of Example 2, p. 445, the theoretical mean slide recoil velocity is 5.77 m/s, taking into account the masses of case and recoil spring, otherwise 5.84 m/s. This difference of approximately 1.2 percent can be considered to be negligible.

For comparison: In a real-world scenario, i.e. when the frame is not rigidly supported but a shot is fired off-hand, relative speed between frame and slide turns out to be approximately 1.5 to 2 m/s less. If one takes the above-mentioned calculation results as a benchmark, this corresponds to a deviation of about 25 to 35 percent. The reason for this lower recoil velocity is to be found in the processes during shooting off-hand. In practice, not only the barrel and slide, but also the frame moves during shooting, which results in a lower relative speed between frame and slide. This is said in advance for a better understanding of the limits of mathematical analytical computing, such as the following theoretical analysis.

a) Straight Blowback

Let us consider a pistol in which the cartridge case immediately moves backward at the very instant where the projectile begins its travel. The casing is backed up by the inertia of a movably mounted slide, which simultaneously experiences acceleration in the longitudinal direction of the barrel. The pistol's barrel is fixed. The type of construction described is known as force-locked, inertia locked, straight blowback, or simple blowback; colloquially blowback.

In order to develop the equations of motion of the horizontally reciprocating slide, we consider the forces acting on the body, namely the inertial and the recoil spring force. Regarding the latter, the spring force F_R is proportional to the displacement x of the recoil spring ($F_R = R\,x$). Regarding inertial force, Newton's second law states: the acceleration a of a body with mass m is proportional to the force acting on it ($F = m\,a$).[701] The net force is zero at any time:

$$ma + Rx = 0 \qquad\qquad \text{A19.12}$$

The velocity v is defined as the first time derivative of the displacement x, the second time derivative of x yields the acceleration a. Hence,

$$a = \frac{d^2x}{dt^2} \;\; := \ddot{x} \qquad\qquad \text{A19.13}$$

We insert A19.13 into A19.12 and obtain a second-order linear homogeneous differential equation with constant coefficients:

$$m\ddot{x} + Rx = 0 \qquad\qquad \text{A19.14}$$

We divide the equation by m and from this follows the equation of motion of the undamped harmonic oscillation:

$$\ddot{x} + \frac{R}{m}x = 0 \qquad\qquad \text{A19.15}$$

To determine the time response of the oscillator, the equation of motion must be solved. We are looking for functions $x = x(t)$, which, when inserted into the equation, fulfill it identically. One receives the solutions with the following function for x:

$$x = e^{\lambda t} \qquad\qquad \text{A19.16}$$

where e is Euler's number, λ is a factor yet to be determined and t stands for time. The first and second derivative of A19.16 is represented as:

$$\dot{x} = \lambda\,e^{\lambda t} \qquad\qquad \ddot{x} = \lambda^2 e^{\lambda t} \qquad\qquad \text{A19.17}$$

Plugging the right hand side from A19.16 and $\ddot{x}$ from A19.17 into A19.15, and excluding $e^{\lambda t}$ yields:

$$e^{\lambda t}\left(\lambda^2 + \frac{R}{m}\right) = 0 \qquad \text{A19.18}$$

We seek solutions for λ, for which the term on the left side of the equation becomes zero. Since the exponential function does not assume the value zero, the expression in parenthesis must disappear. This results in a negative radicand and therefore we get two results for λ:

$$\lambda_1 = j\sqrt{\frac{R}{m}} \qquad \lambda_2 = -j\sqrt{\frac{R}{m}} \qquad \text{A19.19}$$

where j is the imaginary unit. With the abbreviation

$$\omega = \sqrt{\frac{R}{m}} \qquad \text{A19.20}$$

follows:

$$\lambda_1 = j\omega \qquad \lambda_2 = -j\omega \qquad \text{A19.21}$$

We use the solutions for λ separately in A19.16, so that both solutions appear in their respective form

$$x_1 = e^{j\omega t} \qquad x_2 = e^{-j\omega t} \qquad \text{A19.22}$$

One combines both results of A19.22, by summation, considers two arbitrary constants and obtains as general solution for $x(t)$:

$$x(t) = A_1 e^{j\omega t} + A_2 e^{-j\omega t} \qquad \text{A19.23}$$

Taking into account the known Euler identity

$$e^{\pm jx} = cos(x) \pm j\,sin(x) \qquad \text{A19.24}$$

equation A19.23 can be transformed into:

$$x(t) = A_1\big(cos(\omega t) + j\,sin(\omega t)\big) + A_2\big(cos(\omega t) - j\,sin(\omega t)\big)$$

By multiplication this becomes

$$x(t) = (A_1 + A_2)\,cos(\omega t) + j(A_1 - A_2)\,sin(\omega t) \qquad \text{A19.25}$$

Each constant has exactly one value and the bracketed expressions shown as sum or difference can therefore be summarized by replacing them with arbitrary constants C_1 and C_2. The result is the general solution for the differential equation:

$$x(t) = C_1\,cos(\omega t) + C_2\,sin(\omega t) \qquad \text{A19.26}$$

Velocity is the time derivative of displacement. Hence, we get

$$v(t) = -C_1 \omega \sin(\omega t) + C_2 \omega \cos(\omega t)$$

A19.27

We determine the constants by considering the boundary conditions at time zero. For $t = 0$ we have $\sin(\omega 0)$. The sine of zero is zero, the cosine is one. Based on this, we obtain from A19.26 and A19.27:

$$C_1 = x(0)$$

A19.28

$$C_2 = \frac{v(0)}{\omega}$$

A19.29

The zero point of the deflection is the equilibrium position at which the recoil spring is at its free length. In this equilibrium position its free length is L_0. At time $t = 0$ the pistol's slide is in battery position and the compression of the recoil spring is s_1 (i.e. the spring's load height is $L_0 - s_1$). In other words: At $t = 0$ there is a deflection of the spring. From this it can be concluded for constant C_1:

$$C_1 = s_1$$

A19.30

For the determination of C_2 we neglect the pressure build-up at time $t = 0$, which, as a function of time, leads to an acceleration of the projectile that corresponds to a certain time interval (cf. numerical value equation A23.12, on page 478). This is done on the assumption that the time until the projectile leaves the barrel is considerably less than the duration of the total recoil travel of the slide from front to rear stop. We thus simplify the analysis under the theoretical assumption that at time $t = 0$ a sudden increase in recoil velocity from zero to v_{S1} occurs in an infinitely short time. Under this condition we obtain for C_2:

$$C_2 = v_{S1} \frac{1}{\omega}$$

A19.31

The unit of measurement for C_1 and C_2 is meters. As explained at the beginning, this is a straight blowback pistol and for the recoiling mass we take into account the mass of the slide, one third of the recoil spring mass and the mass of the spent cartridge case:

$$m_{rBB} = m_S + \frac{m_R}{3} + m_{CC}$$

A19.32

where: m_{rBB} Recoiling mass, blowback [kg]
m_{CC} Mass, cartridge case [kg]
m_R Mass, recoil spring [kg]
m_S Mass, slide [kg]

Based on the principle of linear momentum for v_{S1} applies:

$$v_{S1} = v_0 \frac{m_P}{m_{rBB}}$$ **A19.33**

where: v_{S1} Slide velocity [m/s], when base of projectile leaves muzzle
m_P Mass, projectile [kg]
v_0 Muzzle velocity [m/s] of projectile

We use A19.20 and consider the recoiling mass m_{rBB}:

$$\omega = \sqrt{\frac{R}{m_{rBB}}}$$ **A19.34**

where: ω Angular frequency [s^{-1}] [702]
R Spring rate, recoil spring [N/m]

It follows for C_2:

$$C_2 = v_0 \frac{m_P}{m_{rBB}} \sqrt{\frac{m_{rBB}}{R}}$$ **A19.35**

With the above parameters we will obtain from A19.26 an equation for the displacement of a straight blowback pistol's slide, in which we take into account the preload s_1 of the recoil spring:

$$x(t) = s_1 \cos\left(\sqrt{\frac{R}{m_{rBB}}}\ t\right) + v_0 \frac{m_P}{m_{rBB}} \sqrt{\frac{m_{rBB}}{R}}\ \sin\left(\sqrt{\frac{R}{m_{rBB}}}\ t\right) - s_1$$ **A19.36**

or, respectively:

$$x(t) = s_1 \cos(\omega t) + \frac{v_{S1}}{\omega}\ \sin(\omega t) - s_1$$ **A19.37**

where: $x(t)$ Displacement [m] at time t
s_1 Deflection [m]; preload of the recoil spring; s_1 represents the amount of shortening by which the spring is compressed when the slide is in battery
t time [s]

An oscillation always occurs about the static rest position of the elastic element; in our case the recoil spring's free length L_0. The recoil spring is pretensioned by s_1, which in the oscillation equations A19.36 and A19.37 corresponds to a zero point offset relative to the free length L_0.[703]

By introducing additional parameters, the oscillation equation A19.36 can be converted into a clearer representation:[704]

$$\hat{x} = \sqrt{s_1^2 + C_2^2}$$ **A19.38**

where: $\hat{x}$ Amplitude [m]

The amplitude $\hat{x}$ represents the theoretical extreme value of the displacement x.[705] In reality, the slide's total recoil travel is structurally limited to the displacement x_d and does not meet the maximum value $\hat{x}$.

$$\varphi = arctan\left(\frac{C_2}{s_1}\right)$$
A19.39

where: φ Phase angle [rad], i.e. initial phase angle[706]

With the above definitions and after summation of the superposed oscillations, the representation of equation A19.36 is simplified:

$$x(t) = \hat{x}\,cos(\omega t - \varphi) - s_1$$
A19.40

Taking into account the correct phase shift along the time axis and solving this equation according to t gives the duration until a displacement x is reached:

$$t(x) = \frac{1}{\omega}\left(\varphi - arccos\left(\frac{s_1 + x}{\hat{x}}\right)\right)$$
A19.41

where: $t(x)$ Duration [s], until displacement x is reached
           x Displacement [m]; recoil travel of slide, where x always
                   satisfies $0 \leq x \leq x_d$, where x_d is the total slide recoil travel

The instantaneous velocity of the slide at a time t is the first time derivative of the oscillator equation A19.40. Hence,

$$v_S(t) = -\hat{x}\omega\,sin(\omega t - \varphi)$$
A19.42

where: $v_S(t)$ Instantaneous velocity [m/s] of the slide at time t

The time t can be eliminated by converting equation A19.36 and its first time derivative by squaring, multiplying by ω^2, adding the equations, applying the trigonometric Pythagorean and square root to an equation of v_S as a function of displacement x:

$$v_S(x) = \omega\sqrt{C_2^2 + s_1^2 - (s_1 + x)^2}$$
A19.43

Alternatively: the law of the conservation of energy (see A19.8) is used to calculate the velocity at the rear stop:

$$v_{S2} = \sqrt{v_{S1}^2 - \omega^2\left((s_1 + x_d)^2 - s_1^2\right)}$$
A19.44

Speed is measured as distance moved over time. Also, our analysis of the straight blowback principle is based on the assumption of an instantaneous increase in slide velocity at $t = 0$ from zero to v_{S1}. Under this condition and by taking A19.41 into account, we can set up the following equation for the theoretical mean slide recoil velocity:

$$\bar{v}_{St_{BB}} = \frac{x_d\,\omega}{\varphi - arccos\left(\frac{s_1 + x_d}{\hat{x}}\right)}$$ **A19.45**

Straight blowback: Slide at rearmost position and during counter-recoil

At the backstop the direction of movement of the oscillator is reversed. The acceleration at the reversal point results from an interaction of the recoil spring force and the
collision at the travel limiter. Due to elastic deformation of the frame, the shooter's
hand, his arm, etc., only a fraction of the mechanical energy present at the reversal
point is returned to the slide.

From a physical point of view, the collision is neither an ideal-elastic nor an ideal-
inelastic impact. The phenomenon occurring here is known as a *partially elastic* or *real
impact*. We simplify the calculation of the rebound at the reversal point by determining
the velocity of the slide after passing the reversal point, i.e. at the start of the counter-
recoil sequence, as follows:[707]

$$v'_{S2} = -k_S v_{S2}$$ **A19.46**

where: v'_{S2} Velocity, slide [m/s] after collision
          k_S Coefficient of restitution [1]; empirical value: $k_S \approx 0$ to 0.2

Note: During the counter-recoil sequence, a negative sign for the slide's velocity results
from the definition made at the beginning. In addition, the mass of a cartridge is taken
into account in the formulas. The slide hits this cartridge shortly after leaving the backstop and feeds it from the magazine into the firing chamber. As a distinguishing feature
to variables used up to now, we introduce the index gamma to differentiate the counter-
recoil sequence from the recoil sequence.

$$x(t_\gamma) = C_\gamma\,sin(\omega_\gamma t_\gamma + \varphi_\gamma) - s_1$$ **A19.47**

Regarding to the coefficients, the following applies:

$$\omega_\gamma = \sqrt{\frac{R}{m'_S + \frac{1}{3}m_R}}$$ **A19.48**

where: m'_S Mass [kg], slide plus one cartridge
          $m'_S = m_S + m_{cartridge}$

$$\varphi_\gamma = arctan\left(\frac{s_1 + x_d}{v'_{S2}}\,\omega_\gamma\right)$$ **A19.49**

Note the following exception: When $v'_{S2} = 0$ then $\varphi_\gamma = \frac{-\pi}{2}$. Example of use: closing the
slide by pressing down on the slide release or when $k_S = 0$.

The constant C_Y is calculated:

$$C_Y = \frac{s_1 + x_d}{\sin \varphi_Y}$$ **A19.50**

Equation A19.47 resolved to t:

$$t_Y(x) = \frac{1}{\omega_Y}\left(arcsin\left(\frac{s_1 + x}{C_Y}\right) - \varphi_Y\right)$$ **A19.51**

For $x = 0$ follows the total duration of the counter-recoil sequence for blowback pistols:

$$t_{Y\,tot} = \frac{1}{\omega_Y}\left(arcsin\left(\frac{s_1}{C_Y}\right) - \varphi_Y\right)$$ **A19.52**

The first time derivative of A19.47 yields the instantaneous velocity,

$$v_Y(t_Y) = C_Y\omega_Y\,cos(\omega_Y t_Y + \varphi_Y)$$ **A19.53**

By forming, the instantaneous velocity can be represented as a function of the deflection x (where x always satisfies $0 \leq x \leq x_R$), independent of t:

$$v_Y(x) = -\omega_Y\sqrt{C_Y^2 - (s_1 + x)^2}$$ **A19.54**

At the position $x = 0$ the slide hits the front stop. At this point the instantaneous velocity is

$$v_{S3} = -\omega_Y\sqrt{C_Y^2 - s_1^2}$$ **A19.55**

To conclude, the following can be said: For straight blowback semi-autos, the recoil velocity v_{S1} is calculated according to A19.33, p. 425, v_{S2} according to Equation A19.44, p. 426, the mean recoil velocity $\bar{v}_{St}$ according to A19.45, p. 427, and the ultimate velocity at the front stop v_{S3} according to Formula A19.55.

The angular frequency ω is a measure of how fast the mass attached to the spring would theoretically oscillate if the system were not limited by design. Reducing the spring rate R and increasing the mass m means a mass on a spring oscillates slower.

The angular frequency is independent of the excitation force. The latter only determines the amplitude, i.e. the distance the mass travels for one cycle.[708]

$$\omega = \sqrt{\frac{R}{m}} \equiv \frac{2\pi}{T} \equiv 2\pi f$$ **A19.56**

where: ω Angular frequency [s^{-1}]
            f Frequency [s^{-1}]
            T Time period [s]

The angular frequency ω differs from the frequency f by the factor 2π. The reciprocal value of f corresponds to the time for one full oscillation, called period T.[709]

$$T = \frac{2\pi}{\omega} \equiv \frac{1}{f}$$ **A19.57**

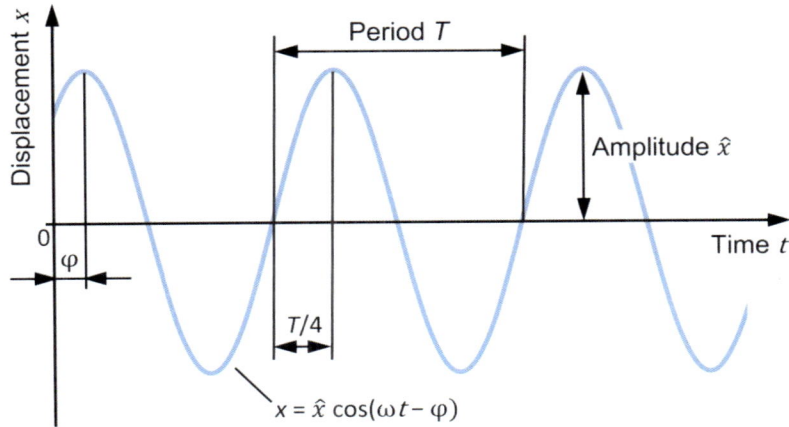

Fig. A19.4: Amplitude and repetition period illustrated in the displacement vs. time diagram.

For comparison purposes, a quarter of the periodic time T is commonly used for reasons of illustration. $T/4$ corresponds to the time from the beginning of the movement until the theoretically maximum deflection of a harmonic motion is reached.

The values $T/4$ and $\hat{x}$ describe the harmonic motion. They are characteristic values that can be used for system comparisons. Similar to engine power and torque, which serve as characteristic values in the world of automotive.[710]

Example 1: Straight Blowback Pistol

Data of a .32 cal. blowback pistol:

$m_S = 0.225$ kg $m_P = 0.0048$ kg $m_R = 0.006$ kg $m_{CC} = 0$ kg
$v_0 = 287$ m/s $x_d = 0.036$ m $R = 670$ N/m $s_1 = 0.046$ m

By applying equation A19.40 a deflection vs. time diagram is obtained. Fig. A19.5 shows $x(t)$ for the gun in standard configuration, also a variant with a recoil spring whose spring rate has been doubled compared to the original and an alternative with a 100 grams heavier slide ($m_S = 0.325$ kg). According to common practice the mass of the cartridge case is neglected; ergo $m_{CC} = 0$ kg.

The limit of the total slide recoil travel is shown in the diagram as a horizontal dashed line. This backstop at x_d represents the reversal point from which the counter-recoil movement starts.

The diagram visualizes the influence of the spring's stiffness and the slide's mass. The effect of both variables can be clearly seen towards the end of the recoil stroke and explains the increasing distance between the curves. A special mention should be made of the lower and upper functional working limits of the gun. The demand for functional reliability and the mutual influence of the design parameters leads in practice to restrictions in the variation of spring rate, recoiling mass, ammunition, etc.

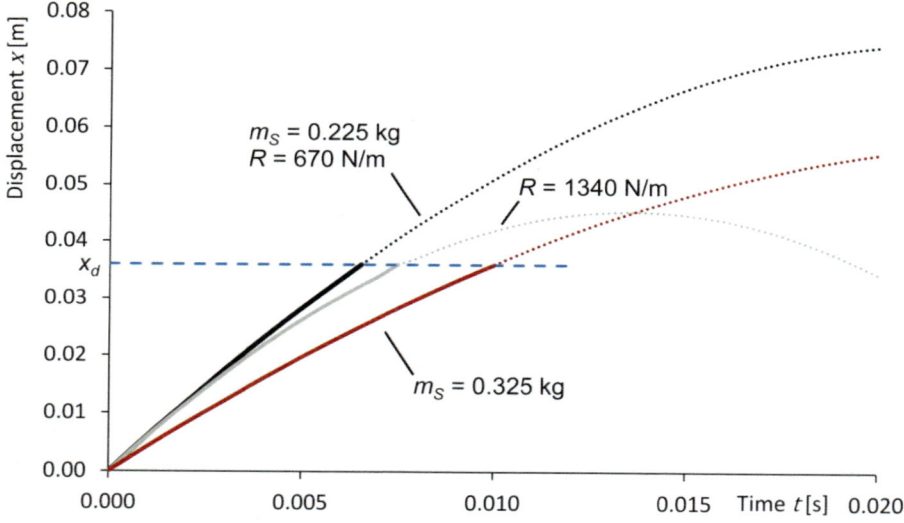

Fig. A19.5: Displacement vs. time diagram for Example 1, .32 cal. blowback pistol.

The continuation of the displacement-time curves, which is only possible in theory because it ignores the backstop, is shown in dots. In the area above x_d the curvature of the sinusoidal functions is noticeable. The increasing curvature results from the harmonic oscillator equation and also signifies a decrease in speed. In the example with doubled spring rate R, maximum of deflection corresponds to that point, where a theoretically free-swinging mass would change its direction of movement and would move back towards the starting position.

Rate of fire, blowback

Under the premise of an identical time duration for recoil and counter-recoil travel of the slide, the following applies:

$$K_{t_{BB}} = \frac{\omega}{2\,arccos\left(\frac{s_1}{s_1+x_d}\right)}$$ **A19.58**[711]

where: K_{tBB} Rate of fire, theoretical, blowback [rounds/s]

The calculation of the rate of fire shown here is based on the approach with an equation of motion. It therefore refers to the duration of the movement sequence during the firing sequence. Since we are dealing with semi-automatic pistols in which the shot is fired from a closed breech and the firing mechanism is released by pulling the trigger, the firing mechanism must react first before the next shot is fired. In this instance the main-spring drives a striking element. It takes approximately 0.003 to 0.01 seconds from the sear release until the primer is struck. This time interval is called lock time.[712]

Furthermore, it is common practice to assume that a pressure of approximately 10 percent of the peak pressure must build up to start the projectile from its seat (cf. End-note 154). The short span of time between when the primer is hit and the gas pressure causes the projectile to move is usually about 0.002 seconds. This is known as ignition time.[713]

Note: Even if lock time and ignition time is taken into account when calculating the rate of fire, the formula usually gives a larger number compared to the actual cyclic rate. For example, it ignores frictional losses; notably the work required to strip the next cartridge from the magazine and to chamber it. Besides, it is substantial whether the gun is fixed in position during the test or whether it is fired off hand. Shooting off-hand reduces the firing rate.

Uncoupling ratio: Support of the casing during firing

Of particular interest is the calculation of the displacement x_ℓ for the instant when the base of the projectile leaves the muzzle and an assessment of whether the cartridge case is adequately supported until the chamber pressure has decreased to a safe level. In the field of weapon technology, various terms are used in this context. For example, *Dannecker* distinguishes between the controlled travel (Sicherheitsstrecke) with refer-ence to blowback-action (Fig. A19.6) and the safeguarded travel (Unterstellstrecke) in systems with coupled barrel and slide.[714] Other sources might use locked dwell time.[715]

This Appendix 19 is devoted to harmonic motion. It uses parlance from classical mechanics and thus the neutral term displacement is used in the following.

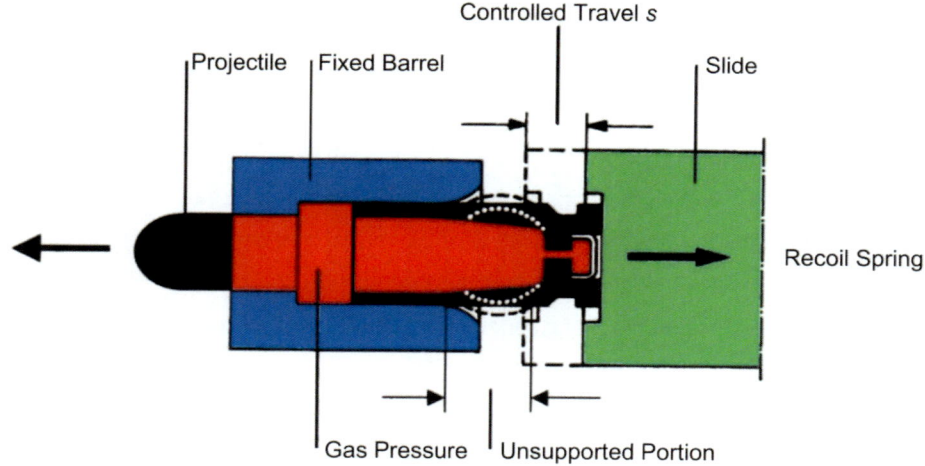

Fig. A19.6: Controlled Travel s of a blowback pistol on the example of *Dannecker*. Furthermore, Dannecker explains that the influence of the recoil spring force is negligible compared to the effect of the inertia of the slide.

© *P. Dannecker*

The TR "Pistol" provides the safeguarded displacement s_{safety} (Sicherheitsweg), which it defines as the slide's displacement until the start of uncoupling.[716] Following our logic, we use the variable x_{u1} for the displacement at which the uncoupling is initiated.

When analyzing blowback pistols, we mainly deal with softer shooting cartridge types than what we would use for a 9 mm Luger. However, in the following we will refer to the hotter 9 mm, as a link to later on following explanations regarding pistols with coupled barrel and slide. Also, the requirements regarding safeguarded recoil travel x_{u1} and cartridge support are identical.

To assess whether the casing is adequately supported ($x_{\ell\,max}$) over a sufficient period of time, one uses the principle of linear momentum, applies the maximum projectile mass $m_{P\,max}$ of commonly used projectiles, doubles it, and usually neglects the masses of the recoil spring and the cartridge case:

$$x_{\ell\,max} = s_\ell \, \frac{2\,m_{P\,max}}{m_{r\,min}}$$

A19.59[717]

or for the further course of our considerations under ordinary conditions:

$$x_\ell = s_\ell \frac{m_P}{m_S + m_B + \frac{1}{3}m_R + m_{CC}}$$ A19.60

$$s_\ell = L_B - L_3 + s_{Ps}$$ A19.61

where: x_ℓ Displacement [m], when base of projectile leaves the muzzle
          L_3 Length, case [m]
          L_B Length, barrel [m]; distance from breech face to muzzle
          m_B Mass, barrel [kg], applies only for pistols with coupled barrel
                 and slide
          m_{CC} Mass, fired case [kg]
          m_P Mass, projectile [kg]
          $m_{P\,max}$ Maximum projectile mass [kg] of commonly used bullets
          $m_{r\,min}$Mass, recoiling [kg], m_S for blowbacks, $m_S + m_B$ for pistols with
                 coupled barrel and slide; use minimum values
          m_R Mass, recoil spring [kg]
          m_S Mass, slide [kg]
          s_ℓ Bore travel [m], projectile base leaves the muzzle
          s_{Ps} Seating depth [m]; i.e. the depth the projectile is placed in
                 the case; ca. 0.005 m for 9 mm Luger with 124 gr projectile

Consider a barrel loaded with a cartridge in the chamber, the case head rests against the breech face. The barrel length is the distance from breech face to muzzle. The bore travel s_ℓ of the projectile's base is the distance between the base and muzzle. Taking into account 5 mm bullet seating depth commonly found on an 8 g FMJ 9 mm Luger means the projectile's base is approximately 14 mm ahead of the breech face. By subtracting 0.014 m from the barrel length L_B of a 9 mm pistol, total projectile bore travel s_ℓ can be obtained straightforwardly; $s_\ell = L_L - 0.014$ m.

The displacement x_ℓ indicates how far the recoiling mass deflects until the projectile's base leaves the muzzle. For blowback pistols, this reveals by how much the case slides out of the chamber until the propellant gas has the chance to escape at the muzzle and thus the pressure in the bore can quickly drop to a tolerable level. In the case of pistols with coupled barrel and slide, x_ℓ is used to assess whether a relative movement between barrel and slide occurs as the projectile moves down the bore. The maximum slide displacement $x_{\ell\,max}$ for these designs must be smaller than the start of the uncoupling at x_{u1}, i.e. the uncoupling process may only be initiated after the projectile has cleared the barrel.

For pistols with a coupled barrel and slide, the uncoupling ratio is also assessed. It corresponds to the relation of the start of uncoupling x_{u1} to the displacement when the projectile's base leaves the muzzle. For this purpose, the calculation of x_ℓ is similar to Formula A19.59, but without doubling the projectile mass:

$$\Phi_{\ell u1} = \frac{x_{u1}\, m_{r\,min}}{s_\ell\, m_{P\,max}}$$

A19.62

where: $\Phi_{\ell u1}$ Uncoupling ratio [1], ratio of displacement x_{u1} to $x_{\ell\,max}$
             x_{u1} Displacement [m], start of uncoupling

For example, $\Phi_{\ell u1} = 2$ expresses a proportion of 2, meaning the uncoupling process begins after a displacement of the slide corresponding to twice the slide recoil travel at which the projectile has left the muzzle. Table 5.5, p. 315, lists $\Phi_{\ell u1}$ of individual pistols.

The time involved when the projectile first starts to move from the case until it leaves the muzzle, the so-called barrel time,[718] can also be calculated. Here, it is necessary to take into account the time course of a projectile velocity that starts from a standstill and increases with intensifying pressure in the barrel (Fig. A23.9, p. 488). For reasons of simplification, we assumed the projectile velocity would rise sharply in an infinitely short time from 0 m/s to the value v_0 in the analysis of the blowback principle, since we focused on the slide's displacement. A calculation according to Formula A19.41 or A19.71 mentioned below using x_ℓ instead of x would result in a too short barrel time. Both formulas are only valid starting from the projectile's passage at the time t_ℓ onward.

While the projectile passes through the barrel, i.e., taking into account the acceleration phase of the projectile and the recoil mass involved, we use Heydenreich's method, explained in Appendix 23, and apply the numerical value equation A23.12, p. 478, to calculate the barrel time t_ℓ:

$$t_\ell = \frac{0.74\, v_0 \left(m_P + \tfrac{1}{2}m_C\right)}{A\{mm^2\}\, p_{max}\{MPa\}} + 1.08\frac{s_\ell}{v_0}$$

(cf. A23.12, p. 478)

The recoiling mass is originally in a state of rest. Its movement is caused by an impact, so to speak, by a disturbance of the state of equilibrium. From a physical point of view, the barrel time t_ℓ is also referred to as impulse force duration or disturbance time. The disturbance begins when the projectile starts to move. In our approach, it ends when the base of the projectile leaves the muzzle.

b) Coupled Barrel and Slide, cocking neglected

The vast majority of recoil operated pistols use a movably mounted barrel which is initially coupled to the slide. After a certain displacement, barrel and slide get uncoupled. The basic arrangement is shown in Fig. A19.9, p. 437.

For a detailed analysis of semi-automatics with coupled barrel and slide, we divide the displacement into four sections. The completion of uncoupling at x_{u2}, the reversal point at x_d and the beginning of coupling when x_{u2} is reached again form the transitions between these segments, and will be referred to below as alpha, beta, gamma and delta.

Fig. A19.7 visualizes the four sections on the example of a displacement-time diagram. It illustrates an example of the sequence of the slide displacement as a function of time. In addition, we find some prominent points of the sequence.

If the actual course of the movement is not of interest and only the recoil velocity is sought instead, the mathematical approach via the energy theorem is recommended. Hence, v_{S2} is calculated by applying A19.64 and A19.65.

$$v_{S1} = v_0 \frac{m_P}{m_S + m_B + \frac{m_R}{3} + m_{CC}} \qquad \text{A19.63}$$

$$v_{\alpha x_{u2}} = \sqrt{v_{S1}^2 - \frac{R}{m_S + m_B + \frac{m_R}{3} + m_{CC}} \left((s_1 + x_{u2})^2 - (s_1 + x_\ell)^2 \right)} \qquad \text{A19.64}$$

$$v_{S2} = \sqrt{v_{\alpha x_{u2}}^2 - \frac{R}{m_S + \frac{m_R}{3} + m_{CC}} \left((s_1 + x_d)^2 - (s_1 + x_{u2})^2 \right)} \qquad \text{A19.65}$$

where: m_B Mass, barrel [kg]
          m_{CC} Mass, cartridge case [kg]
          m_P Mass, projectile [kg]
          m_R Mass, recoil spring [kg]
          m_S Mass, slide [kg]
          R Spring rate, recoil spring [N/m]
          s_1 Deflection [m]; preload of the recoil spring
          v_0 Muzzle velocity [m/s]
          v_{S1} Velocity, slide [m/s], projectile leaves the muzzle
          v_{S2} Velocity, slide [m/s] at backstop
          $v_{\alpha x_{u2}}$ Instantaneous velocity [m/s], at the end of uncoupling
          x_d Displacement, total slide recoil travel [m]
          x_{u2} Displacement, at the end of uncoupling [m]
          x_ℓ Displacement, projectile's base leaves muzzle [m],
                   cf. A19.60, p. 433

Further, v_{S1} follows from A19.63, and the mean slide recoil velocity $\bar{v}_{St}$ yields from Equation A19.89, p. 442.

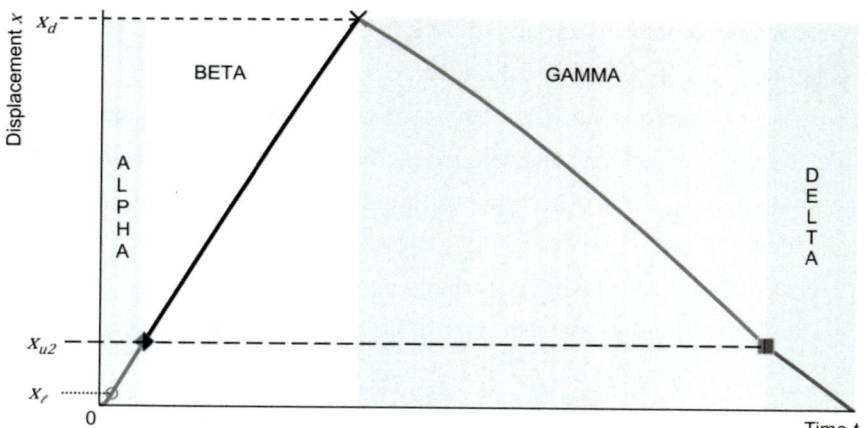

Fig. A19.7: Sections α, β, γ and δ in a displacement vs. time diagram.

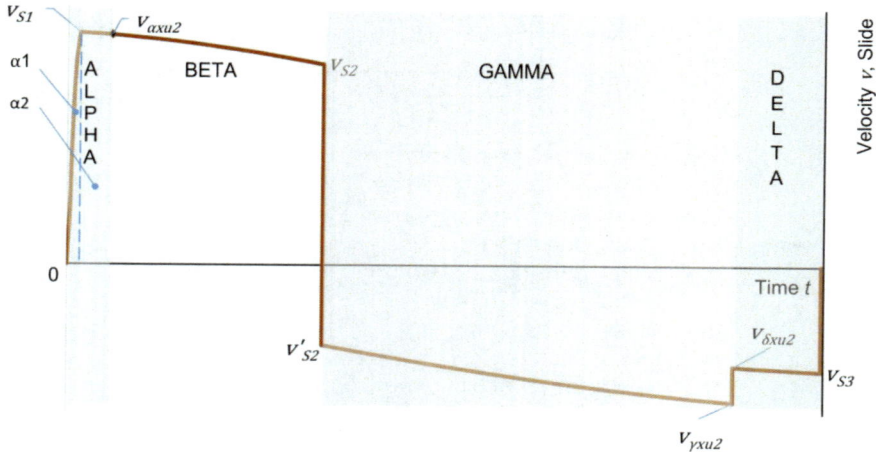

Fig. A19.8: Characteristic sections and velocities in a velocity vs. time diagram.

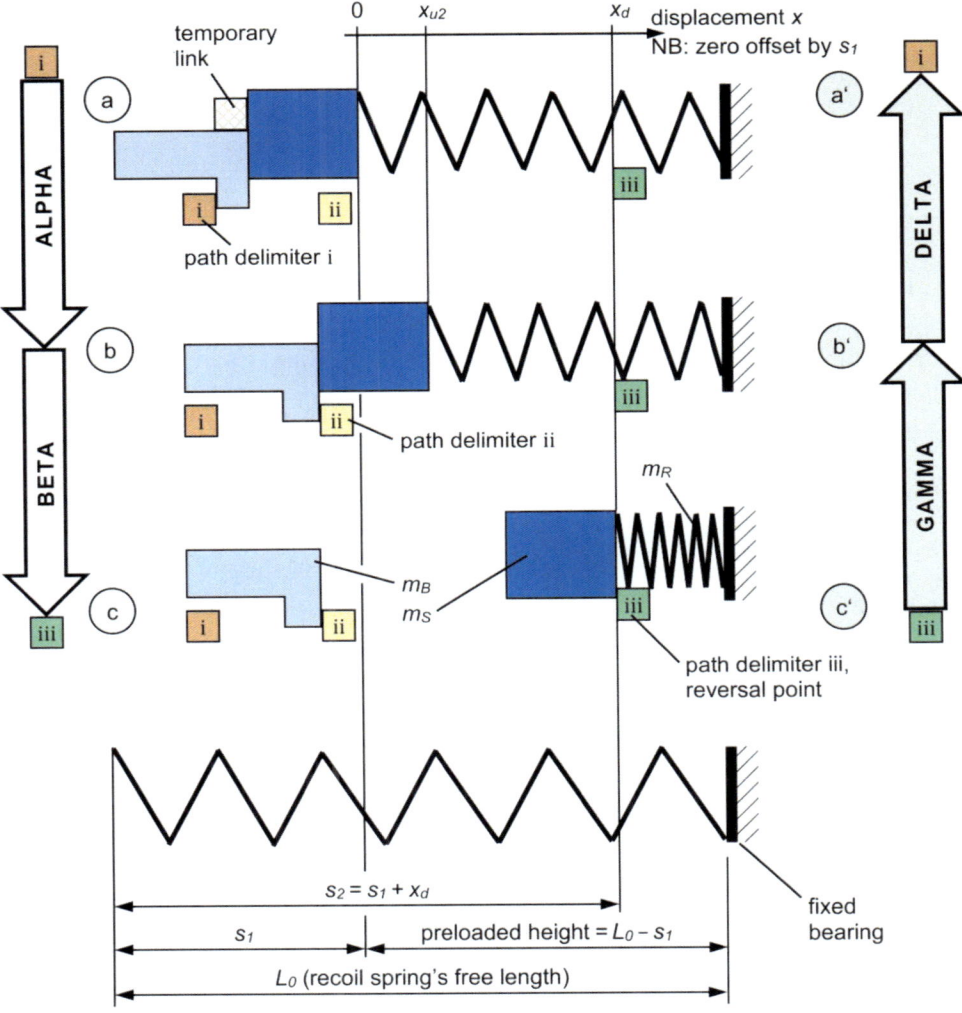

Fig. A19.9: Schematic diagram of the firing cycle using the example of a pistol with coupled barrel and slide.

The illustration abstracts the layout of the coupling and does not show the cartridge and its components or any technical details of the gun.

The cycle is divided into the sections α, β, γ and δ. Their boundaries are defined by: Rest position (a), uncoupling end (b) and slide at the reversal point (c). This corresponds to the sequence during recoil, identified as the sections α and β. Then, from the beginning (c') of the counter-recoil travel, the coupling (b') and the reaching of the rest position (a') are the limits of the sections γ and δ.

Additionally, the deflection of the recoil spring is illustrated at the bottom of the schematic diagram.

Since some parameters, such as the angular frequency, may change at the transitions between the sections, the four parts α, β, γ and δ are analyzed separately. In the course of the calculation, the duration of movement in each of the sections starts with $t = 0$. The displacement x is defined as the instantaneous distance of the slide from the rest position at the front stop. Starting from this zero position, x is counted positively. Consequently, consideration of the amount of a value x does not allow any conclusion to be drawn as to whether the slide is recoiling or closing. For example, a displacement of $x = 0.02$ m merely indicates that the recoiling mass is 0.02 m away from the front stop.

If the slide velocity is plotted against time, a curve shown in Fig. A19.8 is obtained. For this diagram, the increase in slide velocity until v_{S1} is reached was calculated using the Heydenreich/Oerlikon method according to Appendix 23.[719]

The following velocities are relevant here: Initial velocity v_{S1}, instantaneous velocity $v_{\alpha x u2}$ at completion of uncoupling and v_{S2} at the backstop and after reversal of direction (v'_{S2}) respectively, the velocity at the instant of coupling $v_{\gamma x u2}$ and after start of coupling $(v_{\delta x u2})$ respectively as well as the ultimate velocity v_{S3} when the slide reaches the front stop. Fig. A19.8 shows the scale for the velocity on the right ordinate. This was done in preparation of the combination of two diagrams for $x(t)$ and $v(t)$, and using their axis of ordinates on the left and right respectively.

α-Section, $0 \leq x \leq x_{u2}$

The analysis begins with the slide closed. Section alpha extends from the start of the recoil movement to the completion of the uncoupling at x_{u2}. For a detailed observation we divide the section alpha into a fist subsection $\alpha1$, in which the recoil mass is propelled up to the velocity v_{S1} as the projectile runs down the bore, and into a second subsection $\alpha2$. In this one, the velocity of the recoil mass decreases from its maximum at v_{S1}, since the impulsive force is no longer effective after the projectile has left the barrel. Please note, we neglect any accelerating effect of the forward thrust of gas and particles ahead of the muzzle in our examination.

The movement of the projectile starts as soon as the pressure in the case is sufficient to start the projectile from its seat. For the start of the projectile's movement, a gas pressure is generally assumed to be 10 percent of the peak pressure (see Endnote 154). The following applies to the first subsection in alpha: the time $t_{\alpha1} = 0$ is defined as the instant at which the projectile starts to move. When the movement of the projectile begins, the recoil mass is propelled at the same time. In our model, the acceleration ends abruptly when the projectile leaves the muzzle. Functional values for $t_{\alpha 1}$ satisfying $0 < t_{\alpha 1} < t_\ell$ as well as a detailed calculation of the conditions in this time interval prove to be irrelevant for our analysis. Consequently, the course of displacement and velocity in this area is indicated in subsequent diagrams by a linear increase.

First, the initial velocity v_{S1} of the recoiling mass is calculated for the subsection $\alpha 1$ according to Formula A19.63 and the displacement x_ℓ according to Equation A19.60, then the corresponding duration $t_{\alpha 1}$, which is the barrel time t_ℓ (see A23.12, p. 478). The results serve as an entry-point for the oscillation equation. If details of the displacement where x satisfies $0 \le x < x_\ell$ are of interest, see Annex 23.

When the base of the projectile leaves the muzzle, the recoil mass does not experience any further acceleration – characteristic of the time $t_{\alpha 2} = 0$. Due to the effect of the recoil spring, friction, etc., the recoil velocity decreases from this point in time forward. For x satisfying $x_\ell \le x \le x_{u2}$ we start counting the time $t_{\alpha 2}$ starting from zero.

The following parameters are important for the oscillation equation:

$$\omega_\alpha = \sqrt{\frac{R}{m_S + m_B + \frac{1}{3}m_R + m_{CC}}} \qquad\qquad \textbf{A19.66}$$

$$C_\alpha = \frac{v_{S1}}{\omega_\alpha} \qquad\qquad \textbf{A19.67}$$

$$\hat{x}_\alpha = \sqrt{(s_1 + x_\ell)^2 + C_\alpha^2} \qquad\qquad \textbf{A19.68}$$

$$\varphi_\alpha = arctan\left(\frac{C_\alpha}{s_1 + x_\ell}\right) \qquad\qquad \textbf{A19.69}$$

where: C_α Constant [m]
         ω_α Angular frequency [s^{-1}], in section $\alpha 2$
         φ_α Phase angle [rad], i.e. initial phase in section $\alpha 2$
         R Spring rate, recoil spring [N/m]
         s_1 Deflection [m]; preload of the recoil spring
         v_{S1} cf. A19.63, p. 435
         $\hat{x}_\alpha$ Amplitude in section $\alpha 2$ [m]
         x_ℓ Displacement [m], when base of projectile leaves the muzzle

Thus, we modify Equation A19.40, p. 426, and transform it:

$$x(t_{\alpha 2}) = \hat{x}_\alpha \, cos(\omega_\alpha t_{\alpha 2} - \varphi_\alpha) - s_1 \qquad\qquad \textbf{A19.70}$$

Resolved according to $t_{\alpha 2}$ and taking into account the correct phase, this gives the duration required to reach a position x:

$$t_{\alpha 2}(x) = \frac{1}{\omega_\alpha}\left(\varphi_\alpha - arccos\left(\frac{s_1 + x}{\hat{x}_\alpha}\right)\right) \qquad\qquad \textbf{A19.71}$$

where: $t_{\alpha 2}(x)$ Duration [s] to reach displacement x
         x Displacement, slide [m], where x always satisfies $x_\ell \le x \le x_{u2}$

Accordingly, the total duration of the coupled recoil travel $t_{\alpha tot}$ is:

$$t_{\alpha\,tot} = t_{\alpha 1} + t_{\alpha 2\,tot}$$

A19.72

or with A19.71 and where $x = x_{u2}$:

$$t_{\alpha\,tot} = t_\ell + \frac{1}{\omega_\alpha}\left(\varphi_\alpha - arccos\left(\tfrac{s_1 + x_{u2}}{\hat{x}_\alpha}\right)\right)$$

A19.73

where: $t_{\alpha tot}$ Duration [s], from start of recoil to end of uncoupling
         t_ℓ Barrel time [s] (cf. Formula A23.12, p. 478)
         x_{u2} Displacement at the end of uncoupling [m]

The instantaneous velocity of the moving masses at a time t is the first time derivative of $x(t_{\alpha 2})$, i.e. the derivative of A19.70. Hence,

$$v_{\alpha 2}(t_{\alpha 2}) = -\hat{x}_\alpha \omega_\alpha\, sin(\omega_\alpha t_{\alpha 2} - \varphi_\alpha)$$

A19.74

where: $v_{\alpha 2}(t_{\alpha 2})$ Instantaneous velocity [m/s] of the recoiling mass at a time t

Using the transformation mentioned at A19.43, an equation for v as a function of the displacement x can be derived, which describes $v_{\alpha 2}$ for all x that satisfy $x_\ell \leq x \leq x_{u2}$ independent of the time t:

$$v_{\alpha 2}(x) = \omega_\alpha\sqrt{C_\alpha^2 + (s_1 + x_\ell)^2 - (s_1 + x)^2}$$

A19.75

With C_α, ω_α and $x = x_{u2}$ the recoil velocity at position x_{u2} is calculated analogous to A19.75:

$$v_{\alpha x_{u2}} = \omega_\alpha\sqrt{C_\alpha^2 + (s_1 + x_\ell)^2 - (s_1 + x_{u2})^2}$$

A19.76

where: $v_{\alpha x_{u2}}$ Recoil velocity [m/s] at the end of uncoupling
         x_{u2} Displacement at the end of uncoupling [m]

β-Section, $x_{u2} \leq x \leq x_r$

The displacement x_{u2} forms the transition to the second stage of the cycle. The slide has separated from the barrel and sweeps over the beta section. Due to the inertia, the slide moves into the beta section at the velocity $v_{\alpha x_{u2}}$ as it separates from the barrel. The oscillation equation for this range can be derived by calculating the boundary values valid there. Starting from an oscillation equation of the shape:

$$x(t_\beta) = C_\beta\, sin(\omega_\beta t_\beta + \varphi_\beta) - s_1$$

A19.77

where:

$$\omega_\beta = \sqrt{\frac{R}{m_S + \frac{1}{3}m_R + m_{CC}}}$$

A19.78

and the first time derivative of $x(t_\beta)$ yields the velocity v in section beta:

$$v_\beta(t_\beta) = C_\beta \omega_\beta \, cos(\omega_\beta t_\beta + \varphi_\beta) \qquad\qquad \textbf{A19.79}$$

C_β and φ_β can be determined by looking at the initial conditions at the time $t_\beta = 0$. At the beginning of the beta section, the slide is at the position $x = x_{u2}$ and is moving at a velocity $v_{\alpha x_{e2}}$ at that moment. Equation A19.77 and A19.79 thus change to

$$x(0) = C_\beta sin \, \varphi_\beta - s_1 \equiv x_{e2} \qquad\qquad \textbf{A19.80}$$

$$v_\beta(0) = C_\beta \omega_\beta \, cos \, \varphi_\beta \equiv v_{\alpha x_{e2}} \qquad\qquad \textbf{A19.81}$$

This results in the following solutions for φ_β and C_β:

$$\varphi_\beta = arctan \left(\frac{s_1 + x_{u2}}{v_{\alpha x_{u2}}} \omega_\beta \right) \qquad\qquad \textbf{A19.82}$$

$$C_\beta = \frac{s_1 + x_{u2}}{sin \, \varphi_\beta} \qquad\qquad \textbf{A19.83}$$

where: C_β Constant [m]
     φ_β Phase angle [rad], i.e. initial phase in section β
     $v_{\alpha x_{u2}}$ Recoil velocity [m/s] at the end of uncoupling in section α
     ω_β Angular frequency [s^{-1}], in section β

Where x always satisfies $x_{u2} \le x \le x_R$. Solving equation A19.77 for t_β:

$$t_\beta(x) = \frac{1}{\omega_\beta} \left(arcsin \left(\frac{s_1 + x}{C_\beta} \right) - \varphi_\beta \right) \qquad\qquad \textbf{A19.84}$$

By forming, A19.79 can be converted to the instantaneous velocity as a function of the displacement x, which is independent of the time t_β:

$$v_\beta(x) = \omega_\beta \sqrt{C_\beta^2 - (s_1 + x)^2} \qquad\qquad \textbf{A19.85}$$

For $x = x_d$ one obtains the final slide velocity at the rear stop:

$$v_{S2} = \omega_\beta \sqrt{C_\beta^2 - (s_1 + x_d)^2} \qquad\qquad \textbf{A19.86}$$

The interval duration of the second section is analogous to A19.84:

$$t_{\beta \, tot} = \frac{1}{\omega_\beta} \left(arcsin \left(\frac{s_1 + x_d}{C_\beta} \right) - \varphi_\beta \right) \qquad\qquad \textbf{A19.87}$$

where: $t_{\beta \, tot}$ Duration [s], displacement in section β

The total duration of the slide recoil is calculated by summing the time intervals of the alpha and beta sections:

$$t_r = t_{\alpha\, tot} + t_{\beta\, tot} \qquad\qquad\qquad \text{A19.88}$$

where: t_r Duration [s], total duration of slide recoil travel

With this duration the theoretical mean slide recoil velocity can be calculated:

$$\overline{v}_{St} = \frac{x_d}{t_r} \qquad\qquad\qquad\qquad \text{A19.89}$$

γ-Section, $x_{u2} \leq x \leq x_r$

On completion of the recoil stroke, the direction of movement of the slide is reversed. The reversing action occurs almost instantaneously. The slide will rebound from the backstop and is accelerated by the relaxing recoil spring exerting force on it. The velocity of the slide after impact will be:[720]

$$v'_{S2} = -k_S v_{S2} \qquad\qquad\qquad \text{A19.90}^{721}$$

where: v'_{S2} Slide velocity [m/s] after impact
            k_S Coefficient of restitution [1]; empirical values: $k_S \approx 0$ to 0.2

Note: As the slide moves in counter-recoil, i.e. in the gamma and delta sections, a negative sign applies to the slide velocity as defined above. In addition, the mass of a cartridge is taken into account in the calculations for these sections. Shortly after the reversal at the backstop, the cartridge is struck by the slide, stripped from the magazine and fed into the chamber.

The equation of motion is:

$$x(t_\gamma) = C_\gamma\, sin(\omega_\gamma t_\gamma + \varphi_\gamma) - s_1 \qquad\qquad \text{A19.91}$$

As far as the coefficients are concerned:

$$\omega_\gamma = \sqrt{\frac{R}{m'_S + \frac{1}{3}m_R}} \qquad\qquad\qquad \text{A19.92}$$

$$\varphi_\gamma = arctan\left(\frac{s_1 + x_d}{v'_{S2}}\, \omega_\gamma\right) \qquad\qquad \text{A19.93}$$

where: m'_S Mass of slide plus one cartridge [kg]
            $m'_S = m_S + m_{Cartridge}$

Note: For $v'_{S2} = 0$ applies $\varphi_\gamma = \frac{-\pi}{2}$. Application examples: If $k_S = 0$ is chosen when calculating the velocity at the turning point or when the slide is closed manually by releasing it uncontrolled by pressing down on the slide stop lever.

The constant C_γ has the unit meter and is calculated as follows:

$$C_\gamma = \frac{s_1 + x_d}{sin\, \varphi_\gamma} \qquad\qquad\qquad \text{A19.94}$$

The equation A19.91 is resolved to t, where x always satisfies $x_{u2} \le x \le x_R$.

$$t_\gamma(x) = \frac{1}{\omega_\gamma}\left(arcsin\left(\frac{s_1+x}{C_\gamma}\right) - \varphi_\gamma\right)$$

A19.95

From this follows the duration of the process in section gamma:

$$t_{\gamma\,tot} = \frac{1}{\omega_\gamma}\left(arcsin\left(\frac{s_1+x_{u2}}{C_\gamma}\right) - \varphi_\gamma\right)$$

A19.96

The first time derivative of A19.91 yields the instantaneous velocity v_γ. Hence,

$$v_\gamma(t_\gamma) = C_\gamma\omega_\gamma\,cos(\omega_\gamma t_\gamma + \varphi_\gamma)$$

A19.97

By forming, the instantaneous velocity can be represented as a function of the displacement x (x always satisfies $x_{u2} \le x \le x_R$), independent of t:

$$v_\gamma(x) = -\omega_\gamma\sqrt{C_\gamma^2 - (s_1+x)^2}$$

A19.98

At the point $x = x_{u2}$ the slide engages the barrel and the instantaneous velocity of the slide is

$$v_{\gamma x_{u2}} = -\omega_\gamma\sqrt{C_\gamma^2 - (s_1+x_{u2})^2}$$

A19.99

δ-Section, $0 \le x \le x_{u2}$

The barrel is in a state of rest. At the transition from gamma to delta, the slide transmits an impact to the barrel. We take m_S into account at this point and add the mass of the cartridge, which is now chambered, to the mass of the barrel. The three masses are united and continue the rest of the counter-recoil motion united. The velocity of the coupled group after the collision is less than $v_{\gamma x u2}$. It is calculated according to the principle of linear momentum:[722]

$$v_{\delta x_{u2}} = v_{\gamma x_{u2}}\frac{m_S+\frac{1}{3}m_R}{m_r'+\frac{1}{3}m_R}$$

A19.100

where: $v_{\delta x_{u2}}$ Velocity [m/s] of the coupled barrel and slide, instantaneous velocity after unification
         m_r' Recoiling mass [kg], mass of barrel and slide plus one cartridge; $m_r' = m_B + m_S + m_{Cartridge}$ [kg]
         $v_{\gamma x_{u2}}$ Slide velocity [m/s], when impacting the barrel

We get for section delta:

$$x(t_\delta) = C_\delta\,sin(\omega_\delta t_\delta + \varphi_\delta) - s_1$$

A19.101

where

$$\omega_\delta = \sqrt{\frac{R}{m'_r + \frac{1}{3}m_R}} \qquad\qquad \textbf{A19.102}$$

$$\varphi_\delta = arctan\left(\frac{s_1 + x_{u2}}{v_{\delta x_{u2}}}\omega_\delta\right) \qquad\qquad \textbf{A19.103}$$

$$C_\delta = \frac{s_1 + x_{u2}}{sin\,\varphi_\delta} \qquad\qquad \textbf{A19.104}$$

Where x always satisfies $0 \le x \le x_{u2}$. Equation A19.101 resolved to t results:

$$t_\delta(x) = \frac{1}{\omega_\delta}\left(arcsin\left(\frac{s_1+x}{C_\delta}\right) - \varphi_\delta\right) \qquad\qquad \textbf{A19.105}$$

and for the total duration in the delta segment:

$$t_{\delta\,tot} = \frac{1}{\omega_\delta}\left(arcsin\left(\frac{s_1}{C_\delta}\right) - \varphi_\delta\right) \qquad\qquad \textbf{A19.106}$$

From the first time derivative of A19.101 we get the instantaneous velocity:

$$v_\delta(t_\delta) = C_\delta\omega_\delta\,cos(\omega_\delta t_\delta + \varphi_\delta) \qquad\qquad \textbf{A19.107}$$

By forming, the instantaneous velocity can be represented as a function of the displacement, where x always satisfies $0 \le x \le x_{u2}$, independent of time:

$$v_\delta(x) = -\omega_\delta\sqrt{C_\delta^2 - (s_1 + x)^2} \qquad\qquad \textbf{A19.108}$$

At the end of the counter-recoil x satisfies $x = 0$ or with respect to the ultimate speed:

$$v_{S3} = -\omega_\delta\sqrt{C_\delta^2 - s_1^2} \qquad\qquad \textbf{A19.109}$$

Thus, we know the total duration of slide recoil travel t_r, and we can also determine the initial, final, average, and ultimate velocity which occur during the recoil of pistols with coupled barrel and slide.

As explained at the beginning, this is a theoretical approach, which is limited to the essential and is based on highly idealized assumptions.

Rate of fire, pistols with coupled barrel and slide

The cycle time is the sum of the time intervals of all four sections plus lock time and ignition time. The reciprocal value of the cycle time gives the firing rate in rounds per second (cf. Rate of fire, blowback, p. 431).

$$K_t = \frac{1}{t_{\alpha\,tot} + t_{\beta\,tot} + t_{\gamma\,tot} + t_{\delta\,tot} + t_\sigma + t_\tau} \qquad\qquad \textbf{A19.110}[723]$$

where: K_t Rate of fire, theoretical [rounds/s]
             t_σ Lock time [s], ca. 0.003 to 0.01 s
             t_τ Ignition time [s], ca. 0.002 s

Example 2: Coupled Barrel and Slide, hammer-fired

Spreadsheet software is suitable for the application of the above equations. Using the example of a hammer-fired pistol with a short recoiling barrel, calculations were carried out as an illustrative example. A simulation with 50 grams of additional mass at the slide was carried out further, for example to analyze the influence of a removable micro red dot sight. These devices are normally attached to the slide. The values in Table A19.2 were used as the basis for the computations, and Table A19.3 summarizes the results.

Table A19.2: Data, short recoil operated hammer-fired pistol, 9 mm Luger.

Description		Value	Description		Value
Muzzle velocity	v_0 [m/s]	358	Barrel length	L_B [m]	0.1015
Mass, projectile	m_P [kg]	0.008	Spring rate, recoil spring	R [N/m]	720
Mass, slide	m_S [kg]	0.345	Preload, recoil spring	s_1 [m]	0.025
Mass, barrel	m_B [kg]	0.110	Travel, uncoupling, start	x_{u1} [m]	0.0031
Mass, recoil spring	m_R [kg]	0.006	Travel, uncoupling, end	x_{u2} [m]	0.0081
Mass, cartridge	$m_{Cartr.}$ [kg]	0.012	Total slide recoil travel	x_d [m]	0.0491
Mass, propellant	m_C [kg]	0.0004	Coefficient of restitution	k_S [1]	0.1
Mass, fired case	m_{CC} [kg]	0.0036	Mass, additional accessory	m_Δ [kg]	0.05

Table A19.3: Results of Example 2, Table A19.2, without additional accessory.

Miscellaneous

	Value	Reference
x_ℓ [mm]	1.5	(A19.60)
t_ℓ [ms]	0.4	(A23.12)

	Value	Ref.
ζ_{SB}	1:3.1	(A13.5)
$\Phi_{\ell u1}$	1:2.0	(A19.62)

Examination of Recoil Motion

	Equation of Motion						Energy Theorem (rough calculation)	
	Section				Total			
	Alpha		Beta					
	Value	Reference	Value	Reference	Value	Ref.	Value	Ref.
v_{S1} [m/s]	6.2	(A19.63)					6.3	(A19.7)
m_r [kg]	0.461	$m_S + m_B + m_R/3 + m_{CC}$	0.351	$m_S + m_R/3 + m_{CC}$			0.455	$m_S + m_B$
ω [s^{-1}]	39.5	(A19.66)	45.3	(A19.78)				
φ	1.40	(A19.69)	0.24	(A19.82)				
C [m]	0.157	(A19.67)	0.140	(A19.83)				
v [m/s]	6.2	(A19.76)	5.4	(A19.86)			4.4	(A19.8)
$\bar{v}_{St}$ [m/s]					5.8	(A19.89)	5.4	(A19.9)
							$W_{fl} = 3.5$ J	
t [s]	0.0015	(A19.73)	0.0070	(A19.87)	0.0085	(A19.88)		
	Section							
	Gamma		Delta					
	Value	Reference	Value	Reference				
v_S [m/s]	−0.5	(A19.90)	−2.2	(A19.100)				
m [kg]	0.359	$m'_S + m_R/3$	0.469	$m'_r + m_R/3$				
ω [s^{-1}]	44.8	(A19.92)	39.2	(A19.102)				
φ	−1.41	(A19.93)	−0.53	(A19.103)				
C [m]	−0.075	(A19.94)	−0.066	(A19.104)				
v [m/s]	−3.0	(A19.99)	−2.4	(A19.109)				
t [s]	0.0213	(A19.96)	0.0035	(A19.106)	0.0248	$t_\gamma + t_\delta$		

Evaluation of the results of Example 2

In order to obtain certainty about the validity of the mathematical analysis, $\bar{v}_{Ve}$ was determined by evaluating a high-speed video. The empirically determined $\bar{v}_{Ve}$ with a 9 mm Luger, $m_P = 8$ g, was 5.4 ±0.6 m/s. Using the oscillation equations, a theoretical mean recoil velocity $\bar{v}_{St}$ of 5.8 m/s was obtained. The value from the rough calculation according to A19.9 was 5.4 m/s. The result based on the oscillation equations was about 5 percent higher than the value determined by high-speed recording. This deviation is not surprising. As explained at the beginning, computations using oscillation equations neglect the influence of friction and other mechanical influences and therefore provide a higher value for the velocity.

The application of the equations for $x(t)$ and $v(t)$ allows a visualization as a displacement-time or velocity-time diagram. Fig. A19.10, p. 448, shows the characteristic graphs of the slide motion as a function of time. Red lines represent the velocity of the slide, black lines the displacement. The diagram has two ordinates. The left one represents the displacement of the recoil mass. The right one serves as a scale for the velocity of the slide. Solid lines apply to the pistol in its original condition, dashed curves refer to the oscillating body with a mass increased by 50 grams, for example a reflex sight.

The time t is plotted horizontally. As the diagram illustrates, the graph of the displacement x increases continuously from the axis origin until $x = x_d$. When the rear stop is reached, the slide motion reverses, i.e. the characteristic curve drops until the oscillator reaches the initial position at $x = 0$. A slight bend in the line occurs above the abscissa at $t = 0.029$ and 0.032 s respectively. This bend is characteristic for the beginning of the mating of barrel and slide at $x = x_{u2}$.

With regard to the velocity, the following can be said: At time $t = 0$ the slide is in the rest position. With the beginning of the projectile movement, the velocity of the recoiling mass increases to the value v_{S1} within approx. 0.4 ms. This almost instantaneous increase ends when the projectile base leaves the muzzle. At this moment, during theoretical analysis and practical measurements, the velocity of the slide shows the highest value in the entire cycle. Starting from v_{S1}, the velocity decreases until the rear stop is reached. The slide rebounds against this stop. The closing process in this example therefore does not start from a rest position, but at a speed not equal to zero. At the turning point, the velocity suddenly takes on a negative value, since the slide now moves in the opposite direction.

Driven by the recoil spring, the amount of the slide velocity increases. This continuous increase is interrupted at the instance when the slide encounters the barrel and transmits an impact to the barrel. The speed of the oscillator is abruptly reduced. The point can be identified by the short vertical jump in the velocity graph between $t = 0.029$ and 0.032 s.

Three further points are marked in the diagram: A plus symbol marks the displacement x_ℓ of the recoiling slide when the projectile leaves the muzzle. This takes place before the uncoupling begins. Start and end of uncoupling are marked by circles.

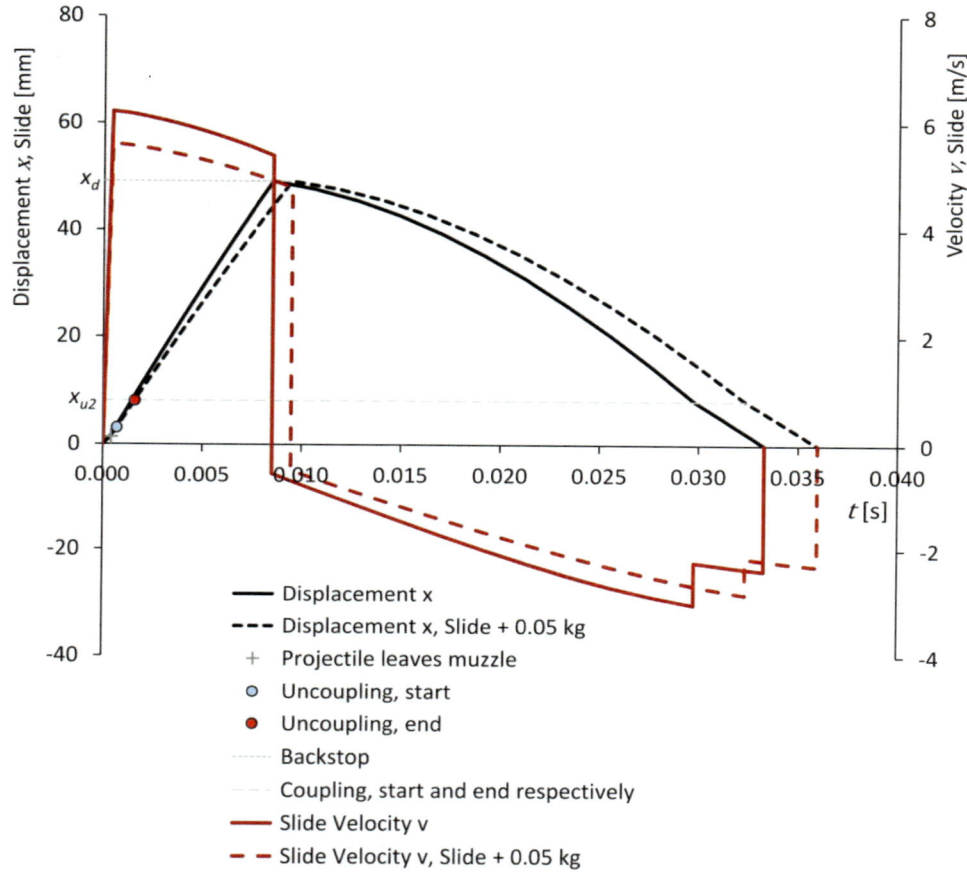

Fig. A19.10: $x(t)$ and $v(t)$ of a hammer-fired pistol, 9 mm Luger, with and without optional slide-mounted accessory.

The diagram shows the influence of additional 50 grams slide-mounted mass on the slide velocity as well as the time required to complete the cycle. The duration of the cycle increases from 0.033 to 0.036 seconds and the ultimate velocity at the end of the cycle decreases from −2.4 to −2.3 m/s as a result of slide-mounted components.

The computation showed that with an identical recoil spring in combination with 50 grams added mass, the theoretical mean recoil velocity is reduced by 0.6 m/s from 5.8 to 5.2 m/s.

A discussion of the equations for the recoil velocities reveals which variables theoretically influence the cycle. The extent of their influence can be easily determined by combining the formulae in a spreadsheet software, for example. The simulation based on a hammer-fired pistol described under Example 2 showed the following effect on the theoretical mean recoil velocity with a ten percent increase of one respective parameter in each case:

- +10 % v_0 Muzzle velocity $\Rightarrow$ +10 % v_{S1},
- ditto $\Rightarrow$ +13 % v_{S2},
- ditto $\Rightarrow$ +11 % $\bar{v}_{St}$,
- +10 % m_P Mass, projectile $\Rightarrow$ +11 % $\bar{v}_{St}$,
- +10 % m_S Mass, slide $\Rightarrow$ −7 % $\bar{v}_{St}$,
- +10 % m_B Mass, barrel $\Rightarrow$ −3 % $\bar{v}_{St}$,
- +10 % R   Spring rate, recoil spring $\Rightarrow$ −0.6 % $\bar{v}_{St}$,
- +10 % s_1 Preload, recoil spring $\Rightarrow$ ca. −0.3 % $\bar{v}_{St}$,
- +10 % x_{u2} Displacement, end of uncoupling $\Rightarrow$ ca. +0.03 % $\bar{v}_{St}$, and
- +10 % x_d Total slide recoil travel $\Rightarrow$ −0.8 % $\bar{v}_{St}$.

The latter is technically limited in order to prevent, among other things, to fully compress the recoil spring to the point where it reaches its solid height. Remarkably, the angular frequency ω of the oscillation process is not influenced by the recoil spring's preload s_1 (see A19.66, p. 439).

The diagram in Fig. A19.10 was based on a hammer-fired pistol and the calculations were performed without taking into account the work needed to force the hammer down into cocked position. The question now literally arises as to how strongly the cocking affects the result, or whether the effect can be neglected. In the following we will look at this detail using the example of a frequently encountered layout: a striker-fired pistol.

Effect of cocking the firing mechanism on striker-fired pistols

The vast majority of striker-fired 9 mm pistols cock the firing mechanism during the last millimeters of the forward stroke. From a physical point of view, the experimental setup is referred to under *horizontal two-spring-mass system*. This arrangement is composed of a single mass attached to two different springs on either side of the mass in a way that the oscillating body is captured between said springs, e.g. recoil spring and main-spring (Fig. A19.11). The mass can slide without friction along a horizontal surface about its equilibrium position. The displacement as a function of time is described and the displacement-time diagram shown below (Fig. A19.12, p. 453).

The tensioning of the striker spring is worked by the recoil spring but it must be strong enough to reliably overcome the force of the mainspring. The latter process is known to reduce the slide's velocity, as the reaction provided by the relaxing recoil spring is used for both, compressing the mainspring and closing.

Regardless of the scope of data already obtained for the analysis, additional infor-mation regarding the mainspring is required. For example, it is necessary to do a test to determine at which point of the counter-recoil

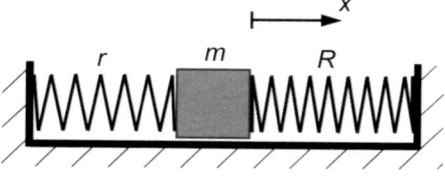

Fig. A19.11: Two-spring-mass system.

motion the cocking of the firing mechanism starts. In practice, this can be determined by assembling the unloaded pistol without the recoil spring and simulating the firing cycle. To do so, point the muzzle downwards, force the slide forward into the battery. Depress the trigger and hold it to the rear until the test is completed: Draw the slide back. Coming from the backstop, guide the slide forward until increased resistance is felt at the point where the striker gets cocked. Now compare this displacement x_p vs. the start of the coupling of barrel and slide at x_{u2}. You may release the trigger now. In the example under consideration we assume that $x_p > x_{u2}$, i.e. the mainspring is getting compressed before the coupling starts, i.e. in section gamma. The equations listed above remain effective in the α and β section, but in γ only for values of x satisfying $x_p < x \leq x_d$. For $0 \leq x \leq x_p$, ergo from the beginning of the cocking process until the front stop is reached, the influence of the mainspring's restoring force has to be taken into account. Formula symbols for this range are marked with an asterisk (*) to distinguish them from previously used notations.

In order for the cocking action to take place, the striker must be retained in position while the slide continues to travel forward. When the cocking process starts, in most of the common designs the mass of the striker is no longer a component of the slide but a part of the frame. We take this into account in the calculation by subtracting the striker mass m_T from the slide mass m_S.

As with the deduction of the oscillation equation at the beginning with reference to the blowback action, the mathematical approach still assumes that the sum of all forces is always zero (cf. A19.12, p. 422). Here, the inertial force and the forces of the recoil spring and the mainspring are effective, with the spring forces acting in opposite directions to each other:

$$F_{inertia} + F_{recoil\ spring} - F_{mainspring} = 0 \qquad \text{A19.111}$$

consequently

$$\left(m_S'^* + \frac{m_R + m_K}{3}\right) a + R(s_1 + x) - r(d_1 + x_p - x) = 0 \qquad \text{A19.112}$$

where: d_1 Preload, mainspring [m]
m_K Mass, mainspring
$m_S'^*$ Mass, slide minus mass of striker plus the weight of one cartridge [kg]; $m_S'^* = m_S - m_T + m_{cartridge}$
r Spring rate [N/m], mainspring
x_p Displacement [m]; start of cocking action

After transformation a second-order non-homogeneous differential equation is obtained:

$$\ddot{x} + \frac{R+r}{m_S'^* + \frac{m_R + m_K}{3}} x = \frac{1}{m_S'^* + \frac{m_R + m_K}{3}}\left(r(d_1 + x_p) - R\,s_1\right) \qquad \text{A19.113}$$

The general solution of this equation consists of the sum of all homogeneous and specific solutions. The homogeneous ones are the same as those derived previously under A19.22, since for their determination the right side of the equation A19.113 is set to zero. For the typical ones we take x as constant A_3 and form their derivatives:

$$x = A_3 \quad \dot{x} = 0 \quad \ddot{x} = 0 \qquad \text{A19.114}$$

We use the approach for the specific solution in A19.113 and resolve to A_3:

$$A_3 = \frac{r(d_1 + x_p) - R\,s_1}{R+r} \qquad \text{A19.115}$$

With the solutions determined, the equation of motion is obtained:

$$x_\gamma^*(t) = A_1 cos(\omega_\gamma^* t) + A_2 sin(\omega_\gamma^* t) + A_3 \qquad \text{A19.116}$$

The integration constants A_1 and A_2 are obtained from the boundary conditions at the beginning of the cocking action. In addition, the respective equations for velocity, time etc. are derived by appropriate forming and for gamma and delta separately. Table A19.6 summarizes the formulas, Table A19.7 the equation symbols (p. 454f).

First, the speed of the slide at the beginning of the cocking action is calculated (v_{xp}). If the cocking starts before the barrel is docked to the slide, the left-hand side of Table A19.6 is used to begin with. Section delta is then entered using Formula A19.117.

If the cocking process starts after or at the beginning of the coupling, i.e. $x_p \leq x_{u2}$, we calculate $v^*_{\delta x_p}$ using A19.108, p. 444, by replacing x by x_p. The result is used to continue the calculation following Table A19.6 in the column for delta.

Example 3: Coupled Barrel and Slide, striker-fired

Table A19.4: Data, short recoil operated striker-fired pistol, 9 mm Luger.

Description		Value	Description		Value
Muzzle velocity	v_0 [m/s]	335	Preload, mainspring	d_1 [m]	0.040
Mass, projectile	m_P [kg]	0.008	Mass, striker	m_T [kg]	0.008
Mass, slide	m_S [kg]	0.354	Mass, mainspring	m_K [kg]	0.002
Mass, barrel	m_B [kg]	0.093	Travel, uncoupling, end	x_{u2} [m]	0.0074
Mass, recoil spring	m_R [kg]	0.010	Total slide recoil travel	x_d [m]	0.046
Spring rate, recoil spring	R [N/m]	430	Coefficient of restitution	k_S [1]	0.2
Preload, recoil spring	s_1 [m]	0.155	Mass, additional accessory	m_Δ [kg]	0.05
Barrel length	L_B [m]	0.1015	Mass, cartridge	$m_{Cartr.}$ [kg]	0.012
Spring rate, mainspring	r [N/m]	600	Mass, propellant	m_C [kg]	0.0004
Cocking action, start	x_p [m]	0.010	Mass, fired case	m_{CC} [kg]	0.0036

Data shown above were used to calculate the ultimate velocity. Table A19.5 compares the results from three scenarios: without cocking, with cocking and the latter with additional mass mounted on the slide. The $v_{\gamma x_p}$ given in Table A19.5 is the velocity of the slide at the start of the cocking action. The figures in percent refer to the difference in comparison to a simulation that does not take into account the cocking action.

Table A19.5: Comparison: Effect of the cocking action on counter-recoil velocity.

	without cocking	with cocking	with cocking + 50 grams
$v_{\gamma x_p}$	−4.0 m/s / 100 %	−4.0 m/s / 100 %	−3.7 m/s / 93 %
$v_{\delta x_{u2}}$	−3.2 m/s / 100 %	−3.1 m/s / 97 %	−3.0 m/s / 94 %
v_{S3}	−3.5 m/s / 100 %	−3.3 m/s / 94 %	−3.2 m/s / 91 %

Fig. A19.12 shows the comparison of the displacement-time curves (black) and the velocity-time curves (red). Solid lines represent the approximate calculation without consideration of the cocking process. The dotted lines are for the same pistol, taking into account the cocking of the striker, and the dashed lines are for the same pistol with its slide mass increased by 50 grams.

The displacement-time and velocity-time graphs of the gun without and with consideration of the cocking action prove to be almost identical. When taking into account an additional 50-gram mass of the slide, a change relative to the basic configuration is evident. This variant is shown in dashed lines.

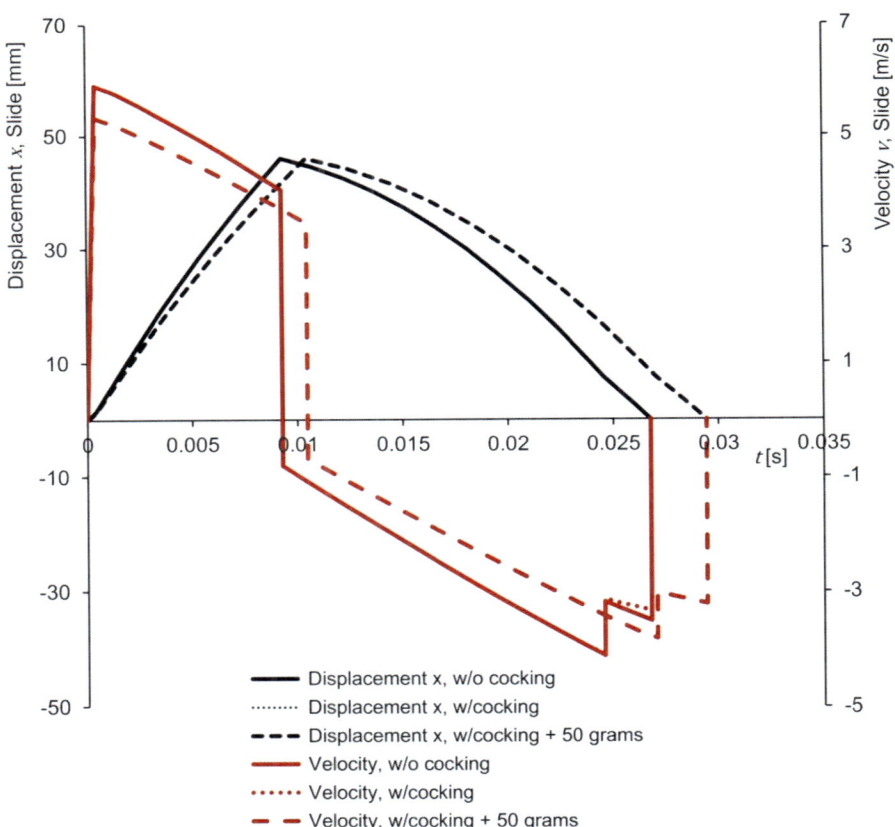

Fig. A19.12: Displacement $x(t)$ and slide velocity $v(t)$ of a striker-fired 9 mm pistol, with and without reflex sight and cocking action.

Table A19.6: Equations with regard to cocking, sections γ and δ.

Gamma Section, for $x_{u2} < x_p \leq x_d$; where x satisfies $x_{u2} < x \leq x_d$		Delta Section, where x satisfies: $0 \leq x \leq x_{u2}$	
$v_{\gamma x_p} = -\omega_\gamma \sqrt{C_\gamma^2 - (s_1 + x_p)^2}$	(A19.98)	$x_p > x_{u2}:\ v_{\delta x_p}^* = v_{\gamma x_{u2}}^* \dfrac{m_S - m_B + \frac{m_R}{3}}{m_R' - m_B + \frac{m_R}{3}}$ A19.117 $x_p \leq x_{u2}:\ v_{\delta x_p}^* = v_\delta(x_p)$ (A19.108)	
$\omega_\gamma^* = \sqrt{\dfrac{R+r}{m_S' - m_T + \frac{m_R + m_K}{3}}}$	A19.118	$\omega_\delta^* = \sqrt{\dfrac{R+r}{m_r' - m_T + \frac{m_R + m_K}{3}}}$	A19.119
$A_{1\gamma} = x_p - A_{3\gamma}$	A19.120	$A_{1\delta} = x_p - A_{3\delta}$ $x_p > x_{u2}:\ x_p := x_{u2}$ $x_p \leq x_{u2}:\ x_p := x_p$	A19.121
$A_{2\gamma} = \dfrac{v_{\gamma x_p}}{\omega_\gamma^*}$	A19.122	$A_{2\delta} = \dfrac{v_{\delta x_p}^*}{\omega_\delta^*}$	A19.123
$A_{3\gamma} = \dfrac{r(d_1 + x_p) - R\,s_1}{R + r}$	A19.124	$A_{3\delta} = \dfrac{r(d_1 + x_p) - R\,s_1}{R + r}$	A19.125
$\hat{x}_\gamma^* = \sqrt{A_{1\gamma}^2 + A_{2\gamma}^2}$	A19.126	$\hat{x}_\delta^* = \sqrt{A_{1\delta}^2 + A_{2\delta}^2}$	A19.127
$\varphi_\gamma^* = \arctan\left(\dfrac{A_{2\gamma}}{A_{1\gamma}}\right)$	A19.128	$\varphi_\delta^* = \arctan\left(\dfrac{A_{2\delta}}{A_{1\delta}}\right)$	A19.129
$x_\gamma^*(t_\gamma^*) = \hat{x}_\gamma^* \cos(\omega_\gamma^* t_\gamma^* - \varphi_\gamma^*) + A_{3\gamma}$	A19.130	$x_\delta^*(t_\delta^*) = \hat{x}_\delta^* \cos(\omega_\delta^* t_\delta^* - \varphi_\delta^*) + A_{3\delta}$	A19.131
$t_\gamma^*(x_\gamma^*) = \dfrac{\arccos\left(\frac{x_\gamma^* - A_{3\gamma}}{\hat{x}_\gamma^*}\right) + \varphi_\gamma^*}{\omega_\lambda^*}$	A19.132	$t_\delta^*(x_\delta^*) = \dfrac{\arccos\left(\frac{x_\delta^* - A_{3\delta}}{\hat{x}_\delta^*}\right) + \varphi_\delta^*}{\omega_\delta^*}$	A19.133
$v_\gamma^*(t_\gamma^*) = -\hat{x}_\gamma^* \omega_\gamma^* \sin(\omega_\gamma^* t_\gamma^* - \varphi_\gamma^*)$	A19.134	$v_\delta^*(t_\delta^*) = -\hat{x}_\delta^* \omega_\delta^* \sin(\omega_\delta^* t_\delta^* - \varphi_\delta^*)$	A19.135
$v_{\gamma x_{u2}}^* = -\omega_\gamma^* \sqrt{A_{1\gamma}^2 + A_{2\gamma}^2 - (x_{u2} - A_{3\gamma})^2}$	A19.136	$v_{S3}^* = -\omega_\delta^* \sqrt{A_{1\delta}^2 + A_{2\delta}^2 - A_{3\delta}^2}$	A19.137

Table A19.7: Formula symbols and units for the equations in Table A19.6, listed separately for the respective section.

Gamma Section		Delta Section	
a	Acceleration [m/s²]		
$A_{1\gamma}$	Constant [m]	$A_{1\delta}$	Constant [m]
$A_{2\gamma}$	Constant [m]	$A_{2\delta}$	Constant [m]
$A_{3\gamma}$	Constant [m]	$A_{3\delta}$	Constant [m]
C_γ	cf. A19.94, p. 442 [m]		
d_1	Deflection, striker spring preset [m]	d_1	Deflection, striker spring preset [m]
F_{res}	Net force [N]		
m_B	Mass, striker [kg]	m_B	Mass, striker [kg]
m_R	Mass, recoil spring [kg]	m_R	Mass, recoil spring [kg]
m_K	Mass, mainspring [kg]	m_K	Mass, mainspring [kg]
m'_S	Mass, slide [kg], incl. one cartridge, cf. A19.93, p. 442	m_S	Mass, slide [kg]
		m'_r	Mass, recoiling [kg], incl. cartridge, cf. A19.100, p. 443
ω_γ	cf. A19.92, p. 442 [s⁻¹]		
ω_γ^*	Angular frequency, during cocking in section γ [s⁻¹]	ω_δ^*	Angular frequency, during cocking in section δ [s⁻¹]
φ_γ^*	Phase angle, section γ, at start of cocking	φ_δ^*	Phase angle, section δ, at start of cocking
r	Spring rate, mainspring [N/m]	r	Spring rate, mainspring [N/m]
R	Spring rate, recoil spring [N/m]	R	Spring rate, recoil spring [N/m]
s_1	Deflection, rec. spr. preset [m]	s_1	Deflection, rec. spr. preset [m]
t	Time [s]	t	Time [s]
t_γ^*	Time [s]	t_δ^*	Time [s]
v_γ^*	Velocity, slide [m/s]	v_δ^*	Velocity, slide [m/s]
$v_{\gamma x_{u2}}^*$	Velocity, slide, at transition from section γ to δ [m/s]	$v_{\delta x_{u2}}^*$	Velocity, coupled barrel and slide at coupling [m/s]
		$v_{\delta x_S}^*$	Velocity, at start of δ or at start of cocking [m/s]
		v_{S3}^*	Velocity, at front stop ($v_{\delta 0}^*$) [m/s]
$v_{\gamma x_S}$	Velocity, start cocking [m/s]		
x	Displacement [m]	x	Displacement [m]
x_{u2}	Displacement [m], start of coupling	x_{u2}	Displacement [m], start of coupling
x_γ^*	Displacement, $x_{u2} < x_\gamma^* \le x_p$ [m]	x_δ^*	Displacement, $0 \le x_\delta^* \le x_p$ [m]
$\hat{x}_\gamma^*$	Amplitude [m]	$\hat{x}_\delta^*$	Amplitude [m]
		$x_{\delta S}$	Displacement, start cocking, $x_p < x_{u2}$ (section δ) [m]
x_p	Displacement, start cocking [m]	x_p	Displacement, start cocking [m]
$\ddot{x}$	second time derivative of the displacement [m/s²]		

As can be seen from Table A19.5, the cocking reduces the velocity v_{S3} at the front stop by approx. 6 % or 9 % respectively. The reference base was the pistol without consideration of the cocking process and without additional payload.

Calculation of v_{S3} in cases where x_p satisfies $x_p > x_{u2}$

If the actual course of movement is not of interest and only the ultimate velocity v^*_{S3} is required, it is recommended to determine it using the principle of energy conservation. The following applies to any analysis where $x_p > x_{u2}$ is satisfied:

$$E_{S\ kin\ x_{u2}} = E_{S\ kin\ x_p} + W_{R\ comp} - W_K \qquad \textbf{A19.138}$$

where: $E_{S\ kin\ xu2}$ Kinetic energy [J], of the slide at start of coupling
$E_{S\ kin\ xp}$ Kinetic energy [J], of the slide at start of cocking
W_K Work, cocking [J],
            from x_p to x_{u2}
$W_{R\ comp}$ Work, compression [J], of the recoil spring
            from x_p to x_{u2}

which is

$$\frac{m'^*_S}{2} v^{*2}_{\gamma x_{u2}} = \frac{m'^*_S}{2} v^2_{\gamma x_p} + \frac{R}{2}\left((s_1 + x_p)^2 - (s_1 + x_{u2})^2\right) - \frac{r}{2}\left(d^2_{x_{u2}} - d^2_1\right) \qquad \textbf{A19.139}$$

Taking a negative sign into account, the following applies to the slide velocity $v^*_{\gamma xu2}$ at the beginning of the coupling:

$$v^*_{\gamma x_{u2}} = -\sqrt{v^2_{\gamma x_p} + \frac{1}{m'^*_S}\left(R\left((s_1 + x_p)^2 - (s_1 + x_{u2})^2\right) - r\left(d^2_{x_{u2}} - d^2_1\right)\right)} \qquad \textbf{A19.140}$$

where: $v^*_{\gamma x_{u2}}$ Instantaneous velocity [m/s], of the slide at start of coupling
            under consideration for cocking action
$d_{x_{u2}} = d_1 + x_p - x_{u2}$
d_1 Preset of mainspring [m]; d_1 represents the amount of shortening by which the mainspring is compressed with the striker in uncocked condition
$m'^*_S = m'_S - m_T + \frac{m_R + m_K}{3}$
m'_S Mass of slide plus one cartridge [kg]
            $m'_S = m_S + m_{cartridge}$
$v_{\gamma x_p}$ Velocity, slide [m/s], at start of cocking action; calculated by setting $x = x_p$ in A19.98, p. 443
x_p Cocking travel [m], displacement at start of cocking action during counter-recoil motion

With $v^*_{\gamma xu2}$ we identify the velocity of the slide when it hits the barrel. The barrel is in the rest position until this instant. At the transition from gamma to delta, the slide transmits an impact to the barrel. We take m_S into account at this point in the calculation and add the mass of the chambered cartridge to the mass of the barrel. The three masses m_S, m_B and $m_{cartridge}$ are united and continue together for the rest of the closing motion. The velocity $v^*_{\delta xu2}$ of the coupled group after the collision is less than v^*_{yxu2}. It is calculated according to the principle of linear momentum theorem. The mass of the mainspring is not listed individually in the associated equation. It is taken into account not only with a share of ⅓, but with 100 percent in the mass of the slide. This was done because in striker-fired pistols the mainspring usually is a component of the slide.

$$v^*_{\delta x u2} = v^*_{\gamma x u2} \frac{m_S - m_T + \frac{1}{3}m_R}{m'_r - m_T + \frac{1}{3}m_R}$$ A19.141

where: $v^*_{\delta x u2}$ Velocity [m/s] of coupled slide and barrel
right after start of coupling
m'_r Moving mass [kg], including one cartridge;
$m'_r = m_S + m_B + m_{cartridge}$
m_T Mass, striker [kg]
$v^*_{\gamma x u2}$ Velocity [m/s] of slide when impacting the barrel at start of coupling

As the coupled barrel and slide group approaches the front stop, the cocking action continues. The kinetic energy of the moving mass returning to its forward stop, $E_{S\ kin\ s1}$, is equal to the kinetic energy $E_{S\ kin\ xu2}$ which the moving mass initially had at displacement x_{u2}, plus the work $W_{R\ comp}$ performed by the recoil spring minus the work W_K performed for cocking:

$$E_{S\ kin\ s_1} = E_{S\ kin\ x_{u2}} + W_{R\ comp} - W_K$$ A19.142

which is

$$v^*_{S3} = -\sqrt{v^{*2}_{\delta x u2} + \frac{1}{m'_r - m_T + \frac{m_R + m_K}{3}}\left(R\big((s_1 + x_{u2})^2 - s_1^2\big) - r\left((d_1 + x_p)^2 - d^2_{x_{u2}}\right)\right)}$$ A19.143

where: v^*_{S3} Velocity [m/s] of slide at front stop
when taking the cocking action into account
m_K Mass, mainspring [kg]
$d_{x_{u2}} = d_1 + x_p - x_{u2}$

Calculation of v_{S3} in cases where x_p satisfies $x_p \leq x_{u2}$

The calculation is simplified for striker-fired pistol designs in which the cocking of the firing mechanism takes place exclusively in the delta section. In this case, where $x_p \leq x_{u2}$ is satisfied, the following applies:

$$v_{S3}^* = -\sqrt{v_{\delta x_p}^{*2} + \frac{1}{m_r' - m_T + \frac{m_R + m_K}{3}} \left(R\left((s_1 + x_p)^2 - s_1^2 \right) - r\left((d_1 + x_p)^2 - d_1^2 \right) \right)} \qquad \textbf{A19.144}$$

where: v_{S3}^* Velocity [m/s] of slide at front stop
                                  when taking the cocking action into account
                     m_r' Moving mass [kg], including one cartridge;
                                  $m_r' = m_S + m_B + m_{cartridge}$
                     $v_{\delta x_p}^*$ Velocity, slide [m/s], at start of cocking action; calculated by
                                  setting $x = x_p$ in A19.108, p. 444

Hindsight

For a rough estimate within the context of a feasibility study, the procedure using Formula A19.9, p. 416, is appropriate. Although this method only provides the value of the theoretical mean slide recoil velocity, this should be sufficient in practice. One reason for this is that meaningful proof of the expected durability must be provided by endurance tests later in in the design process of a new pistol model.

  For the sake of simplification, we had assumed in our considerations of the straight blowback principle that the recoil velocity would rise sharply in an infinitely short time from zero to velocity v_{S1}. In reality, the recoil velocity approaches its maximum within a certain time interval. Displacement x_ℓ and recoil velocity v_{S1} can be determined quickly and accurately via the principle of linear momentum. The calculation of the barrel time requires additional effort. Heydenreich's method is recommended for an analysis of the temporal course of displacement x satisfying $0 \leq x \leq x_\ell$. It allows a closer look at the recoil as a function of time or projectile bore travel. Starting from the instant when the base of the projectile leaves the muzzle, a description by means of oscillation equations permits a detailed discussion of the dynamic course of the firing cycle.

  In reality, we observe a backward movement of the frame during discharge. This recoil is the result of forces exerted by components subject to recoil, for example via the recoil spring to the frame or by the uncoupling process. It is characteristic of insufficient support of the frame during the shot. If the frame and the slide move together in the same direction, the velocity of the slide relative to the frame is reduced as a consequence. Influences and losses of this kind are not taken into account in the calculation. Bear in mind that a gun can show a weak ejection pattern despite theoretically satisfactory computation results.

  The slide velocity is a decisive criterion for the functional reliability of the feeding cycle. The faster the slide, the less time the magazine has to present the next cartridge for feeding. For a short moment, while the breech face is behind the magazine, the feedway is clear for the next cartridge. Within this short time window, the magazine spring must set the cartridge column in motion to move the top cartridge into the loading area and position it correctly in alignment until the slide strips it from the magazine. If the next cartridge is not in proper orientation within the feedway, failures to feed may occur. This is caused by the interaction of the slide velocity near the rear stop at x_d and the tension of the magazine spring. The recommendation to maintain a mean recoil velocity in the range of 6 to 8 m/s, which has been mentioned previously, is also considered to be promising for this task with regard to the avoidance of impairments of the functional sequence.

In applications where the recoil spring has to cock the striker during counter-recoil, this spring deserves special attention. As the slide returns to its forward position, the recoil spring is simultaneously needed for stripping a cartridge from the magazine under the tension of the magazine spring and feeding it into the chamber, mastering the coupling process and performing the work required to cock the striker. The resistance of the extractor claw, and possibly of safety devices, or the inertia of additional weight attached to the barrel could also be considerable. Besides all this, the recoil spring has to exert enough force on the slide to keep it reliably in battery. This must be done taking into account the effect of the spring forces of all springs involved and any conceivable application scenarios. For example, also in the case of a pistol with double action trigger when the trigger is almost completely pulled through, moments before firing with muzzle pointing up. Slide and barrel must not get forced away from their rest position in this scenario. As a practical value for the locking force, approximately 30 Newton can be taken as an empirical magnitude.

No less interesting: the solid length L_c of the recoil spring. The recoil motion of the slide must be limited by a defined stop, and in no case by the effective spring coils going solid. In mathematical terms, this is expressed as follows:

$$L_c < L_0 - (s_1 + x_d) \qquad\qquad\qquad \textbf{A19.145}$$

where: L_c Length [m], solid length of recoil spring
                 L_0 Length [m], free length of recoil spring
                  s_1 Deflection [m], preset of recoil spring
                  x_d Displacement [m], total slide recoil travel

Appendix 20   MSRP and Street Prices

The comparison of the manufacturer's suggested retail price (MSRP) with the street prices allows some conclusions on how much a consumer is willing to pay in reality for the products being offered. MSRP's were researched on the websites of the manufacturers. Actual prices, commonly called *street prices*, were obtained online on September 4, 2014. The following online dealers were surveyed:

- CheaperThanDirt.com
- ImpactGuns.com
- GanderMountain.com
- Grabagun.com
- BudsGunShop.com
- Firearms4u.com

At the time of this survey, no street price could be found for the BERETTA *Pico*. Apparently, the *Pico* wasn't available yet. Same applied for the TARA *TM-9* pistol, neither MSRP nor street price could be obtained.

The results are summarized in Table A20.1. It shows the street prices, their calculated average, as well as the minimum street price and MSRP.

Note: Purchasers need to have a Federal Firearms License (FFL) to purchase firearms and have them shipped directly in the United States, or they need to have the firearm delivered to an FFL holder such as a local U.S. shop.

When shipping to a shop the buyer will be required to undergo the usual NICS background check as if the firearm were purchased from said shop.

Table A20.1: MSRP and Street Price as of September, 2014.

	Street Prices, as of September 4, 2014 [USD]								MSRP [USD]
	Cheaper Than Dirt	Impact Guns	Gander Mountain	Grab a Gun	Buds Gun Shop	Firearms 4u	Average	Minimum	
BERETTA Nano	369.00	374.00	449.99	372.08	398.00		392.61	369.00	445.00
BERETTA Pico							N/A	N/A	398.00
CARACAL F		459.99					459.99	459.99	625.00
GLOCK G19 Gen4		537.99	699.99	539.00	539.00	539.00	571.00	537.99	649.00
HECKLER & KOCH VP9		599.99				575.00	587.50	575.00	719.00
HI-POINT C9	121.76		179.99	152.07	152.00		151.46	121.76	189.00
KEL-TEC PF-9	228.30	256.99	349.99	254.92	257.00	275.00	270.37	228.30	333.00
RUGER LC9	321.36	346.99	429.99		351.00	328.95	355.66	321.36	449.00
SCCY CPX-1	275.00	269.99	339.99	269.56	275.00	276.00	284.26	269.56	334.00
SCCY CPX-2	253.70	246.99	299.99	247.11	275.00		264.56	246.99	314.00
SIG-SAUER P320 Carry	519.00	564.99	649.99	554.98	549.00	541.00	563.16	519.00	713.00
S&W M&P9, 4.25"	442.00	469.99	529.99	469.48	478.00	467.00	476.08	442.00	569.00
SPRINGFIELD XD^M, 4.5"	536.54	563.99		550.77	551.00		550.58	536.54	699.99
TARA TM-9							N/A	N/A	N/A
TAURUS 24/7 G2	395.15	403.99		423.31	408.00		407.61	395.15	492.00
WALTHER PPX	354.75	369.99	499.99	359.00	388.00	382.00	392.29	354.75	449.00

Appendix 21   Technical Data of Sample Pistols

Table A21.1: CARACAL *F*, 9 mm Luger, technical data.

	Criteria	Manufacturer's Specifications	Survey
Pistol	Serial Number		LA 516
	Dimensions *L×W×H*, with (w/o) Magazine [mm]	193 × 28 × 135	192.2 × 31.7 × 137.8 (131.4)
	Weight, with (w/o) Mag [g]	790	763 (676)
	Fire-control Mechanism	Short Double Action	striker-fired, part. cocked
	Trigger Pull / Trigger Travel / Trigger Work	2.2 kg / 8 mm	24 N / 5 mm / 69 Nmm
	Slide Racking Force / Work	N/A	92 N / 3.8 J
	Copper Crusher Indent / max. Diameter [mm]	N/A	0.24 / 1.26
	Automatic Safeties: Striker Safety Trigger Safety Other Features: Loaded Ch. Indicator Others	Firing Pin Trigger, Drop Cocking Indicator	Striker Safety Trigger Safety Viewport Cocking Indicator
	Number of Magazines [pcs]	2	2
	Magazine capacity [rounds]	18	18
	Uncoupling start / -end, Total Slide Recoil Travel [mm]	N/A	3.2 / 5.4 / 50.1
	v_0/E_0 (RUAG *Sintox* 8.0 g FMJ)	N/A	351 m/s / 492 J
	$\overline{v}_{St}$ / $\overline{v}_{Se}$	N/A	5.3 m/s / 4.8 m/s
	Magazine Release	ambidextrous	Push-Button, ambidextrous
	Slide Stop		left side
	MSRP, Street Price [USD]	625 [a]	460 [a]
Slide	Dimensions *L×W×H*, with (w/o) recoil lug [mm]	N/A	178.6 × 28.0 × 31.0 (17.4)
	Gross Weight	N/A	337 g
	Manufacturing Technology	N/A	milling, MIM'ed insert
	Retracted Striker	N/A	no
Barrel	Weight	N/A	110 g
	Barrel Length	104 mm	104.3 mm
	Barrel Profile	250 mm twist rate	6 lands/grooves, RHT
	Manufacturing Technology	N/A	milling
Frame	Gross Weight	N/A	217 g
	Manufacturing Technology	synthetic	injection molding

---

[a] As of Sept. 2014

Table A21.2: GLOCK *G19 Gen4*, 9 mm Luger, technical data.

	Criteria	Manufacturer's Specifications	Survey
Pistol	Serial Number		WSB758
	Dimensions *L×W×H*, with (w/o) Magazine [mm]	185 × 30 × 127	184.9 × 31.8 × 128.1 (120.6) L = 188.7 w/Beavertail
	Weight, with (w/o) Mag [g]	670 (595)	669 (600)
	Fire-control Mechanism	Safe Action	striker-fired, part. cocked
	Trigger Pull / Trigger Travel / Trigger Work	25 N /12.5 mm	41 N / 8 mm / 122 Nmm
	Slide Racking Force / Work	N/A	115 N / 4.0 J
	Copper Crusher Indent / max. Diameter [mm]	N/A	0.34 / 0.60 × 0.96
	Automatic Safeties: Striker Safety Trigger Safety Passive Safeties: Out-of-battery disc. S. Other Features: Loaded Ch. Indicator Others	Striker Safety Trigger Safety LCI Drop Safety	Striker Safety Trigger Safety out-of-battery disconnect LCI Guided Trigger Bar
	Number of Magazines [pcs]	2	2
	Magazine capacity [rounds]	15	15
	Uncoupling start / -end, Total Slide Recoil Travel [mm]	2.5 mm [724] / – / –	2.5 / 5.0 / 46.9
	v_0/ E_0 (RUAG *Sintox* 8.0 g FMJ)	350 m/s / 490 J	358 m/s / 513 J
	$\bar{v}_{St}$ / $\bar{v}_{Se}$	N/A	5.4 m/s / 4.7 m/s
	Magazine Release	reversible	Push-Button, reversible
	Slide Stop	N/A	left side
	MSRP, Street Price [USD]	649 [a]	538 [a]
Slide	Dimensions *L×W×H*, with (w/o) recoil lug [mm]	N/A	171.2 × 25.5 × 32.9 (21.4)
	Gross Weight	N/A	347 g
	Manufacturing Technology	N/A	milling
	Retracted Striker	N/A	lance-shaped striker nose
Barrel	Weight	N/A	104 g
	Barrel Length	102 mm	102.0 mm
	Barrel Profile	hexagonal	polygonal, 6 "lands", RHT
	Manufacturing Technology	cold-hammer-forged[725]	milling
Frame	Gross Weight	N/A	130 g
	Manufacturing Technology	HiTech polymer w/o glass fibers	injection molding

---

[a] As of Sept. 2014

Table A21.3: HECKLER & KOCH *VP9*, 9 mm Luger, technical data.

	Criteria	Manufacturer's Specifications	Survey
Pistol	Serial Number		224-001989
	Dimensions *L×W×H*, with (w/o) Magazine [mm]	186.5 × 33.5 × 137.5	186.9 × 33.5 × 138.1 (133.1)
	Weight, with (w/o) Mag [g]	753 (660)	753 (660)
	Fire-control Mechanism	Striker Fired	striker-fired, SA
	Trigger Pull / Trigger Travel / Trigger Work	24 N / 6 mm	24 N / 6 mm / 67 Nmm
	Slide Racking Force / Work	N/A	91 N / 3.5 J
	Copper Crusher Indent / max. Diameter [mm]	N/A	0.31 / 1.23
	Automatic Safeties: Striker Safety Trigger Safety Passive Safeties: Out-of-battery disconn. Other Features: Loaded Ch. Indicator Others	Striker Safety Trigger Safety LCI Cocking Indicator	Striker Safety Trigger Safety out-of-battery disconnect LCI Cocking Indicator
	Number of Magazines [pcs]	N/A	2
	Magazine capacity [rounds]	15	15
	Uncoupling start / -end, Total Slide Recoil Travel [mm]	N/A	4.6 / 7.7 / 47.1
	v_0 / E_0 (RUAG *Sintox* 8.0 g FMJ)	N/A	365 m/s / 533 J
	$\bar{v}_{St}$ / $\bar{v}_{Se}$	N/A	5.6 m/s / 5.0 m/s
	Magazine Release	Paddle, ambidex.	Paddle, ambidex.
	Slide Stop	ambidextrous	ambidextrous
	MSRP, Street Price [USD]	719 [a]	575 [a]
Slide	Dimensions *L×W×H*, with (w/o) recoil lug [mm]	N/A	181.5 × 28.8 × 33.5 (22.6)
	Gross Weight	N/A	352 g
	Manufacturing Technology	N/A	milling
	Retracted Striker	N/A	yes
Barrel	Weight	N/A	98 g
	Barrel Length	104 mm	104.3 mm
	Barrel Profile	polygonal, 6 "lands", cold hammer forged, 250 mm RHT	polygonal, RHT
	Manufacturing Technology	Hammer forged	milling
Frame	Gross Weight	N/A	183 g
	Manufacturing Technology	reinforced polyamide	injection molding

[a] As of Sept. 2014

Table A21.4:　KEL-TEC *PF-9*, 9 mm Luger, technical data.

	Criteria	Manufacturer's Specifications	Survey
Pistol	Serial Number		SOF52
	Dimensions $L \times W \times H$, with (w/o) Magazine [mm]	149 × 22 × 109	151.1 × 25.0 × 111.9 (109.4)
	Weight, with (w/o) Mag [g]	360 (279)	415 (359)
	Fire-control Mechanism	Hammer Fired, DAO	hammer-fired, part. cocked
	Trigger Pull / Trigger Travel / Trigger Work	23 N	25 N / 16 mm / 280 Nmm
	Slide Racking Force / Work	N/A	89 N / 3.3 J
	Copper Crusher Indent / max. Diameter [mm]	N/A	0.23 / 1.10
	Automatic Safeties: Hammer block Other Features: Strips w/Mag. Assembles w/Mag.	Hammer-Lock Safety	Hammer Block yes yes
	Number of Magazines [pcs]	1	1
	Magazine capacity [rounds]	7	7
	Uncoupling start / -end, Total Slide Recoil Travel [mm]	N/A / 6 / N/A	2.9 / 5.2 / 41.3
	v_0 / E_0 (RUAG *Sintox* 8.0 g FMJ)	540 J	343 m/s / 470 J
	$\bar{v}_{St} / \bar{v}_{Se}$	N/A	11.0 m/s / 8.8 m/s
	Magazine Release	N/A	Push-Button, left side
	Slide Stop	N/A	left side
	MSRP, Street Price [USD]	333 [a]	228 [a]
Slide	Dimensions $L \times W \times H$, with (w/o) recoil lug [mm]	N/A	143.3 × 22.2 × 28.3 (20.5)
	Gross Weight	N/A	171 g
	Manufacturing Technology	N/A	milling
	Retracted Striker	N/A	yes
Barrel	Weight	N/A	58 g
	Barrel Length	79 mm	78.9 mm
	Barrel Profile	N/A	6 lands/grooves, RHT
	Manufacturing Technology	N/A	milling
Frame	Gross Weight	N/A	126 g
	Manufacturing Technology	N/A	injection molding

---

[a] As of Sept. 2014

Table A21.5:   RUGER *LC9*, 9 mm Luger, technical data.

	Criteria	Manufacturer's Specifications	Survey
Pistol	Serial Number		320-43011
	Dimensions *L×W×H*, with (w/o) Magazine [mm]	152.4 × 22.9 × 114.3	152.4 × 27.2 × 113.6 (111.9) Height w/Extension: 125.8
	Weight, with (w/o) Mag [g]	485	485 (429)
	Fire-control Mechanism	DAO	hammer-fired, part. cocked
	Trigger Pull / Trigger Travel / Trigger Work	N/A	35 N / 17 mm / 380 Nmm
	Slide Racking Force / Work	N/A	89 N / 3.0 J
	Copper Crusher Indent / max. Diameter [mm]	N/A	0.26 / 1.13
	Manual Safeties: Manual Safety Locking Device Automatic Safeties: Striker Safety Mag Disconnect Other Features: Loaded Ch. Indicator Strips w/Mag. Assembles w/Mag.	Manual Safety Internal Lock Mag Disconnect LCI	Manual Safety Internal Lock Striker Safety Mag Disconnect LCI yes yes
	Number of Magazines [pcs]	1	1
	Magazine capacity [rounds]	7	7
	Uncoupling start / -end, Total Slide Recoil Travel [mm]	N/A	3.5 / 5.2 / 42.6
	v_0 / E_0 (RUAG *Sintox* 8.0 g FMJ)	N/A	339 m/s / 459 J
	$\overline{v}_{St} / \overline{v}_{Se}$	N/A	8.7 m/s / 8.0 m/s
	Magazine Release	unilateral	Push-Button, left side
	Slide Stop	unilateral	left side
	MSRP, Street Price [USD]	449 [a]	321 [a]
Slide	Dimensions *L×W×H*, with (w/o) recoil lug [mm]	N/A	147.8 × 22.8 × 28.5 (20.6)
	Gross Weight	N/A	227 g
	Manufacturing Technology	N/A	milling
	Retracted Striker	N/A	yes
Barrel	Weight	N/A	57 g
	Barrel Length	79.3 mm	79.7 mm
	Barrel Profile	N/A	6 lands/grooves, RHT
	Manufacturing Technology	N/A	milling
Frame	Gross Weight	N/A	138 g
	Manufacturing Technology	N/A	injection molding

[a] As of Sept. 2014

Table A21.6:   SMITH & WESSON *M&P9*, 9 mm Luger, technical data.

	Criteria	Manufacturer's Specifications	Survey
Pistol	Serial Number		DST7991
	Dimensions L×W×H, with (w/o) Magazine [mm]	194 × 30 × 139	193.8 × 33.0 × 141.7 (135.3) W = 41.2 w/Manual Safety
	Weight, with (w/o) Mag [g]	680	793 (701)
	Fire-control Mechanism	Striker Fired DA	striker-fired, SA
	Trigger Pull / Trigger Travel / Trigger Work	28.9 N / 7.6 mm	25 N / 7 mm / 77 Nmm
	Slide Racking Force / Work	N/A	84 N / 3.2 J
	Copper Crusher Indent / max. Diameter [mm]	N/A	0.32 / 1.05
	Manual Safeties: Manual Safety Locking Device Automatic Safeties: Striker Safety Trigger Safety Passive Safeties: Out-of-battery disconn. Other Features: Loaded Ch. Indicator	optional optional	Manual Safety Striker Safety Trigger Safety out-of-battery disconnect Viewport
	Number of Magazines [pcs]	2	2
	Magazine capacity [rounds]	17	17
	Uncoupling start / -end, Total Slide Recoil Travel [mm]	N/A	2.0 / 4.8 / 47.4
	v_0 / E_0 (RUAG *Sintox* 8.0 g FMJ)	N/A	361 m/s / 520 J
	$\bar{v}_{St} / \bar{v}_{Se}$	N/A	5.2 m/s / 4.4 m/s
	Magazine Release	Reversible Magazine Release	Push-Button, reversible
	Slide Stop	ambidextrous	ambidextrous
	MSRP, Street Price [USD]	569 [a]	442 [a]
Slide	Dimensions L×W×H, with (w/o) recoil lug [mm]	N/A	182.4 × 27.4 × 32.8 (21.7)
	Gross Weight	N/A	368 g
	Manufacturing Technology	N/A	milling
	Retracted Striker	N/A	yes
Barrel	Weight	N/A	103 g
	Barrel Length	108 mm	108.0 mm
	Barrel Profile	N/A	6 lands/grooves, RHT
	Manufacturing Technology	N/A	milling
Frame	Gross Weight	N/A	207 g
	Manufacturing Technology	Zytel Polymer Frame	injection molding

---

[a] As of Sept. 2014

Table A21.7: SPRINGFIELD *XD^M*, 9 mm Luger, technical data.

	Criteria	Manufacturer's Specifications	Survey
Pistol	Serial Number		MG957178
	Dimensions $L{\times}W{\times}H$, with (w/o) Magazine [mm]	193 × 30 × 146	200.7 × 33 × 145.7 (140.5)
	Weight, with (w/o) Mag [g]	822	801 (715)
	Fire-control Mechanism	N/A	striker-fired, SA
	Trigger Pull / Trigger Travel / Trigger Work	24.5 N to 34 N	24 N / 7 mm / 68 Nmm
	Slide Racking Force / Work	N/A	84 N / 3.6 J
	Copper Crusher Indent / max. Diameter [mm]	N/A	0.32 / 1.17
	Automatic Safeties: Striker Safety Grip Safety Trigger Safety Passive Safeties: Out-of-battery disconn. Other Features: Loaded Ch. Indicator Strips w/Mag. Others	Striker Block Grip Safety Ultra Safety Assurance Trigger System LCI Striker Status Indic.	Striker Safety Grip Safety Trigger Safety out-of-battery disconnect LCI yes Cocking Indicator
	Number of Magazines [pcs]	3	3
	Magazine capacity [rounds]	19	19
	Uncoupling start / -end, Total Slide Recoil Travel [mm]	N/A	3.1 / 6.4 / 46.8
	v_0 / E_0 (RUAG *Sintox* 8.0 g FMJ)	N/A	371 m/s / 550 J
	$\overline{v}_{St} / \overline{v}_{Se}$	N/A	5.6 m/s / 5.2 m/s
	Magazine Release	Ambidextrous Magazine Release	Push-Button, ambidextrous
	Slide Stop	N/A	left side
	MSRP, Street Price [USD]	700 [a]	537 [a]
Slide	Dimensions $L{\times}W{\times}H$, with (w/o) recoil lug [mm]	N/A	185.9 × 26.3 × 36.1 (23.3)
	Gross Weight	N/A	349 g
	Manufacturing Technology	Melonite	milling
	Retracted Striker	N/A	yes
Barrel	Weight	N/A	107 g
	Barrel Length	114 mm	116.7 mm
	Barrel Profile	N/A	6 lands/grooves, RHT
	Manufacturing Technology	N/A	milling
Frame	Gross Weight	N/A	227 g
	Manufacturing Technology	polymer	injection molding

---

[a] As of Sept. 2014

Table A21.8: TAURUS *PT 24/7 PRO*, 9 mm Luger, technical data.

	Criteria	Manufacturer's Specifications	Survey
Pistol	Serial Number		TZJ22447
	Dimensions $L \times W \times H$, with (w/o) Magazine [mm]	181 × 32 × 140	184.8 × 31.5 × 146.1 (137.5)
	Weight, with (w/o) Mag [g]	825	795 (710)
	Fire-control Mechanism	SA/DA	striker-fired, SA/RC[a]
	Trigger Pull / Trigger Travel / Trigger Work		SA: 21 N / 11 mm / 63 Nmm RC: 27 N / 11 mm / 155 Nmm
	Slide Racking Force / Work	N/A	88 N / 3.0 J
	Copper Crusher Indent / max. Diameter [mm]	N/A	SA: 0.31 / 1.10 RC: N/A [b]
	Manual Safeties: Manual Safety Locking Device Automatic Safeties: Striker Safety Trigger Safety Other Features: Loaded Ch. Indicator	Ext. Safety Latch Key Lock Firing Pin Block Trigger Safety LCI	Manual Safety Internal Lock Striker Safety Internal Trigger Safety LCI
	Number of Magazines [pcs]		2
	Magazine capacity [rounds]	17	17
	Uncoupling start / -end, Total Slide Recoil Travel [mm]	N/A	3.4 / 5.2 / 41.8
	v_0 / E_0 (RUAG *Sintox* 8.0 g FMJ)	N/A	356 m/s / 508 J
	$\bar{v}_{St} / \bar{v}_{Se}$	N/A	4.9 m/s / 4.4 m/s
	Magazine Release	N/A	Push-Button, reversible
	Slide Stop	N/A	left side
	MSRP, Street Price [USD]	492 [c]	395 [c]
Slide	Dimensions $L \times W \times H$, with (w/o) recoil lug [mm]	N/A	179.6 × 25.6 × 35.0 (23.6)
	Gross Weight	N/A	383 g
	Manufacturing Technology	Stainless Steel	milling
	Retracted Striker	N/A	yes
Barrel	Weight	N/A	103 g
	Barrel Length	107 mm	108.4 mm
	Barrel Profile	6 grooves, RHT, 250 mm	6 lands/grooves, RHT
	Manufacturing Technology	N/A	milling
Frame	Gross Weight	N/A	223 g
	Manufacturing Technology	Ribber Grip Overlay	injection molding

---

[a] Restrike Capability (RC)
[b] Pistol's trigger action would not allow to take this measurement (see Chapter 5.10, p. 280ff).
[c] As of Sept. 2014

Table A21.9: WALTHER *PPX*, 9 mm Luger, technical data.

	Criteria	Manufacturer's Specifications	Survey
Pistol	Serial Number		FAM7387
	Dimensions *L×W×H*, with (w/o) Magazine [mm]	186 / 34 / 143	186.0 × 33.1 × 142.7 (141.7)
	Weight, with (w/o) Mag [g]	765 (675)	763 (672)
	Fire-control Mechanism	Hammer Fired, DA pre-set	hammer-fired, SA
	Trigger Pull / Trigger Travel / Trigger Work	29 N / 7 mm	22 N / 7 mm / 84 Nmm
	Slide Racking Force / Work	N/A	111 N / 3.8 J
	Copper Crusher Indent / max. Diameter [mm]	N/A	0.33 / 1.26
	Automatic Safeties: Striker Safety	Striker Safety	Striker Safety
	Passive Safeties: Other passive safeties	Drop Safety	two Drop Safeties
	Other Features: Loaded Ch. Indicator Strips w/Mag. Assembles w/Mag.	LCI	Viewport yes yes
	Number of Magazines [pcs]	2	2
	Magazine capacity [rounds]	16	16
	Uncoupling start / -end, Total Slide Recoil Travel [mm]	N/A	3.1 / 8.1 / 49.1
	v_0/E_0 (RUAG *Sintox* 8.0 g FMJ)	500 J	358 m/s / 513 J
	$\bar{v}_{St}$ / $\bar{v}_{Se}$	N/A	5.4 m/s / 5.4 m/s
	Magazine Release	Button, reversible	Push-Button, reversible
	Slide Stop	N/A	left side
	MSRP, Street Price [USD]	449 [a]	355 [a]
Slide	Dimensions *L×W×H*, with (w/o) recoil lug [mm]	N/A	184.4 × 29.1 × 37.0 (24.0)
	Gross Weight	N/A	345 g
	Manufacturing Technology	N/A	milling with polymer insert
	Retracted Striker	N/A	yes
Barrel	Weight	N/A	110 g
	Barrel Length	102 mm	101.5 mm
	Barrel Profile	6 lands/grooves	6 lands/grooves, RHT
	Manufacturing Technology	N/A	Barrel Tube, MIM'ed Barrel Block, MIM'ed Feed Ramp
Frame	Gross Weight	N/A	206 g
	Manufacturing Technology	polymer	injection molding

---

[a] As of Sept. 2014

Appendix 22   Technical Data, Summary

Table A22.1: Technical data, summary.

	CARACAL F	GLOCK G19 Gen4	HECKLER & KOCH VP9	KEL-TEC PF-9	RUGER LC9	SMITH & WESSON M&P9	SPRINGFIELD XD^M	TAURUS PT 24/7 PRO	WALTHER PPX	Average
Trigger Pull Force [N]	24	41	24	25	35	25	24	21	22	27
Trigger Travel [mm]	5	8	6	16	17	7	7	11	7	9
Trigger Work [Nmm]	69	122	67	280	380	77	68	63	84	134
Copper Crusher Indent, Depth [mm]	0.24	0.34	0.31	0.23	0.26	0.32	0.32	0.31	0.33	0.30
Copper Crusher Indent, Diameter [mm]	1.26	0.6 × 0.96	1.23	1.1	1.13	1.05	1.17	1.1	1.26	1.16*
Slide Racking Force [N]	92	115	91	89	89	84	84	88	111	94
Uncoupling, start [mm]	3.2	2.5	4.6	2.9	3.5	2.0	3.1	3.4	3.1	3.1
Uncoupling, end [mm]	5.4	5.0	7.7	5.2	5.2	4.8	6.4	5.2	8.1	5.9
Slide Recoil Travel, total [mm]	50.1	46.9	47.1	41.3	42.6	47.4	46.8	41.8	49.1	45.9
Gross Weight, Slide (m_s) [g]	337	347	352	171	227	368	349	383	345	320
Weight, Barrel (m_B) [g]	110	104	98	58	57	103	107	103	110	94
Barrel Length [mm]	104.3	102.0	104.3	78.9	79.7	108.0	116.7	108.4	101.5	100.4
Barrel Profile (Lands/Grooves, Polygonal)	L/G	P	P	L/G	L/G	L/G	L/G	L/G	L/G	N/A
v_0 [m/s]	351	358	365	343	339	361	371	356	358	356
E_0 [J]	492	513	533	470	459	520	550	508	513	507
Recoiling Mass ($m_s + m_B$) [g]	447	451	450	229	284	471	456	486	455	414
V_{St} [m/s]	5.3	5.4	5.6	11.0	8.7	5.2	5.6	4.9	5.4	6.3
V_{Se} [m/s]	4.8	4.7	5.0	8.8	8.0	4.4	5.2	4.4	5.4	5.6
V_{Se}/V_{St}	91%	87%	89%	80%	92%	85%	93%	90%	100%	90%
Width (W), Slide [mm]	28.0	25.5	28.8	22.2	22.8	27.4	26.3	25.6	29.1	26.2
Height (H), Slide [mm]	31.0	32.9	33.5	28.3	28.5	32.8	36.1	35.0	37.0	32.8
Cross-Sectional Area (W · H), Slide [mm²]	868	839	965	628	650	899	949	896	1077	863
Magazine Capacity [rounds]	18	15	15	7	7	17	19	17	16	15
Magazine Release	ambidextrs.	reversible	reversible	unilateral	unilateral	reversible	ambidextrs.	reversible	reversible	N/A
Slide Stop Lever	unilateral	unilateral	ambidextrs.	unilateral	unilateral	ambidextrs.	unilateral	unilateral	unilateral	N/A
Consistent Trigger Pull for each shot	yes	yes	yes	yes	yes	yes	yes	yes	yes	N/A
Hammer-/Striker-Fired	Striker Fired	Striker Fired	Striker Fired	Hammer Frd.	Hammer Frd.	Striker Fired	Striker Fired	Striker Fired	Hammer Frd.	N/A
Field Stripping w/o pulling Trigger	no	no	yes	yes	yes	yes	yes	no	yes	N/A
Strips w/Mag seated	no	no	no	yes	yes	no	yes	no	yes	N/A
Reassembles w/Mag seated	no	no	no	yes	yes	no	no	no	yes	N/A
MSRP [USD]	625	649	719	333	449	569	700	492	449	554
Street Price [USD]	460	538	575	228	321	442	537	395	355	428
Street Price/MSRP	74%	83%	80%	69%	72%	78%	77%	80%	79%	77%

* w/o GLOCK

Appendix 23   Heydenreich's Method

In 1908 *Heydenreich*[726] publishes a quickly comprehensible approximate calculation of internal ballistic data. It is based on generalized characteristic curves of gas pressure, projectile velocity and the time of the projectile travel. Taking into account the projectile mass, muzzle velocity, peak bore pressure, caliber and barrel length, *Heydenreich* resizes standardized curves, the so-called normal curves[727].

First of all, it is necessary to calculate the mean pressure. This is a consistent pressure level theoretically high enough to propel the mass of the projectile and half of the propellant weight to the same muzzle velocity as is achieved in reality as the projectile is accelerated down the bore with a variable pressure curve (cf. Fig. A23.9, p. 488):

$$\bar{p} = \frac{\left(m_P + \frac{1}{2}m_C\right)v_0^2}{2\,s_\ell\,A\{mm^2\}} \qquad \textbf{A23.1}$$

where: $\bar{p}$ Mean pressure [MPa]
         A Bore cross-sectional area [mm²], (cf. A23.12, p. 478)
         m_C Mass, propellant [kg]
         m_P Mass, projectile [kg]
         s_ℓ Bore travel [m], projectile base leaves the muzzle,
               (cf. A19.61, p. 433)
         v_0 Muzzle velocity [m/s]

Heydenreich relates this mean pressure to the actually measured peak pressure. The spike, p_{max}, is to be determined by pressure testing. If firing pressure measurements cannot be accomplished, reloading tables or the Maximum Average Pressure p_{Tmax} from the C.I.P. pressure specifications for the respective cartridge type can be used instead (see A23.12, p. 478).

$$\eta = \frac{\bar{p}}{p_{max}} \qquad \textbf{A23.2}$$

where: η Pressure ratio [1]
         $\bar{p}$ Mean pressure [MPa]
         p_{max} Peak pressure [MPa]

Once the pressure ratio η is known, the projectile base travel $s_{p\,max}$, instantaneous barrel time $t_{p\,max}$ and projectile velocity $v_{p\,max}$ at pressure peak can be determined. In addition, the muzzle pressure p_ℓ and the barrel time t_ℓ can be determined for the instant at which the projectile's base leaves the muzzle.

The following relationships apply:[728]

$$s_{p\,max} = s_\ell\, \Sigma(\eta)$$ A23.3

$$t_{p\,max} = \frac{2s_\ell}{v_0}\, \Theta(\eta)$$ A23.4

$$v_{p\,max} = v_0\, \Phi(\eta)$$ A23.5

$$p_\ell = \bar{p}\, \Pi(\eta)$$ A23.6

$$t_\ell = \frac{2s_\ell}{v_0}\, T(\eta)$$ A23.7

Whereby the multipliers Σ, Θ, Φ, Π, and T depend on the pressure ratio η and, if necessary, can be obtained by interpolation for each argument η satisfying $0.25 \leq \eta \leq 0.55$ (see Table A23.1).[729]

For a linear interpolation, Table A23.1 shows the nearest pressure ratio below (η_0) or above (η_1) of a given argument η computed according to Formula A23.2 and its associated multipliers $Y(\eta_0)$ and $Y(\eta_1)$. The intermediate magnitude $Y(\eta)$ is

$$Y(\eta) = \frac{(\eta - \eta_0)}{\eta_1 - \eta_0}\, (Y(\eta_1) - Y(\eta_0)) + Y(\eta_0)$$ A23.8

Table A23.1: Multipliers for conditions at pressure peak or when projectile base leaves the muzzle respectively; according to *Oerlikon*.[730]

η	$\Sigma(\eta)$	$\Theta(\eta)$	$\Phi(\eta)$	$\Pi(\eta)$	$T(\eta)$
0.25	0.0313	0.139	0.324	0.216	0.725
0.30	0.0402	0.172	0.333	0.242	0.762
0.35	0.0500	0.207	0.343	0.278	0.800
0.40	0.0608	0.244	0.354	0.304	0.836
0.45	0.0729	0.284	0.366	0.340	0.873
0.50	0.0875	0.326	0.380	0.382	0.910
0.55	0.1059	0.370	0.396	0.454	0.950

For Example 2 and 3 on pages 441 and 448, respectively, the following approximate numbers are obtained:

$s_{p\,max}$ = 5 mm, $t_{p\,max}$ = 0.1 ms, $v_{p\,max}$ = 120 m/s, p_ℓ = 30 MPa, t_ℓ = 0.4 ms

Comment on barrel time t_ℓ

The duration from the beginning of projectile motion until the projectile leaves the muzzle depends on how strongly the projectile is accelerated in the barrel, or how great the resulting average projectile velocity is (Fig. A23.9, p. 488). *Brukner* uses the generally accepted formula for calculating the velocity:

$$v = \frac{s}{t} \qquad\qquad \textbf{A23.9}$$

This equation states that the velocity, in our case the mean velocity of the projectile, is calculated as the ratio of the bore travel to barrel time.

Brukner provides a user-friendly rough calculation for the determination of the barrel time. He assumes that the average velocity of the projectile running down the bore is approximately ⅔ of the muzzle velocity. With this assumption we obtain:

$$\frac{2}{3} v_0 = \frac{s_\ell}{t_\ell} \qquad\qquad \textbf{A23.10}$$

Solved after barrel time, a constant of $3/2$ is obtained as a factor:

$$t_\ell = {}^3/_2 \, \frac{s_\ell}{v_0} \qquad\qquad \textbf{A23.11}[731]$$

where: t_ℓ Barrel time [s]
                 s_ℓ Bore travel [m], projectile base leaves the muzzle,
                      (cf. A19.61, p. 433)
                 v_0 Muzzle velocity [m/s]

Equation A23.11 serves well as a rough estimate. In comparison, however, we will take a closer look at the barrel time using Heydenreich's method.

In principle, Heydenreich uses a similar approach to formula A23.11 for the calculation of t_ℓ, but in addition his method takes into account internal ballistic conditions. Consequently, the algorithm contains propellant weight m_C, bore cross-sectional area A and peak pressure p_{max}.

The various calculations for A23.7 are summarized and given as a numerical value equation:

$$t_\ell = \frac{0.74\, v_0 \left(m_P + \frac{1}{2} m_C\right)}{A\{\text{mm}^2\}\, p_{max}\{\text{MPa}\}} + 1.08\, \frac{s_\ell}{v_0} \qquad\qquad \textbf{A23.12}^{732}$$

where: t_ℓ Barrel time [s]

           A Bore cross-sectional area [mm²]

               $A = \frac{\pi}{4} D_B^2 + \frac{D_G - D_B}{2} W_G n_G$ using bore diameter D_B, groove diameter D_G, groove width W_G and total number of grooves n_G.[733]
If the internal dimensions are not known, the C.I.P. tables can be used for reference. They give values for A. These correspond to the minimum value of the bore cross-sectional area; e.g. 62.61 mm² for 9 mm Luger[734]

           m_C Mass, propellant charge [kg]

           m_P Mass, projectile [kg]

           p_{max} Peak pressure [MPa]
If chamber pressure measurements are not available, values from the reloading literature or the Maximum Average Pressure p_{Tmax} from C.I.P. tables can be used as an alternative; e.g. 235 MPa for 9 mm Luger (cf. Endnote 734)

           s_ℓ Bore travel [m], projectile base leaves the muzzle, (cf. A19.61, p. 433)

           v_0 Muzzle velocity [m/s]

In the practical example of Table A19.2, p. 445, and by applying *Heydenreich* we have a barrel time of $t_\ell = 0.41$ ms. The counter test with *Brukner* gives 0.36 ms. The results are obviously not too far apart. Using Heydenreich's method, a multiplier of 1.684 is calculated in this example, whereas Brukner calculates with ³/₂, i.e. 1.5 (cf. A23.11).

In the outset of this discussion it is worthwhile to note that firing 10,000 rounds under the conditions mentioned in the above example results in a total barrel operating time of about 4 seconds. This is calculated by multiplying the barrel time by the number of shots.[735]

The output power can also be impressive. The heat engine under consideration transmits a certain amount of energy to the projectile within the aforementioned barrel time. The output power P of the machine is calculated by

$$P = \frac{E_0}{t_\ell} \qquad\qquad \textbf{A23.13}^{736}$$

where: P Output power [W]

           E_0 Projectile's muzzle energy [J], see A13.2, p. 374

           t_ℓ Barrel time [s]

The 9 mm Luger in our example produces roughly 1240 kW or 1700 horsepower.

Pressure-time and velocity-time curves

It is possible to determine the pressure curve and projectile velocity as a function of time or the projectile base travel by using further multipliers. The latter are determined as a function of λ. The relative distance lambda is calculated by dividing any respective instantaneous projectile base travel s_P by the bore travel at pressure peak ($s_{p\,max}$, see A23.3):

$$\lambda = \frac{s_P}{s_{p\,max}} \qquad\qquad \textbf{A23.14}$$

Heydenreich provides the following relationships for pressure p, velocity v and time t as a function of projectile base travel s_P:[737]

$$p(s_P) = p_{max}\,P(\lambda) \qquad\qquad \textbf{A23.15}$$

$$v(s_P) = v_{p\,max}\,\Psi(\lambda) \qquad\qquad \textbf{A23.16}$$

$$t(s_P) = t_{p\,max}\,\Omega(\lambda) \qquad\qquad \textbf{A23.17}$$

Associated multipliers depending on λ are shown in Table A23.2, p. 480. Intermediate values are computed by interpolation using A23.8.

The displacement of the slide when the projectile exits the barrel is calculated according to A19.60, p. 433. Intermediate values of the displacement x depending on any instantaneous projectile base travel s_P are calculated using

$$x\,m_r = s_P\,m_P \qquad\qquad \textbf{A23.18}$$

hence

$$x = s_P\,\frac{m_P}{m_r} \qquad\qquad \textbf{A23.19}$$

and, where appropriate, slide displacement x is expressed as a dependency of λ using A23.14:

$$x = \lambda\,s_{p\,max}\,\frac{m_P}{m_r} \qquad\qquad \textbf{A23.20}$$

where: x Displacement, slide [m]
         m_P Mass, projectile [kg]
         m_r Mass, recoiling [kg], m_S for blowback pistols,
                 $m_S + m_B$ for pistols with coupled barrel and slide
         s_P Instantaneous projectile base travel [m]

Note: While retaining the magnitude for the peak pressure and the results from A23.3 to A23.5, the influence of any optional barrel length on pressure ratio, muzzle pressure, muzzle velocity and barrel time can be roughly predicted. A23.14 and A23.16 form the entry point, followed by A23.1, A23.2, A23.6 and A23.7.

Table A23.2: Multipliers for the computation of pressure, velocity and time specified as a function of the relative distance λ (by *Oerlikon*).[738]

λ	P(λ)	Ψ(λ)	Ω(λ)
0.25	0.741	0.392	0.610
0.50	0.912	0.635	0.780
0.75	0.980	0.834	0.903
1	1	1	1
1.25	0.989	1.140	1.081
1.50	0.965	1.262	1.154
1.75	0.932	1.366	1.219
2.00	0.898	1.468	1.282
2.5	0.823	1.632	1.394
3.0	0.747	1.763	1.495
3.5	0.675	1.875	1.589
4.0	0.604	1.983	1.682
4.5	0.546	2.068	1.769
5	0.495	2.140	1.851
6	0.403	2.269	2.012
7	0.338	2.363	2.163
8	0.284	2.445	2.309
9	0.248	2.509	2.451
10	0.220	2.566	2.589
11	0.199	2.615	2.725
12	0.181	2.659	2.858
13	0.164	2.702	2.988
14	0.150	2.740	3.116
15	0.137	2.777	3.253
16	0.125	2.811	3.390
17	0.117	2.837	3.502
18	0.109	2.862	3.618
19	0.102	2.887	3.740
20	0.096	2.910	3.861
25	0.073	3.003	4.455
30	0.058	3.075	5.031
35	0.048	3.162	5.657

Description of the multipliers by function equations

Heydenreich's normal curves or their description via the normal chart (Table A23.2) are particularly suitable for the uncomplicated solution of internal ballistic tasks. Please note: The calculation method is based on the standardization of the data determined empirically by Heydenreich and later by OERLIKON, which are based on recoil tests.

If one wants to use the values from Table A23.1 and Table A23.2 by means of spreadsheet software, the possibility of interpolation via program-controlled access to the corresponding value pairs remains. However, the method of reconstruction of entire rational functions can also be used. The result of the latter option describes the normal curves mathematically. One uses polynomial functions of the form

$$P(x) = a_n x^n + \cdots + a_3 x^3 + a_2 x^2 + a_1 x + a_0$$

Each *a* represents a coefficient of the term. The *n* stands for the number of coefficients or the number of powers. The degree of the function is equal to that of the polynomial. It corresponds to the highest exponent *n*.

At points where the reconstruction leads to extreme or turning points in the function equation, the following description omits the exact consideration of all values of the data series. Complete coverage would increase complexity without increasing the precision of the results. One should remember the empirical character of the database.[739]

The presentation of some of the coefficients is in scientific format. The E mentioned there, as in 1.234E–5, represents "multiplied by ten to the power of", in our example this means $1.234*10^{-5}$ and thus 0.00001234. In other words: "–5" stands for a shift of the decimal point by 5 places to the left.

Coefficients for polynomial functions as substitute for Table A23.1

The coefficients for the polynomials for respective values at pressure peak, for muzzle pressure, and barrel time (Table A23.1) are listed in Table A23.3.

Table A23.3: Coefficients for polynomial functions of multipliers for $0.25 \leq \eta \leq 0.55$.

n	a_n for $\Sigma(\eta)$	a_n for $\Theta(\eta)$	a_n for $\Phi(\eta)$	a_n for $\Pi(\eta)$
12	0	0	0	$\dfrac{3.57887969264285\text{E}25}{4{,}714{,}992{,}056{,}248{,}420}$
11	0	0	0	$\dfrac{-1.63141302844222\text{E}26}{4{,}602{,}819{,}418{,}031{,}550}$
10	0	0	0	$\dfrac{2.76168375555064\text{E}26}{3{,}664{,}595{,}791{,}498{,}830}$
9	0	0	$-33{,}280{,}000/2079$	$\dfrac{-3.36670310159993\text{E}25}{348{,}960{,}669{,}969{,}523}$
8	0	0	$12{,}944{,}000/297$	$\dfrac{7.25230219408452\text{E}26}{8{,}755{,}811{,}471{,}486{,}110}$
7	$-69{,}760/231$	$2{,}739{,}200/2079$	$-104{,}700{,}800/2079$	$\dfrac{-2.2287278323852\text{E}25}{443{,}615{,}090{,}405{,}761}$
6	$23{,}504/33$	$-937{,}280/297$	$9{,}517{,}040/297$	$\dfrac{6.63216764237353\text{E}+25}{3{,}004{,}031{,}690{,}773{,}170}$
5	$-7540/11$	$920{,}960/297$	$-327{,}976/27$	$\dfrac{-2.1950598498428\text{E}25}{3{,}099{,}307{,}676{,}736{,}770}$
4	$19{,}011/55$	$-474{,}932/297$	$8{,}190{,}461/2970$	$\dfrac{1.47410091969018\text{E}25}{8{,}954{,}510{,}848{,}817{,}310}$
3	$-1{,}591{,}951/16{,}500$	$3{,}389{,}396/7425$	$-18{,}395{,}413/51{,}975$	$\dfrac{-8.13295464279229\text{E}23}{3{,}007{,}840{,}567{,}650{,}010}$
2	$2{,}363{,}063/165{,}000$	$-5{,}052{,}833/74{,}250$	$2{,}208{,}167/99{,}000$	$\dfrac{3.37341736506988\text{E}21}{113{,}234{,}640{,}267{,}329}$
1	$-148{,}263/192{,}500$	$64{,}753/13{,}860$	$-117{,}457/231{,}000$	$\dfrac{-260281657310377\text{E}6}{131{,}655{,}664{,}305{,}223}$
0	0	0	$81/250$	$\dfrac{518838530648811\text{E}6}{8{,}682{,}032{,}369{,}689{,}590}$

For $T(\eta)$, select the value pairs $(0.25|0.725)$ and $(0.5|0.91)$ from Table A23.1. This results in the linear function

$$T(\eta) = 0.74\eta + 0.54 \qquad\qquad \textbf{A23.21}$$

Coefficients for polynomial functions as substitute for Table A23.2

The reconstruction of the graphs for the pressure curve, the velocity and the time for any λ that satisfies $0 \leq \lambda \leq 0.55$ is done by seamlessly merging curve pieces with minimal overall curvature.

The number of splines required to describe the graphs is *i*, i.e. the curve is divided into *i* segments for lambda. Each of these sections is described by a polynomial of the form

$$f(\lambda) = a_3(\lambda - L_i)^3 + a_2(\lambda - L_i)^2 + a_1(\lambda - L_i) + a_0$$

Table A23.4: Coefficients of the polynomial function for the pressure curve; $P(\lambda)$.

i	λ	L_i	a_3	a_2	a_1	a_0
1	$15 < \lambda \le 35$	15	−4.497E−6	2.698E−4	−0.008047	0.137
2	$8 < \lambda \le 15$	8	−2.258E−4	0.005012	−0.04502	0.284
3	$4 < \lambda \le 8$	4	−9.335E−4	0.01621	−0.1299	0.604
4	$1.75 < \lambda \le 4$	1.75	0.004074	−0.01128	−0.1410	0.932
5	$1.5 < \lambda \le 1.75$	1.5	0.09906	−0.08556	−0.1168	0.965
6	$1.25 < \lambda \le 1.5$	1.25	−0.009562	−0.07841	−0.07580	0.989
7	$1 < \lambda \le 1.25$	1	0.1952	−0.2248	1.146E−7	1
8	$0.75 < \lambda \le 1$	0.75	0.002696	−0.2268	0.1129	0.985908
9	$0.5 < \lambda \le 0.75$	0.5	0.03236	−0.2511	0.2324	0.943
10	$0.25 < \lambda \le 0.5$	0.25	8.206	−6.405	1.896	0.741
11	$0 < \lambda \le 0.25$	0	−8.54	0	3.498	0

Table A23.5: Coefficients of the polynomial function for the projectile velocity curve; $\Psi(\lambda)$.

i	λ	L_i	a_3	a_2	a_1	a_0
1	$20 < \lambda \le 35$	20	1.265E−5	−5.692E−4	0.02249	2.91
2	$10 < \lambda \le 20$	10	6.215E−5	−0.002434	0.05252	2.566
3	$5 < \lambda \le 10$	5	8.203E−4	−0.01474	0.1384	2.14
4	$1.5 < \lambda \le 5$	1.5	0.00497	−0.06693	0.4242	1.262
5	$1 < \lambda \le 1.5$	1	0.2653	−0.4648	0.6901	1

Where λ satisfies $0 \le \lambda \le 1$ follows

$$\Psi(\lambda) = \frac{96}{125}\lambda^5 - \frac{1096}{375}\lambda^4 + \frac{538}{125}\lambda^3 - \frac{2491}{750}\lambda^2 + \frac{543}{250}\lambda \qquad \textbf{A23.22}$$

Table A23.6: Coefficients of the polynomials for instantaneous barrel time; $\Omega(\lambda)$.

i	λ	L_i	a_3	a_2	a_1	a_0
1	$10 < \lambda \le 35$	10	1.109E−5	−8.315E−4	0.1363	2.589
2	$5 < \lambda \le 10$	5	2.746E−4	−0.00495	0.1655	1.851
3	$2 < \lambda \le 5$	2	0.001037	−0.01428	0.2232	1.282
4	$1 < \lambda \le 2$	1	0.04454	−0.1479	0.3854	1
5	$0.5 < \lambda \le 1$	0.5	−0.07724	−0.03203	0.4786	0.78
6	$0.25 < \lambda \le 0.5$	0.25	5.707	−4.312	1.561	0.57
7	$0 < \lambda \le 0.25$	0	−5.75	0	2.639	0

Using the polynomials, the diagrams illustrated in Fig. A23.9 and Fig. A23.10 (page 488) are obtained, as well as the slide recoil speed curve in section $\alpha 1$ of Fig. A19.8 (p. 436).

The comparison between the chart values from *Oerlikon* or their respective graphs vs. the values of the reconstructed polynomial functions is shown from Fig. A23.1 through Fig. A23.8. Intermediate values for the graphs based on the numbers taken from *Oerlikon* were obtained by interpolation according to A23.8.

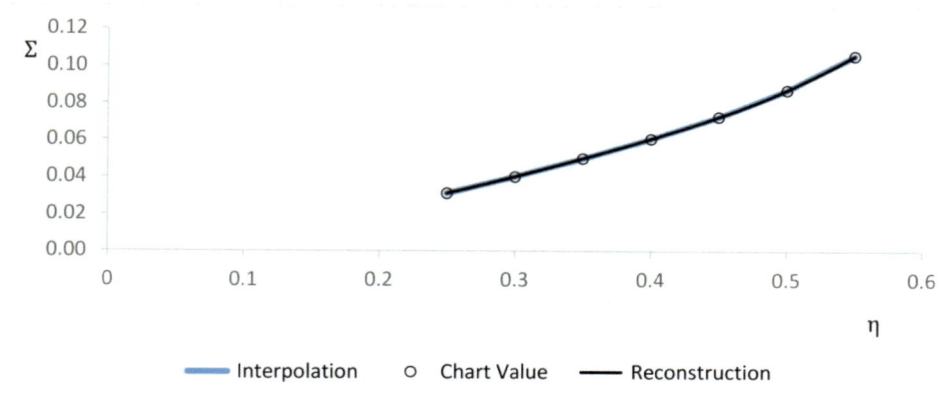

Fig. A23.1: Multiplier "Sigma"; *Oerlikon* chart values vs. polynomial function.

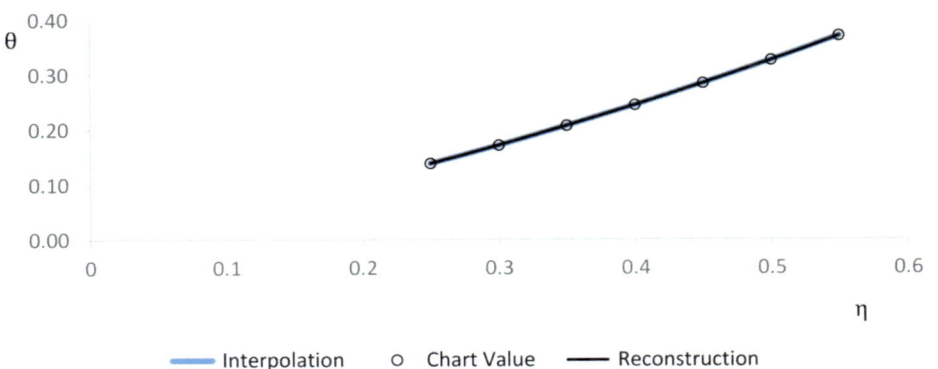

Fig. A23.2: Multiplier "Theta"; *Oerlikon* chart values vs. polynomial function.

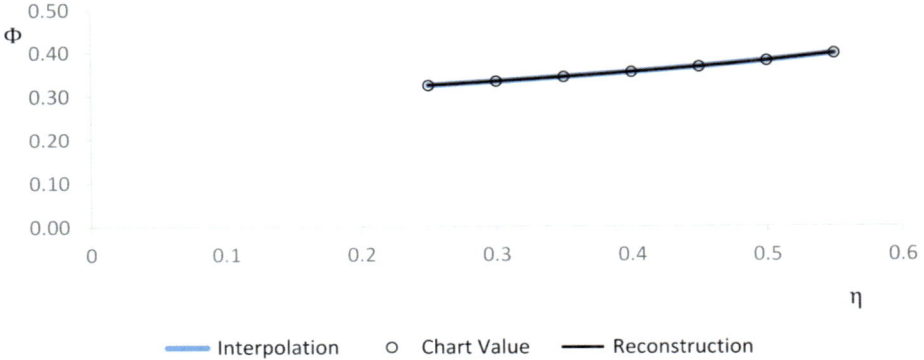

Fig. A23.3: Multiplier "Phi"; *Oerlikon* chart values vs. polynomial function.

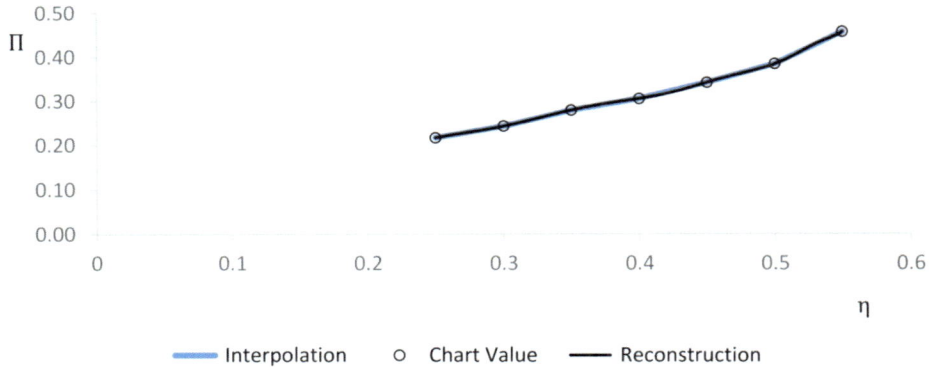

Fig. A23.4: Multiplier "Pi"; *Oerlikon* chart values vs. polynomial function.

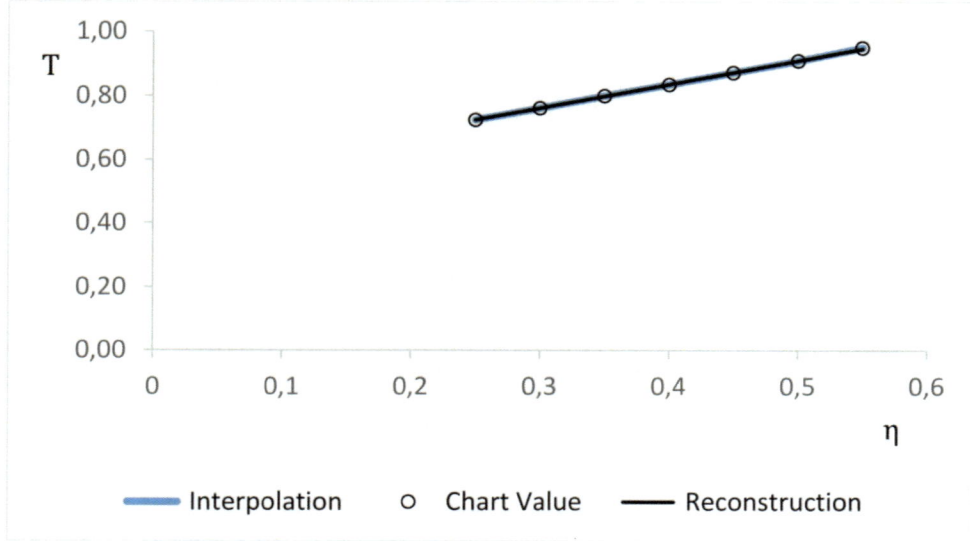

Fig. A23.5: Multiplier "Tau"; *Oerlikon* chart values vs. polynomial function.

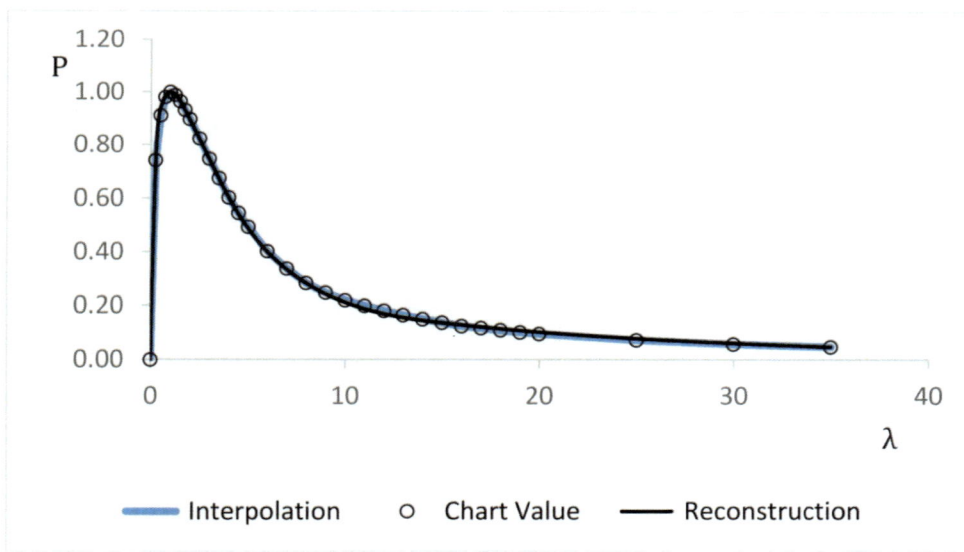

Fig. A23.6: Normalized Pressure Curve; *Oerlikon* chart values vs. polynomial function.

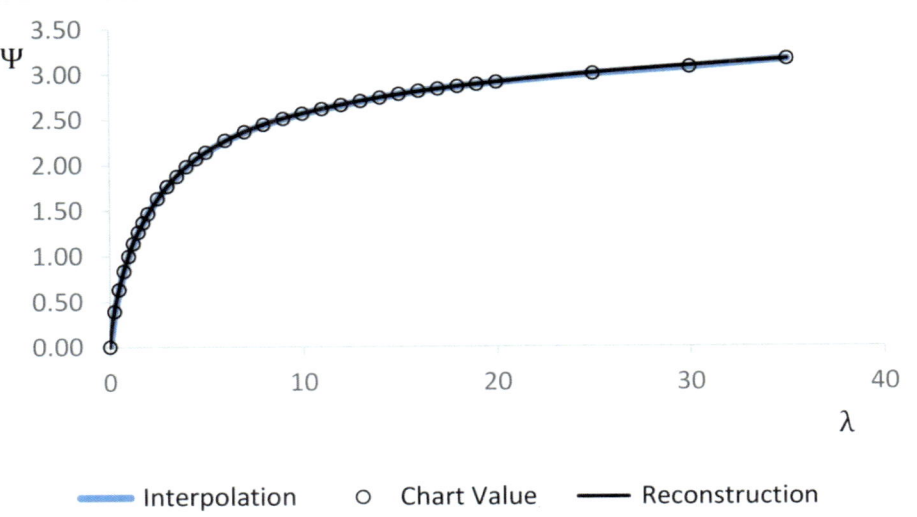

Fig. A23.7: Normalized Velocity Curve; *Oerlikon* chart values vs. polynomial function.

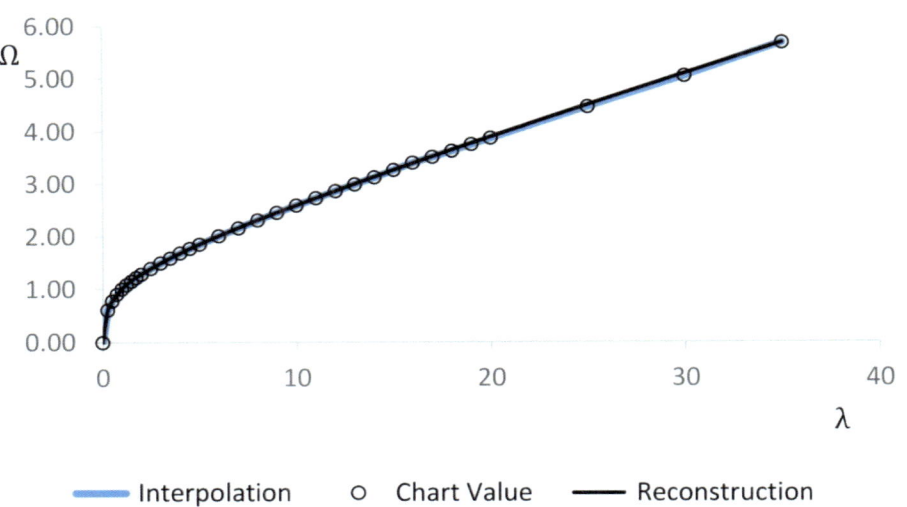

Fig. A23.8: Normalized Time Curve; *Oerlikon* chart values vs. polynomial function.

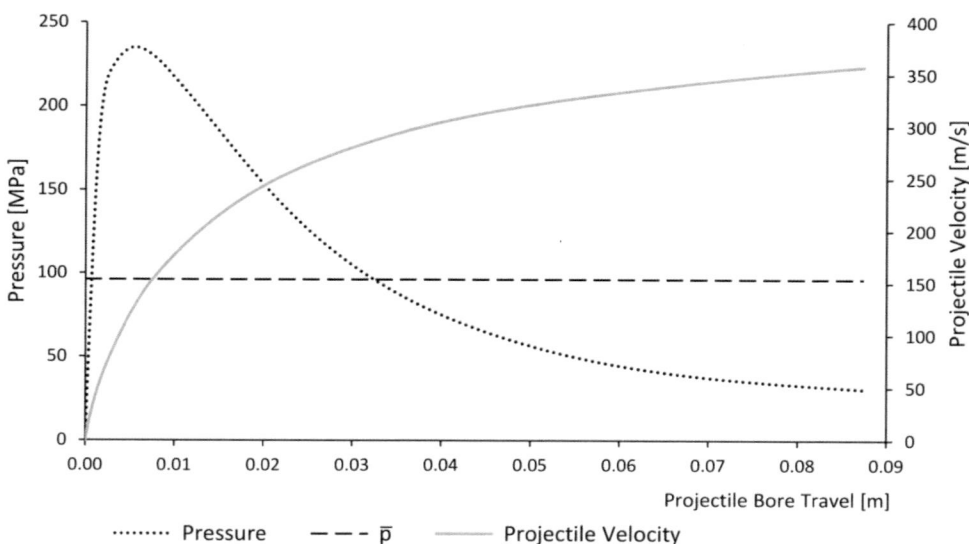

Fig. A23.9: Pressure, mean pressure and projectile velocity as a function of projectile base travel for Example 2, Table A19.2, p. 445, 87.5 mm bore travel.

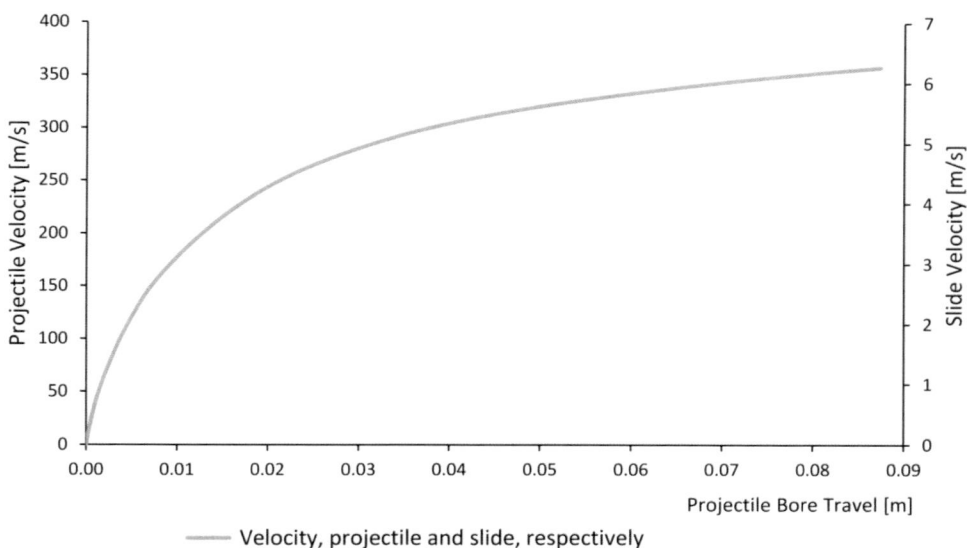

Fig. A23.10: Comparison of the curves of projectile and slide velocity of an 8 gram projectile of a 9 mm Luger running down the bore (using the data from Table A19.2, p. 445, as an example, bore travel is 87.5 mm). Both curves are identical. Their respective magnitude differs by the factor of the quotient of projectile mass to recoil mass.

Appendix 24   Lock Time and Striker Energy, Striker-Fired Pistols

The time between the release of the striker by the sear and the moment the primer is struck is called the lock speed, action time or lock time.[740] The calculation of this time interval requires the derivation of the oscillation equation. Appendix 19 deals with this topic. The calculation is of the same basic pattern as for section gamma, which is described in detail starting from page 442. For this reason, we do not derive formulas here and look at the practical case immediately.

The mechanical setup of striker fired actions corresponds to that of the horizontal spring-mass oscillator (cf. p. 419). A mass m_T is propelled forward by a compression spring whose spring rate is r. Before the striker nose starts to protrude from the breech face, its acceleration by the striker spring ends when the relaxing spring runs against a stop (Fig. A24.1) in the slide's striker channel. At this moment the striker has reached its maximum velocity v. From this point on it is in free flight to bridge the gap to the primer. For reasons of simplicity we assume that the motion continues from hereon at constant velocity v. The effect of a retract spring is neglected.

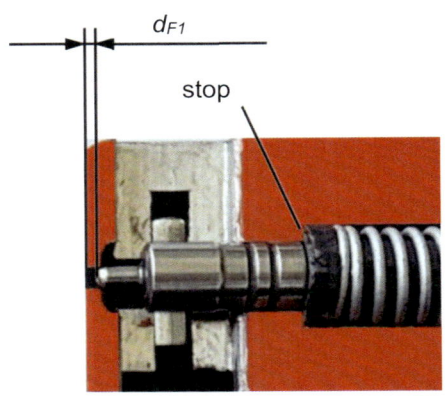

Fig. A24.1: Retracted striker, clearance d_{F1}.

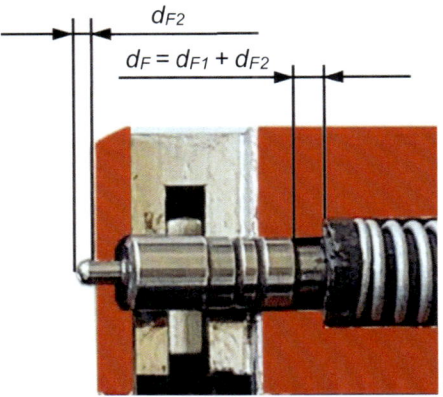

Fig. A24.2: Striker protrusion d_{F2} (exaggerated for clarity).

The free flight d_F is determined by the striker tip clearance d_{F1} from striker nose to breech face of the retracted striker and the distance d_{F2} it travels past the breech face. The protrusion d_{F2} can be roughly assumed to be 0.2 mm. This value takes into account the headspace, i.e. the distance between the breech face and the primer; in addition, the indentation until the impact-sensitive priming compound is set off.

The characteristic values of the process are computed as follows:

$$\omega = \sqrt{\frac{r}{m_T + \frac{1}{3}m_K}}$$ **A24.1**

$$\varphi = -\frac{\pi}{2}$$ **A24.2**

$$C = \frac{d_1 + x_p}{\sin\varphi}$$ **A24.3**

$$t_1 = \frac{\arcsin\left(\frac{d_1}{C}\right) - \varphi}{\omega}$$ **A24.4**

$$v = \omega\sqrt{C^2 - d_1^2}$$ **A24.5**

$$d_F = d_{F1} + d_{F2}$$ **A24.6**

$$t_2 = \frac{d_F}{v}$$ **A24.7**

$$t_{tot} = t_1 + t_2$$ **A24.8**

$$E_{kin} = \frac{m_T}{2}v^2$$ **A24.9**

where: ω Angular frequency [s^{-1}]
           φ Phase angle [rad]
           C Constant [m]
           d_1 Preload [m], mainspring; difference between free length L_0
              and length with striker in uncocked condition
           d_F Travel [m], free flight
           d_{F1} Clearance [m], retracted striker
           d_{F2} Travel [m], headspace and indent, approximately 0.0002 m
           E_{kin} Kinetic energy, striker [J]
           L_0 Free length [m]; mainspring
           m_K Mass, mainspring [kg]
           m_T Mass, striker [kg]
           r Spring rate, mainspring [N/m]
           t_1 Time, acceleration [s]
           t_2 Time, free flight [s]
           t_{tot} Lock time [s]
           v Velocity, striker [m/s], free flight
           x_p Displacement [m]; striker tension: difference between the
              spring length with striker in uncocked condition and its length
              when the sear disengages

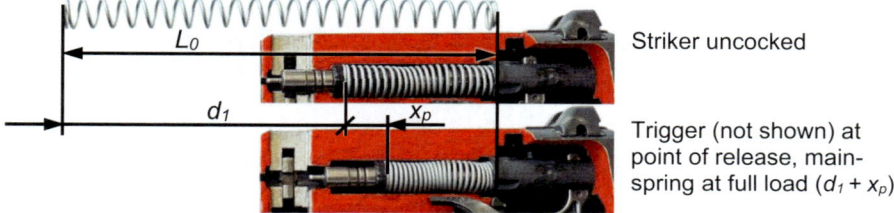

Fig. A24.3: Measurements L_0, d_1 and x_p.

The striker travel x shall be 0 at the point of release. At the end of the mainspring's relaxation $x = x_p$ applies. The mainspring exerts a force on the striker where x satisfies $0 \leq x \leq x_p$. The instantaneous velocity as a function of decompression travel x is

$$v(x) = \omega \sqrt{C^2 - (d_1 + x_p - x)^2} \qquad \qquad \textbf{A24.10}$$

The acceleration time t as a function of the mainspring's decompression x is obtained with

$$t(x) = \frac{1}{\omega}\left(arcsin\left(\frac{d_1 + x_p - x}{C}\right) - \varphi\right) \qquad \qquad \textbf{A24.11}$$

The instantaneous velocity as a function of time t is calculated from

$$v(t) = -C\omega\, cos(\omega t + \varphi) \qquad \qquad \textbf{A24.12}$$

Example: Lock time, striker-fired

Table A24.1: Specs, striker-fired firing mechanism.

Description	Value	Description	Value
Spring rate, mainspring	r [N/m]	600	Mass, striker	m_T [kg]	0.008
Striker tension	x_p [m]	0.0098	Mass, mainspring	m_K [kg]	0.002
Preload, mainspring	d_1 [m]	0.040	Clearance, retracted	d_{F1} [m]	0.001

Results based on values from Table A24.1 and Formulae A24.1 to A24.9:

$\omega = 263\ \mathrm{s}^{-1}$	$C = -0.05\ \mathrm{m}$	$t_1 = 2.4\ \mathrm{ms}$	$t_{tot} = 2.5\ \mathrm{ms}$
$\varphi = -1.57$	$v = 7.8\ \mathrm{m/s}$	$t_2 = 0.1\ \mathrm{ms}$	$E_{kin} = 0.244\ \mathrm{J}$

This is a theoretical consideration under ideal conditions. In reality, frictional losses and other adverse influences occur, which can reduce the impact energy and consequently increase the lock time. In the above example, t_{tot} is rounded up to whole milliseconds, hence $t_{tot} = 3$ ms is obtained.

If the striker was made of titanium (density: 4.506 g/cm³) instead of steel (7.87 g/cm³), it would accelerate to 10 m/s. The figures obtained are $t_{tot} = 2$ ms and $E_{kin} = 0.231$ J.

NB: Calculating the kinetic energy on the basis of the energy released by the mainspring, e.g. calculated according to Formula A19.10, p. 416, results in a larger value for E_{kin} than according to A24.9. The difference is explained by the consideration of the mass of the mainspring and the mass of the striker.

Fig. A24.4 illustrates the velocity curve and the impact energy as a function of the striker travel based on the data of Table A24.1.

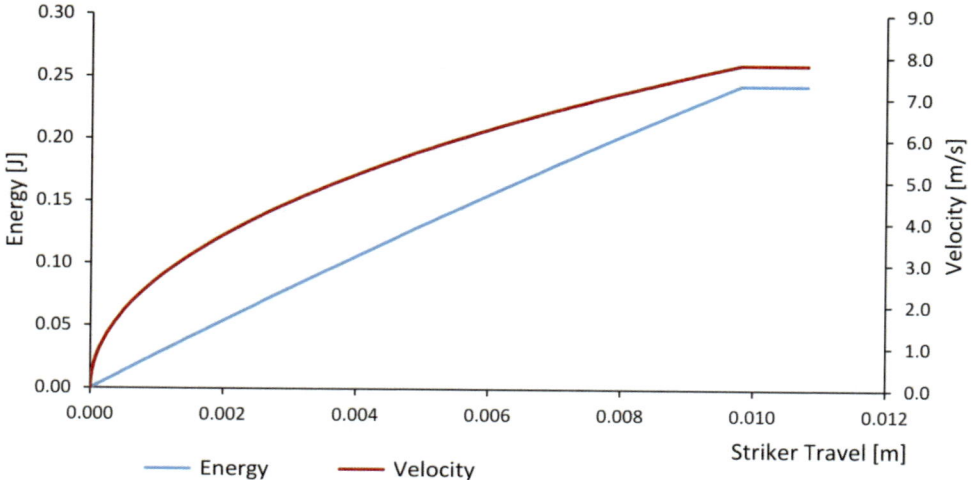

Fig. A24.4: Energy-travel and velocity-travel curves from start of striker motion until the end of motion based on the example of Table A24.1.

Primer-striking energy and copper crusher impression

The ignition sensitivity and the ignition insensitivity of primed casings is tested with a run-down test and a drop mass of 55 grams. The nose of the firing pin for this test has a hemispherical shape with a 1 mm radius.

The safety limit of ammunition for service pistols of the German police is defined at a drop height of 75 mm, which means no ignition is accepted at this drop level. When the functional limit is reached at 350 mm, no misfire may occur.[741]

The TR "Patrone" names the corresponding indent depths in copper crusher cylinders, which were determined under the same test conditions. Table A24.2 shows the values given there. In addition, a comparison of the impact energy is made, which is calculated from the drop height of the falling mass.

As can be seen from Table A24.2, the falling mass provides an impact energy of approximately 0.189 J at the functional limit and causes an indent depth of circa 0.3 mm. The E_{kin} of our example (0.244 J) exceeds the functional limit by 29 percent. Consequently, a good ignition reliability can be assumed for the striker action of our example.

Table A24.2: Kinetic energy E_{kin} and primer strike in copper crusher cylinders
as a function of the drop height of a 55 gram falling mass.[742]

	Drop Height [mm]	E_{kin} [J]	Indent Depth [mm]
	70	0.038	0.128
Safety Limit	75	0.040	N/A
	76	0.041	0.136
	80	0.043	0.146
	275	0.148	0.270
	300	0.162	0.282
	325	0.175	0.294
Functional Limit	350	0.189	0.304
	375	0.202	0.313
	400	0.216	0.319
	425	0.229	0.330
	450	0.243	0.341

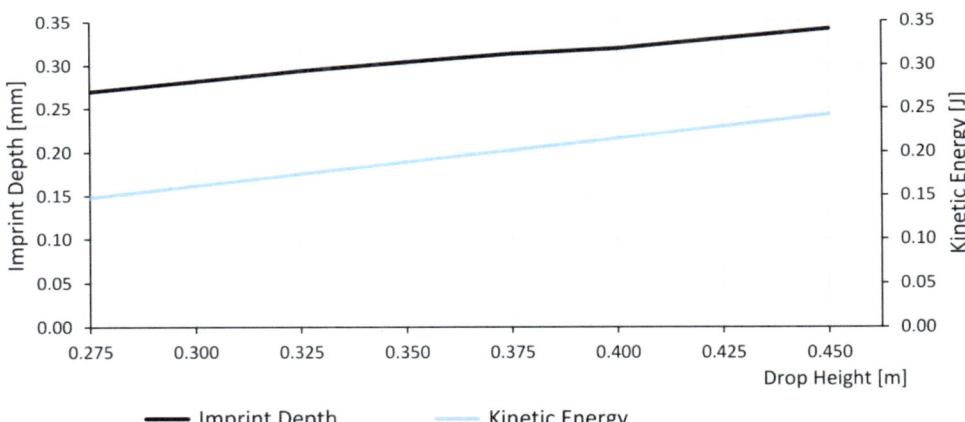

Fig. A24.5: Imprint in copper crusher cylinders and impact energy
as a function of the drop height of a 55 grams falling mass.

Note: The above data are "laboratory values". If the shape and diameter of the firing
pin nose differ from those specified for the run-down test, different results may be obtained for the imprint depth.

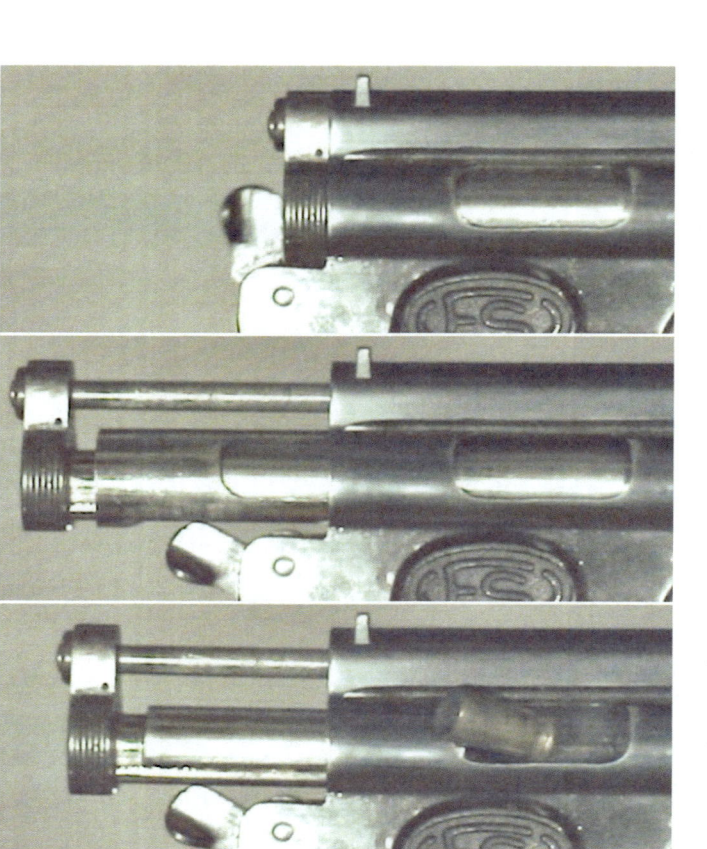

Hammer drops

Barrel and bolt
at backstop

Bolt is held back
temporarily as barrel
returns forward and
ejects the fired case

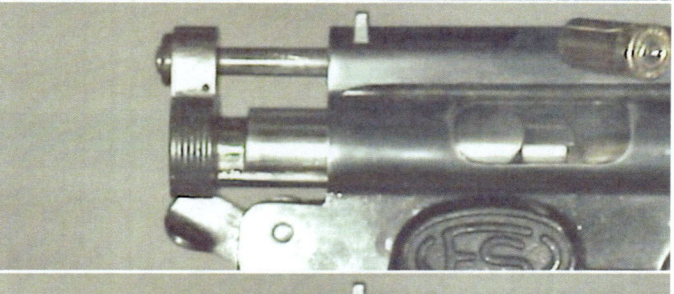

Bolt snaps forward and
feeds cartridge

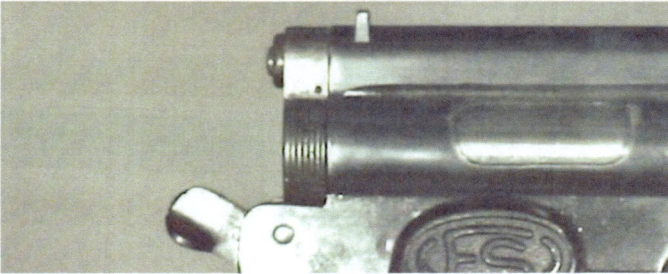

End of cycle

Long recoil operation cycle
shown on the example of a FROMMER *Stop*, cal. 7.65 Frommer

Bibliography

18 USC § 922. 2014. 18 US Code § 922 : Unlawful Acts. *Legal Information Institute.* [Online] Cornell University Law School, 2014. [Cited: October 27, 2014.] http://www.law.cornell.edu/uscode/text/18/922.

Aicher, Hans. 1984. Russische Sportpistole ij35. [ed.] Emil Schwend. *Deutsches Waffen-Journal.* Juli 1984.

—. **1983.** SIG-Sauer P226. [ed.] Emil Schwend. *Deutsches Waffen-Journal.* December 1983.

Allsop, D. F. and Toomey, M. A. 1999. *Small Arms : General Design.* London : Brassey's (UK) Ltd., 1999. 1-85753-250-3.

Arbeitsgruppe "Pflichtenheft Faustfeuerwaffen". 1975. Pflichtenheft Faustfeuerwaffen : Katalog der Konstruktions- und Funktionsmerkmale einer Faustfeuerwaffe für den Polizeidienst. June 19, 1975.

Armatix. 2011. *21st Century Gun Safety.* [Online] Armatix, 2011. [Cited: September 23, 2014.] http://www.armatix.de/?L=0.

Assmann, Bruno and Selke, Peter. 2011. *Technische Mechanik 3 – Kinematik und Kinetik.* 15. Auflage. München : Oldenbourg Wissenschaftsverlag GmbH, 2011. 978-3-486-59751-6.

ATF AFMER. 2018. Annual Firearms Manufacturing and Export Report. *ATF.* [Online] January 4, 2018. [Cited: February, 2018.] https://www.atf.gov/resource-center/data-statistics.

ATF Commerce Report. 2017. Firearms Commerce in the United States : Annual Statistical Update : 2017. *ATF.* [Online] 2018. [Cited: February, 2018.] https://www.atf.gov/resource-center/data-statistics.

ATF Factoring Criteria. 2008. Factoring Criteria for Weapons : ATF Form 4590 (5330.5). *U.S. Department of Justice : Bureau of Alcohol, Tobacco, Firearms and Explosives.* [Online] Revision: March, 2008. [Cited: March 13, 2013.] http://www.atf.gov/forms/download/atf-f-5330-5.pdf.

ATF Guidebook. 2009. Guidebook : Importation & Verification of Firearms, Ammunition, and Implements of War. *U.S. Department of Justice : Bureau of Alcohol, Tobacco, Firearms and Explosives.* [Online] 2009. [Cited: April 10, 2013.] http://www.atf.gov/files/firearms/guides/importation-verification/download/firearms-imporation-verification-guidebook--complete.pdf.

ATF Mission. 2016. Bureau of Alcohol, Tobacco, Firearms and Explosives. [Online] September 22, 2016. [Cited: April 27, 2017.] https://www.atf.gov/about.

ATF Reference Guide. 2005. Federal Firearms Regulations Reference Guide : ATF Publication 5300.4. *U.S. Department of Justice : Bureau of Alcohol, Tobacco, Firearms and Explosives.* [Online] September, 2005. [Cited: May 24, 2013.] http://www.atf.gov/files/publications/download/p/atf-p-5300-4.pdf.

ATF State Laws. 2010. State Laws and Published Ordinances : Firearms, 2010–2011 : ATF Publication 5300.5. *U.S. Department of Justice : Bureau of Alcohol, Tobacco, Firearms and Explosives.* [Online] 31st Edition, 2010. [Cited: May 24, 2013.] https://www.atf.gov/files/publications/download/p/atf-p-5300-5-31st-editiion/2010-2011-atf-book-final.pdf.

Awiszus, Birgit, et al. 2016. *Grundlagen der Fertigungstechnik.* 6. Auflage. München : Hanser Verlag, 2016. 978-3-446-44779-0.

Ayoob, Massad. 2013. Glocks On Duty : Why The Glock Has Long Since Become The Preeminent Police Pistol. [ed.] Stanley R. Harris. *Glock Autopistols Annual.* Vol. 19 No. 1, 2013.

—. **2007.** *The Gun Digest Book of Combat Handgunnery.* 6th Edition. Iola : Krause Publication, 2007. 978-0-89689-525-6.

Bahners, Patrick. 2012. *Verfassungsgemäß bewaffnet unterwegs.* [ed.] Werner D'Inka, et al. Frankfurt am Main : Frankfurter Allgemeine Zeitung GmbH, 2012.

Barnes, Frank C. 2012. *Cartridges of the World.* [ed.] Richard A. Mann. 13th Edition. Iola : Krause Publications, 2012. 978-1-4402-3059-2.

Barrett, Paul M. 2012a. Buying Power : The Obama Gun Surge. [Online] Bloomberg Businessweek, October 11, 2012. [Cited: August 2, 2013.] http://www.businessweek.com/articles/2012-10-11/the-obama-gun-surge.

—. **2012b.** *Glock : The rise of America's gun.* New York : Crown Publishing Group, 2012. 978-0-307-71993-5.

BAuA. 2014. Bundesanstalt für Arbeitsschutz und Arbeitsmedizin: Gefährdungsbeurteilung.de. *Thermische Gefährdungen; heiße Medien/Oberflächen.* [Online] 2014. [Cited: June 26, 2014.] http://www.gefaehrdungsbeurteilung.de/de/ gefaehrdungsfaktoren/thermische_gefaehrdungen/heiss.

Beeley, P. R. and Smart, R. F. 2008. *Investment Casting.* Paperback Edition of the 1995 1st Edition. Leeds : Maney Publishing, 2008. 978-1-906540-57-9.

Beretta. 2014. *Accessories for Semi Automatic Shotguns : Gun Pod Instructions.* [Online] 2014. [Cited: April 4, 2014.] http://www.berettausa.com/products/berettagunpodunit/.

—. **2013.** *Beretta Press Kit i-PROTECT.* [DVD-ROM] Gardone : Areacom 51, 2013. CI0013701403.

Berz, Peter. 2001. *08/15 : Ein Standard des 20. Jahrhunderts.* München : Wilhelm Fink Verlag, München, 2001. 3-7705-3507-3.

Blank. Feinguss-Blank. *Feingussverfahren und Feingussprozess im Überblick.* [Online] Feinguss Blank GmbH. [Cited: April 26, 2014.] http://www.feinguss-blank.de/de/technologien-loesungen/feinguss/feingussverfahren-feingussprozess-keramiktauchen/.

Boberg. 2014. *Boberg XR9-L : Power, Prestige, Performance.* [Prospekt]. White Bear Lake : Boberg Arms Corporation, 2014. p. 6.

Bock, Gerhard. 1941. *Moderne Faustfeuerwaffen und ihr Gebrauch.* 3. Auflage. Berlin-Friedenau : Verlag J. Neumann Neudamm, 1941.

Bock, Gerhard, et al. 1989. *Handbuch der Faustfeuerwaffen.* 8. Auflage. Melsungen : Neumann-Neudamm, 1989. 3-7888-0497-1.

Böge, Alfred and Böge, Wolfgang. 2019. *Technische Mechanik : Statik – Reibung – Dynamik – Festigkeitslehre – Fluidmechanik.* 33. Auflage. Wiesbaden : Springer Vieweg, 2019. 978-3-658-25723-1.

Böhlein, Karl. 1967. Die Patronenzündung und ihre Wirkung in Munition und Waffe. [ed.] Emil Schwend. *Deutsches Waffen-Journal.* August 1967.

Bolotin, David Naumovich. 1990. *Sovetskoe Strelkovoe Oruzhie [Sovjet Firearms].* 3rd Edition. Moskau : Voenisdat, 1990. 5-203-00631-8.

—. **1995.** *Soviet Small-Arms and Ammunition.* [ed.] John Walter and Heikki Pohjolainen. [trans.] Igor F. Naftul'eff. Hyvinkää : Suomen Asemuseosäätiö Finish Arms Museum Foundation, 1995. 951-97184-1-9.

Boßlet, Joachim. 2009. *Tenifer-/QPQ-Verfahren.* V.5. Mannheim : Durferrit GmbH, 2009.

—. **2007.** *Tenifer-QPQ-Verfahren.* Mannheim : Durferrit, 2007.

Brandeis, Friedrich. 1881. *Die moderne Gewehrfabrikation : Praktisches Hand- und Lehrbuch.* [ed.] Gesellschaft von Künstlern, technischen Schriftstellern und Fachgenossen. Weimar : Bernhard Friedrich Voigt, 1881. Vol. 131.

Brauer, Jurgen. 2013. *The US Firearms Industry : Production and Supply.* Geneva : Small Arms Survey, 2013. p. 98. 978-2-9700856-0-7.

Brukner, Bruno. 2003. *Faustfeuerwaffen : Technik und Schießlehre.* Schwäbisch Hall : DWJ Verlags GmbH, 2003. 3-936632-26-x.

—. **1983.** *Faustfeuerwaffen : Technik und Schießlehre.* Melsungen : Verlag J. Neumann-Neudamm, 1983. 3-7888-0394-0.

Brunner, Sybille and Kehrle, Karl. 2014. *Volkswirtschaftlehre.* 3. Auflage. München : Verlag Franz Vahlen GmbH, 2014. 978-3-8006-4769-9.

Bundesministerium der Justiz. 2001. Gesetze im Internet. *Strahlenschutzverordnung.* [Online] July 20, 2001. [Cited: February 26, 2014.] http://www.gesetze-im-internet.de/bundesrecht/strlschv_2001/gesamt.pdf.

Burhop, Carsten. 2011. *Wirtschaftsgeschichte des Kaiserreichs : 1871–1918.* Göttingen : Vandenhoeck & Ruprecht GmbH & Co. KG, 2011. 978-3-8252-3454-6.

Bussard, Michael. 2017. *Ammo Encyclopedia.* [ed.] John B. Allen, David Kosowski and Charles F. Priore. 6th Edition. Minneapolis : Blue Book Publications, Inc., 2017. 978-1-936120-91-8.

Buxton, Warren H. 1978. *The P.38 Pistol : The Walther Pistols 1930–1945.* Second Printing 1999. Dallas : Taylor Publishing Company, 1978. Vol. 1. 0-87833-303-7.

—. **2009.** WaltherForums. *Ejection Port, Post #10.* [Online] March 9, 2009. [Cited: June 27, 2014.] http://www.waltherforums.com/forum/p5/9409-ejection-port.html.

C.I.P. 2011. *Commission Internationale Permanente pour l'Epreuve des Armes à Feu portatives.* [Datenträger] [ed.] Bureau Permanent de la Commission Internationale Permanente. Edition 2011, Brüssel : s.n., 2011.

California Code of Regulations. 2013. CCR : Title 11 (Law) : Division 5 (Firearms Regulations) : Chapter 5 (Laboratory Certification and Handgun Testing) : Article 4 (Operational Requirements) : § 4059 (Which Handguns Must Be Tested, Who May Submit Handguns, [...]). *State of California : Department of Justice : Office of the Attorney General : Bureau of Firearms : California Office of Administrative Law.* [Online] 2013. [Cited: March 27, 2013.] http://government.westlaw.com/linkedslice/default.asp?SP=CCR-1000.

California Firearms Laws. 2007. CFL 2007. *California Department of Justice : Edmund G. Brown Jr. : Attorney General.* [Online] January 1, 2007. [Cited: March 27, 2013.] https://oag.ca.gov/sites/all/files/pdfs/firearms/forms/Cfl2007.pdf.

California Penal Code. 2013. CPC : State of California : Department of Justice. *Office of the Attorney General : Bureau of Firearms.* [Online] 2013. [Cited: March 28, 2013.] http://www.leginfo.ca.gov/cgi-bin/displaycode?section=pen&group=31001-32000&file=32000-32030.

Caranta, Raymond. 1986. The Extraordinary Glock. [ed.] Dan Shideler and Corrina Peterson. *GunDigest Book of Classic Combat Handguns.* Iola : F+W Media, Inc., 1986. 2011. 978-1-4402-2384-6.

Cerakote. 2014. Cerakote Firearms Coatings. [Online] 2014. [Cited: May 22, 2014.] http://www.cerakoteguncoatings.com/.

Chinn, George Morgan. 1955. *The Machine Gun : Design Analysis of Automatic Firing Mechanisms and Related Components.* Washington D.C. : s.n., 1955. Vol. 4.

—. 1987. *The Machine Gun : Development of Full Automatic Machine Gun Systems, High Rate of Fire Power Driven Cannon, and Automatic Grenade Launchers by the United States and her Allies, following World War II, Korean Police Action, and the Vietnam Conflict.* USA : s.n., 1987. Vol. 5.

—. 1952. *The Machine Gun : History, Evolution and Development of Manual, Automatic, and Airborne Repeating Weapons by the Soviet Union and her Satellites.* 1952. Vol. 2.

Christensen, Clayton M. 1997. The Innovator's Dilemma. Translation of the 1997 1st Edition [trans.] Kurt Matzler and Stephan Friedrich von den Eichen. München : Verlag Franz Vahlen GmbH, 1997. 2011. 978-3-8006-3791-1.

CIP. 2011. *Commission Internationale Permanente pour l'Epreuve des Armes à Feu portatives.* [CD-ROM] [ed.] Bureau Permanent de la Commission Internationale Permanente. Edition 2011, Brüssel : s.n., 2011.

Code of Federal Regulations. 2008. Code of Federal Regulations : Alcohol, Tobacco and Firearms : Title 27 CFR Part 478 : Commerce in Firearms and Ammunition : Subpart F : Conduct of Business : § 478.92 : How must licensed manufacturers and licensed importers identify firearms [...]. [Online] 2008, Government Printing Office, 2008. [Cited: June 11, 2013.] http://www.ecfr.gov/cgi-bin/text-idx?c=ecfr&SID=8d44544b33b353a347b721e7d500a141&rgn=div8&view=text&node=27:3.0.1.2.3.6.1.2&idno=27.

Code of Massachusetts Regulations 501 CMR 7.02. 2007. The Official Website of the Attorney General of Massachusetts : Code of Massachusetts Regulations : 501 CMR 7.02 : Approved Weapon Rosters. [Online] 2007. [Cited: May 10, 2013.] http://www.mass.gov/eopss/docs/chsb/firearms/501-cmr-7.pdf.

Code of Massachusetts Regulations 940 CMR 16.00. 1998. Code of Massachusetts Regulations : 940 CMR 16.00 et seq. : Attorney General's Handgun Sales Regulations : Handgun Sales. *The Official Website of the Attorney General of Massachusetts.* [Online] 1998. http://www.mass.gov/ago/government-resources/ags-regulations/940-cmr-1600.html.

Congressional Research Service. Cornell University Law School. *Bearing Arms, Second Amendment.* [Online] [Cited: February 26, 2014.] http://www.law.cornell.edu/anncon/html/amdt2_user.html#amdt2_hd1.

Consumer Product Safety Act. 1972. 15 USC §§ 2051–2089. [Online] 1972. [Cited: July 18, 2013.] http://www.cpsc.gov/PageFiles/105435/cpsa.pdf.

Cowgill, James P. 1981. Walther's P5 Pistol. [ed.] Dan Shideler and Corrina Peterson. *GunDigest Book of Classic Combat Handguns.* Iola : F+W Media, Inc., 1981. 2011. 978-1-4402-2384-6.

Cranz, Carl, Poppenberg, O. and Eberhard, O. von. 1926. *Lehrbuch der Ballistik : Innere Ballistik : Die Bewegung des Geschosses durch das Rohr und ihre Begleiterscheinungen.* [ed.] Carl Cranz. Nachdruck der Ausgabe von 1926. Berlin : Julius Springer, 1926. Vol. 2. 978-3-642-52558-2.

Dangerous Weapons Control Law. 2008. DWC : §§ 12125–12133 : Handgun Safety Testing : Chapter 1.3 : Unsafe Handguns. *State of California : Department of Justice : Bureau of Firearms.* [Online] 2008. [Cited: May 17, 2013.] http://ag.ca.gov/firearms/dwcl/dwc.pdf.

Dannecker, Peter and Schulz, Walter. 2017. Ganz neue Verriegelung. [ed.] Walter Schulz. *Deutsches Waffen-Journal.* October 2017.

Dannecker, Peter. 2016. *Verschlusssysteme von Feuerwaffen.* 4. Auflage. Blaufelden : dwj Verlags-GmbH, 2016. 978-3-936632-97-2.

—. **2015.** Vertikalblock auf Italienisch. [ed.] Walter Schulz. *Deutsches Waffen-Journal.* Mai 2015.

de Vries, Guus and Martens, Bas J. 2004. *Die Maschinenpistole MP 38, 40, 40/1 und 41.* [trans.] Torsten Verhülsdonk. Herne : VS-Books Carl Schulze & Torsten Verhülsdonk, 2004. Vol. 2. 3-932077-23-7.

Deichsel, Alexander and Schmidt, Manfred. 2010. *Jahrbuch Markentechnik 2011/2012.* [ed.] Manfred Schmidt. Wiesbaden : Gabler Verlag, 2010. 978-3-8349-2533-6.

Dern, Heinz-Jürgen and Schlünken, Sigrid. 2012. *Phenolharze (Phenoplaste, PF-Harze).* [ed.] Peter Elsner, Peter Eyerer and Thomas Hirth. 8. Auflage. Heidelberg : Springer-Verlag, 2012. p. 1494. 978-3-642-16172-5.

Diller, Herman. 2000. *Preispolitik.* 3. Auflage. Stuttgart : Verlag W. Kohlhammer, 2000. 3-17-016670-0.

Dobrinski, Paul, Krakau, Gunter and Vogel, Anselm. 2010. *Physik für Ingenieure.* 12. Auflage. Wiesbaden : Vieweg+Teubner, 2010. 978-3-8348-0580-5.

Dockery, Kevin. 2007. *Future Weapons.* New York, NY, USA : The Berkley Publishing Group, 2007. 978-0-425-21750-4.

—. **2012.** *The M60 Machine Gun.* [ed.] Martin Pegler. Oxford : Osprey Publishing Ltd., 2012. Vol. WPN020. 978-1-84908-844-2.

Doege, Eckart and Behrens, Bernd-Arno. 2010. *Handbuch der Umformtechnik : Grundlagen, Technologien, Maschinen.* 2. Auflage. Berlin : Springer-Verlag Berlin Heidelberg, 2010. 978-3-642-04248-5.

Dynamit Nobel. 1988. *Wiederladen : Ein praktisches Handbuch für Jäger und Schützen.* 3. Auflage. Troisdorf : s.n., 1988. Nr. 029975 / 3 / 0588 / KÖ.

Ebell, Max. 1921. *Wilhelm Mauser – ein deutscher Erfinder : Sein Leben an Hand seiner Briefe.* München : C. H. Beck'sche Verlagsbuchhandlung, 1921.

Eckardt, Werner and Morawietz, Otto. 1973. *Die Handwaffen des brandenburgisch-preußisch-deutschen Heeres 1640–1945.* 2. Auflage. Hamburg : Helmut Gerhard Schulz Verlag, 1973.

Eco, Umberto. 2010. *Wie man eine wissenschaftliche Abschlußarbeit schreibt :
Doktor-, Diplom- und Magisterarbeit in den Geistes- und Sozialwissenschaften.*
[trans.] Walter Schick. 13. Auflage. Wien : facultas.wuv, 2010. 978-3-8252-1512-5.

Edwards, Chris. 2013. New Gen4 Models. [ed.] Stanley R. Harris.
Glock Autopistols Annual. Vol. 19 No. 1, 2013.

Egelko, Bob. 2013. Gun control : Cartridge ID law to take effect : Patent expires,
clearing hurdle for micro-stamping technology : NRA threatens to sue. *San Francisco
Chronicle.* [Online] Hearst Communications Inc., May 18, 2013. [Cited: July 19,
2013.] http://www.sfchronicle.com/news/article/Gun-control-Cartridge-ID-law-to-take-
effect-4527165.php?t=87572811fc.

Ehrenfried, M. 1984. Laufprofile : Ein Vergleich des Zug-Feld-Profils mit dem
Polygonprofil. [ed.] Emil Schwend. *Deutsches Waffen-Journal.* August 1984.

Eilers, Konrad. 1938. *Handbuch der praktischen Schusswaffenkunde und
Schießkunst für Jäger und Sportschützen.* 4. Auflage. Berlin : Verlag von Paul Parey,
1938.

Einstein, Albert and Infeld, Leopold. 1938. *Die Evolution der Physik.* 2014,
Nachdruck der deutschen Ausgabe von 1950. New York : Anaconda Verlag GmbH,
1938. 978-3-7306-0086-3.

Elbe, Ronald E. 1975. *External Barrel Temperature of the M16A1 Rifle.*
Rock Island, IL : Rock Island Arsenal, 1975. p. 25.

Emri, Igor and Gonzalez-Gutierrez, Joamin. 2014. Improving Powder Injection
Molding : An Opportunity for the Aerospace Industry. *Science & Education of the
Bauman MSTU.* s.l. : Bauman Moscow State Technical University, 2014.
ISSN 1994-0448.

ER "Pistolen". 2008. Erprobungsrichtlinien zur Technischen Richtlinie : Pistolen im
Kaliber 9 mm x 19. *Münster : Polizeitechnisches Institut der Deutschen Hochschule
der Polizei : Waffen-/Gerätetechnik.* [Online] Revision: January, 2008.
[Cited: August 10, 2012.] http://www.pfa.nrw.de/PTI_Internet/pti-intern.dhpol.local/
WG/Regelungen/Pistolen/TR_01_08/ER-Pistole_mit_Anlagen_31-01-08.pdf.html.

Ezell, Edward Clinton. 1981. *Handguns of the World : Military Revolvers and Self-
Loaders from 1870 to 1945.* London : Arms and Armour Press, 1981. 0-85368-504-5.

—. **1983.** *Small Arms of the World.* 12th Edition. Houston : Barnes & Noble, Inc.,
1983. 0-88029-601-1.

—. **1984.** *The Great Rifle Controversy.* Harrisburg : Stackpole Books, 1984.
0-85368-686-6.

Ferner, Matt. 2013. The Huffington Post: Bill That Bans High-Capacity Magazines :
Limits Them To 15 Rounds Passes Colorado Senate Committee. [Online] March 5,
2013. [Cited: January 27, 2014.] http://www.huffingtonpost.com/2013/03/04/bill-that-
bans-high-capac_n_2808548.html.

Field, Leslie E. and Martens, Bas J. 2011. *Ott-Helmuth von Lossnitzer : Technical
Director of the Mauser Company : 1933-1945 : An Oral Recollection.*
[ed.] Bas J. Martens. Arnheim : SAM Wapenmagazine, 2011. 978-9-08-173780-7.

Field, Robert. 1999. Getrennte Eltern : Walther KSP200. [ed.] Michael Schwend and
Gerhard Wirnsberger. *Deutsches Waffen-Journal.* Januar 1999.

Firearms Industry. 2013. Pistol Production by U.S. Manufacturers 2011. *U.S. Firearms Industry.* [Online] 2013. [Cited: August 15, 2013.] http://www.shootingindustry.com/u-s-firearms-industry-today-2013/.

FirmenABC. 2019. Glock Gesellschaft m.b.H. [Online] 2019. [Cited: April 30, 2019.] https://www.firmenabc.at/glock-gesellschaft-m-b-h_HTxx.

Fjestad, Steve P. 2017. *Blue Book of Gun Values.* 38th Edition. Minneapolis : Blue Book Publications, Inc., 2017. 978-1-936120-90-1.

Fleck, A. 1901. Die Entwickelung der Faustfeuerwaffe. [ed.] E. Hartmann. *Kriegstechnische Zeitschrift.* 1901, 4. Jahrgang, p. 572.

FN Herstal. *FN SmartCore.* [Online] FNH. [Cited: May 25, 2014.] http://www.fnherstal.com/primary-menu/products-capabilities/shot-counter/general.html.

Franck, Adolf. 2000. *Kunststoff-Kompendium.* 5., überarbeitete Auflage. Würzburg : Vogel Verlag und Druck, 2000. 3-8023-1855-2.

Frankonia. 1990. Gesamtjahres-Katalog 90/91. Würzburg : s.n., August 1990. p. 600.

Gebhardt, Andreas. 2016. *Additive Fertigungsverfahren : Additive Manufacturing und 3D-Drucken für Prototyping – Tooling – Produktion.* 5. Auflage. München : Carl Hanser Verlag, 2016. 978-3-446-44401-0.

Gebhardt, Andreas, Kessler, Julia and Thurn, Laura. 2016. *3D-Drucken : Grundlagen und Anwendungen des Additive Manufacturing (AM).* 2. Auflage. München : Carl Hanser Verlag, 2016. 978-3-446-44672-4.

Geek, Jerry The. 2013. California finds "Perfect Storm" to drown the Second Amendment. [Online] Blogger.com, June 21, 2013. [Cited: July 19, 2013.] http://jerrythegeek.blogspot.de/2013_06_16_archive.html.

Gerthsen, Christian. 2015. *Gerthsen Physik.* [ed.] Dieter Meschede. 25. Auflage. Bonn : Springer Verlag GmbH, 2015. 978-3-662-45976-8.

Glock IM. 2013. Glock Safe Action Pistolen, alle Modelle. [Owner's Manual]. Deutsch-Wagram, Austria : Glock, 2013. p. 58. Glock 33250 / 08 13.

Glock, Gaston. 2014. Annual Message from the Founder. [ed.] Stanley R. Harris. *Glock Autopistols Annual.* 2014, Vol. 20 No. 1.

—. 2013. *Glock Autopistols Annual.* [ed.] Stanley R. Harris. New York : Harris Publications, Inc., 2013.

Goldratt, Eliyahu M. and Cox, Jeff. 1984. *Das Ziel : Ein Roman über Prozessoptimierung.* Newly revised and extended 2013 edition. Frankfurt : Campus Verlag GmbH, 1984. 978-3-593-39853-2.

Görtz, Joachim and Sturgess, Geoffrey. 2011. *The Borchardt & Luger Automatic Pistols : A Technical History for Collectors from C93 to P.08.* Galesburg : Brad Simpson Publishing & G. I. Sturgess, 2011. Vol. III. 978-0-9727815-7-2.

Görtz, Joachim. 1981. Das Patronenlager der Pistole 08 : Der seltsame Ring. [ed.] Emil Schwend. *Deutsches Waffen-Journal.* April 1981.

—. 1988. *Die Pistole 08.* 2. Auflage. Dietikon-Zürich : Verlag Stocker-Schmid AG, 1988. 3-7276-7065-7.

—. 2004. *Die Pistole 08.* 2. Auflage. Dietikon-Zürich : Verlag Stocker-Schmid AG, 2004. 3-7276-7065-7.

Götz, Hans-Dieter. 1981. *Die deutschen Militärgewehre und Maschinenpistolen 1871–1945.* 3. Auflage. Stuttgart : Motorbuch Verlag, 1981. 3-87943-350-X.

—. **1996.** *Militärgewehre und Pistolen der deutschen Staaten 1800–1870.* 2. Auflage. Stuttgart : Motorbuch Verlag, 1996. 3-87943-533-2.

—. **1979.** *Waffenkunde für Sammler : Vom Luntenschloss zum Sturmgewehr.* 5. Auflage. Stuttgart : Motorbuch Verlag, 1979. 3-87943-303-8.

Greene, Mark. 2013. *A Review of Gun Safety Technologies.* Office of Justice Programs, National Institute of Justice. Washington DC : National Institute of Justice, 2013. NCJ242500.

Greener, William Wellington. 1910. The Gun and its Development. 9th Edition (2002 Reprint) London : Cassell & Co., 1910. The Lyons Press, 2002. 1-58574-734-3.

Greiner, Walter, Neise, Ludwig and Stöcker, Horst. 1993. *Theoretische Physik : Thermodynamik und Statistische Mechanik.* 2. Auflage. Thun : Verlag Harri Deutsch, 1993. Vol. 9. 3-8171-1262-9.

Gross, Dietmar, et al. 2019a. *Technische Mechanik 1 : Statik.* 14. Auflage. Berlin : Springer Vieweg, 2019. 978-3-662-59156-7.

—. **2019b.** *Technische Mechanik 3 : Kinetik.* 14. Auflage. Berlin : Springer Vieweg, 2019. 978-3-662-59550-3.

Gun Test Team. 2012. Options for Concealed Carry : Two Nines vs. a Forty Wheelgun. [ed.] Todd Woodward. *Gun Test.* 2012, July.

Günther, Reinhold. 1900. *Bergmann's Rückstoßlader.* EOD Reprint. Berlin : Militär-Verlagsanstalt, 1900.

H.E.F. Groupe. 2002. *Vakuumbeschichtungen.* Edition no. 2. Andrézieux-Bourthéon Cedex : s.n., 2002. p. 12.

Handgunlaw.us. 2014. State Restrictions on Magazines, Chemical Sprays and Stun Guns. [Online] January 6, 2014. [Cited: January 7, 2014.] http://www.handgunlaw.us/documents/NoHiCapChemSpray.pdf.

Handrich, Hans-Dieter. 1993. *Vom Gewehr 98 zum Sturmgewehr.* [ed.] Wolfram Funk, et al. Herford : Verlag E.S. Mittler & Sohn GmbH, 1993. Vol. 8. 3-8132-0409-X.

Hartmann, E., [ed.]. 1901. *Kriegstechnische Zeitschrift : Für Offiziere aller Waffen : Zugleich Organ für kriegstechnische Erfindungen und Entdeckungen auf allen militärischen Gebieten.* Berlin : Ernst Siegfried Mittler und Sohn, 1901. p. 572. Vol. 4.

Hatcher, Julian S. 1966. *Hatcher's Notebook.* 3rd Edition. Harrisburg : The Stackpole Company, 1966. 0-8117-0795-4.

Hecht, Eugene. 2009. *Optik.* [trans.] Dr. Anna Schleitzer. 5. Auflage. München : Oldenbourg Wissenschaftsverlag GmbH, 2009. 978-3-486-58861-3.

Heckler & Koch. 1970. *Der beweglich abgestützte Rollenverschluß, seine Kinematik und sein konstruktiver Aufbau.* 4. Auflage 1985, inhaltlich identischer Nachdruck der 1970 erschienenen 1. Auflage. Oberndorf : HK, 1970.

—. **2014.** VP9 Operator's Manual. [Owner's Manual]. Oberndorf/Neckar : s.n., 2014. p. 34. HK USA-05282014.

Heydenreich. 1908. *Die Lehre vom Schuß für Gewehr und Geschütz.* Berlin : Ernst Siegfried Mittler und Sohn, 1908.

Heym. 2014. Heym-Läufe. *Lauffertigung.* [Online] 2014. [Cited: May 6, 2014.] http://www.heym-läufe.de/public/laufherstellung.html.

HK Archives. 2014. Heckler & Koch Archives. *VP9 ... Striker Fired Perfection.* [Online] 2014. [Cited: June 30, 2014.] http://www.gandrtactical.com/images/archive/HK/VP9%20Product%20Sheet%20JUN E.pdf. HK-USA06022014.

Hlebinsky, Ashley. 2016. Glock Makes History. [ed.] Stanley R. Harris. *Glock Autopistols Annual.* 2016, Vol. 22 No. 1.

Hoenow, Gerhard and Meißner, Thomas. 2012. *Konstruktionspraxis im Maschinenbau.* 3. Auflage. München : Carl Hanser Verlag, 2012. 978-3-446-43082-2.

Hoffmann, Henning. 2012. *Feuerkampf & Taktik : Taktischer Schusswaffengebrauch im 21. Jahrhundert.* 3. Auflage. Blaufelden : dwj Verlags-GmbH, 2012. 978-3-936632-71-2.

Hogg, Ian Vernon. 2001. *German Handguns : The Complete Book of the Pistols and Revolvers of Germany, 1869 to the Present.* London : Greenhill Books, 2001. 1-85367-461-3.

Horsch, Florian. 2014. *3D-Druck für alle : Der Do-it-yourself-Guide.* 2. Auflage. München : Carl Hanser Verlag, 2014. 978-3-446-44261-0.

Hounshell, David A. 1985. *From the American System to Mass Production, 1800– 1932 : The Development of Manufacturing in the United States.* Baltimore : The Johns Hopkins University Press, 1985. 0-8018-3158-X.

Hüttel, Klaus. 2007. Marktsegmentierung durch produktpolitische Maßnahmen. [book auth.] Werner Pepels. [ed.] Werner Pepels. *Marktsegmentierung : Erfolgsnischen finden und besetzen.* 2. Auflage. Düsseldorf : Symposion Publishing GmbH, 2007.

IDPA. 2019. IDPA.COM. *2017 IDPA Rulebook.* [Online] 2019. [Cited: April 3, 2019.] https://www.idpa.com/wp-content/themes/idpa/assets/match-files/2017_Rule_ Book.pdf.

Information Bulletin 2013-BOF-03. 2013. Information Bulletin No. 2013-BOF-03 : Certification of Microstamping Technology. *Information Bulletins : State of California : Department of Justice.* [Online] May 17, 2013. [Cited: July 19, 2013.] https://oag.ca.gov/sites/all/files/agweb/pdfs/firearms/infobuls/2013-BOF-03.pdf?.

IPSC. 2017. Handgun Competition Rules. [Online] January 2017 Edition, 2017. [Cited: June 10, 2017.] http://www.ipsc.org/.

Jentzsch, Robert B. 1902. *Ist "Kaisertreu" Wahrheitstreu?* Wien : Selbstverlag des Verfassers, 1902. p. 85.

Johnson, Steve. 2014. The Firearm Blog. *Beretta PX4i Storm (i-Protect System).* [Online] April 22, 2014. [Cited: May 25, 2014.] http://www.thefirearmblog.com/blog/2014/04/22/beretta-px4i-storm-i-protect-system/.

Johnston, Gary Paul and Nelson, Thomas B. 2010. *The World's Assault Rifles.* 2nd Edition. Lorton, Virginia, USA : Ironside International Publishers, Inc., 2010. 978-0935554007.

Kahaner, Larry. 2007. *AK-47 : The Weapon That Changed The Face Of War.* Hoboken : John Wiley & Sons, Inc., 2007. 978-0-470-16880-6.

Kaisertreu. 1902. *Die principiellen Eigenschaften der automatischen Feuerwaffen : Eine Studie über die neuesten Errungenschaften der Waffentechnik für Officiere aller Waffen.* Wien : Verlag Wilh. Braumüller & Sohn, k. u. k. Hof- und Universitäts-Buchhandlung, 1902. (Karel Krnka, Ps. Kaisertreu).

Kamiske, Gerd F. and Brauer, Jörg-Peter. 2006. *Qualitätsmanagement von A bis Z : Erläuterungen moderner Begriffe des Qualitätsmanagements.* 5. Auflage. München : Carl Hanser Verlag, 2006. 3-446-40284-5.

Kasler, Peter Alan. 1992. *GLOCK : The New Wave in Combat Handguns.* Boulder, Colorado, USA : Paladin Press, 1992. 0-87364-649-5.

Keane, Larry. 2013. NSSFBLOG.com. *New NSSF Report Shows Americans are skeptical of Smart Guns.* [Online] November 13, 2013. [Cited: February 15, 2014.] http://www.nssfblog.com/americans-skeptical-of-smart-guns-oppose-their-legislative-mandate-national-poll-finds/.

Keller, Julia. 2008. *Mr. Gatling's terrible marvel : The gun that changed everything and the misunderstood genius who invented it.* New York : Viking Penguin, 2008. 978-0-14-311564-9.

Kel-Tec. 2014a. *Kel-Tec Product Overview.* [Prospekt]. Cocoa : Kel-Tec CNC, 2014. p. 8.

—. 2014b. FAQ. *What is the expected life of a Kel-Tec firearm.* [Online] 2014. [Cited: July 30, 2014.] http://www.keltecweapons.com/faq/.

—. 2014c. *About Us.* [Online] Kel-Tec, 2014. [Cited: September 14, 2014.] http://www.keltecweapons.com/about/.

—. 2011. Kel-Tec Warranty Update : Second Owner Responsibilities. [Online] Kel-Tec, August 4, 2011. [Cited: September 14, 2014.] http://www.keltecweapons.com/news/warranty-update-second-owner-responsibilities/.

—. 2006. *PF-9 Pistol : Safety, Instruction, and Parts Manual.* Rev. 0606. Cocoa : KEL-TEC, 2006.

Kersten, Manfred and Schmid, Walter. 1999. *Heckler & Koch : Die offizielle Geschichte der Oberndorfer Firma Heckler & Koch : Einblicke in die Historie, Beschreibung der Waffenmodelle, Darstellung der Technik.* Wuppertal : Verlag Udo Weispfennig, 1999. 3-00-005091-4.

—. 2008. *Mauser-Waffenforschungsanstalt : Die Monatsberichte.* Kleve : service K, 2008. Art.-Nr. 01-07-01.

Kim, W. Chan and Mauborgne, Renée. 2005. *Blue Ocean Strategy : How to create uncontested market space and make the competition irrelevant.* Boston : Harvard Business School Publishing Corporation, 2005. 978-1-59139-619-2.

Klemm, Friedrich. 1998. *Geschichte der Technik : Der Mensch und seine Erfindungen im Bereich des Abendlandes.* 3. Auflage. Stuttgart : B. G. Teubner Verlagsgesellschaft Leipzig, 1998. 3-8154-2512-3.

Kneubuehl, Beat Paul. 2018. *Ballistik : Theorie und Praxis.* Berlin : Springer-Verlag GmbH, 2018. 978-3-662-58299-2.

—. 2013. *Geschosse : Gesamtausgabe.* Dietikon-Zürich : Verlag Stocker-Schmid AG, 2013. 978-3-7276-7176-0.

Kneubühl, Fritz Kurt. 1994. *Repetitorium der Physik.* 5. Auflage. Stuttgart : Teubner Studienbücher : Physik, 1994. 3-519-43012-6.

Korn, R. H. 1908. *Mauser-Gewehre und Mauser-Patente.* [ed.] Waffenfabrik Mauser A.G. Berlin : Ecksteins biographischer Verlag, 1908.

Korwin, Alan. 2009. *Gun Laws of America.* Scottsdale, AZ : Bloomfield Press, 2009. 978-1-889632-24-7.

Lamb, Vernon A. and Young, John P. 1960. *Experimental Plating of Gun Bores to Retard Erosion.* Washington : National Bureau of Standards, 1960. NBS-TN-46.

Law Center to Prevent Gun Violence. 2014. Regulating Guns in America. *A Comprehensive Analysis of Gun Laws Nationwide.* [Online] May 2014. [Cited: June 12, 2014.] http://smartgunlaws.org/wp-content/uploads/2014/10/RGIA-For-Web.pdf.

Leupold, Andreas and Glossner, Silke. 2016. *3D-Druck, Additive Fertigung und Rapid Manufacturing : Rechtlicher Rahmen und unternehmerische Herausforderung.* München : Verlag Franz Vahlen GmbH, 2016. 978-3-8006-5149-8.

Lewisch, R. 1971. Bemerkungen zur Verschlußdynamik. [ed.] Emil Schwend. *Deutsches Waffen-Journal.* November 1971.

Lindner, Helmut. 1993. *Grundriss der Atom- und Kernphysik.* 17. Auflage. Leipzig; Köln : Fachbuchverlag Leipzig - Köln, 1993. 3-343-00840-0.

Linnenkohl, Hans. 1990. *Vom Einzelschuss zur Feuerwalze.* Koblenz : Bernhard & Graefe Verlag, 1990. 3-7637-5866-6.

Lipson, Hod and Kurman, Melba. 2013. *Fabricated : The New World of 3D Printing.* Indianapolis : John Wiley & Sons, Inc., 2013. 978-1-118-35063-8.

Lott, John R. 2000. *More guns, less crime : Understanding crime and gun-control laws.* 2nd Edition. Chicago : The University of Chicago Press, 2000. 0-226-49364-4.

Lugs, Jaroslav. 1982. *Handfeuerwaffen : Systematischer Überblick über die Handfeuerwaffen und ihre Geschichte.* [trans.] Rudolf Winkler. 7. Auflage. Berlin : Militärverlag der Deutschen Demokratischen Republik, 1982. Tschechischer Originaltitel: Rucni palne zbrane, Prag, 1956. Bestellnummer 745 1570.

LumiNova AG. *Information Super-LumiNova.*

Macharzina, Klaus and Wolf, Joachim. 2012. *Unternehmensführung : Das internationale Managementwissen : Konzepte – Methoden – Praxis.* 8. Auflage. Wiesbaden : Springer Gabler, 2012. 978-3-8349-3412-3.

Maier, Karl Wilhelm. 1994. Biography. Boynton Beach, FL, USA : s.n., November 1994. p. 2. Letter from Karl W. Maier to Henk Visser.

—. 1993a. Zur Entwicklung der 20 mm Flugzeugbordwaffe MK 213 C der Abteilung Vs. [Document]. Bad Liebenzell : s.n., June 1993. p. 4. Letter from von Karl W. Maier to Henk Visser (see endnote II, Field, et al., 2011 p. 105).

—. **1993b.** Zur Entwicklung von Selbstladegewehren bei Mauser Oberndorf. [Document]. Bad Liebenzell : s.n., June 1993. p. 10. Letter from Karl W. Maier to Henk Visser (see endnote II, Field, et al., 2011 p. 105).

Malloy, John. 1989. Early Rivals of the Model 1911 45 Automatic. [ed.] Dan Shideler and Corrina Peterson. *GunDigest Book of Classic Combat Handguns.* Iola : F+W Media, Inc., 1989. 2011. 978-1-4402-2384-6.

Marquardt, Erik. 2014. *VDI Statusreport : Additive Fertigungsverfahren.* [ed.] VDI. Düsseldorf : s.n., 2014. p. 26.

Martens, Bas. 2019. Het Duitse Jäger pistool. *SAM-Wapenmagazine.* February 2019, Vol. 217.

Maryland Handgun Roster. 2013. Maryland Handgun Roster. *Department of Maryland State Police.* [Online] 2013. [Cited: July 4, 2013.] http://mdsp.maryland.gov/Organization/Pages/CriminalInvestigationBureau/Licensing Division/HandgunRoster.aspx.

Massachusetts General Laws, C. 269, § 11E. 2013. Massachusetts Laws : Chapter 269 : Section 11E : Serial identification numbers on Firearms. *The 188th General Court of The Commonwealth of Massachusetts : Massachusetts Laws.* [Online] 2013. [Cited: July 19, 2013.] https://malegislature.gov/Laws/GeneralLaws/PartIV/TitleI/Chapter269/Section11E.

Massachusetts General Laws, C. 140, § 121. 2013. Massachusetts Laws : Chapter 140 : Section 121 : Firearms sales; definitions; antique firearms; application of law; exceptions. *The 188th General Court of The Commonwealth of Massachusetts : Massachusetts Laws.* [Online] 2013. [Cited: June 15, 2013.] https://malegislature.gov/Laws/GeneralLaws/PartI/TitleXX/Chapter140/Section121.

Massachusetts General Laws, C. 140, § 123. 2013. Massachusetts Laws : Chapter 140 : Section 123 : Conditions of licenses. *The 188th General Court of The Commonwealth of Massachusetts : Massachusetts Laws.* [Online] 2013. [Cited: June 15, 2013.] https://malegislature.gov/Laws/GeneralLaws/PartI/TitleXX/Chapter140/Section123.

Mathews, J. Howard. 1973. *Firearms Identification.* Second Printing. Springfield : Charles C Thomas Publisher, 1973. 0-398-02355-7.

Matschoß, Conrad, Haßler, Friedrich and Bihl, Adolf. 1939. *50 Jahre Deutsche Waffen- und Munitionsfabriken Aktiengesellschaft.* Berlin : VDI-Verlag GmbH, 1939. Denkschrift, Herausgegeben aus Anlaß des 50-jährigen Bestehens der DWM als AG.

—. **1938.** *Geschichte der Mauser Werke.* Berlin : VDI-Verlag GmbH, 1938. Denkschrift, Herausgegeben aus Anlaß des 125-jährigen Bestehens der Gewehrfabrik in Oberndorf a. N.

McNab, Chris. 2015. *Glock : The World's Handgun.* London : Amber Books Ltd, 2015. 978-1-78274-256-2.

Meissner, Manfred and Schorcht, Hans-Jürgen. 2007. *Metallfedern : Grundlagen, Werkstoffe, Berechnung, Gestaltung und Rechnereinsatz.* 2. Auflage. Berlin, Heidelberg : Springer-Verlag, 2007. 978-3-540-49868-1.

Meissner, Manfred and Wanke, Klaus. 1993. *Handbuch Federn : Berechnung und Gestaltung im Maschinend Gerätebau.* 2., bearbeitete Auflage. Berlin; München : Verlag Technik GmbH, 1993. 3-341-01087-4.

Meyer, Klaus-Dieter. 1986. *Handbuch für den Wiederlader.* 3. Auflage. Schwäbisch Hall : Journal-Verlag Schwend GmbH, 1986.

Miller, Emily. 2014. The Washington Times: Smith & Wesson to stop selling guns in California due to microstamping law. *D.C. passes emergency bill to postpone the firearm law until 2016.* [Online] January 22, 2014. [Cited: January 27, 2014.] http://www.washingtontimes.com/news/2014/jan/22/smith-wesson-stop-selling-guns-california-due-micr/.

Miniter, Frank. 2014. *The Future of the Gun.* Washington, DC : Regnery Publishing, 2014. 978-1-62157-240-4.

ModernArms. 2013. Modern Arms Blog. *California Approves Sale of Smart Gun.* [Online] October 30, 2013. [Cited: February 27, 2014.] http://modernarms.net/blog/california-approves-sale-of-smart-gun/.

Moll, Friedrich Wilhelm. 2009. *Die Mauser HSP.* 2. Auflage. Kleve : service K, 2009. Vol. 1. Art.-Nr. 01-04-02.

Mootz, Werner. 1989. *Geschichte und Technik der Selbstladepistole : Von den Anfängen bis zur Gegenwart.* [ed.] Elmar W. Caspar, et al. Herford : Verlag E. S. Mittler & Sohn GmbH, 1989. Vol. 5. 3-8132-0305-0.

Morawietz, Otto. 1968. Waffenkunde. [ed.] Deutsche Gesellschaft für Heereskunde e. V. *Einführung in die Heereskunde : Beilage der Zeitschrift für Heereskunde : Folge 13.* 1985, unveränderter Nachdruck der im Januar 1968 erschienenen Auflage, 1968, 32. Jahrgang.

Möser, Kurt. 2002. *Geschichte des Autos.* Frankfurt : Campus Verlag GmbH, 2002. 3-593-36575-8.

Mötz, Josef and Schuy, Joschi. 2013. *Die Weiterentwicklung der Selbstladepistole I : Selbstladepistolen in Österreich-Ungarn bzw. Österreich von 1914 bis heute : Österreichische Pistolen.* Laxenburg bei Wien : Herausgegeben in den Selbstverlagen der Verfasser, 2013. Vol. 2. 978-3-9502342-2-0.

—. 2015. *Die Weiterentwicklung der Selbstladepistole II : Maschinenpistolen, Pistolentaschen, Ergänzungen zu Band 1 und 2 : Österreichische Pistolen.* Laxenburg bei Wien : Herausgegeben in den Selbstverlagen der Verfasser, 2015. Vol. 3. 978-3-9502342-3-7.

—. 2007. *Vom Ursprung der Selbstladepistole : Repetiend Selbstladepistolen in Österreich-Ungarn 1884 bis 1918 : Österreichische Pistolen.* Laxenburg bei Wien : Herausgegeben in den Selbstverlagen der Verfasser, 2007. Vol. 1. 978-3-9502342-0-6.

MRI. 2016. History. [Online] 2016. https://www.magnumresearch.com/MRI-history.asp.

NICS. 2016. FBI National Instant Criminal Background Check System. *NICS Firearm Background Checks : Year by State/Type.* [Online] 2016. [Cited: June 10, 2017.] https://www.fbi.gov/file-repository/nics_firearm_checks_-_year_by_state_type.pdf/view.

NIJ. 2016. *Baseline Specifications for Law Enforcement Service Pistols with Security Technology.* [ed.] U.S. Department of Justice. Washington, DC : National Institute of Justice, 2016. p. 23. NCJ 250377.

Nitz, Stefan. 2015. *3D-Druck : Der praktische Einstieg.* Bonn : Galileo Press, 2015. 978-3-8362-2875-6.

nks. 2012. *Obama: Keine Entschuldigung für Untätigkeit.* [ed.] Werner D'Inka, et al. Frankfurt am Main : Frankfurter Allgemeine Zeitung GmbH, 2012.

NRC. 1998. United States Regulatory Commission. *NUREG-1556, Vol. 8, "Consolidated Guidance About Materials Licenses: Program-Specific Guidance about Exempt Distribution Licenses".* [Online] September 1998. [Cited: February 27, 2014.] http://www.nrc.gov/.

NSSF. 2014. *2013–2014 Industry Reference Guide : A Compilation of Firearm and Ammunition Industry Data.* [ed.] National Shooting Sports Foundation. Newtown, CT, USA : s.n., 2014.

—. 2013. NSSF.org Fact Sheets. *"Smart Gun" Technology.* [Online] 2013. [Cited: February 15, 2014.] http://nssf.org/factsheets/.

nssfnews. 2014. NSSF.org. *Gun Industry Sales Reflect a 'New Normal'.* [Online] August 11, 2014. [Cited: August 20, 2014.] http://www.nssfblog.com/gun-industry-sales-reflect-a-new-normal/.

O'Connor, Patrick D. T. 1990. *Zuverlässigkeitstechnik : Grundlagen und Anwendungen.* Weinheim : VCH Verlagsgesellschaft, 1990. 3-527-26890-1.

Oehler, Gerhard and Kaiser, Fritz. 1993. *Schnitt-, Stanz- und Ziehwerkzeuge.* 7. Auflage. Berlin : Springer-Verlag, 1993. 3-540-56700-3.

Oerlikon. 1981. *Oerlikon Taschenbuch.* 2. Auflage. Zürich : Oerlikon-Bührle AG, 1981.

Ortmeier, Gerhard. 2016. Verkleinert : Taschenpistole Walther Modell 7. [ed.] Walter Schulz. *Deutsches Waffen-Journal.* July 2016.

Pawlas, Karl R. 1971–2001. *Waffen-Revue.* Nürnberg : Publizistisches Archiv für Militär und Waffenwesen, 1971–2001. Vierteljährlich, 123 Bände.

Pegler, Martin. 2013. *The Vickers-Maxim Machine Gun.* Oxford : Osprey Publishing Ltd., 2013. Vol. WPN025. 978-1-78096-382-2.

Pietzner, Wolfgang. 2005. *Waffenkunde Teil II.* Hilden : Verlag Deutsche Polizeiliteratur GmbH Buchvertrieb, 2005. 3-8011-0501-6.

Ponsford, Matthew and Glass, Nick. 2014. CNN Business. *The night I invented 3D printing.* [Online] Februar 14, 2014. [Cited: Mai 25, 2019.] https://edition.cnn.com/2014/02/13/tech/innovation/the-night-i-invented-3d-printing-chuck-hall/index.html.

Popenker, Maxim and Williams, Anthony G. 2007. *Modern Combat Pistols.* [ed.] Anthony G. Williams. Wiltshire : The Crowood Press Ltd, 2007. 10-186126-894-7.

Porter, Michael E. 1985. *Competitive Advantage : Creating and Sustaining Superior Performance.* New York : Free Press, 1985. 978-0-684-84146-5.

—. 1999. *Wettbewerbsstrategie (Competitive Strategy) : Methoden zur Analyse von Branchen und Konkurrenten.* Frankfurt/Main : Campus Verlag GmbH, 1999. 3-593-36177-9.

—. 2010. *Wettbewerbsvorteile (Competitive Advantage) : Spitzenleistungen erreichen und behaupten.* 7. Auflage. Frankfurt/Main : Campus Verlag GmbH, 2010. 978-3-593-38850-2.

PTOOMA Productions. 2004. *The Complete Glock Reference Guide.* [ed.] PTOOMA Productions LLC. 2004. p. 290.

Raber, Michael S., et al. 2009. *Forge of Innovation : An Industrial History of the Springfield Armory, 1794–1968.* [ed.] Richard Colton. Springfield : Eastern National, 2009. 978-1-59091-100-6.

Rheinmetall. 1982. *Handbook on Weaponry.* 2nd Edition. Düsseldorf : Brönners Druckerei Breidenstein GmbH, Frankfurt a. M., 1982.

Rinker, Robert A. 2011. *Understanding Firearms Ballistics.* 6th Edition. Clarksville : Mulberry House Publishing, 2011. 978-0-9645598-5-1.

Riordan, Dennis. 1980. Model 1911 Colt : Six Decades Of Service. [ed.] Dan Shideler and Corrina Peterson. *GunDigest Book of Classic Combat Handguns.* Iola : F+W Media, Inc., 1980. 2011. 978-1-4402-2384-6.

Rosenberger, Manfred R. and Hanné, Katrin. 1993. *Vom Pulverhorn zum Raketengeschoß : Die Geschichte der Handfeuerwaffen-Munition.* Stuttgart : Motorbuch Verlag, 1993. 3-613-01541-2.

Rosenwald, Michael S. 2014. The Washington Post. *'We need the iPhone of guns' : Will smart guns transform the gun industry?* [Online] February 18, 2014. [Cited: February 28, 2014.] http://www.washingtonpost.com/local/we-need-the-iphone-of-guns-will-smart-guns-transform-the-gun-industry/2014/02/17/6ebe76da-8f58-11e3-b227-12a45d109e03_story.html.

Rüb, Matthias. 2012. *Einzigartiger Ausdruck der amerikanischen Kultur.* [ed.] Werner D'Inka, et al. Frankfurt am Main : Frankfurter Allgemeine Zeitung GmbH, 2012.

Ruger. 2013a. RUGER LC9 and LC380. [Owner's Manual]. Southport, CT : Sturm, Ruger & Co., Inc., 2013.

—. 2013b. The Ruger LC9 Centerfire Pistol. [Online] 2013. [Cited: December 17, 2013.] http://www.ruger.com/products/lc9/models.html.

SAAMI : Setting the Standard. 1998. Setting the Standard : Safety and Technical Standards for Firearms and Ammunition. *SAAMI Publication #241.* [Online] 2/98, 1998. [Cited: July 16, 2013.] http://www.saami.org/specifications_and_information/publications/download/SAAMI_ITEM_241-Setting_the_Standard_Safety_and_Technical_Standards_for_Firearms_and_Ammunition.pdf.

SAF. 2013. Second Amendment Foundation. *News Release: Glock Files Amicus Brief Supporting SAF's California Case.* [Online] November 6, 2013. [Cited: November 8, 2013.] http://www.saf.org/viewpr-new.asp?id=461.

Saleh, Bahaa E. A. and Teich, Malvin Carl. 2008. *Grundlagen der Photonik.* [trans.] Dr. Michael Bär. 2. Auflage. Weinheim : Wiley-VCH Verlag GmbH & Co. KGaA, 2008. 978-3-527-40677-7.

SB-281. 2013. Maryland Legislature. *The President, et. al.: Senate Bill 281 (3lr0154): Firearms Safety Act of 2013.* [Online] 2013, April 4, 2013. [Cited: July 11, 2013.] http://mgaleg.maryland.gov/2013RS/bills/sb/sb0281E.pdf.

SB-293. 2013. California Legislative Information. *De Saulnier : Senate Bill 293 Firearms : Owner-authorized handguns.* [Online] May 24, 2013. [Cited: February 27, 2014.] http://leginfo.legislature.ca.gov/faces/billStatusClient.xhtml.

Scarlata, Paul. 2008. *Das Gewehr 88 : Deutschlands erstes modernes Militärgewehr.* [trans.] Bernd Rolff. Dietikon-Zürich : Verlag Stocker-Schmid AG, 2008. 978-3-7276-7167-8.

SCCY. 2014. *SCCY Firearms.* [Product Information]. Daytona Beach : SCCY Industries, 2014.

SCCY Industries, LLC. 2014. *Handguns.* [Online] 2014. [Cited: August 28, 2014.] http://www.sccy.com/products/handguns.

Schmid, Walter. 2003. *Die Entwicklungsgeschichte der Mauser-Flugzeugbordwaffe MG / MK 213 C : Vorläufer der BK 27.* Oberndorf a. N. : DWJ Verlags-GmbH, 2003. 3-936632-38-3.

Schmit, Cajetan. 1972. Die Heckler & Koch-Pistole VP 70. [ed.] Emil Schwend. *Deutsches Waffen-Journal.* September 1972.

Schreiber, Edzard. 1978. Der Seltsame Ring. [ed.] Emil Schwend. *Deutsches Waffen-Journal.* August 1978.

—. 1981. Der Seltsame Ring : Die Lösung des Rätsels. [ed.] Emil Schwend. *Deutsches Waffen-Journal.* January 1981.

Schuh, Günther. 2012. *Innovationsmanagement : Handbuch Produktion und Management 3.* 2. Auflage. Berlin : Springer Vieweg, 2012. 978-3-642-25049-1.

Schumpeter, Joseph Alois. 1947. Kapitalismus, Sozialismus und Demokratie. 8. Aufl. 2005, Nachdruck der 1947 erschienenen 1. Aufl. Tübingen : A. Franke UTB, 1947. 2005. 3-8252-0172-4.

—. 1934. Theorie der wirtschaftlichen Entwicklung : Eine Untersuchung über Unternehmensgewinn, Kapital, Kredit, Zins und des Konjunkturzyklus. 8. Aufl. 1993, unveränderter Nachdruck der 1934 erschienenen 4. Aufl. Berlin : Duncker und Humblot, 1934. 1993. 3-428-07725-3.

Schwab, Klaus. 2016. *The Fourth Industrial Revolution.* Geneva : Portfolio Penguin, 2016. 978-0-241-30075-6.

Secubit. 2014. GSC™ Secubit Ltd. Shot Counter. *About GSC™.* [Online] Secubit, 2014. [Cited: May 25, 2014.] http://www.secubit-ltd.com/.

Seel, Wolfgang. 1993. *Die G11 Story.* Schwäbisch Hall : Journal Verlag Schwend GmbH, 1993.

Shaydurov, Ilya. 2010. *Russische Schusswaffen : Typen, Technik, Daten.* Stuttgart : Motorbuch Verlag, 2010. 978-3-613-03187-6.

Shepherd, R. V., et al. 1945. Visit to Mauser-Werke A. G. Oberndorf am Neckar and Mauser Personnel at Lager Haiming, Ötztal, near Innsbruck. *Supreme Headquarters Allied Expeditionary Force. G-2 (Intelligence) Division. Operational Intelligence Sub-Division. 2/13/1944-7/14/1945.* s.l. : Combined Intelligence Objectives Sub-Committee, 1945.

SIG-Sauer. 2014. *P320 : Meeting the Demands of Today's Armed Professionals.* [Brochure]. Newington : SIG-Sauer, Inc., 2014. p. 4. MKT0082.

Simmons, Donald M. 1988. A Second Look At The Glock 17. [ed.] Dan Shideler and Corinna Peterson. *GunDigest Book of Classic Combat Handguns.* Iola : F+W Media, Inc., 1988. 2011. 978-1-4402-2384-6.

Smith, Walter H. B. 1960. *The Book of Rifles : An Encyclopedic Reference Work.* 2nd Edition. Harrisburg : The Stackpole Company, 1960.

—. **1953.** *The N.R.A. Book of Small Arms, Pistols and Revolvers : Volume 1 : Pistols and Revolvers.* 4th Edition. Washington, D. C. : The Military Service Publishing Company, 1953.

Speck, Christoph. 2009. *Automatisierte Auswertung forensischer Spuren auf Patronenhülsen.* Schriftenreihe Nr. 014. Karlsruhe : Universitätsverlag Karlsruhe, 2009. 978-3-86644-365-5.

Sterett, Larry S. 1961. Mars Automatic Pistols. [ed.] Dan Shideler and Corinna Peterson. *GunDigest Book of Classic Combat Handguns.* Iola : F+W Media, Inc., 1961. 2011. 978-1-4402-2384-6.

Stevens, R. Blake. 2006. *Full Circle : A Treatise on Roller Locking.* Ontario : Collector Grade Publications Incorporated, 2006. 0-88935-400-6.

Stiefel, Dieter. 1984. Kunststoff und Stahl : Die Glock 17. [ed.] Emil Schwend. *Deutsches Waffen-Journal.* August 1984.

Stock, Kyle. 2014. People Aren't Buying Guns. [Online] Bloomberg Businessweek, August 27, 2014. [Cited: August 29, 2014.] http://www.businessweek.com/articles/2014-08-27/gun-sales-are-down.

Storz, Dieter. 2012. *Deutsche Militärgewehre : Schußwaffen 88 und 91 sowie Ziel- und Fechtgewehre, Seitengewehre und Patronentaschen.* [ed.] Ansgar Reiß. Kataloge des Bayerischen Armeemuseums Ingolstadt. Wien : Verlag Militaria, 2012. Vol. 9. 978-3-902526-55-7.

Stroppe, Heribert, et al. 2018. *Physik für Studierende der Natur- und Ingenieurwissenschaften.* 16. Auflage. München : Fachbuchverlag Leipzig, 2018. 978-3-446-45533-7.

Sweeney, James P. 2009. Casing-code issues snag handgun law. *The San Diego Union-Tribune.* [Online] The San Diego Union-Tribune, August 10, 2009. [Cited: March 28, 2013.] http://www.utsandiego.com/news/2009/aug/10/1n10guncodes235322-casing-code-issues-snag-handgun/?page=2#article-copy.

Sweeney, Patrick. 2010. *1911 : The First 100 Years.* Iola : Krause Publications, 2010. 978-1-4402-1115-7.

—. **2013.** *Glock : Deconstructed.* Iola : Krause Publications, 2013. 978-1-4402-3278-7.

—. **2008.** *The Gun Digest Book of the Glock.* 2nd Edition. Iola : Krause Publications, 2008. 978-0-89689-642-0.

TARA. 2014. *TARA TM-9 PISTOL.* [Brochure]. Mojkovac, Montenegro : TARA Perfection D.O.O., 2014. p. 8.

Tarr, James. 2014. SIG Strikes Back. [ed.] Chris Agnes. *Handguns.* December 2014, Volume 29, Number 6.

The Economist. 2013. The Economist : Technology Quarterly. *Smart Weapons : Kill switches and safety catches.* [Online] November 30, 2013. [Cited: September 5, 2014.] http://www.economist.com/news/technology-quarterly/21590764-arms-control-new-technologies-make-it-easier-track-small-arms-and-stop-them.

Thomson West. USA Government Documents : California Code of Regulations. [Online] Vol. 15. [Cited: January 30, 2014.] https://archive.org/details/gov.ca.ccr.11.

ToddG. 2010. HKpro.com. *Forum's member "ToddG" in Thread "Striker fired 'P40'",* *page 9, Post #85.* [Online] December 20, 2010. [Cited: January 11, 2011.] http://www.hkpro.com/forum/showthread.php?131124-Striker-fired-quot-P40-quot/page9.

TR "Patrone". 2009. Technische Richtlinie : Patrone 9 mm x 19, schadstoffreduziert. *PTI Online.* [Online] Revision: September, 2009. [Cited: August 10, 2012.] http://www.pfa.nrw.de/PTI_Internet/pti-intern.dhpol.local/WG/Regelungen/Munition/ TR_Munition_mit_Anlagen1-13_09-09.pdf.html.

TR "Pistolen". 2008. Technische Richtlinie : Pistolen im Kaliber 9 mm x 19. *PTI Online.* [Online] Revision: January, 2008. [Cited: August 10, 2012.] http://www.pfa.nrw.de/PTI_Internet/pti-intern.dhpol.local/WG/Regelungen/Pistolen.

Uhlig, Albrecht. 1964. *Untersuchungen über die Bewegungen und Kräfte beim Rundkneten.* Dissertation. Hannover : Technische Hochschule Hannover, 1964.

United States District Court. 2013. United States District Court : Western District of New York. *Case 1:13-cv-00291-WMS : Document 140.* [Online] December 31, 2013. [Cited: January 7, 2014.] http://www.handgunlaw.us/documents/agopinions/NYDisCtRulingOn7RdMags.pdf.

USPSA. 2014. USPSA.ORG. *Handgun Rules.* [Online] February 2014. [Cited: October 18, 2014.] http://www.uspsa.org/document_library/rules/2014/ Feb%202014%20Handgun%20Rules.pdf.

Vahs, Dietmar and Brem, Alexander. 2013. *Innovationsmanagement : Von der Idee zur erfolgreichen Vermarktung.* 4. Auflage. Stuttgart : Schäffer-Poeschel Verlag Stuttgart, 2013. 978-3-7910-2857-6.

Vickery, W. F. 1940. *Advanced Gunsmithing.* Kingsport : Kingport Press, Inc., 1940.

Vizzard, William J. 2000. *Shots in the dark : The policy, politics, and symbolism of gun control.* Lanham : Rowman & Littlefield Publishers, Inc., 2000. 0-8476-9559-X.

Wacker, Albrecht. 1993. *Das System Adalbert : Der K98k.* Düsseldorf : Barett Verlag GmbH, 1993. 3-924753-57-1.

Walter, John. 2004. *Guns of the Third Reich.* [ed.] Hugh Schoenmann. London : Greenhill Books, 2004. 1-85367-598-9.

—. **2008.** *The Handgun Story : A Complete Illustrated History.* Yorkshire : Frontline Books, 2008. 978-1-84832-500-5.

Walther. 1984. PP/PPK. [Owner's Manual]. Ulm : Carl Walther, June 1984. p. 32. Item No. 2187256.

Walz, Karlheinz. 1944. *Federfragen : Insbesondere aus dem Gebiet der Zeitfestigkeit : Bei sinusförmiger und schlagartiger Beanspruchung.* [ed.] Mauser. Oberndorf/Neckar : Buchdruckerei Gebrüder Knöller, 1944.

Weiss, Douglas R. 1996. *SANDIA REPORT : Smart Gun Technology Project Final Report.* Sandia National Laboratories. Albuquerque, New Mexico : National Technical Information Service, US Department of Commerce, 1996. SAND96-1131.

Welter, Patrick. 2013. *Amerika gesundet.* [ed.] Werner D'Inka, et al. Frankfurt am Main : Frankfurter Allgemeine Zeitung GmbH, 2013.

Wille, Richard. 1897. *Mauser-Selbstlader.* EOD Reprint. Berlin : R. Eisenschmidt, 1897. 3-226-01331-9.

—. **1902.** *Selbstlader-Fragen : Eine Studie.* EOD Reprint. Berlin : R. Eisenschmidt, 1902. 3-226-01332-7.

—. **1896.** *Selbstspanner (Automatische Handfeuerwaffen).* EOD Reprint. Berlin : R. Eisenschmidt, 1896. 3-226-01452-8.

Willemsen, Mathieu. 2012. *Erprobung und Versuch : Prototypen und Versuchsstücke militärischer Handfeuerwaffen des 19. und frühen 20. Jahrhunderts.* [ed.] Legermuseum, Heleen Bronder. Wien : Verlag Militaria GmbH, 2012. 978-3-902526-53-3.

Wintermute, Garen J. 1997. Violence Prevention Research Program : California's Guns and Crime : New Evidence. [Online] May 1997, 1997. [Cited: July 2, 2013.] http://www.ucdmc.ucdavis.edu/vprp/publications/cagunweb.html.

Wirtgen, Arnold. 2004. *Die preußischen Handfeuerwaffen : Modelle, Manufakturen, Gewehrfabriken 1814–1856, Steinschloß- und Perkussionswaffen.* Bonn : Bernard & Graefe Verlag, 2004. 3-7637-6250-7.

Wirtgen, Rolf. 1987. *Geschichte und Technik der automatischen Waffen in Deutschland : Von den Anfängen bis 1871 : Teil 1.* [ed.] Elmar W. Caspar, et al. Herford : Verlag E. S. Mittler & Sohn, 1987. Vol. 1. 3-8132-0262-3.

Wirtgen, Rolf, et al. 1991. *Das Zündnadelgewehr : Eine militärtechnische Revolution im 19. Jahrhundert.* Herford : Verlag E.S. Mittler & Sohn GmbH, 1991. 3-8132-0380-8.

Wittel, Herbert, et al. 2019. *Roloff/Matek Maschinenelemente : Normung, Berechnung, Gestaltung.* 24. Auflage. Wiesbaden : Springer Vieweg, 2019. 978-3-658-26279-2.

Woll, Artur. 2011. *Volkswirtschaftlehre.* 16. Auflage. München : Verlag Franz Vahlen GmbH, 2011. 978-3-8006-3835-2.

Womack, James P., Jones, Daniel T. and Roos, Daniel. 1990. *The Machine That Changed the World.* New York : Rawson Associates, 1990. 0-89256-350-8.

Woods, Steve. 2014. Glock's Polygonal Rifling. [ed.] Stanley R. Harris. *Glock Annual Autopistols.* 2014, Vol. 20 No. 1.

Zimring, Franklin E. and Hawkins, Gordon. 1992. *The Citizen's Guide to Gun Control.* New York : Macmillan Publishing Company, 1992. 0-02-897505-7.

Zimring, Franklin E. 1975. Firearms and Federal Law : The Gun Control Act of 1968 : Gun Rights Policy Conference. *Journal of Legal Studies : Vol. 4 : 1975.* [Online] 1975. [Cited: July 10, 2013.] http://www.saf.org/lawreviews/zimring68.htm.

ZOLLERN GmbH & Co. KG. 2009. Zollern Feinguss. *Giessereitechnik.* Sigmaringen : Zollern, June 2009. Z427.

Manufacturing imperfection: .380 ACP case without extractor groove.

Endnotes

A chemical mixture in so vanishingly small doses, where the total of five to eight rifle loads equals the weight of a simple letter, is forced to transfer so much force and energy into a small metal plunger of a few grams weight, that this tiny body is able to smoothly penetrate thick tree trunks, strong walls, steel plates, and may put a man down at thousands of meters. Indisputably, this is a formidable achievement.

But far more it apparently wants to say that this little demon, this sparse amount of powder, was made subservient with lots of creativity and artfulness, by a relatively simple mechanism, to become a hardworking and nimble henchman of the shooter and forced him, besides and simultaneously into performing his tasks as slayer and annihilator to fulfil the work of opening, cocking, feeding, and closing, briefly the overall operation of the weapon except for aiming, pulling the trigger, and loading the magazine, performing this with unbeatable perfection and speed. Without a doubt, this is a first-rate technological advance, which again delivers a full proof of the inexhaustible invention and design gift of preferred spirits, a progress, which practical recognition and exploitation will be hardly denied in perpetuity.

(Wille, 1896 p. 106) [translated by PD]

ENDNOTES

[1] These days the term "innovation" is used universally and frequently, although the concept of innovation is clearly described by innovation researchers of all disciplines. Generally, the term "innovation" derives from the Latin "innovare," meaning "renew." This process of renewing is preceded by conception or invention. Without invention, innovation is not possible. (Macharzina, et al., 2012 pp. 735–738)

Economist Joseph A. Schumpeter is of particular importance among innovation researchers. At the beginning of the 20th century Schumpeter was engaged in the research of economic development and innovation and he described the concept of leadership. This term is characterized by the will and power to deal with very specific issues and by the ability to move forward alone where no antitype can be found, as well as the effect on others that emanates from this leadership. A leader must thereby not necessarily find new opportunities or create, but he will make them live, real, and enforce them.

Furthermore, Schumpeter uses the term new combinations which has been superseded in modern parlance by "innovation." Schumpeter stated that the peculiarity was the enforcing of new combinations and not finding or inventing them. In addition, he identifies three characteristics of innovation: (i) uncertainty, i.e. the usual data-base on which decision-making could be based does not exist, (ii) the behavior of the leader to do something which appears to be difficult, or something which is different or new, and (iii) the resistance of the social environment, which usually arises if someone wants to do something new. In the economic environment, the latter could be the resistance of groups, which might feel threatened by the changes as well as the resistance of various individuals on which the potential leader depends on, but who will not cooperate, and finally: the resistance in the form of consumer's inertia. (Schumpeter, 1934 pp. 124–129)

Christensen uses the concept of *disruptive innovation* in a context which has been referred to by Schumpeter as creative destruction. Schumpeter coined the term *creative destruction* for the capability of innovation, technical and economic progress, as well as the associated displacement of existing products offered on the market through innovative new products, which can be seen as the primary engine of economic progress. Schumpeter puts it this way: The competition which arises by the introduction of a new product, a new technology, a new supply source, a new type of business organization, in other words, any competitor who has a decisive cost or quality advantage and who challenges not just the profit margin and production limits of the existing companies, but who shatters their foundations, their actual life mark, is so much more effective, just like a bombardment would be when compared to breaking down a door. (Schumpeter, 1947 pp. 134–142)

Currently, innovation is understood as a commercially successful implementation of a product or process, which not just differs significantly from the known state of the art, but for which technology is also at a level for full commercial deployment, and once in full production, successfully marketed. According to this definition, just a good idea or invention is not considered as an innovation; only through the success of a product can it be considered to be an innovation. (Vahs, et al., 2013 p. 21)

Throughout this book the term "innovation" is used as per Schumpeter's definition, i.e. for any novelty that has actually been introduced on the market and gained importance.

[2] *Bussard*'s technical definition of a firearm: "A firearm is a heat engine that converts the chemically stored energy in the propellant into kinetic energy by accelerating the bullet or shot pellets to high velocity." (Bussard, 2017 p. 21)

SAAMI specifies under its "Definitions":

"Handgun – A firearm designed to be held and fired with one hand.

Pistol –   A handgun in which the chamber is part of the barrel.

Revolver – A firearm, usually a handgun, with a cylinder having several chambers so arranged as to rotate around an axis and be discharged

ENDNOTES

successively by the same firing mechanism." (ANSI/SAAMI Z299.5-2016 p. 1)

If not otherwise specified, the term pistol will be used throughout this book for self-loading pistols, i.e. semi-automatic handguns, which can be discharged successively by pulling the trigger, and in which the chamber is part of the barrel.

[3] (Chinn, 1987 pp. 560–587) and (Chinn, 1987 Preface), ibid.: Chinn distinguishes between five principles: (1) short recoil, (2) long recoil (cf. p. 494), (3) gas-operation, (4) blowback, and (5) muzzle blast actuation. Chinn's work focuses on machine guns and he points out, gas-operation would have not limit in its application.

[4] (Lugs, 1982 p. 524)

[5] Karel Krnka (1858–1926) published under the pseudonym Kaisertreu. Depending on the source, as a given name also Karl or Carl can be found; Kaisertreu chose the latter spelling.

Kaisertreu's series of essays on contemporary issues of small arms technology had been first published from March 1900 to November 1901 in *Danzer's Armee-Zeitung* [Danzer's army weekly; used to be an Austrian military journal; PD]. The development of modern self-loading pistols had begun at that time with constructions by Borchardt, Browning, Krnka, Luger, von Mannlicher, Mauser and Schmeisser. One of the challenges of the days was which locking system should be chosen for a service pistol. Krnka was an early proponent of recoil-operated pistols with positive locking. However, similar to how the name of the gun designer Schmeisser was linked to the manufacturers BERGMANN and HAENEL, Krnka's name was connected to ROTH, and so he had chosen to publish under a pseudonym. When Kaisertreu published a compilation of his "pro ROTH" papers in 1902, Robert Jentzsch would publish an analysis of Kaisertreu's compilation, revealing the true identity of the author.

The introduction of Jentzsch's replica is written in a very emotional style, but the main part of the book tries to focus on technical facts in spite of emotions and is based on reliable sources; in contrast to Kaisertreu, who explained just his own point of view. In any case, both sources give a picture of where pistol technology was standing at the beginning of the 20th century and what challenges the design engineers had to meet. The source Kaisertreu is therefore referred to several times. Jentzsch turns out to be a treasure trove of patents, which at the same time sets out the history of technology during the beginnings of modern weapons.

[6] The citations in the endnotes allow the reader to place the referred content into a context, such as to the origin of the statement or the age of a source. In weapons technology, as well as in other established fields, real innovations are rare. But sometimes our thoughts complement each other and something new is created that could not have been so without reading a particular book or an inspiring communication with third parties. In this regard, providing sources is comparable to paying one's dues. (Eco, 2010 p. 213)

[7] (McNab, 2015 p. 15)

[8] (Mötz, et al., 2015 pp. 235–240) Walter mentions "spray casting" as an example for a new method of barrel production. (Walter, 2008 pp. 229–235) Apparently, this method was not successful, even though the basic idea is more popular than ever: additive manufacturing is paving its own way. (Marquardt, 2014 pp. 4–7) A similar idea can be found in patent AT 142,529, filed March 17, 1934. Anton Krathky discloses a method which can be used to manufacture metal-ceramic barrels with lands and grooves rifling.

[9] Automatic firearms would not work without the invention of self-contained cartridges. The initial idea of a "cartridge", however, was to improve loading, i.e. to reduce the time to load a firearm. Today, due to the long history of firearms, the word cartridge has a variety of different meanings. For example: "A muzzle loading cartridge is a paper, gutta percha, linen, wood, or other non-metallic material container which holds a pre-measured charge of black powder. It may also contain a bullet (a conical bullet or a round ball)." (Bussard, 2017 p. 21) In the following we are going to use cartridge for a self-contained combination of projectile,

ENDNOTES

propellant charge, and primer, all of them carried by a cartridge case. Self-contained cartridges are a special kind of ammunition:

"Ammunition is the collective term for all projectiles which are thrown, slung, or fired by means of energy stored in any form, and released by a triggering process. The word was borrowed at the beginning of the 16th century from the French 'munition de guerre' for the requisites of war, and narrowed to gunnery requirements." (Rheinmetall, 1982 p. 506)

Also known is caseless ammunition, which is commonly not used in handguns.

Cased ammunition has two distinct features worth mentioning: During firing the cartridge case serves as an obturator by sealing securely against the wall of the barrel's chamber, so that no combustion gases can get through to the breech, (Rheinmetall, 1982 p. 506) and the case will absorb some heat which will be removed together with the casing from the chamber during the auto-loading cycle. (Kneubuehl, 2013 p. 54)

Bussard's technical definition of a cartridge: "A cartridge is a complete round of self-contained ammunition consisting of, at minimum, a case, primer, propellant powder, and bullet or shot charge." And he explains, "a modern cartridge provides all components necessary for a firearm to convert stored chemical energy into kinetic energy and transfer that kinetic energy, via the projectile, on/into the target." (Bussard, 2017 p. 21) Bussard continues, "a cartridge is often mistakenly referred to as a bullet. [...] This is incorrect as a bullet is only one of the component parts of a cartridge." (Bussard, 2017 p. 22) In this book, we are going to use projectile in most cases, trying to avoid the word "bullet."

10 (Görtz, et al., 2011 p. 1483) Görtz explains in his book *Pistole 08* the history of the 9 mm cartridge in more detail. According to this source, a Parabellum-Pistol chambered in 9 mm was offered in 1902 to the British "Small Arms Committee," however a gun was not available any sooner than January 1903. (Görtz, 2004 p. 260)

A concise history of pistol cartridge milestones: .30 Borchardt and .30 Mauser (1893), .32 ACP and 7.65 Parabellum (1899), 9 mm Luger (1902), 8 mm Nambu (1904), .25 ACP and .45 ACP (1905), .380 ACP and 9 mm Mauser (1908), 9 mm Steyr (1910). (Brukner, 2003 p. 81)

11 (Barnes, 2012 p. 341), ibid.:

"The 9mm Luger, or 9mm Parabellum, was introduced, in 1902, with the Luger automatic pistol. It was adopted first by the German Navy, in 1904, and then by the German Army, in 1908. Since that time, it has been adopted by the military of practically every non-Communist power. It has become the world's most popular and widely used military handgun and submachine gun cartridge. In the United States, Colt's, Smith & Wesson, Ruger, and many others chamber the 9mm, as do many foreign-made pistols. In 1985, the 9mm Luger was adopted as the official military cartridge by U.S. Armed Forces, along with the Beretta Model 92-F (M-9) 15-shot semi-auto pistol."

12 (Görtz, et al., 2011 pp. 1487–1490) Until then, bottle-necked cartridges had been the industry standard, furnishing a relatively large body for the propellant powder, an external shape allowing consistent feeding from the magazine into the barrel's chamber, and offering good obturation, i.e. ductile materials which will give way during discharge, providing an expanding neck and shoulder which serve as a breech seal. (Görtz, 2004 p. 259) and (Smith, 1960 pp. 36, 43)

In 1903 Georg Luger himself had presented the 9 mm cartridge to the U.S. Ordnance Department, which then ordered 50 sample pistols. These pistols were shipped in 1904. (Görtz, 2004 p. 260) In Germany, this ammunition type was officially referred to as "Pistolenpatrone 08", abbreviated "Pist. Patr. 08." In the early years until 1914 the officials in Germany would refer to the cartridge as "9 mm cartridge." This was explicit enough in the field of German military during those days, because there was no other officially introduced service round than the *Pistolenpatrone 08*. The German Navy sometimes referred to the same cartridge as "Patrone 1904." (Görtz, 2004 p. 259)

[13] (Wirtgen, 1987 p. 67), ibid.: Muzzle loaders with rifled barrels were used as well. Also, early types of single-shot breech-loading rifles chambered for metallic cartridges, and, in the final phase of the war, repeating rifles, such as the *Spencer* repeating rifle. The latter can already be considered to have represented a high technology level of rapid fire weaponry, which was only outperformed at that time by crank-operated weapons.

Ezell mentions before the Civil War the rifle manufacturers concentrated on improvements in production technology, "major innovation was neither sought nor encouraged." The Ordnance establishment thought the U.S. Model 1862 rifle-musket was a much deadlier weapon at ranges up to 250 meters than the flintlock smooth-bores, and hence, and it was argued the U.S. Army "could not afford to adopt breech-loading rifles in the middle of a shooting war [...]." After Civil War, Ezell explains, "America lagged in the development of repeating military breech-loaders." (Ezell, 1984 pp. 5–10)

[14] (Pegler, 2013 pp. 5, 10), foremost the 1862 Gatling gun, but also Vandenburgh volley gun, the Ager coffee-mill gun and the Billinghurst-Requa battery gun. (Keller, 2008 p. 23)

[15] The Dreyse needle gun was the first successful military breechloader adopted that fired a self-contained cartridge. With its rifled barrel the accuracy was better than known from smoothbore muskets, and its high rate of fire, compared to the standard at that time, resulted from this breech-loader's increased speed in loading.

For correct spelling of Nicolaus von Dreyse's first name refer to: (Wirtgen, et al., 1991 pp. 18, 45f endnote 28)

In the old days, when armies with muzzle loaders marched up to the battle-duel, one tolerated the low accuracy of the individual musket. A target wall, the column target, was common. It measured 30 meters in width and had a man-sized height (1.8 m). (Götz, 1996 p. 31) In the armies of Europe, however, growing expectations were placed on accuracy. The objective was to increase the effectiveness of infantry combat by means of rapid, precise and long-range fire. For this purpose, the standard issue rifle was to be equipped with a rifled barrel.

From the point of view of the time, this was a major challenge. Smoothbore muzzle loaders did not deliver the required accuracy, rifled muzzle loaders required more carefulness when loading, i.e. the loading speed was too low. A technological progress was needed that would make it possible to circumvent this dilemma: the transition to the breech loader with paper cartridges. (Wirtgen, et al., 1991 p. 39) Although this system was known to have weaknesses, it clearly demonstrated its superiority over muzzle loaders in the field. It was not long before breech loaders for metal cartridges and bolt action repeating rifles followed. (Willemsen, 2012 pp. 14–17, 20–24) This turning point was to initiate an improvement of propellant powders as well as a long-lasting trend toward smaller projectile calibers. (Götz, 1981 pp. 72–80) For example, the Dreyse needle gun with its sub-caliber projectile (barrel caliber 15.43 mm, projectile diameter 13.6 mm) was ballistically inferior to the Chassepot needle rifle in 11 mm caliber, which was 25 years younger. (Götz, 1981 p. 9) Regarding the long-range effect, *Ebell* notes:

> "In the meantime, the Franco-Prussian War had broken out in which the superiority of the French Chassepot rifle over our needle rifle clearly proved itself. We recall the bloody and useless sacrifice of the Prussian Guard at St. Privat, who due to the greater long-range effect of the Chassepot rifle could not make use of their own weapon at all and stormed the high French position in the densest bullet rain and uncovered terrain." (Ebell, 1921 p. 44) [translated by PD]

ENDNOTES

Brief comment on the nomenclature of officially introduced weapons in Germany: In the transition from the infantry rifle *I.G. Mod. 71/84* to the *Gew. 88*, the "Mod.", short for "model" in model names rifles was abandoned. (Storz, 2012 p. 45) *Handrich* explains, it was common practice in the German Army to specify the type of the service weapon (e.g. P: pistol, G: rifle, MG: machine gun, MP: submachine gun, Stg: assault rifle) in the name, followed by a numeric designation indicating the year in which the design was completed, or the year when the model was officially accepted as a service weapon (e.g. *P08*). After the First World War, the German military deviated from this principle. To muddy the water when the Treaty of Versailles would not allow the design of new military arms, they would use the number under which the blueprints were registered. The new light machine gun by DREYSE for example had the drawing number 13 and when it was introduced in 1930 it was named *MG13*. In 1934 the German military returned to denote their small arms after the year of construction/introduction. This was continued until the end of World War II. (Handrich, 1993 p. 55)

The German Luftwaffe (air force) used a different system beginning 1935. The first part of the model number consisted of the caliber, followed by the number of the construction pattern of the weapon model, i.e., the first model would have been indicated by the number "1" in the designation. Hence, the *MG81* was an 8 mm caliber (7.92 mm) machine gun, and the first model. The *MG131* had the caliber 13 mm, the *MG151* 15 mm caliber. The system was replaced in 1942 for all air force machine guns and machine cannons with calibers of less than 20 mm by a system which included the development company: "1" stood for RHEINMETALL-BORSIG, Unterlüß, "2" for MAUSER, Oberndorf "3" for KRIEGHOFF, Suhl, and "4" for KRUPP, Essen, with the first digit in the name of the weapon for the development company, and the following numbers and letters would indicate the specific type of weapon, such as the *MK108* by RHEINMETALL-BORSIG (= 1). (Eckardt, et al., 1973 p. 306f)

After WWII consecutive model numbers were introduced. The *P38* pistol would become the *P1* pistol, and in the seventies the count was up to five, six, and seven, when the German police introduced their WALTHER *P5*, SIG SAUER *P6*, and HECKLER & KOCH *P7* duty pistols. The model numbers are given by BAAINBw K6.2. (Helmut Kimmerle, per e-mail, 2015)

[16] (Götz, 1981 pp. 47–51)

[17] (Lugs, 1982 pp. 79–82), ibid.: Pauly creates the first self-contained cartridge (France, 1812), improvement of the rimfire ignition (USA, ca. 1857), 1864 Edward M. Boxer's cartridge for British Rifles, and ca. 1865 Hiram Berdan's cartridge (USA).

[18] (Götz, 1981 p. 25) Ruptured casings as well as failures to extract were common.

[19] (Pawlas, 1971 Vol. 1 pp. 57–63)

[20] (Götz, 1981 p. 72f)

[21] (Linnenkohl, 1990 p. 11f)

[22] Four problems were faced: (i) During feeding the extractor would engage the cartridge's extractor groove too late, allowing double feeds and causing unintentional discharges, (ii) in the event of a ruptured case hot gases would be discharged under pressure from the rifle, (iii) ruptured barrels, (iv) extreme barrel wear. (Scarlata, 2008 p. 64f)

[23] (Berz, 2001 pp. 62–76) Starting May 1915, a reorganization in the field of infantry weapon manufacturing was initiated. The decision was to not build more production facilities, but to outsource production instead; known as de-centralization in manufacturing. The basis for this de-centralization was that the military would change their attitude toward what was seen classified information and would focus on modern production technology as well. This did not directly lead to an interchangeability of components made by different suppliers. With the blueprints and gauges handed over in May 1915 it was not possible for the suppliers to manufacture interchangeable parts. First standardized drawings had to be created and then gauges had to be designed which were based on the new blueprints.

ENDNOTES

[24] (Berz, 2001 p. 55) *Hounshell* explains, the interchangeability of components presented a considerable challenge with the introduction of mass production. Before the introduction of moving line assembly at FORD MOTOR COMPANY two conditions had to be met: firstly, Henry Ford became a specialist for the interchangeability of parts after he had learned from the practices in the arms industry, when it came to the introduction of production with high manufacturing accuracy. Secondly, Henry Ford adopted sheet steel punch and press work. (Hounshell, 1985 p. 10)

According to *Klemm* the French gunsmith Honoré Blanc was a pioneer of the use of interchangeable parts. In 1785 Blanc started to use gauges and filing jigs to bring duplicate parts to interchangeability. *Klemm* continues, Eli Whitney was inspired by Blanc and started in 1801 to pursue the idea of interchangeable parts for small arms. The concept of interchangeability could only be fully achieved after improvements in machining operations were available. The new accuracy achieved in manufacturing processes allowed the production of standardized components which were truly interchangeable. (Klemm, 1998 pp. 165–168)

Long before Henry Ford, starting from 1815, U.S. armory managers realized the need for interchangeability of component parts. At first, they introduced a production method which was based on gauging, however interchangeability wasn't fully achieved initially. (Raber, et al., 2009 pp. 78–94)

[25] (Berz, 2001 pp. 73–76)

[26] (Lugs, 1982 p. 82) [Quote translated by PD] According to *Smith*, the principle of self-loaders had been theorized centuries ago but could only be realized after progress in metallurgy, the availability of metal cartridges, and smokeless powder with reduced fouling were available. Starting from approximately 1877 the time was right to begin with serious research and the development of self-loading guns.

[27] (Ezell, 1981 p. 133)

[28] (Smith, 1953 p. 26)

[29] (Smith, 1953 pp. 27, 178f) *Smith* refers to it as the model *M98* pistol, *Hogg* as M96. (Hogg, 2001 p. 211f) *Ezell* points out, Schwarzlose's pistol *1898* would have been the first commercially available pistol with a slide stop feature. (Ezell, 1981 p. 153)

[30] (Lugs, 1982 p. 464f)

[31] (Rüb, 2012 p. 3)

[32] (Welter, 2013 p. 17)

[33] (Rüb, 2012 p. 3) Rüb also describes, based on estimations it would be 55 Guns per 100 citizens in Yemen, 46 in Switzerland, and 34 in Iraq.

[34] (Law Center to Prevent Gun Violence, 2014 p. 279): District of Columbia v. Heller

[35] (Congressional Research Service)

[36] (Rüb, 2012 p. 3)

[37] (NSSF, 2014 p. 181), ibid.:

> "Firearms retailers have been reporting that nearly 25 percent of their customers in recent years were purchasing their first firearm. To learn more about these new customers, NSSF commissioned this study to better understand this important segment. The study reveals that first-time gun buyers are largely active in one or more shooting activities and that women are motivated to purchase their first firearm predominately for personal defense."

Ibid.: A survey conducted during President Obama's tenure about the most frequent reasons for first gun purchase reported (multiple answers possible): home-defense (87.3 percent), self-defense (76.5 percent), wanted to go shooting with my family and/or friends (73.2 percent), want to be self-sufficient (69.6 percent), always wanted a gun (63.8 percent).

[38] (Barrett, 2012a)

[39] (Porter, 2010 p. 643)

[40] (nssfnews, 2014) and (Stock, 2014)

ENDNOTES

[41] (Porter, 2010 p. 29ff)
[42] (Porter, 1999 p. 75)
[43] (Porter, 1999 p. 78ff)
[44] (Brauer, 2013 pp. 11–13, 54)
[45] (Brunner, et al., 2014 pp. 7, 13)
[46] (Burhop, 2011 p. 137f) and (Brunner, et al., 2014 pp. 7, 13, 224–231)
[47] (Burhop, 2011 p. 43) and (Woll, 2011 p. 372f)
[48] (McNab, 2015 p. 24f)
[49] (Brunner, et al., 2014 p. 310f)
[50] (Brunner, et al., 2014 pp. 274–281)
[51] Also: (Goldratt, et al., 1984 p. 383)
[52] (Womack, et al., 1990 pp. 21–26)
[53] (Womack, et al., 1990 pp. 26f, 276)
   American manufacturers of sewing machines, revolvers, and harvesting machines had adopted the idea of interchangeable parts long before automakers. Cadillac was the first U.S. carmaker demonstrating the interchangeability of its component parts during a reliability test in 1908. Cadillac participated in the 1908 interchangeability test where three of its vehicles were disassembled, the components mixed, then the cars were reassembled and driven for 500 miles. However, the truly revolutionary increase in productivity was achieved when Henry Ford combined the idea of interchangeable components with the moving assembly line. (Möser, 2002 p. 156f) The assembly line was first tried on April 1, 1913. Within eighteen months of the first experiments continuous assembly lines were used in most assemblies. (Hounshell, 1985 p. 10)

[54] (Womack, et al., 1990 p. 55f)
   Henry Ford coined the term *mass production* in 1926. (Womack, et al., 1990 p. 26 and 280)

[55] (Womack, et al., 1990 p. 81)
[56] (Womack, et al., 1990 pp. 48–69)
[57] (Schwab, 2016 pp. 6–13)
[58] (Hoffmann, 2012 p. 14) [Quote translated by PD]
[59] (Deichsel, et al., 2010 p. 16)
[60] (Diller, 2000 p. 366)
[61] (Lugs, 1982 p. 521f) [Quote translated by PD]
[62] (Beeley, et al., 2008 p. 9ff)
[63] *Gebhardt* explains the distinction used in the Anglo-Saxon literature between subtractive, formative and additive manufacturing processes. Subtractive processes create the desired geometry by defined removal of material, formative methods reshape the volume of the part but does not change the volume of it, and the latter create the desired shape by adding volume elements. Figuratively speaking: similar to creating a sculpture by connecting existing components.
   Additive manufacturing is an automated fabrication process based on layer technology. In contrast to the mere joining of prefabricated geometries, these processes will generate the geometry of an object and the material properties. Liquids or powders are used, which are applied and solidified in individual layers. The process is also referred to as generative manufacturing but more commonly known as additive manufacturing or 3D printing. (Gebhardt, 2016 pp. 1–3, 60)

[64] (Gebhardt, 2016 pp. 1–3)
[65] (Gebhardt, 2016 p. 13) As Gebhardt mentions, the product development lead time is of decisive importance for the market success of a product. The time that elapses between the decision to develop and produce a product and its introduction into the market is referred

ENDNOTES

to as *time to market*. Gebhardt explains that if the budgeted development costs were exceeded by 50 %, this would reduce the budgeted profits by only 3.5 %, whereas postponing a product launch by six months would reduce the planned profit by 33 % even if any other requirements of the project were met. Frontloading, i.e. defining binding requirements at an early stage, would be conducive to minimize time to market. (Gebhardt, 2016 pp. 348–352)

[66] (Lipson, et al., 2013 pp. 30–39)

[67] (Ponsford, et al., 2014), (Gebhardt, 2016 pp. 47–51)
1986 3D SYSTEMS, INC. is founded; the first 3D printing company. Co-founder is Charles Hull. (Horsch, 2014 p. 30)

[68] (Gebhardt, 2016 pp. 60–67) Gebhardt explains that the term laser sintering (LS) or selective laser sintering (SLS) is preferably used in context of the processing of polymers. Whereas laser melting or selective laser melting (SLM) would refer to metallic materials. (Gebhardt, et al., 2016 pp. 46, 49)

[69] (Gebhardt, 2016 p. 6f)

[70] (Gebhardt, 2016 pp. 26–29) The STL file format was introduced in 1989. (Horsch, 2014 p. 67) Depending on the literature used, the abbreviation STL is explained to be short for: Stereolithography Language (Awiszus, et al., 2016 p. 324), Standard Transformation Language (Gebhardt, 2016 p. 35), Surface Tesselation Language (description of just the surface by triangles), Standard Triangulation Language (Horsch, 2014 p. 67), or Standard Tesselation Language (Lipson, et al., 2013 p. 101). *Lipson* explains that the possibilities offered by STL may reach their limits in the future and that more modern file formats are required, which, for example, take into account combinations of colors and materials as well as other physical and technological properties. Lipson proposes the Additive Manufacturing Format (AMF). STL does not supply the content in question, as this file format only contains geometric information. (Lipson, et al., 2013 p. 101f) and (Gebhardt, 2016 pp. 37, 45–46)

[71] (Horsch, 2014 pp. 67f, 70)

[72] (Leupold, et al., 2016 pp. 27–31)

[73] (Leupold, et al., 2016 pp. 1–23)

[74] (Lipson, et al., 2013 p. 229f), (Leupold, et al., 2016 pp. 57–63) As a feasible way of protection *Leupold* recommends the registration of a mark as a trademark in the register kept by the trademark office under *wares or service*, Class 9:
> "Trademark owners who cannot rule out the possibility that they will offer their own files for their products in the future should extend the protection of their trade marks to Class 9 and register it for 'downloadable image files' and 'electronic publications (downloadable)' in order to be (also) able to take action from their trade mark against unauthorized trade in 3D artwork for their products." (Leupold, et al., 2016 p. 100f)

[75] (Lipson, et al., 2013 pp. 222–230)

[76] (Awiszus, et al., 2016 p. 322), (Gebhardt, 2016 p. 24)

[77] (Lipson, et al., 2013 pp. 231–236)
Do-it-yourself 3D printing, hobbyists and the respective Internet community enjoy a high status at *Horsch*. (Horsch, 2014 pp. 31–36)

[78] *Nitz* introduces the reader to printers for beginners: *Builder* (3DPRINTER4U), *Ultimaker 2* (ULTIMAKER) and *BeeTheFirst* (BEEVERYCREATIVE). (Nitz, 2015 pp. 37–59) *Horsch* mentions *Replicator* (MAKERBOT), *Renkforce RF1000* (CONRAD ELECTRONIC) and *ZEUS* (AIO ROBOTICS). Suitable for professional use would be *Replicator Z18* (MAKERBOT) as well as *RepRap Industrial* (KÜHLING & KÜHLING). (Horsch, 2014 pp. 208–210) Also known are fabricators from ANYCUBIC, FORMLABS and DREMEL.

[79] (Lipson, et al., 2013 pp. 217–220)

[80] (Awiszus, et al., 2016 pp. 124–127)

[81] (Awiszus, et al., 2016 pp. 146–163)

526

ENDNOTES

[82] Bakelite is a thermosetting phenol formaldehyde resign with fabric fibers or sawdust.

In 1907 Leo Baekeland filed for the first patent for the production of synthetic material. He developed the manufacture of thermosetting molding compounds by condensation of phenol and formaldehyde, i.e. the production of phenolic resins. In 1910, the first industrial production was carried out by the BAKELITE GMBH in Erkner near Berlin, Germany. Until the middle of the 20th century items such as bayonet- and pistol grip panels, as well as parts of telephones, radios, light switches etc. were made of Bakelite. (Dern, et al., 2012 pp. 1032–1034)

The addition of fillers and reinforcement materials, as well as sliding and release agents, and dyes influenced the physical properties of the later substance. Usual reinforcements are wood flour, powdered cellulose, olive seed flour, and different shell flours. Wood flours in different fineness represent the most important filler. Inorganic, mineral fillers such as chalk, kaolin, mica and graphite increase the heat resistance and reduce the shrinkage behavior of the substance. (Dern, et al., 2012 pp. 1040–1043)

[83] (Franck, 2000 p. 169ff)

[84] (Hoenow, et al., 2012 p. 67ff)

[85] (Beeley, et al., 2008 p. 23)

[86] (ZOLLERN GmbH & Co. KG, 2009 p. 4)

[87] (Beeley, et al., 2008 p. 111f)

[88] This information was taken from the image DVD of ECRIMESA. See also: (Blank) and (ZOLLERN GmbH & Co. KG, 2009)

[89] (Doege, et al., 2010 p. 623)

[90] (Sweeney, 2013 p. 48)

[91] (Doege, et al., 2010 pp. 496ff, 622)

[92] (Emri, et al., 2014 p. 402)

[93] See patent Raymond E. Wiech US 4,197,118 dated April 12, 1976:

[94] This information was taken from the image DVD of MIMECRISA, and a non-published description of the MIM-process by CM-PULVERSPRITZGUSS GMBH (1994).

[95] This information is from a non-published description of the MIM-process by CM-PULVERSPRITZGUSS GMBH (1994).

[96] See disclosure DE 10,2010,004,106 A1, filed January 7, 2010.

[97] (Handrich, 1993 p. 69) and (Wacker, 1993 p. 66)

[98] (Awiszus, et al., 2016 pp. 106–122)

[99] (Oehler, et al., 1993 p. 65ff)

[100] (Oehler, et al., 1993 p. 112ff)

[101] Dr. Joachim Boßlet, DURFERRIT, per e-mail, October 7, 2013.

[102] (Kersten, et al., 2008 p. 74)

[103] (Boßlet, 2009 pp. 1–3)

[104] (Boßlet, 2009 p. 12)

[105] (Boßlet, 2009 p. 14)

[106] Also, when workpieces are treated while on stands, nitrogen will not penetrate where workpieces are in contact with the stands. Joachim Boßlet per e-mail, September 23, 2016.

[107] (Boßlet, 2007 p. 3)

[108] (Boßlet, 2009 p. 2)

[109] (Boßlet, 2009 p. 6)

[110] (Boßlet, 2009 p. 4)

[111] (Boßlet, 2009 p. 6)

[112] (Boßlet, 2009 p. 9)

[113] (Boßlet, 2009 p. 2)

[114] (Boßlet, 2007 p. 2)

[115] Boßlet per e-mail, October 7, 2016.

ENDNOTES

[116] Boßlet per e-mail, September 23, 2016

[117] (H.E.F. Groupe, 2002 pp. 1, 4) and information provided by Roger Pingel of H.E.F. DURFERRIT.

[118] Ilaflon is made by INDUSTRIELACK AG, Wangen, Switzerland.

[119] *DuraCoat* is distributed by HOUTS ENTERPRISES LLC, Dayton, Nevada, USA.

[120] (Cerakote, 2014)

[121] (Seel, 1993 pp. 156–159)

[122] (Rheinmetall, 1982 p. 290) Apart from this, *Chinn* describes the requirements on the stability of the charge, e.g. to clear the chamber, as well as the risk of an ignition of the content of a magazine in the event where a primer gets hit. (Chinn, 1987 pp. 581–587)

To better protect the nose of the firing pin from erosion, it has been suggested to use some form of electronic ignition (Shepherd, et al., 1945 pp. 81–113), which is used on the VOERE *VEC 91* rifle. (Dockery, 2007 pp. 86–88) Now, VEORE offers an interchangeable bolt for their VOERE *X3* bolt action precision rifle, which sets off the primer via *LaserIgnition*.

[123] In 1963 Gaston Glock formed GLOCK KG, a company that produced polymer and metal parts. (Hlebinsky, 2016 p. 12) In the following years this company from Deutsch-Wagram, just outside of Vienna, Austria, won contracts to make machine gun belt links, field knives, and training grenades for the Austrian Armed Forces. (Mötz, et al., 2013 p. 530)

On January 9, 1981, the actual pistol manufacturer was founded, i.e. GLOCK GES.M.B.H. (FirmenABC, 2019) On October 29, 1982, GLOCK GES.M.B.H. had been put on notice that the Austrian Defense Ministry decided on Gaston Glock's handgun with an initial order for 25,000 *P80* pistols. (Mötz, et al., 2013 p. 539)

[124] (Chinn, 1955 p. 3)

[125] (Wille, 1896 pp. 21–26)

[126] (Kaisertreu, 1902 p. 25f)

[127] (Kneubuehl, 2013 pp. 67–70)

[128] (Gerthsen, 2015 p. 17)

[129] (Stroppe, et al., 2018 p. 51f)

[130] (Dobrinski, et al., 2010 p. 49f) *Einstein* tries to convey the term force as follows:

"The connection between force and change of velocity – and not, as one might intuitively believe, the connection between force and velocity itself – is the basis of classical mechanics of Newtonian coinage. [...]

What is force? Intuitively, we think we know what is meant by this expression. The term originated from the activity of pushing, throwing or pulling or from the muscle excitation associated with all these actions. [...] Whenever and wherever we perceive a change in speed, we must, generally speaking, blame an external force for it. Newton writes in his 'Principia':

An external force is an influence exerted on a body in order to change its state, either that of rest or that of uniformly straightforward forward motion.

This force appears only as an active influence and does not adhere to the body when the influence ceases to be active, for each body remains in each new state it enters, solely by virtue of its vis inertiae. External forces can be of various origins, such as impact, pressure and centripetal force." (Einstein, et al., 1938 p. 20f) [translated by PD]

Einstein also mentions: "The actual physics began with the creation of the terms 'mass', 'force' and [..]. These terms are all pure abstractions. They formed the basis for mechanical thinking." (Einstein, et al., 1938 p. 314) [translated by PD]

In the following endnotes various theories concerning recoil-actuation and gas-actuation are presented. But first a clarification of the systematics:

The task of the discipline of classical mechanics is to describe and predetermine the motion of bodies and the forces associated with these movements. A sub-discipline, dynamics, deals with the interaction of forces and movements: Forces are physical quantities that cause a change in the state of motion of bodies. In this context, the discipline of kinetics studies movements of bodies under the effect of forces. Kinetics was scientifically founded by Isaac Newton, who gave the first formulation of the laws of motion in 1687. With the help of these idealizations many technical problems of mechanics can be solved. Basically, the considerations and calculations in mechanics are always based on forces. (Gross, et al., 2019b p. XIIIf) In the context of pressures, the effect of pressure is always described by the force that acts per unit area. In other words: classical mechanics reduces pressures to the forces they exert. As is well known, every force produces a counterforce. Forces thus always occur in pairs; in our application, the effect of force action results in the recoil effect. Consequently, in this book we will refer to all semi-automatic pistols that use the recoil effect as recoil operated. In comparison, the term gas-operated typifies systems in which gas pressure is tapped from the barrel and its force effect is used outside the barrel.

[131] (Dobrinski, et al., 2010 p. 76)

An impact is the sudden collision of two bodies and the resulting change in movement. An impact is characterized by the fact that a very large force is applied over a very short period of time, the so-called impact duration. The mass undergoes a sudden change in speed and the same amount of force is applied to the bodies. The sum of the impulses of both bodies remains constant at every moment of the impact. The principle of linear momentum is often used for impact processes. The exact course of the force during the impact is usually not of interest. (Böge, et al., 2019 p. 242) and (Gross, et al., 2019b p. 51) Three assumptions apply: The impact duration t_S is so small that the change in position of the bodies involved during t_S can be neglected. The forces occurring at the point of contact of the bodies are so large that during t_S all other forces can be neglected compared to them. The deformations of the bodies are so small that they can be neglected with respect to the movement of the bodies as a whole; i.e. the bodies are regarded as rigid in the laws of motion. (Gross, et al., 2019b p. 93f)

[132] (Dobrinski, et al., 2010 p. 78)

[133] (Rheinmetall, 1982 pp. 427–433), (Mootz, 1989 p. 126), (Kneubühl, 1994 p. 40f)

[134] (Dobrinski, et al., 2010 p. 71)

[135] (Dobrinski, et al., 2010 p. 176f)

[136] (Greiner, et al., 1993 pp. 2–5, 17), (Rinker, 2011 pp. 28, 56), (Kneubuehl, 2013 pp. 38–40)

[137] (Chinn, 1955 p. V)

[138] (Mootz, 1989 p. 33)

[139] (Brukner, 2003 p. 107)

[140] (Kneubuehl, 2013 p. 56f) Kneubuehl differentiates between actuation by recoil and gas-actuation. The latter uses the feature of a gas porting to activate the unlocking mechanism on the outside of the barrel. Based on this approach, we are going to classify blowback pistols as recoil-actuated handguns.

[141] (Johnston, et al., 2010 p. 390f) *Jentzsch* explains, a gas-operated mechanism is not driven by recoil, but it will use gas pressure to power the mechanism. (Jentzsch, 1902 p. 27)

[142] A pistol's locking system lies in the field of classic mechanical machine components and within the question of how to join objects to prevent them from moving when a load is applied. Gun locking systems are non-permanent joints, in that they may be fastened and unfastened repeatedly. Two objects can be temporarily joined to each other by *force locking* or *form locking*. The former includes inertia in combination with spring loads, the latter is based on shaped components to prevent movement of objects relative to each other while transmitting large forces.

[143] (Dannecker, 2016 pp. 22, 33) *Force locked mechanism* is an umbrella term borrowed by Dannecker from the technical language, such as the field of connection technology of machine elements, i.e. non-positive connections. Under this umbrella would be common terms, such as "unlocked breech," "inertia type breech block," "inertia locked breech," "semi-

locked," "blowback," "simple blowback," "straight blowback," "plain blowback," "retarded blowback" or "delayed blowback."

Force locked mechanisms can be implemented in the form of fixed or temporary connections. For our application, a temporary connection is of interest and Dannecker describes a force locked mechanism could be achieved by inertia, spring force, friction, the power of combustion gases, or displacement of media, such as liquids or gases.

[144] A *form locked connection* or *positive locking* is a permanent connection of components which are interlocked by a certain shape. (Dannecker, 2016 p. 21) *Form locked connections* are designed to transmit forces, whereby the transmission is done on the surfaces by surface pressure. A form locked connection is created by interlocking of the involved components. The connection remains locked with or without the transmission of force. Instead of "form locked" or "positively locked" often the word "locked" is colloquially used. For a clear conceptual distinction between (1) a form locked group consisting of a barrel and a slide, which will start to reciprocate as a unit when a shot is fired, and (2) a breech, which does not move relative to other components of the gun, i.e. remains in steady position until it gets unlocked, we are going to use for the former the term *coupled barrel and slide*, and for the latter *form locked breech*.

Terms such as *coupled* or *paired* can be found as early as in 1893 in patent descriptions of locked breeches. Examples are Bergmann's patent DE 76,571, dated May 9, 1893, and patents by *Mauser*, CH 11,943, dated January 9, 1896, and DE 159,157, dated November 6, 1902.

[145] The chart is based on a diagram published by Morawietz, (Morawietz, 1968 p. C22).

[146] Enormous forces and stresses during recoil and cycling of the slide have always been and still are a challenge for pistol design engineers. (Jentzsch, 1902 p. 15)

[147] (Chinn, 1955 p. 65)

[148] Short-recoil-operated and long-recoil-operated self-loading mechanisms are known. The terminology describes how far the coupled barrel and slide group will be reciprocated before they get separated. Most modern pistols are short-recoil-operated; the group stays coupled for about 4 mm. In a long-recoil-operated action, they would remain coupled for a distance which is the same length like the cartridge or even longer. (Kaisertreu, 1902 pp. 46f, 58), (Chinn, 1955 p. 66), (Hatcher, 1966 p. 53), (Lugs, 1982 p. 304f)

[149] (Rheinmetall, 1982 pp. 263, 417f), (Chinn, 1955 p. 3), (Lewisch, 1971 p. 1055)

[150] (Dannecker, 2016 pp. 10–22)

[151] (Shaydurov, 2010 p. 17)

[152] Sometimes, ISRAEL MILITARY INDUSTRIES (IMI) and SACO DEFENSE are mentioned in connection with the *Desert Eagle* pistol. IMI, as well as SACO DEFENSE would be purely manufacturing companies, explains MAGNUM RESEARCH, INC. (MRI) on its website and emphasizes the intellectual rights would belong to MRI exclusively:

> "Magnum Research's founders, Jim Skildum and John Risdall had been involved with the company since 1979. Both men oversaw the ascent of the Desert Eagle® Pistol from a concept on paper to a pop culture icon including a Playboy magazine front cover featuring Pamela Anderson showing off a Pink Desert Eagle pistol.
>
> Over the past 25, years MRI has been responsible for the design and development of the Desert Eagle pistol. The design was refined and the actual pistols were manufactured by Israel Military Industries until 1995, when MRI shifted the manufacturing contract to Saco Defense in Saco, Maine. In 1998, MRI moved manufacturing back to IMI, which later reorganized under the name Israel Weapon Industries. Both Saco and IMI/IWI were strictly contractors: all of the intellectual property, including patents, copyrights and trademarks are the property of Magnum Research. Since 2009, the Desert Eagle Pistol has been produced in the USA at MRI's Pillager, MN facility." (MRI, 2016)

ENDNOTES

[153] (Kersten, et al., 1999 p. 244) Dwell time is the amount of time that the gas system is charged with high pressure, i.e. the amount of time it takes the projectile's base to get from the gas port to the muzzle is the amount of time the gas system is charged with gas pressure to cycle the weapon.

[154] The theoretical definition of the start of the projectile's movement is when the gas pressure level reaches 10 percent of the peak pressure (short shot). The time elapsing from the point when the firing pin strikes the primer until short shot, or start of the projectile's movement respectively, is referred to as ignition delay time or *ignition time*. The time from the start of projectile movement until the base of the projectile exits the muzzle is referred to as *barrel time*. The time between the firing pin hitting the primer and the projectile base leaving the muzzle, i.e. the sum of ignition time and barrel time, is the *action time*. (Rinker, 2011 pp. 1, 63, 394), (Hatcher, 1966 p. 396), (Rheinmetall, 1982 p. 82f)

[155] The kinetic energy of propellant gases depends upon the powder type used, the charge weight, grain size, shape, volumetric mass, and density, and the diameter of the flash hole in the cartridge case and even the primer used. The projectile's mass, bearing surface, shape, diameter, and material all have an influence on pressure and muzzle velocity. Bullet pull (the projectile must be retained with sufficient strength in the case mouth) affects the force required to overcome the crimp. Bullet seating, chamber, forcing-cone, and bore dimensions determine the ease with which the bullet starts its journey up the barrel. Also, the temperature of ammunition and gun, as well as how the ammunition has been stored since it was manufactured. (Rinker, 2011 p. 45)

[156] (Cranz, et al., 1926 pp. 2–4)

[157] (Cranz, et al., 1926 p. 185)

[158] (Kneubuehl, 2013 p. 68f)

[159] (Chinn, 1955 p. 4)

[160] (Allsop, et al., 1999 p. 71)

[161] (Kneubuehl, 2013 p. 78)

[162] (Chinn, 1955 p. 4), (Allsop, et al., 1999 pp. 42, 71)

[163] (Bock, et al., 1989 p. 13f)

[164] (Rinker, 2011 p. 127)

[165] (Hatcher, 1966 p. 428)

[166] (Rinker, 2011 p. 402)

[167] *Rinker* explains forcing cone is the tapered section in front of a shotgun chamber but is not used correctly for a rifle or handgun. "The final *e* spelling on *leade* is a colloquialism and used only in regard to ballistics and firearms. This spelling is not found in a standard dictionary." (Rinker, 2011 pp. 402, 406)

[168] (Kneubuehl, 2013 p. 67)

[169] In contrast to the dynamics of "crimp jump" where the projectile creeps out of the case mouth due to its inertia during the heavy recoil of some guns.

[170] (Rinker, 2011 pp. 135–137), (Meyer, 1986 p. 359)

[171] (Schreiber, 1978 p. 1072ff), (Schreiber, 1981 p. 42ff), (Görtz, 1981 p. 479), (Görtz, 2004 p. 261f), (Görtz, et al., 2011 pp. 1490–1492)

Sometimes the question is asked why pistol barrels were shortened between WWI and WWII in Germany. The answer on barrel length and caliber restrictions is simple:

"The restrictions go back to the Treaty of Versailles and the rulings based on it. On July 27, 1927, the law 'Gesetz über Kriegsgerät' (RGBl. 1927, Teil I, pp. 239–242) was passed. §§ 1 and 2 banned the manufacturing, storing and exportation of military equipment. In § 3, the military equipment in question is listed. According to § 3, Number 6: 'Pistols and revolvers, automatic or with self-loading device, with a barrel length of more than 9.8 cm or with a caliber of more than 8 mm'."

(Oberstleutnant Michael Peter M.A., ZMSBw, per e-mail, June 19, 2017) [translated by PD]

ENDNOTES

[172] (Lugs, 1982 p. 102)

[173] (CIP, 2011), ibid. Appendix A4, Section 4.1.

[174] (CIP, 2011), ibid. Appendix A4, Section 4.1.

[175] Ernst Voigt, patent DE 356,817, dated September 21, 1920. Button rifled barrels were used by MAUSER in WWII. (Shepherd, et al., 1945 p. 40), (Field, et al., 2011 p. 103)

[176] (Field, et al., 2011 p. 103f) Before, since 1939, canons with squeeze bores had been tested and were then put in service by the Germans, i.e. barrels which forged the projectile down in diameter. As mentioned by Pawlas, the manufacture of tapered bores was a challenge because it demanded a different production technology. In a study dated November 26, 1940, by German Ordnance Directorate, Heereswaffenamt Wa. Prüf, Referat 2 V a, it is reported that hammer forging was used to manufacture tapered bores. Three companies, GUSTAV APPEL, LÜBECKER MASCHINENFABRIK and MANNESMANN-WERKE, Remscheid, were involved in developing a non-cutting shaping technique for the manufacture of bores. According to the study, the first tapered bore hammer forged barrel was tested on November 13, 1940. (Pawlas, 1979 Vol. 33 p. 5265–5269)

[177] (Uhlig, 1964 pp. 21, 23f)

[178] Patent DE 916,485 re-dated October 11, 1944 (filed January 7, 1954), by Kralowetz, Grill, and Ribback. Also, patent DE 923,103, re-dated February 16, 1945 (filed May 20, 1954), inventors Laswitz, Bethke, Ribback. In both cases, the assignee mentioned is GUSTAV APPEL MASCHINENFABRIK, Berlin-Spandau.

[179] Kralowetz was part of the Austrian Ordnance and worked for a division of the German Ordnance Directorate during WWII. "This division sponsored and controlled Gustav Appel's swaging developments." (Field, et al., 2011 p. 104)

[180] (Field, et al., 2011 p. 104) *Uhlig* presumes Bruno Kralowetz and Franz Grill developed the rotary swaging process in 1940. (Uhlig, 1964 p. 22)

[181] (Heym, 2014) Regarding hammer forging a barrel profile and the chamber, see patent DE 768,149, dated December 24, 1941, assignee GUSTAV APPEL MASCHINENFABRIK, Berlin-Spandau, inventors Franz Grill, Bruno Kralowetz, and Otto Stelle.

[182] (Uhlig, 1964 p. 21)

[183] (Heym, 2014)

[184] (Ehrenfried, 1984 p. 982)

[185] "A distinction is drawn between tube wear due to the abrasive effect of the projectiles, in particular the driving bands, and tube erosion due to the washing out effect of the hot rapidly flowing propellant gases, and chemical processes taking place between the tube wall and these gases. [...] An effective means of reducing tube wear is hard chrome plating of the bore surface." (Rheinmetall, 1982 p. 362f)

[186] An example is the U.S. *M14* rifle. With the adoption of this rifle by the U.S. Army in 1957, the Army decided that all subsequent automatic rifles should have chromium-plated barrels and chambers, because experimentation demonstrated that plated barrels lasted longer and had fewer extraction problems. (Ezell, 1984 p. 197) ARMALITE, the manufacturer of the *M16*, ignored this requirement in the beginning. (Ezell, 1984 p. 333 endnote 45) but after corrosion in the chamber was found to be the reason for extraction problems, in an interim step, the company just chrome plated the chamber to protect it from corrosion. Later, the entire bore was chrome plated as per Army requirement. (Ezell, 1983 p. 47)

[187] "The beginning of the rifling is the hardest hit by wear and erosion. Entire lands can be torn away as the number of rounds fired increases. Furthermore, the caliber is enlarged." (Rheinmetall, 1982 p. 362)

[188] (Shepherd, et al., 1945 p. 117), (Rheinmetall, 1982 pp. 362f, 572–576)

[189] (Rheinmetall, 1982 p. 575)

[190] (Shepherd, et al., 1945 p. 117)

[191] (Rinker, 2011 p. 134)

ENDNOTES

[192] (Cranz, et al., 1926 pp. 359–374), (Shepherd, et al., 1945 p. 117), (Rinker, 2011 p. 134)

[193] (Shepherd, et al., 1945 p. 118)

[194] (Shepherd, et al., 1945 pp. 38, 118)

[195] (Schmid, 2003 p. 14), (Shepherd, et al., 1945 p. 121), (Field, et al., 2011 p. 76), (Rinker, 2011 p. 134), (Lamb, et al., 1960 p. 29), (Rheinmetall, 1982 pp. 282–284)

[196] (Ehrenfried, 1984 p. 982ff), (Greener, 1910 pp. 634–637)

[197] (CIP, 2011), ibid.: *see Texts, Table of Contents, Decisions, 4. Individual proof of weapons, 4.2 Carrying out of individual proofs -Rejection before firing, Article 7 Rejection before firing, 7.6 dimensions other than those stipulated in CIP Standards*

[198] (ER "Pistolen", 2008 p. 26)

[199] (ER "Pistolen", 2008 p. 26)

[200] (Kaisertreu, 1902 p. 28)

[201] *Martens* explains that the use of formed sheet metal parts on Franz Eduard Jäger's pistol is not proven. On November 16, 1909, however, Franz Pfannl had submitted the application "Verfahren zur Herstellung des Gehäuses für selbsttätige Pistolen." This would describe a frame made of sheet metal. The application had been issued as Austrian patent no. 45314 on December 10, 1910, and in one the illustrations Pfannl's *Kolibri* pistol is shown. (Martens, 2019 pp. 58–65)

[202] (Kersten, et al., 1999 p. 47)

[203] (Aicher, 1983 p. 1556f)

[204] For headspace dimensions refer to CIP. (CIP, 2011)

[205] (Chinn, 1955 pp. 5–6)

[206] The extractor fulfills two tasks: (1) when clearing a pistol, it extracts the cartridge from the chamber, (2) it can be used to achieve a consistent ejection pattern for every shot.

The extractor does improve the reliability of a pistol, but in general the spent case will get pushed out of the chamber by the residual pressure, without being pulled by the extractor claw. (Wille, 1897 p. 30) What the extractor does is, it will make sure the case's head hits the ejector properly. (Brukner, 2003 p. 87f) The *Bergmann 1894* pistol did not have an extractor, and just used the residual pressure. (Lugs, 1982 p. 413) More information on this pistol can be found in patent DE 78,500, dated June 10, 1893, page 2, and Fig. 3.

[207] Normally, casings get ejected to the right, however exceptions exist: The *C96* pistol and the *Pistole 08* eject to the top, WALTHER *Model 3*, *Model 4*, *P38/P1* and *P5* eject to the left. Sometimes, it is said in the early days of pistols that left side ejection was chosen because it was supposed to be more ergonomic when unloading a pistol. Forums member searcher451 published on *WaltherForums* the following correspondence with Warren Buxton (†2016), which is cited here because sometimes information gets lost on the Internet:

"Dear Warren:

I'm wondering if you know why Walther decided on a left-hand ejection port for the P.38 and its continuation models through the P5. I know that the Modell 3 and the Modell 4 both had a left-hand ejection port, but the later models did not, nor did the PP and PPK series. So why switch back to the left-side system?

Sincerely,

Here's his reply:

Hello, William:

There is no definitive or verified answer to your question referencing why Walther put a left-hand ejection port in the Mod. 3. It may have been done based on a trivial aspect of the initial design process of 'why not' because when the Mod. 3 was introduced, there was as yet no firm standard, implied or otherwise, as to which was better or more preferred: a left, right, or top ejection path. Since most people are right-handed, it was natural for a designer to use a right ejection path or

ENDNOTES

possibly a top path; but if the designer were left-handed, he may have naturally thought in terms of a left path.

I don't know if any of the Walthers, i.e. Carl (the boss) or any of the four brothers, were left-handed. The left-hand path was transferred to the Mod. 4 simply because the same tooling and machine setup used to make the Mod. 3 was also used to manufacture the Mod. 4. The next two major tooling changes involved the Mod. 6 and Mod. 7, and both of them had the right ejection path, so the 'left-handed Walther' idea falls apart with them, or so it appears.

All of the caliber 9mm Parabellum double-action Mod. MP pistols that were based on the Mod. PP platform had the right ejection path, as of course did the Mod. PP itself. However, the next radical design change after the Mod. PP-style Mod. MP pistols was the Mod. MP concealed hammer pistols. They had a left-hand ejection path because the ejector was mounted in the right side of the frame and the extractor in the left side of the breech face.

Once again, there is no known definitive or verified answer as to why the ejection path with these new designs reverted back to the left. With these guns, it was best to put the ejector on the right side of the magazine well, due to the position of the hammer drop lever in the frame and the hammer block pin, also located in the frame. But since these guns were new designs, it seems it would have been just as simple to design a right ejection path into the gun as a left. Anyway, the evolution of the concealed hammer guns then went to the Mod. AP with no major changes, once the type and style of Mod. MP was decided upon.

Another design change was made that resulted in the Mod. MP/H (as I call it), or a Mod. MP with the exposed hammer. Still, however, the left side path was retained; but by then to have changed that path would have meant a completely new design, and the funds, time, and military backing would not have allowed that. Also, as far as I know, there was never any complaint from the military or commercial customers about the left-hand ejection path, so there was no incentive for Walther to change it. The Mod. MP/H then morphed into the Mod. HP and then into the P.38.

The P5 was developed in the 1970s in response to a German police request for a new pistol, and three new designs were eventually adopted: SIG P220, H&K P7, and Walther P5. The P5 was simply a P1 redesigned to conform to the characteristics set down by the German police request-for-quotation rulings. The full length P5 slide could be made much stronger than the older P1 slide and would easily handle hot ammunition. The P1 slide, although it had been strengthened in the 1970s just for that purpose, was still considered marginal in that aspect (although it was eventually found that was of no concern) and also had the possibility of allowing the entrance of too much dirt into the pistol due to its open configuration. In addition, the P1 safety mechanism would not conform to the new RFQ [request for quotation; PD] rules, and an entirely new mechanism had to be designed, and it was thought that a fully enclosed slide would work better with it.

Some of the prototype P5s had the open slide like the P1, but the design quickly evolved into the closed version eventually put on the market. The P5 did gain itself an outstanding reputation for durability and reliability, even when fired with a [sic] large quantities of hot ammunition, so the factory made the correct choice there. The reason why the left-hand path was transferred from the P1 to the P5 was simply to cut design costs. As many of the original design features as possible of the P1 slide were incorporated into the new P5 slide, and that included the position and location of the extractor. The P5's frame is little changed from the P1, including the location of the ejector in the right of the magazine well, and therefore no extensive design changes were required.

The more cost-cutting the factory could do, the better, since police contracts are often non-profitable, or very close to it. The design rules for the new pistol required a magazine with about eight rounds in a single line. Walther was not too enthusiastic about this since pistols with double row magazines was what the commercial market and most other police or military markets wanted, so trying to peddle a single-row 9mm pistol was not too good an idea at that time (or now, I suppose). So, once again, whatever cost-cutting could be done was done.

So, the reason why the P5 was designed with a left ejection path is known, but that still doesn't answer the basic question of why the Mod. 3 was designed with it in the first place.

Have a nice day.
Warren Buxton" (Buxton, 2009)

[208] Brisance is the shattering capability of a high explosive. *Rheinmetall* also explains, how brisance was measured in the 1980s: "The brisance (shattering power) of an explosive is determined experimentally by placing an explosive charge against a copper cylinder (KAST) or a lead cylinder (HESS) and measuring the amount of compression." (Rheinmetall, 1982 p. 43f)

[209] "**Lead trinitroresorcinate** (Lead styphnate, Lead tricinate, Tricinate) [...] is mixed with a small amount of tetrazene to form a component of the "sinoxide" charges." (Rheinmetall, 1982 p. 40)

[210] (Sweeney, 2013 p. 52f) Sweeney recommends to pay attention to the thin walled breech face of the slides. Primer cups of high-pressure ammunition hit the breech face around the striker hole, peening may appear, and pierced primers could cause erosion. (Sweeney, 2008 p. 107f)

[211] (Götz, 1979 p. 150) See also Paul Mauser's Belgian patent no. 120,477 dated March 12, 1896. (Korn, 1908 p. 289f)

[212] (Bock, et al., 1989 pp. 17, 653)

[213] (Bundesministerium der Justiz, 2001) According to the German Radiation Protection Ordnance, Appendix III, Table 1, Column 2, the total activity must not exceed 1E+9 Bq.

[214] (Saleh, et al., 2008 p. 621)

[215] (Hecht, 2009 p. 946f)

[216] (Hecht, 2009 p. 974)

[217] (Saleh, et al., 2008 p. 621ff)

[218] DIN 67510, *Photoluminescent pigments and products*, (part 1 and part 2) defines photoluminescent pigments as products that can be stimulated by UV or visible radiation, and which will emit light for more than thirty minutes with a luminance of at least 3 mcd/m² after stimulation. The duration time is defined by DIN 67510 as the time within which the luminance drops to 0.3 mcd/m²; where said 0.3 mcd/m² is approximately 100 times above the limit of visibility.

[219] (Hecht, 2009 pp. 14, 91ff) Light is absorbed and emitted in the form of photons. Photons are elementary particles, the quantum of the electromagnetic field including electromagnetic radiation such as light. Photons are stable, have zero mass, no electric charge, and move at the speed of light. We cannot directly see photons, but the effects of the interaction of light with its surroundings, and such effects occur only if photons are generated or disappear. For the birth or the death of a photon, charged particles are needed. Usually, photons are emitted or absorbed by electrons, the latter are part of the electron shells of atoms.

Quantum mechanics describes the manner in which light is absorbed and emitted from atoms.

[220] (LumiNova AG p. 1f)

ENDNOTES

[221] (LumiNova AG p. 3)

[222] (LumiNova AG p. 3)

[223] (Lindner, 1993 p. 112)

[224] Information provided by TRITEC, Switzerland.

[225] (Bundesministerium der Justiz, 2001 p. 49)

[226] (NRC, 1998 pp. 8.1–9.10, Appendix O)

[227] (NRC, 1998 pp. 8.1–9.10), testing requirements are listed ibid., Appendix O.

[228] (Frankonia, 1990) The GLOCK pistol is listed for the first time in this catalog in 1990. In order to get comparable prices, it was decided to use the prices shown in this catalog: BERETTA p. 147, GLOCK p. 166, HECKLER & KOCH p. 145, SIG SAUER p. 146, and WALTHER p. 144.

[229] (Schmit, 1972 p. 936) [Quote translated by PD]

[230] (Popenker, et al., 2007 p. 34), ibid.: First trials with aluminum alloy frames were carried out by J. P. SAUER & SOHN and WALTHER in the late 1930s. A broader interest was created, when COLT and SMITH & WESSON introduced pistols and revolvers with aluminum alloy in the 1950s. Companies, such as RUGER, were able to reduce the machining effort of investment cast components made of steel or aluminum alloy.

[231] (Kasler, 1992 pp. 4–14)

[232] (Mötz, et al., 2013 p. 609)

[233] (Kasler, 1992 pp. 20–25)

[234] (Mötz, et al., 2013 p. 609) According to the media, GLOCK allegedly had pistols delivered to Libya. Because of these reports, the gun advanced to a "terrorists' weapon" and dedicated gun for plane hijackers. (Mötz, et al., 2013 p. 609f)

[235] (de Vries, et al., 2004 p. 16)

[236] (Dockery, 2012 p. 22)

[237] Johnston mentions with regard to the development of the AR-15 rifle fiberglass stock as well as synthetic stock. (Johnston, et al., 2010 pp. 1014–1024) The stock of the M16 was made of black fiberglass. (Johnston, et al., 2010 p. 1044)

[238] (Sweeney, 2013 p. 30)

[239] (Johnston, et al., 2010 p. 63) The stock of the STEYR AUG is made of glass fiber reinforced polyamide 66.

[240] (Kersten, et al., 1999 pp. 52–56)

[241] (Dockery, 2007 pp. 154–156) Dockery explains, the VP70 had been the first commercially produced pistol with a polymer frame. A metal insert was cast into the polymer frame. This insert had slide rails and helped to "give the pistol an expected service life of over 30,000 rounds."

[242] (Kersten, et al., 1999 p. 72) The frame of the VP70 pistol was made of ABS polymer.

[243] See disclosure DE 1,728,251, assignee HECKLER & KOCH GMBH, dated September 18, 1968, page 25, claim 1.

[244] See disclosure DE 1,728,251, assignee HECKLER & KOCH GMBH, dated September 18, 1968, page 6f: The sliding trigger of the VP70 would allow to avoid pivoting points in the frame, which might have had to carry high loads and would have required some reinforcement measures.

[245] Patent AT 368,807 by Gaston Glock, dated April 30, 1981, page 10, claim 16, describes the firing mechanism assembly fits into an opening inside the frame. And claim 17 explicitly mentions the one-piece frame would be preferably made of polymer.

[246] (PTOOMA Productions, 2004 p. 22) Sweeny doubts the aforementioned 80 seconds: "Each frame is formed when the rails and serial number plate are placed in the mould, the mould closed and the polymer fed into it. [...] the details on temperature, feed rate, production rate and curing are closely-held production secrets. I do know that each one spends something just over a minute in the mould. Everyone I talk to keeps repeating the same figure: 80 seconds. Color me cynical, but when

ENDNOTES

everyone has such a precise and consistent number, I have to wonder if it really is '80 seconds' or if the number was selected for some reason and once it was uttered by a single source has been repeated ever since." (Sweeney, 2008 p. 45)

Typically, the cycle time would be between 70 and 90 seconds, depending on the number of inserts to be placed in the mold.

[247] (Popenker, et al., 2007 p. 38) Q.v. disclosure "Handgriff für eine Schußwaffe," DE 195,05,829 A1, dated February 21, 1995.

[248] (Hüttel, 2007 p. 168f)

[249] NIJ demands a reversible push-button magazine release for LE Pistols with Security Technology and requires it to release with a minimum of 4 lbs of pressure and no more than 7 lbs [18–31 N; PD]. (NIJ, 2016 p. 9)

[250] *Kaisertreu* discusses as early as in 1901 in *Danzer's Armee-Zeitung* the issue of an ideal cartridge storage and feeding device. His opinion is to safe costs and more than one magazine per pistol would be a waste of money. If more than one magazine were to be provided, the soldiers were required to not lose them and would have to put empty magazines in their pockets when performing a mag change. Kaisertreu's other concern is, if it were just one magazine per pistol, the combatant would be in trouble when he run out of ammo and was forced to re-load this magazine during a fight. Kaisertreu also mentions the magazine spring might set if the magazines were continuously carried loaded. Hence, Kaisertreu concludes, the same cartridge clips as used for rifles would be the only reasonable solution, and therefore, pistols would have a fixed magazine. (Kaisertreu, 1902 pp. 92–95) *Jentzsch* doubts this discussion and thinks Kaisertreu is just whitewashing the design of the ROTH pistol. (Jentzsch, 1902 pp. 3f, 47–70)

The first test pistols provided by COLT for the pistol trials (1907–1911) did not have a button magazine release. "The 45-caliber Luger pistol was an enlarged version of the 30-caliber 1900 Parabellum pistol [...]. It was the only test pistol submitted that had a grip safety, and one of only two that allowed ejection of the magazine by the shooting hand. These features were viewed favorably by the board and, indeed, the final victor – the Colt 1911 – incorporated bothof [sic] them." (Malloy, 1989 p. 52)

Even during the development of the Glock pistol, this was still an issue. In this case, the intention was that the magazines would not drop free from the pistol on release. (Kasler, 1992 p. 216)

[251] (Allsop, et al., 1999 p. 209), (Brandeis, 1881 p. 220f)

[252] The pros and cons of hammer-fired vs. striker-fired were discussed as early as self-loading pistols were becoming popular. In a publication from 1902, Richard Wille concludes hammer-fired pistols should be the preference of the military. (Wille, 1902 pp. 22–40)

[253] (Allsop, et al., 1999 p. 209)

[254] (Bock, et al., 1989 p. 15)

[255] (Smith, 1953 p. 37) Smith writes: "In general, strikers in pistols require fewer parts than hammer mechanisms but are not as reliable." This might not be true anymore in modern pistols, and it must be noted that Smith considers just the number of parts of a firing mechanism. But it should be taken into account how the components are manufactured, i.e. how intricate and how demanding they are in manufacturing.

[256] (Bock, et al., 1989 p. 15)

[257] (Brukner, 2003 p. 140) According to Brukner, the pocket pistols made during the first decades of the 20th century were mainly striker-fired, because this allowed for a more compact pistol design. He explains, that however, hammer-fired designs were supposed to be safer and sometimes faster to use, and this would have been the reason why the demand for striker-fired designs slowed down.

[258] (Görtz, 2004 pp. 10, 24f, 32, 38) Since the success of Glock pistols, striker-fired designs without cocking indicator are widely accepted. Some might wonder, why users still prefer

pistol designs with exposed hammer. *Bolotin* answers this question in his review on early pistols from the 19th century: "[...] the reliability of the earliest pistols fell far short of time-proven revolvers. Concealed firing-pins were widely regarded as less safe than external hammers, which allowed the state of cocking to be seen at a glance." (Bolotin, 1995 p. 12) *Fleck*'s discussions from the year 1901 prove Bolotin's point. (Fleck, 1901 p. 84) The Swiss army had adopted the Luger pistol in 1900 and *Kaisertreu* discusses the missing loaded chamber indicator of this pistol and why it was not possible to cock the firing mechanism in the event of a misfire. (Kaisertreu, 1902 p. 16f)

[259] (Görtz, 2004) Luger's pistol was first introduced by the Swiss army in 1900, next by the Imperial German Navy (*Pistole 04*), and then a model with shorter barrel was adopted by the German Army, known as the *Pistole 08*.

[260] (Smith, 1953 p. 266) Smith describes the model *American Savage 1910*. SAVAGE manufactured also a .45 cal. model, i.e. SAVAGE Model 1907, which they submitted for the U.S. pistol trials. "In the 1907 pistol trials, it was neck-and-neck between the Savage and Colt pistols. In fact, the Savage beat the Colt in the dust test and was not far behind in the rust test." (Sweeney, 2010 p. 39)

[261] (Götz, 1979 pp. 41–44), (Lugs, 1982 p. 226)

[262] (Lugs, 1982 p. 59f)

[263] (Lugs, 1982 p. 226) Hammer-fired revolvers with rotating cylinders appeared already in the 17th century and samples can be found in the National Army Museum in Great Britain.

[264] (Rheinmetall, 1982 p. 39) "To determine *sensitivity to impact*, the drop hammer method is used. Here, a hammer of definite weight is dropped on the explosive to be tested from increasingly greater heights. The height from which the hammer is dropped when the reaction begins, is multiplied by the weight of the hammer, to give the value of sensitivity to impact." (Rheinmetall, 1982 p. 42)

[265] (NIJ Standard-0112.03, 1999 p. 3), ibid.:
> "3.16 Single Action
> A mode of operation that uses the trigger to fire the pistol only."

Smith gives the following definition for single action: "A weapon in which the hammer must be cocked before pressure on the trigger will fire the gun. Each act (coking hammer and pressing trigger) is distinct from the other." (Smith, 1953 p. 603)

[266] (Görtz, 2004 p. 14) Approximately 3,000 *C93* pistols were built. (Görtz, et al., 2011 p. 117) During loading or when the firearm discharges, a "self-cocking" of the firing mechanism is accomplished during the cycle of operation. The next shot can be fired by just pulling the trigger, i.e. in single-action. This explains why sometimes pistols were called self-cockers back in the days of early semi-autos, particularly in Europe. (Wille, 1896 p. 1)

[267] (Lugs, 1982 p. 226), ibid.: "In 1834, Ethan Allen of Grafton, Mass., invented a very simple but practical firing mechanism with restrike trigger and first used it on the single-shot pistol with bolt-on barrel."
Lugs elaborates:
> "The strongest competitor to Colt's percussion revolver was the one made by Robert Adams. Adams' revolver is a gun with restrike trigger, i.e., by pressing the trigger, the cylinder rotates, the hammer is cocked and released. Although this was not new to pistols with rotatable barrels, but rather very common around 1830, but Adams' lock was nevertheless an advance over the earlier designs, especially because Adams achieved a lighter trigger action." (Lugs, 1982 p. 247) [translated by PD]

[268] (Smith, 1953 p. 38) Smith describes the repeat strike capability of this trigger action:
> "*Cocking Systems* of the double-action type have been highly developed in Europe. (Walther PPK, Sauer Model 38 and Little Tom, Caliber .32 ACP). [...] Instead of thumb-cocking the hammer or pulling the slide back as is normally

ENDNOTES

required to fire an automatic pistol under these conditions, in the double action types it is necessary only to press the trigger to raise and drop the hammer. If the cartridge fires, the slide as it is blown back will recock the hammer. If the cartridge should misfire, a second pull on the trigger will again raise and drop the hammer, usually discharging the average defective cartridge."

Other sources ignore the restrike capability of this trigger, e.g. for compliance with ATF Form 4590 a partially cocked trigger, such as found on GLOCK pistols, would be classified as DA.

[269] (Götz, 1979 p. 213)

[270] (Fleck, 1901 p. 4)

[271] (Fjestad, 2017 p. 2385) In 1953, *Smith* explained Double-Action triggers on pistols as follows: "[…] Some autoloading pistols are equipped with a type of double-action trigger which goes through the cycle of cocking and firing the first shot by continuous pressure on the trigger […]." (Smith, 1953 p. 603) *Smith's* description does not go as much into detail as Fjestad's explanation, who's definition of Double-Action is as follows:

"The principle in a revolver or auto-loading pistol wherein the hammer can be cocked and dropped by a single pull of the trigger. Most of these actions also provide capability for single action fire. In auto loading pistols, double action normally applies only to the first shot of any series, the hammer being cocked by the slide for subsequent shots."

and Double-Action-Only:

"A firearm which cannot be operated in single action mode. Many newer DAO firearms are either hammerless, or their hammers and triggers cannot be positioned in a single action status."

[272] (NIJ Standard-0112.03, 1999 p. 1) See definitions of Double-Action and Striker-Fire-Action

[273] (Smith, 1953 pp. 36f, 311)

[274] (NIJ Standard-0112.03, 1999 p. 4)

[275] (Ayoob, 2007 p. 22) Whether pre-cocked actions affect the safety of a pistol is also of concern in the USA., ibid.:

"Safety? There was no manual safety per se. All safeties were internal and passive. 'Point gun, pull trigger,' just like the revolver. When BATF declared the Glock pistol to be double-action only in design, the argument about cocked guns being dangerous went out the window, too."

[276] (TR "Pistolen", 2008), l.c. Anlage 5, Glossar, 6. Feuerbereitschaft:

"A pistol can be regarded as uncocked if the energy stored in the system is not sufficient to fire a round, even if the firing mechanism is partially cocked. The maximum permissible energy in the firing mechanism is 41 mJ. This is the amount of energy which, according to the *Technical Guidelines for Cartridge 9 mm × 19*, must not be able to ignite a 9 mm × 19 cartridge when a test ball weighing 55 ± 0.57 g is dropped on it from a height of 76 mm. In cases of doubt, a comparative study must be made of impacts in a copper crusher cylinder. In these tests, a penetration depth of 0.11 mm must not be exceeded." [translated by PD]

[277] (TR "Pistolen", 2008 p. 7), l.c.: "If a decocking of the action is needed for the handling of the pistol, decocking must be possible without pulling the trigger (also when field-stripping it for maintenance)."

[278] (Mootz, 1989 p. 117) Mootz explains, it had been decided to not automatically cock the hammer to avoid negligent discharges. However, this particular pistol did not have a reciprocating slide, it had a forward moving barrel instead.

An unintentional discharge can happen to anyone and at any time. In a letter dated June 30, 1869, Wilhelm Mauser describes what happened to his brother: "Paul shot himself through the right index finger with a revolver. He himself does not know how it happened.

ENDNOTES

At first it was believed that he would lose his finger, but the course was then very favorable, so that he would only have a stiff finger for a long time." (Ebell, 1921 p. 29) [translated by PD]

[279] (Pawlas, 1971 Vol. 2 pp. 241–248) According to Pawlas, this model is supposed to be the first pistol with restrike/SA-trigger.

[280] Patent AT 36 387, 1908, p. 1, ibid.:

"The invention relates to a self-loading firearm with restrike trigger, in which the cartridge magazine can be inserted between the trigger and the hammer. This is made possible by the fact that the hammer, in contrast to those of the known trigger actions, does not come into direct engagement with the trigger, but is located at a certain distance from it [...].

The pistol described [in the patent specification; PD] belongs to the category of hammer-fired self-loading pistols, the hammer of which is cocked only for the first shot by hand or by means of the trigger, but subsequently by means of the slide which is driven back by the gas pressure." [translated by PD]

By 1927, WIENER WAFFENFABRIK produced over 27,000 copies of the *Little Tom*. (Mötz, et al., 2013 pp. 272, 288)

[281] (Ayoob, 2007 p. 9) Ayoob explains:

"Today's Double-Action Autos

Walther popularized the double-action auto with a de-cocking feature in the 1930s. It was seen at the time as a 'faster' auto, the theory being that with a single-action auto like the Colt or Browning, you had to either move a safety lever, or cock a hammer, or jack a slide before firing. With the DA auto, it was thought, one could just carry it off safe and pull the trigger when needed, like a revolver.

At the time, most of America felt that if they wanted an auto that worked like a revolver, they would just carry one of their fine made-in-USA *revolvers*, thank you very much. In the middle of the 20th century, 1911 flag-bearer Jeff Cooper applied an engineer's phrase that would stick to the double-action auto forever after. The concept was, he said: 'an ingenious solution to a non-existent problem.'" (Ayoob, 2007 p. 25)

[282] (Brukner, 2003 p. 164)

[283] (NIJ Standard-0112.3, 1999 p 5), (Vickery, 1940 p. 241)

[284] (Wille, 1896 p. 26)

[285] (Brukner, 2003 pp. 144–149)

[286] (Allsop, et al., 1999 p. 216)

"The trigger system must be designed to ensure disconnection of the trigger mechanism[13,14] when the slide is not completely closed.

For **locked breech systems**, the trigger mechanism must be disconnected before unlocking. The travel of moving parts up to disconnection of the trigger mechanism must not exceed 75 % of the *safeguarded displacement* s_{safety}.

For **blowback systems**, the trigger mechanism must be disconnected after no more than 1 mm rearward travel of the slide." (TR "Pistolen", 2008 p. 11)

"[13] **Trigger mechanism**

The trigger mechanism is the mechanical connection from the trigger to the striker.

[14] **Disconnection of the trigger mechanism**

Separation of trigger mechanism during reloading process and disconnection as long as the breech is not adequately locked.

The trigger mechanism is considered disconnected if the striker cannot hit the primer, especially when the slide moves forward." (TR "Pistolen", 2008 p. 27) [translated by PD]

[287] (Rheinmetall, 1982 pp. 80–83, 637–639), (Kneubuehl, 2013 p. 67)

ENDNOTES

[288] (TR "Pistolen", 2008 p. 11) The dimensions of the adapter cartridge are shown in Appendix 4 of the ER "Pistolen," 2008. For the requirement of 0.3 mm refer to: (TR "Patrone", 2009 p. 12) and l.c. Annex 4 where copper crusher indentations are depicted graphically.

[289] (ER "Pistolen", 2008 p. 12)

[290] (Kneubuehl, 2013 pp. 52, 67)

[291] (TR "Pistolen", 2008 p. 11)

[292] (Eilers, 1938 p. 38) [Quote translated by PD]

[293] Austrian standard ÖNORM S 1200, April 1, 1997, p. 2, lists for centerfire handgun striker noses a diameter range from 1.5 to 2.0 mm [0.059 to 0.079"; PD], a hemispheric tip, and a striker hole diameter, which may be 0.15 mm [0.006"; PD] wider than the diameter of the striker.

[294] (Allsop, et al., 1999 p. 210)

[295] (U.S. 4,825,744, 1988 Column 4)

[296] (TR "Patrone", 2009 p. 12)

[297] (Shepherd, et al., 1945 p. 91)

[298] (Mathews, 1973)

[299] (Stiefel, 1984 p. 966) Also, European patent EP 0,154,356 B1, dated April 29, 1982, where a lance-shaped striker nose is disclosed. The shape of the nose will allow the base of the cartridge to cam the striker nose back during feeding.

[300] (ER "Pistolen", 2008 p. 8)

[301] This trend was undoubtedly started by the success of GLOCK pistols in recent decades. It still persists today and more and more manufacturers are following.

[302] (Meissner, et al., 2007 p. 3)

[303] (Rheinmetall, 1982 p. 288)

[304] (TR "Pistolen", 2008 p. 7), ibid.:
"The pistol must have a service life of 10,000 shots. During this time, the operational reliability of the pistol must be assured and there must be no damage to the main components relevant to operation or safety, aside from normal wear.

In contrast to the *main components*[2] of a pistol, the *other components* (wear parts) must have a service life of at least 5,000 shots." [translated by PD]

L.c., Annex 5 (Anlage 5):
"[2] Main components and other components of the pistol
The components of the pistol are divided into main components and other components, depending on their wear characteristics.

Main components
The main components of the pistol are the breech (and slide, if separate), barrel and frame. These components must not show wear over the service life of 10,000 rounds (or 5,000 load cycles in the case of dry fire testing) to the extent that replacement is necessary.

Other components are wear parts.
Components of this type may be replaced after 5,000 rounds, but only because of damage, and only once. Preventive replacement during trials is not permitted." [translated by PD]

Conclusions: minimum service life of slide, barrel, and frame is 10,000 shots, and 5,000 shots for other components.

[305] (Walz, 1944 p. 16) [Quote translated by PD]

[306] (Walz, 1944 p. 3)

[307] (Schmid, 2003 p. 14)

[308] (Walz, 1944 p. 16) [Quote translated by PD]

ENDNOTES

[309] A first prototype of the *ShKAS* was tested in 1930 and the machine gun got introduced later in 1932, named *ShKAS* Model 1932. It's rate of fire was 2,000 rounds per minute and service life was just 1,500 to 2,000 shots. It took until 1935 and a variety of improvements, e.g. the introduction of a stranded spring, until guns from full production would reach a service life of 15,000 rounds. From 1936 on, this new design was standard equipment in fighter planes and had its baptism of fire in the Spanish War. (Bolotin, 1995 pp. 200–204)

  Chinn explains, the different models of the *ShKAS* were all classified as type 426, but depending on the revision of the design, they were named *KM-35*, *KM-36*, and *Model 41*. (Chinn, 1952 p. 72f)

[310] (Bolotin, 1990 pp. 200–204)

[311] (Lugs, 1982 p. 330f)

[312] Even though this is a rimmed cartridge, it was fed from a disintegrating belt. (Bolotin, 1995 p. 201)

[313] (Bolotin, 1990 p. 274)

[314] (Field, et al., 2011 p. 63) *Bolotin* verifies what is mentioned in Field: First, Shpitalny designed the stranded spring for the *ShKAS* machine gun, which was later used during the Spanish War in the I-16 fighter plane. It would take until after this war, when stranded springs appeared in German and American made automatic firearms. (Bolotin, 1995 p. 203)

[315] "All progress on the weapon's development was kept in close security. In 1936, Russian reports of its use in the Spanish Civil War referred to it simply as 'special machine gun.'" (Chinn, 1952 p. 72f)

[316] (Field, et al., 2011 p. 63)

[317] (Field, et al., 2011 p. 63)

[318] (Field, et al., 2011 p. 64)

[319] (Walz, 1944 p. 30f)

[320] (Meissner, et al., 1993 p. 213)

[321] (Meissner, et al., 2007 p. 166)

[322] (Meissner, et al., 2007 p. 166)

[323] (Rheinmetall, 1982 p. 288), ibid.:

  "*Weapon springs* are often stressed by sudden shocks caused by high velocities. This stress cannot be evaluated using static considerations. So the results of traveling wave theory have to be used as a basis [1]. Only in this way can the comparatively short service lives of certain springs, for example, the return springs, be explained. It shows that a helical spring, made from round wire, sustains a stress increase of about 3.5 N/mm² for each 1 m/s of impact speed, which under certain circumstances – at least in one section of the spring – is even doubled. For this reason, with impact velocities of over 12 m/s, it is not possible to manufacture such springs with satisfactory durability from the usual materials. By using stranded wire springs, the limit of the stress capacity can be increased to around 14 m/s, with an acceptable service life."

[324] (TR "Pistolen", 2008 p. 10)

[325] (NIJ, 2016 p. 11)

[326] (DIN 2090, 1971 p. 1)

[327] (Wittel, et al., 2019 p. 369)

[328] (Meissner, et al., 2007 p. 143)

[329] (Meissner, et al., 2007 p. 144)

[330] (Meissner, et al., 2007 p. 144)

[331] (Meissner, et al., 2007 p. 144)

[332] Information provided by BAUMANN AG, Switzerland

[333] (Meissner, et al., 2007 p. 13f)

[334] (Meissner, et al., 2007 pp. 10, 73–76)

ENDNOTES

[335] (Meissner, et al., 2007 p. 39)

[336] (Meissner, et al., 2007 pp. 14, 39–43)

[337] (Meissner, et al., 1993 p. 80)

[338] (Meissner, et al., 2007 p. 69ff)

[339] (Meissner, et al., 1993 p. 81)

[340] (NIJ Standard-0112.3, 1999 p. 13)

[341] (Ayoob, 2007 pp. 84–88)

[342] ("Arbeitsgruppe Pflichtenheft Faustfeuerwaffen", 1975 p. 3, 8), ibid.:

> "1.2 Handling, Operation, User Safety
> […] Also dropping or bumping must not cause a discharge. Drop tests may render the handgun out of service.
>
> 1.3 Readiness to Fire
> The handgun loaded with a round in the chamber can be carried and deployed safely.
>   It must be possible to immediately fire a shot by pulling the trigger without prior manipulation of other components (e.g. safeties).
>   Cocking the hammer must be provided in a fast and safe manner." (Arbeitsgruppe "Pflichtenheft Faustfeuerwaffen", 1975 p. 3) [translated by PD]

And regarding safeties:

> "2.2.5 Safeties
> The safety system must allow to safely carry and handle the loaded weapon.
>   Preferably, the handgun is equipped with an automatic striker safety, without any external components, e.g. grip- or magazine-safeties. The striker must be safetied even with hammer retracted.
>   The safety of the loaded weapon, as well as with hammer cocked, is to be proven by drop tests according to the testing guidelines." (Arbeitsgruppe "Pflichtenheft Faustfeuerwaffen", 1975 p. 8) [translated by PD]

[343] (Kaisertreu, 1902 p. 105f) [Quote translated by PD]

This quote was first published in *Danzer's Armee-Zeitung* on October 31, 1901. The issue was, if pistols should have automatic safeties. Kaisertreu opines, the *C93* pistol was the first self-loading pistol with an automatic safety. His conclusion is, self-loading pistols should preferably be equipped with restrike/SA trigger action. (Kaisertreu, 1902 pp. 101, 105f)

[344] For example, the broad term *vehicle safety* refers to the reduced likelihood that drivers and other road users might get injured by an accident. In contrast to other content in this book the terminology of active safety and passive safety is quite young. It was proposed in 1964 by Luigi Locati, the director of the Fabbrica Italiana Automobili Torino (FIAT) research center and rapidly found its way into the field of automotive safety concepts. (Möser, 2002 p. 261)

[345] (Smith, 1953 p. 38f), ibid.: "Safeties fall into two general classes, manual and automatic."

[346] (Wille, 1902 p. 35) [Quote translated by PD]

[347] (Cowgill, 1981 p. 163) Cowgill describes the WALTHER *P5* pistol and its safeties in the summary of his article: "Thus there are multiple, independent, and redundant safety features incorporated into the P5's design. However, even the safest weapon can be dangerous through improper operation." He differentiates between automatic safeties (independent) and passive safeties (redundant). But alas, the *P5* pistol built in 1981 did not have any passive safety yet, except for a disconnector, but it sure had an automatic safety, which separated the firing pin from the hammer, and it featured an automatic hammer block. Both of them were safeties which would be engaged in default position. It took the *P5* pistol until the 1990s to get equipped with a passive drop safety, which was located underneath the rear sight.

ENDNOTES

[348] (TARA, 2014 p. 3), ibid.: TARA describes the possibility to field-strip the pistol without having to pull the trigger and calls this feature an "additional passive safety" of their *TM-9* pistol.

Sweeney uses "passive safety" with reference to GLOCK pistols (Sweeney, 2013 p. 28), and so does *Ayoob*: "Safety? There was no manual safety per se. All safeties were internal and passive." (Ayoob, 2007 p. 22) Based on the latter, it can be seen how "passive" is used instead of "automatically working" in normal course.

Let us focus on the definitions used in the general field of safety technology: active safeties help to keep an accident from happening, and passive safeties reduce the effects in the event where an accident is actually happening or a failure occurs. A passive safety device is supposed to prevent an accidental discharge in the event of an "accident," so to speak.

[349] (Brukner, 1983 p. 122) [Quote translated by PD]

[350] (NIJ, 2016 p. 13)

[351] (Simmons, 1988 p. 179)

[352] (Walter, 2004 p. 75)

[353] (Lugs, 1982 p. 100), ibid.: Trigger safeties do not engage the firing mechanism, and based on this fact, Lugs concluded trigger safeties would be less reliable in preventing an unintentional discharge.

[354] (TR "Pistolen", 2008 p. 27)

[355] (Kahaner, 2007 pp. 178–183)

[356] (Law Center to Prevent Gun Violence, 2014 pp. 249–253)

[357] (Geek, 2013), blog article in *The San Francisco Chronicle* on the introduction of microstamping.

[358] (Sweeney, 2009)

[359] (Egelko, 2013), l.c. California Attorney's General Kamala Harris for the Introduction of microstamping.

[360] (Information Bulletin 2013-BOF-03, 2013), see appendix 6.

[361] (Miller, 2014), ibid.:

"Smith & Wesson announced it will stop selling its handguns in California rather than manufacture them to comply with the new microstamping law. The other publicly traded firearms manufacturer in the U.S., Sturm, Ruger, also said this month that it will stop new sales to California. [...] Smith & Wesson President and CEO James Debney said, 'As our products fall off the roster due to California's interpretation of the Unsafe Handgun Act, we will continue to work with the NRA and the NSSF to oppose this poorly conceived law which mandates the unproven and unreliable concept of microstamping and makes it impossible for Californians to have access to the best products with the latest innovations.' [...] The company reported that all M&P pistols (other than the M&P Shield) will fall off the roster by August because of performance enhancements, which will make them subject to the microstamping regulation. The M&P9c has already been taken off the list, and several more M&P models will be unavailable for Californians to purchase by the end of January."

Revolvers and pistols which are already listed on the roster, and hence do not need to have microstamping technology, may still be sold.

[362] (SAF, 2013 p. 1), see also appendix 7.

[363] (Greene, 2013 p. 7)

[364] (The Economist, 2013), (Miniter, 2014 pp. 227–234)

[365] (Keane, 2013)

[366] (SB-293, 2013), ibid:

"This bill would define an owner-authorized handgun as a handgun that has a permanent feature that renders the handgun incapable of being fired except when activated by the lawful owner or owners of the handgun. The bill would specify requirements that an owner-authorized handgun would be required to meet, and would require a manufacturer that has developed an owner-authorized handgun meeting those requirements to submit the handgun for testing, at the manufacturer's expense, before the handgun may be placed on the roster of handguns determined not to be unsafe. If two owner-authorized handguns have been placed on the roster, the bill would, commencing two years from the date that the second handgun was placed on the roster, prohibit the Department of Justice from placing a handgun on the roster that is not an owner-authorized handgun."

SB-293 was rejected on November 30, 2014: "Inactive Bill – Died."

[367] SB-293 suggests the following features for owner-authorized handguns in California:

"In addition to complying with the provisions of [section 31910; PD] subdivisions (a) and (b), as applicable, owner-authorized handguns shall comply with the following minimum performance standards:

(1) The firearm shall not fail to recognize the authorized user, and shall not falsely recognize an unauthorized user, more than one time per thousand recognition attempts.

(2) The time from first contact to use recognition and firearm enablement shall be no more than [..] 0.3 seconds.

(3) The time from loss of contact with the authorized user to firearm disablement shall be no more than [..] 0.3 seconds.

(4) When the firearm is enabled, the 'ready' condition shall be indicated by a visible indicator.

(5) If the recognition technology on the firearm is battery operated, the firearm shall be equipped with a low power indicator that emits an audible signal.

(6) If the user is not recognized, or if the power supply fails, the firearm shall be inoperable.

(7) Enabling authorized user information shall be stored in the firearm as permanent memory that is restored when power is restored.

(8) The firearm shall be capable of use by more than one authorized user and, if the firearm uses hand recognition technology, it shall recognize either of the authorized user's hands.

(d) As used in this section, an 'owner-authorized handgun' means a handgun that has a permanent programmable biometric or other permanent programmable feature as part of its original manufacture that renders the handgun incapable of being fired except when activated by the lawful owner or other users authorized by the lawful owner, and that cannot be readily deactivated.

(1) An owner-authorized handgun shall only be programmed by a licensed firearms dealer.

(2) Biometric data collected for purposes of programming the owner-authorized handgun shall not be used for any purpose other than programming the owner-authorized handgun.

(3) The Department of Justice shall not retain any biometric data that may be stored in an owner-authorized handgun." (SB-293, 2013)

NIJ's *Baseline Specifications for LE Service Pistols with Security Technology* requires that a pistol must permit the operator to fire the gun when an attempt to block the authorization process is detected, e.g. as a countermeasure to electromagnetic interference. It also answers the question if the gun should fire or not in case where "smart technology" fails: "If the security device malfunctions, it shall default to a state to allow the pistol to fire." (NIJ, 2016 p. 13f)

[368] (ModernArms, 2013)

[369] (Rosenwald, 2014)

ENDNOTES

[370] (NSSF, 2013)

[371] (NIJ, 2016 pp. 1, 14), ibid.: p. 14, 4.18.12

[372] (Greene, 2013 pp. 14–16)

[373] (Greene, 2013 p. 8)

[374] (Weiss, 1996 p. 2)

[375] (Weiss, 1996 p. 34)

[376] (Greene, 2013 p. 19)

[377] (Greene, 2013 p. 22f)

[378] (Greene, 2013 pp. 24–27), see also disclosure DELSY, DE 200,13,901 U1, dated August 13, 2000.

[379] (Greene, 2013 pp. 24–27), see also disclosure DELSY, DE 200,13,901 U1, dated August 13, 2000

[380] (Greene, 2013 p. 25)

[381] (Greene, 2013 p. 25f)

[382] (Greene, 2013 p. 26)

[383] (Greene, 2013 p. 26f)

[384] (Greene, 2013 p. 27)

[385] (Greene, 2013 p. 3f)

[386] (Greene, 2013 p. 43f)

[387] (Greene, 2013 p. 18)

[388] (Armatix, 2011)

[389] (The Economist, 2013)

[390] (Beretta, 2014)

[391] (Beretta, 2013)

[392] (Johnson, 2014)

[393] MIL-STD-1913 accessory rail, standardized by Picatinny Arsenal: "Military Standard : Dimensioning of Accessory Mounting Rail for Small Arms Weapons," colloquially Picatinny Rail.
(U.S. Army ARDEC Standardization Office. Quarterbore.com. *MIL-STD-1913 : Military Standard : Dimensioning of Accessory Mounting Rail for Small Arms Weapons.* [Online] February 2, 1995. [retrieved: January 15, 2013] http://www.quarterbore.com/library/pdf_files/mil-std-1913.pdf)

[394] (Secubit, 2014)

[395] (Dockery, 2007 pp. 177–180)

[396] One of the inventors named in this patent is Ernst Mauch. Mauch changed jobs later from HK to ARMATIX GMBH.

[397] (NIJ, 2016 p. 14)

[398] (Günther, 1900 p. 69f) [Quote translated by PD]

[399] *Rinker* mentions, the term *Saturday night special* is now used by anti-gun groups to describe any handgun, regardless of condition or value. (Rinker, 2011 p. 412)

[400] (Vizzard, 2000 p. 14)

[401] (Barrett, 2012b p. 245f), ibid.: In Barrett's opinion, crime levels started to proliferate in the early 1960s. It was believed that this was caused by "a combination of demography (rebellious baby boomers hitting prime crime-committing years), sociology (waves of heroin- and, later, cocaine-related criminality), and racially tinged history (urban riots in the late 1960s, followed by years of decay in inner cities)."

[402] (Vizzard, 2000 p. 14)

[403] (Barrett, 2012b pp. 77f, 112f, 137)
"In June 1992, the US Bureau of Alcohol, Tobacco and Firearms reported the top eighteen models of the nearly fifty-seven thousand handguns seized by law enforcement and traced during 1990 and 1991. [...] The most common crime gun [...]

ENDNOTES

was the .38-caliber Smith & Wesson revolver, almost certainly because it had been on the market for generations, and millions of the guns circulated on the legitimate used market and on the black market. Filling out the top five, in descending order, were: a cheap and unreliable .25-caliber pistol made by Raven Arms; an inexpensive Davis Industries .380 (like the Raven, a type of handgun often referred to as a Saturday Night Special); the nine-millimeter Smith & Wesson Model 3904 semi-automatic [...]; and the heavy-duty Colt .45, another model that had been sold for many decades." (Barrett, 2012b p. 112f)

And regarding law suits filed against GLOCK:

"Initially, the targets of these [..] suits were manufacturers and retailers of inexpensive, unreliable 'Saturday Night Specials': revolvers and pistols that could be purchased for as little as $29 U.S. and were favorites of stickup artists, drug dealers, and cash-strapped residents of inner-city neighborhoods who feared those criminals. Lawyers representing accident and crime victims argued that Saturday Night Specials had no redeeming social value; they couldn't plausibly be marketed for target shooting, hunting, or police work. By their very nature, according to this view, cheap handguns were meant only to kill people and therefore were 'unreasonably hazardous.'" (Barrett, 2012b p. 137)

[404] (Wintermute, 1997)
[405] (Law Center to Prevent Gun Violence, 2014 p. 206)
[406] (ATF Reference Guide, 2005 p. 4ff)
[407] (Vizzard, 2000 p. 24)
[408] (Zimring, et al., 1992 p. 86)
[409] (Vizzard, 2000 p. 14)
[410] (Lott, 2000 p. 239)
[411] (Zimring, 1975 p. 8)
[412] (ATF Reference Guide, 2005 p. 23)
[413] (Korwin, 2009 p. 37ff)
[414] (18 USC § 922, 2014), Title 18 USC § 922(l): "[...] it shall be unlawful [...] to import [...] into the United States or any possession thereof any firearm or ammunition; [...]."
[415] (ATF Reference Guide, 2005 p. 76) THE NATIONAL FIREARMS ACT, TITLE 26, UNITED STATES CODE, CHAPTER 53, INTERNAL REVENUE CODE, § 5845(a) Firearm.
[416] (ATF Reference Guide, 2005 p. 76) THE NATIONAL FIREARMS ACT, TITLE 26, UNITED STATES CODE, CHAPTER 53, INTERNAL REVENUE CODE, § 5845(f) Destructive device.
[417] (ATF Reference Guide, 2005 p. 1)
[418] (ATF Mission, 2016), ibid. see "Mission"
[419] (ATF Reference Guide, 2005 p. 167f)
[420] (ATF Factoring Criteria, 2008), refer to Appendix 3.
[421] (ATF Guidebook, 2009 *Policies and Procedures* p. 5), ibid.: "The factoring criteria apply only to complete firearms, not actions, frames or receivers."
[422] (ATF Guidebook, 2009 *Policies and Procedures* p. 7)
[423] (ATF Reference Guide, 2005 p. 167f)
[424] (ATF Guidebook, 2009 *Policies and Procedures* p. 5)
[425] (Code of Federal Regulations, 2008 *Title 27 CFR 478.92(a)*)
[426] (ATF Reference Guide, 2005 *§478.92(a)(5) Measurement of height and depth of markings*)
[427] (ATF Reference Guide, 2005 *§478.92(a)(5) Measurement of height and depth of markings*), ibid.: "[...] flat surface of the metal [...]."
[428] (ATF Guidebook, 2009 *Firearms Identification 27 CFR 478.92(a)(4) Exceptions (i) Alternate means of identification.*)

ENDNOTES

[429] (ATF Reference Guide, 2005 *Firearms Identification 27 CFR 478.92(a)(2) Firearm frames or receivers.*)

[430] (ATF Reference Guide, 2005 *27 CFR § 478.112(d)(2)*)

[431] (Dockery, 2007 p. 134f)

[432] (ATF Reference Guide, 2005 p. 26), for general information regarding assault weapons refer to: (Law Center to Prevent Gun Violence, 2014 pp. 140–148)

[433] (Miniter, 2014 p. 63)

[434] (18 USC § 922, 2014), see Title 18 USC § 922(p)(1)(B), § 922(p)(2)(B), § 922(p)(2)(C)(i)

[435] (18 USC § 922, 2014), see Title 18 USC § 922(p)(1)(A), § 922(p)(2)(A), § 922(p)(1)(B), § 922(p)(2)(C)

[436] (Law Center to Prevent Gun Violence, 2014 pp. 206–210)

[437] (nks, 2012 p. 1ff), ibid.: On December 14, 2012, 20-year-old Adam Lanza had shot and killed six adults and 20 children at Sandy Hook Elementary School in Newtown, Connecticut. The mass shooter used a semi-automatic rifle and two pistols, which he had taken from his 52-year-old mother whom he had killed with four shots to her head. According to media reports, after the shooting, the Federal Government examined several options in an attempt at tightening gun laws. One consideration was the ban of large capacity magazines. From 1994 to 2004, certain types of semi-automatic rifles were banned in America, the former Act also banned magazines with a capacity of more than 10 rounds. President Obama had pronounced he would implement this law again, although this did not come to fruition. In his first term, he held back with concrete initiatives even though the shooting at Columbine High School (1999) and the bloodshed during a film premiere in Aurora (July 2012) had happened before the events at Sandy Hook Elementary (December 2012).

[438] (Law Center to Prevent Gun Violence, 2014 pp. 212–229)

[439] (Bahners, 2012 p. 3), (Law Center to Prevent Gun Violence, 2014 pp. 8–12, 58–60, 117)

[440] (Dangerous Weapons Control Law, 2008 § 12001)

[441] (California Firearms Laws, 2007 p. 4)

[442] (ATF State Laws, 2010, p. 262, 750.222(b))

[443] (Dangerous Weapons Control Law, 2008 § 12001(a))

[444] (Dangerous Weapons Control Law, 2008 § 12125)

[445] (California Penal Code, 2013 Section 32010)

[446] (Dangerous Weapons Control Law, 2008, Section 12131)

[447] (California Firearms Laws, 2007 pp. 11, 34f), (Dangerous Weapons Control Law, 2008 § 12132 (h)(1))

[448] (Dangerous Weapons Control Law, 2008 § 12020.3)

[449] (Dangerous Weapons Control Law, 2008 § 12020(a)(1))

[450] (Dangerous Weapons Control Law, 2008 § 12020(c)(22))

[451] (California Firearms Laws, 2007 p. 49), Penal Code Section 12087.6

[452] (Dangerous Weapons Control Law, 2008 § 12088.1)

[453] (California Firearms Laws, 2007 p. 5), Penal Code Section 12020(a)(2)

[454] (California Firearms Laws, 2007 pp. 4, 7), Penal Code Section 12020(c)(25)

[455] (Dangerous Weapons Control Law, 2008 § 12276.1(a)(4))

[456] (California Firearms Laws, 2007 p. 10), Penal Code Section 12276.1(a)(5)

[457] (Dangerous Weapons Control Law, 2008 §§ 12125 to 12133), (California Penal Code, 2013 *Section 32010*)

[458] (Dangerous Weapons Control Law, 2008 § 12126)

[459] (Dangerous Weapons Control Law, 2008 § 12127)

[460] (Dangerous Weapons Control Law, 2008 § 12128)

[461] (Dangerous Weapons Control Law, 2008 § 12131.5(a))

[462] (Dangerous Weapons Control Law, 2008 § 12131.5(b))

[463] (Dangerous Weapons Control Law, 2008 § 12131)

ENDNOTES

[464] (Thomson West), ibid.: California Administrative Code, Title 11, Section 4070

[465] (Thomson West), ibid.: California Administrative Code, Title 11, Section 4070 and 4071

[466] (California Penal Code, 2013 Section 32030(a))

[467] (California Penal Code, 2013 Section 32030(a))

[468] (Miller, 2014), ibid.: SMITH & WESSON refers to the changes carried out as "performance enhancements."

[469] (Dangerous Weapons Control Law, 2008 Section 12131)

[470] (Dangerous Weapons Control Law, 2008 § 12131(c))

[471] (Dangerous Weapons Control Law, 2008 § 12131.5(f))

[472] (Dangerous Weapons Control Law, 2008 § 12131.5(g))

[473] (California Code of Regulations, 2013 11 CCR § 4059(a)–(d))

[474] (California Code of Regulations, 2013 11 CCR § 4059)

[475] (California Code of Regulations, 2013 11 CCR § 4059)

[476] (California Code of Regulations, 2013 11 CCR § 4060(a))

[477] (California Penal Code, 2013 Section 31910(a)(1))

[478] (California Penal Code, 2013 Section 31910(b)(1))

[479] (California Code of Regulations, 2013 11 CCR § 4060(b)(1))

[480] (California Code of Regulations, 2013 11 CCR § 4060(b)(2))

[481] (California Code of Regulations, 2013 11 CCR § 4060(c))

[482] (California Code of Regulations, 2013 11 CCR § 4060(d)(1))

[483] (California Penal Code, 2013 Section 31910(5))

[484] (California Code of Regulations, 2013 11 CCR § 4060(d)(2))

[485] (California Code of Regulations, 2013 11 CCR § 4060(f))

[486] (Dangerous Weapons Control Law, 2008 § 12130)

[487] (California Code of Regulations, 2013 11 CCR § 4060(a) and (e))

[488] (California Code of Regulations, 2013 11 CCR § 4060(g))

[489] (California Code of Regulations, 2013 11 CCR § 4060(h))

[490] (California Code of Regulations, 2013 11 CCR § 4052)

[491] (Dangerous Weapons Control Law, 2008 § 12127(a))

[492] (California Code of Regulations, 2013 11 CCR § 4060(f))

[493] (California Code of Regulations, 2013 11 CCR § 4060(f)(3))

[494] (California Code of Regulations, 2013 11 CCR § 4060(f)(4))

[495] (California Code of Regulations, 2013 11 CCR § 4060(f)(1))

[496] (Dangerous Weapons Control Law, 2008 § 12127(c))

[497] (Dangerous Weapons Control Law, 2008 § 12127(a)(1))

[498] (Dangerous Weapons Control Law, 2008 § 12127(a)(2))

[499] (Dangerous Weapons Control Law, 2008 § 12127(b))

[500] (California Code of Regulations, 2013 11 CCR § 4060(f)(5))

[501] (Dangerous Weapons Control Law, 2008 § 12128), (California Code of Regulations, 2013 11 CCR § 4060(k))

[502] (California Code of Regulations, 2013 11 CCR § 4060(i)(1))

[503] (Dangerous Weapons Control Law, 2008 § 12128)

[504] (California Code of Regulations, 2013 11 CCR § 4060(i)(2))

[505] (California Code of Regulations, 2013 11 CCR § 4060(i)(3))

[506] (California Code of Regulations, 2013 11 CCR § 4060(i)(4))

[507] (California Code of Regulations, 2013 11 CCR § 4060(i)(5))

[508] (California Code of Regulations, 2013 11 CCR § 4060(i)(6))

[509] (California Code of Regulations, 2013 11 CCR § 4060(i))

[510] (Code of Massachusetts Regulations, 1998 940 CMR 16.01)

[511] (Massachusetts General Laws, 2013 Title XX, Chapter 140, § 121, "Large capacity feeding device")

ENDNOTES

[512] (Massachusetts General Laws, 2013 Title XX, Chapter 140, § 121, "Firearm")

[513] (Massachusetts General Laws, 2013 Title XX, Chapter 140, § 121, "Assault Weapon")

[514] (Massachusetts General Laws, 2013 Chapter 269, Section 11E "Serial identification numbers on Firearms")

[515] (Code of Massachusetts Regulations, 1998 940 CMR 16.03 "Tamper Resistant Serial Numbers")

[516] (Code of Massachusetts Regulations, 1998 940 CMR 16.04 "Sale of Handguns Made From Inferior Materials")

[517] (Massachusetts General Laws, 2013 Title XX, Chapter 140, § 123(18)) and (Code of Massachusetts Regulations, 1998 940 CMR 16.04(3))

[518] (Code of Massachusetts Regulations, 1998 940 CMR 16.05(1)), (Massachusetts General Laws, 2013 Title XX Chapter 140 § 131(K))

[519] (Code of Massachusetts Regulations, 1998 940 CMR 16.05(2) "Childproofing")

[520] (Code of Massachusetts Regulations, 1998 940 CMR 16.05(4) and 16.01 "Definitions: Hammer deactivation device")

[521] (Code of Massachusetts Regulations, 1998 940 CMR 16.05(3) and 16.05(4))

[522] (Massachusetts General Laws, 2013 Title XX, Chapter 140, § 121, "Firearm"), (Code of Massachusetts Regulations, 2007 501 CMR 7.02)

[523] (Massachusetts General Laws, 2013 Title XX, Chapter 140, § 123(18))

[524] (Code of Massachusetts Regulations, 1998 501 CMR 7.00)

[525] (Massachusetts General Laws, 2013 Title XX, Chapter 140 § 123(18))

[526] (Massachusetts General Laws, 2013 Title XX, Chapter 140, § 123(19))

[527] (Massachusetts General Laws, 2013 Title XX, Chapter 140 § 123(19))

[528] (Massachusetts General Laws, 2013 Title XX, Chapter 140, § 123(20))

[529] (Code of Massachusetts Regulations, 1998 940 CMR 16.06(3))

[530] (Massachusetts General Laws, 2013 Title XX, Chapter 140, § 123(21))

[531] (Code of Massachusetts Regulations, 1998 940 CMR 16.06(1))

[532] (Code of Massachusetts Regulations, 1998 940 CMR 16.06(1))

[533] (Code of Massachusetts Regulations, 1998 940 CMR 16.06(2))

[534] (ATF State Laws, 2010, p. 220, § 5-101(n))

[535] (ATF State Laws, 2010, p. 226, § 5-406(a))

[536] (ATF State Laws, 2010, p. 226, § 5-405(a)(1))

[537] (ATF State Laws, 2010, p. 226, § 5-405(b))

[538] (ATF State Laws, 2010, p. 226, § 5-405(c))

[539] (ATF State Laws, 2010, p. 226, § 5-405(d))

[540] (SB-281, 2013 p. 10)

[541] (SB-281, 2013 pp. 6f)

[542] (ATF State Laws, 2010 p. 224 § 5-132(c))

[543] (ATF State Laws, 2010 p. 224 § 5-131(a)(6))

[544] (ATF State Laws, 2010 p. 224 § 5-132(d))

[545] (ATF State Laws, 2010 p. 328 § 265.00(3))

[546] (ATF State Laws, 2010 p. 329 § 265.00(21))

[547] (ATF State Laws, 2010 p. 326 § 396-ee(2))

[548] (ATF State Laws, 2010 pp. 326f § 396-ee(1))

[549] New York Secure Ammunition and Firearms Enforcement Act, Senate Bill S2230

[550] (ATF State Laws, 2010, p. 329, § 265.00(23))

[551] Autoloading Pistols for Police Officers : NIJ Standard-0112.03. Office of Justice Programs : National Institute of Justice : Law Enforcement and Corrections Standards and Testing Program : U.S. Department of Justice. [Online] Rev. a, 1999. https://www.justnet.org/pdf/NIJSTD011203REVA.pdf.

[552] (NIJ Standard-0112.03, 1999 p. iii)

ENDNOTES

[553] (NIJ Standard-0112.03, 1999 p. 4)

[554] (NIJ Standard-0112.03, 1999 p. 8)

[555] (NIJ Standard-0112.03, 1999 p. 5)

[556] (NIJ Standard-0112.03, 1999 p. 6)

[557] (NIJ Standard-0112.03, 1999 p. 6)

[558] (NIJ Standard-0112.03, 1999 p. 6)

[559] (NIJ Standard-0112.03, 1999 p. 6)

[560] (NIJ Standard-0112.03, 1999 pp. 1, 18)

[561] (NIJ Standard-0112.03, 1999 pp. 8–11)

[562] (NIJ Standard-0112.03, 1999 pp. 11f)

[563] (NIJ Standard-0112.03, 1999 p. 12)

[564] (NIJ Standard-0112.03, 1999 p. 13)

[565] (NIJ Standard-0112.03, 1999 pp. 13f)

[566] (NIJ Standard-0112.03, 1999 pp. 14f)

[567] (NIJ Standard-0112.03, 1999, p. 5)

[568] Section 3(e) of Public Law 94-284 Section 644; May 11, 1976.

[569] (Consumer Product Safety Act, 1972)

[570] (SAAMI : Setting the Standard, 1998 p. 5)

[571] (SAAMI : Setting the Standard, 1998 p. 2ff)

[572] (ANSI/SAAMI Z299.5-2016 *American National Standard*)

[573] (ANSI/SAAMI Z299.5-2016 *Members*)

[574] (ANSI/SAAMI Z299.5-2016 *Foreword*)

[575] ANSI/SAAMI Z299.5-2016 : American National Standard : Voluntary Industry Performance Standards : Criteria for Evaluation of New Firearms Designs Under Conditions of Abusive Mishandling for the Use of Commercial Manufacturers.

[576] (ANSI/SAAMI Z299.5-2016 p. 1)

[577] (ANSI/SAAMI Z299.5-2016 p. 2f)

[578] (ANSI/SAAMI Z299.5-2016 p. 3f)

[579] (ANSI/SAAMI Z299.5-2016 p. 4f)

[580] (ANSI/SAAMI Z299.5-2016 p 5)

[581] (IPSC, 2017 p. 56)

[582] (IPSC, 2017 p. 56)

[583] (IPSC, 2017 p. 11)

[584] Trivia: A compact size pistol with small grip handle and typically chambered for a large caliber cartridge, was invented by Henry Deringer, Philadelphia, and it was named after him (spelled with one "R"). Copies of this pistol design are often referred to as "Derringer," i.e. spelled with double "R". (Smith, 1953 p. 612)

"In general, post 1/1/2001, any handgun to be sold in CA must be listed on the CA Roster of Handguns. There are limited exceptions to this ruling, Private Party transfers, curio or relic handguns, certain single-action revolvers (amended the single shot exemption 1/1/2015), and pawn/consignment returns. CA amended the single shot exemption which went into effect on 1/1/2015, which exempted certain single-action revolvers from inclusion into the roster of handguns and shall not apply to a single-shot pistol with a break top or bolt action and a barrel length of not less than six inches and that has an overall length of at least 10 ½ inches when the handle, frame, or receiver, and barrel are assembled.

But to address the question of Derringers and California, yes, it is possible that a Derringer can be legally sold in CA under the certain single-action revolver exemption so long as it meets physical characteristics listed in the exemption." (Jason Haddock, per e-mail, June 26, 2017)

[585] (ATF Commerce Report, 2016 p. 1)

ENDNOTES

[586] The increase in exports is remarkable when absolute figures are considered or if one takes a closer look at the relevant curve in Fig. A10.1, p. 353. The concerned reader may draw his own conclusions after looking at these numbers: 333 266, 275.424, 172.408, 140.787 (2018, 2017, 2016, 2015).

[587] For this survey, numbers were compiled from ATF AFMER, Year 2018 and earlier.

[588] (ER "Pistolen", 2008 p. Anlage 3)

[589] (TR "Pistolen", 2008 p. 19)

[590] (Greener, 1910 pp. 310–312), (Cranz, et al., 1926 pp. 72–91), (Bock, et al., 1989 pp. 638–642), (Kneubuehl, 2013 pp. 318–320)

[591] (Walther, 1984 p. 24)

[592] (Bock, et al., 1989 p. 15)

[593] (Pietzner, 2005 p. 16)

[594] (CIP, 2011)

[595] In the event where the ballistic of the projectile in front of the muzzle is affected by propellant gases, these effects will be taken into account if the muzzle velocity is measured in a distance from the muzzle. (Hartmann, 1901 p. 97) CIP specifies 2.5 m, whereas ER "Pistolen" uses 3 meters (ER "Pistolen", 2008 p. 13).

[596] (Stroppe, et al., 2018 p. 97)

[597] (Brukner, 2003 p. 388f)

[598] See (Brukner, 2003 p. 389f) and (Kneubuehl, 2013 p. 80f). The effect of the weight of the propellant gases could be taken into consideration if we would include half of the powder charge's mass into the formula.

[599] (Bock, et al., 1989 p. 17)

[600] For more on differentiation see: (Porter, 1985 pp. 14f, 119–127)

[601] (Ayoob, 2013 p. 8)

[602] (PTOOMA Productions, 2004 p. 14) GLOCK delivered samples in 1983 to the U.S. Department of Defense. Later, in July of 1985, the GLOCK G17 was submitted to the BATF for evaluation. Initially, there were concerns by the BATF since the gun was made of "plastic." In 1986, the ATF approved suitability for importation. (PTOOMA Productions, 2004 pp. 15, 17) Sweeney explains, the G17 was not heavy enough to obtain the 75 points needed for compliance with ATF Form 4590. Gaston Glock then developed adjustable rear sights, which gave him additional points to compensate for the light polymer frame. (Sweeney, 2008 p. 18)

[603] (Glock, 2013 p. 4)

[604] (Glock, 2014 p. 6)

[605] (Edwards, 2013 p. 9) Generation 2 was introduced in 1989, generation 3 in 1997. (PTOOMA Productions, 2004 p. 22)

The first encounter with a GLOCK G22 Gen4 Sweeney commented in a description of a picture as follows: "The first Gen4 Glock I laid eyes on, a G22, back in January of 2010. I was underwhelmed." (Sweeney, 2013 p. 267) GLOCK had perhaps oversold the improvements of the Gen4.

[606] GLOCK was frugal and wanted to use as many components of the .40 cal. pistol as possible for the weaker 9 mm Luger, Sweeney recounts. In the case of the Gen4, initially the same recoil spring assembly was used for the 9 mm pistol as well as for the .40 cal. To improve the life of the recoil spring for the .40 S&W cartridge, GLOCK had changed the spring wire from flat to round, in combination with a recoil spring assembly consisting of two springs. After the introduction of the pistol, this recoil spring assembly proved to be too heavy for the 9 mm Luger round and GLOCK started an exchange of spring assemblies for guns sold before July 22, 2011. (Sweeney, 2013 p. 269)

[607] The previous mag release button was smaller than the Gen4 mag release button and GLOCK justified the poor accessibility of the original magazine release with the argument that this

ENDNOTES

would serve as a protection against accidental loss of the magazine on previous pistol generations. (Kasler, 1992 p. 222f)

[608] (Sweeney, 2013 p. 272), ibid.: The approach of GLOCK differs fundamentally from other gun manufacturers. For instance, SMITH & WESSON uses a palmswell insert to cover the otherwise open sides and the rear of the grip handle. For users, it is essential, that a gun remains to be functional, even if exterior parts were detached accidentally from the grip handle. (Riordan, 1980 p. 29)

[609] (Glock IM, 2013 p. 36f)

[610] The adjustable rear sight compensates for the light polymer frame in getting the ATF approval. (Sweeney, 2008 p. 18)

[611] (Mötz, et al., 2013 p. 538f) Allegedly, the former salt-bath treated slides had been phosphated in addition to the tenifer treatment, to obtain an attractive black color. (PTOOMA Productions, 2004 p. 23) This statement could be found only at PTOOMA. *Sweeney* speaks of "black oxide coating," which would correspond to a treatment in the oxidation bath, which is part of the tenifer process anyway. (Sweeney, 2008 p. 45)

[612] (PTOOMA Productions, 2004 p. 130f) Early GLOCK *G17* pistol slides did not have this striker channel liner and normally, hardly anyone notices this liner inside the striker channel, because it is pressed into the channel and it has the same color as the slide. (PTOOMA Productions, 2004 pp. 23, 74)

[613] (Sweeney, 2013 pp. 92f, 272) The magazine body of early GLOCK *G17* pistols used to be polymer, with the exception of the feed lips. (Caranta, 1986 p. 176) Starting from 1990, a complete metal body was molded into the polymer to improve toughness. (PTOOMA Productions, 2004 p. 33f) Sweeney notes that the polymer sometimes starts to come off the metal insert in the area of the feed lips. (Sweeney, 2013 p. 272) Also *Kasler* discusses the wear of the edge of the magazine's engagement planes. (Kasler, 1992 pp. 220, 222)

[614] For details see user manual: (Glock IM, 2013 p. 38)

[615] (Woods, 2014 p. 42f), ibid.: the polygonal profile of the barrels is cold hammer forged, and Woods explains, this procedure would provide a profile with minimal variations of the caliber and produces a hard surface at the same time, which is good for accuracy and durability. The polygonal profile would enable the projectile to smoothly adjust to the barrel profile and to tighten the bore avoiding gas leakage, which leads to more velocity of the projectile and thus, a greater muzzle energy. Avoiding sharp-edged grooves would weaken the barrel less and tensions inside the barrel could be reduced. Both would have a positive effect on the life of the barrel. Also, a polygonal profile was easier to clean because of the smooth transitions of the surfaces.

[616] (NIJ, 2016 p. 5f) Compact, L × H × W: ≤ 8" × ≥ 4.75" × ≤ 1.35", ≤ 35 ounces. Full Size, L x H x W: ≤ 9" × ≤ 6" × ≤ 1.35", ≤ 42 ounces. (NIJ, 2016 p. 7f)

[617] The *G46* is the first GLOCK that can be field-stripped without pulling the trigger. (Dannecker, et al., 2017 pp. 68–73)

[618] (HK Archives, 2014)

[619] (TR "Pistolen", 2008 p. 11) TR "Pistolen" demands 0.15 J minimum trigger work, i.e. 150 Nmm.

[620] (HK Archives, 2014)

[621] (Heckler & Koch, 2014 p. 9)

[622] (Heckler & Koch, 2014 p. 11)

[623] (HK Archives, 2014)

[624] (Heckler & Koch, 2014 p. 30)

[625] George Kellgren established KEL-TEC CNC. in 1991. According to him, the unequaled design of his guns is the reason why the company is among the top five manufacturers. KEL-TEC has 200 employees, and is focused on innovative products, in combination with polymer technology and modern manufacturing technologies. (Kel-Tec, 2014c)

ENDNOTES

KEL-TEC provides original owners with a lifetime warranty on all of the company's firearms. (Kel-Tec, 2011)

[626] (Kel-Tec, 2006 p. 4), ibid.: "The PF-9 pistol is the lightest and flattest 9mm Luger caliber ever made."

[627] (Kel-Tec, 2006 p. 6)

[628] (Kel-Tec, 2006 p. 4)

[629] (Kel-Tec, 2006 p. 4)

[630] (Kel-Tec, 2006 p. 12)

[631] (Kel-Tec, 2006 p. 10)

[632] (Kel-Tec, 2014c)

[633] (Ruger, 2013b) Technical data was retrieved from the company's website.

[634] (Ruger, 2013b) Technical data was retrieved from the company's website.

[635] (Ruger, 2013a p. 7)

[636] (Ruger, 2013a p. 2ff), ibid.:

California:
"Children are attracted to and can operate firearms that can cause severe injuries or death. Prevent child access by always keeping guns locked away and unloaded when not in use. If you keep a loaded firearm where a child obtains and improperly uses it, you may be fined or sent to prison."

Connecticut:
"UNLAWFUL STORAGE OF A LOADED FIREARM MAY RESULT IN IMPRISONMENT OR FINE."

Florida:
"IT IS UNLAWFUL, AND PUNISHABLE BY IMPRISONMENT AND FINE, FOR ANY ADULT TO STORE OR LEAVE A FIREARM IN ANY PLACE WITHIN THE REACH OR EASY ACCESS OF A MINOR UNDER 18 YEARS OF AGE OR TO KNOWINGLY SELL OR OTHERWISE TRANSFER OWNERSHIP OR POSSESSION OF A FIREARM TO A MINOR OR A PERSON OF UNSOUND MIND."

Maine:
"ENDANGERING THE WELFARE OF A CHILD IS A CRIME. IF YOU LEAVE A FIREARM AND AMMUNITION WITHIN EASY ACCESS OF A CHILD, YOU MAY BE SUBJECT TO FINE, IMPRISONMENT OR BOTH. KEEP FIREARMS AND AMMUNITION SEPARATE. KEEP FIREARMS AND AMMUNITION LOCKED UP. USE TRIGGER LOCKS."

Maryland:
"WARNING: Children can operate firearms which may cause death or serious injury. It is a crime to store or leave a loaded firearm in any location where an individual knew or should have known that an unsupervised minor would gain access to the firearm. Store your firearm responsibly!"

Massachusetts:
"WARNING FROM THE MASSACHUSETTS ATTORNEY GENERAL: This handgun is not equipped with a device that fully blocks use by unauthorized users. More than 200,000 firearms like this one are stolen from their owners every year in the United States. In addition, there are more than a thousand suicides each year by younger children and teenagers who get access to firearms. Hundreds

more die from accidental discharge. It is likely that many more children sustain serious wounds, or inflict such wounds accidentally on others. In order to limit the chance of such misuse, it is imperative that you keep this weapon locked in a secure place and take other steps necessary to limit the possibility of theft or accident. Failure to take reasonable preventive steps may result in innocent lives being lost, and in some circumstances may result in your liability for these deaths."
And:
"IT IS UNLAWFUL TO STORE OR KEEP A FIREARM, RIFLE, SHOTGUN OR MACHINE GUN IN ANY PLACE UNLESS THAT WEAPON IS EQUIPPED WITH A TAMPER-RESISTANT SAFETY DEVICE OR IS STORED OR KEPT IN A SECURELY LOCKED CONTAINER."

New Jersey:
"IT IS A CRIMINAL OFFENSE TO LEAVE A LOADED FIREARM WITHIN EASY ACCESS OF A MINOR."

New York City:
"THE USE OF A LOCKING DEVICE OR SAFETY LOCK IS ONLY ONE ASPECT OF RESPONSIBLE FIREARM STORAGE. FOR INCREASED SAFETY FIREARMS SHOULD BE STORED UNLOADED AND LOCKED IN A LOCATION THAT IS BOTH SEPARATE FROM THEIR AMMUNITION AND INACCESSIBLE TO CHILDREN AND OTHER UNAUTHORIZED PERSONS."

North Carolina:
"IT IS UNLAWFUL TO STORE OR LEAVE A FIREARM THAT CAN BE DISCHARGED IN A MANNER THAT A REASONABLE PERSON SHOULD KNOW IS ACCESSIBLE TO A MINOR."

Texas:
"IT IS UNLAWFUL TO STORE, TRANSPORT, OR ABANDON AN UNSECURED FIREARM IN A PLACE WHERE CHILDREN ARE LIKELY TO BE AND CAN OBTAIN ACCESS TO THE FIREARM."

Wisconsin:
"IF YOU LEAVE A LOADED FIREARM WITHIN THE REACH OR EASY ACCESS OF A CHILD YOU MAY BE FINED OR IMPRISONED OR BOTH IF THE CHILD IMPROPERLY DISCHARGES, POSSESSES, OR EXHIBITS THE FIREARM."

[637] Patrick Sweeney in *Guns & Ammo*, "Spurs & Strikers – Ruger's LCRx and LC9s feature new bits." (G&A, October 2014 p. 66)

[638] (Sweeney, 2010 p. 30), ibid.: The .38 Special cartridge type was introduced at the same time with the *M&P* revolver.

[639] (Sweeney, 2013 p. 232f)

[640] (Sweeney, 2013 p. 232)

[641] (US 7 810 269 B2, Column 4f)

[642] (WO 2007/128 252 A1, p. 4)

[643] Springfield Armory was the primary center for the manufacture of U.S. military firearms from 1795 until 1968. (Raber, et al., 2009 p. 10) Based in Springfield, Massachusetts, it was where the first American designed musket was produced beginning in 1795, and through the next two centuries the Springfield Armory was where some landmark firearms including Springfield "Trapdoor," the Springfield *Model 1903* bolt action rifle and notably the *M1 Garand* were first developed and produced. When the USA decided to introduce the *M16* rifle

ENDNOTES

made by ARMALITE, the U.S. Department of Defense determined that the Springfield Armory should be shut down.

The final rifle developed at the Springfield Armory was the *M14* rifle, which was replaced by the *AR-15/M16* soon after. The name lived on when Elmer C. Balance of San Antonio began using the name "Springfield Armory" through his company LH MANUFACTURING. It introduced the first civilian production of the *M14*, branded as the *M1A*. Robert Reese acquired the company including the brand name and founded SPRINGFIELD ARMORY, INC. in Geneseo, Illinois. Initially the newly founded company would continue to manufacture the *M14* rifle, later the *Model 1911* pistol was added to the product portfolio.

The company more and more moved away from gun production and focused on the importation and distribution of firearms instead. Their great break through started in the year 2001 with the *XD* pistol. This model is based on the *HS2000* pistol made by the Croatia based HS PRODUKT D.O.O. HS PRODUKT would establish a new production line for the *XD* pistol parallel to their production of the *HS2000* pistol to satisfy the demand of SPRINGFIELD ARMORY, INC. In 2007 an *XD^M*-line was added to the *XD*-series.

Currently the importation and the successful distribution of the *XD* family of pistols are the basis of the company's success. In the years 2018 / 2017 / 2016 / 2015 295,107 / 326,653 / 574,486 / 338,535 pistols were imported from Croatia into the USA. (ATF Commerce Report, 2019/2018/2017/2016 p. 9)

[644] (Sweeney, 2013 p. 236f)

[645] (Dannecker, 2015 pp. 36–39), (Dannecker, 2016 pp. 274–280)

[646] Patent application US 2011/0 119 979 A1 refers expressly to a "Picatinny Rail" (paragraph 0048 through 0051), avoiding sockets or outer electric contacts (paragraph 0085), a shot counter (claim 44 through 57) and the possibility of a one-piece design of a module and firearm combination (paragraph 0078 and claim 82).

[647] (Boberg, 2014)

[648] (Aicher, 1984 pp. 862–866), (Field, 1999 pp. 40–45)

[649] (SCCY Industries, LLC, 2014)

[650] (Gun Test Team, 2012 p. 10ff)

[651] (SCCY, 2014)

[652] Patrick Sweeney in *Guns & Ammo*, "This is a gun – SIG Sauer's new P320 has multiple personalities. We meet one of them." (G&A, May 2014 p. 46)

[653] (Tarr, 2014 p. 59)

[654] (SIG-Sauer, 2014)

[655] (Tarr, 2014 p. 61)

[656] In 2009, TARA PERFECTION D.O.O. was established in Montenegro. The parent company is TARA GROUP. TARA PERFECTION D.O.O. is their Small Arms Division, other branches are TARA PRECISION WORKS, and TARA AEROSPACE. TARA PERFECTION has the objective to offer solutions for law enforcement and military customers. The company offers the *TM-9* pistol and the *TM-4* assault rifle. (TARA, 2014 p. 2)

According to them, the leading product engineer originally came from GLOCK via CARACAL to TARA.

[657] (TARA, 2014 p. 4)

[658] (TARA, 2014 p. 6)

[659] (TARA, 2014 p. 5)

[660] (Kel-Tec, 2014b) KEL-TEC offers a lifetime guarantee, but limits it to the original buyer. Users of pre-owned firearms will have to carry $40 U.S. flat for shipping and handling, plus costs for the spares, if they want to have their KEL-TEC repaired. (Kel-Tec, 2011)

[661] (Christensen, 1997 p. 101)

[662] (Christensen, 1997 p. 7)

[663] (Christensen, 1997 p. 110f)

ENDNOTES

664 (Christensen, 1997 p. 6)
665 (Christensen, 1997 p. 6)
666 (Schumpeter, 1947 pp. 134–142), (Christensen, 1997 pp. 6f, 16f, 125)
667 (Christensen, 1997 pp. III, 6, 13)
668 (Christensen, 1997 pp. 6, 45)
669 (Christensen, 1997 pp. 74–76)
670 (Christensen, 1997 pp. 13, 36)
671 (Kim, et al., 2005)
672 (Rheinmetall, 1982 p. 415)
673 (Dobrinski, et al., 2010 p. 76)
674 (Dobrinski, et al., 2010 p. 77)
675 (TR "Pistolen", 2008 p. 13)

Occasionally, the question is asked about the effective range of the 9 mm Luger. Older technical performance specifications for accuracy and effective range requirements may be consulted to provide orientation in answering this question:

According to the acceptance conditions for the *Pistolenpatrone 08*, accuracy was to be tested at a distance of 50 m (Görtz, 1988 p. 212), while the younger STANAG 4090 (NATO Standardization Agreement) demands 46 m (50 yards) when fired from a 199 mm proof barrel. (STANAG 4090, Edition 2, Annex C, PRECISION)

A general definition of effective range is mentioned in *Rheinmetall* with regard to infantry rifle cartridges: "Defined as the effective range is that range at which a 1.31 mm thick plate having a tensile strength of 1670 N/mm² (helmet steel plate) can just be penetrated." (Rheinmetall, 1982 pp. 547, Table 1105, Note 3) Furthermore, in STANAG with regard to the 9 mm cartridge: "The ammunition must be capable of inflicting a fatal wound on person-nel protected by steel helmets and body armour at a range of 23m (the steel helmet is defined as the United States Helmet M1 and the body armour as the United States Body Armour M1952)." (STANAG 4090, Edition 2, Annex C, TERMINAL EFFECTS)

676 (Brukner, 2003 p. 424)
677 (Barnes, 2012 p. 342) Data listed in Barnes is given in *United States Engineering Units*, e.g. in *grain* (gr), which makes it necessary to convert the values to ISO-units. The conversion from *grain* to *gram* is carried out by dividing the mass given in *grain* by 15.43 gr/g.
678 (Barnes, 2012 p. 342) Data listed in Barnes is given in *United States Engineering Units*, e.g. in *feet per second* (fps), which makes it necessary to convert the values to ISO-units. The conversion of the projectile's muzzle velocity from *fps* to *meters per second* is done by multiplying the velocities given in *fps* by 0.3048 m/ft.
679 (Brukner, 2003 p. 413f) Brukner relates his parameters to the observation of the change in momentum. He states that the momentum of the slide changes with $m_S / (m_B + m_S)$, the energy changes with $m_S / (m_B + m_S)^2$.
680 (TR "Pistolen", 2008 pp. 7, 24)
681 (TR "Pistolen", 2008 p. 18), ibid.: The malfunction rate is evaluated by shooting 10,000 rounds each with three pistols. The result must be 0.2 percent or less per pistol, i.e. 20 malfunctions or less per 10,000 rounds.
682 (O'Connor, 1990 p. 22)
683 (Kamiske, et al., 2006 p. 383)
684 (Stroppe, et al., 2018 p. 241)
685 (Kneubuehl, 2013 p. 54)
686 A similar effect was found when *M16A1* rifles were tested. A lower temperature than on the forward end of the barrel was measured at the chamber end of the barrel (Elbe, 1975 p. 7)
687 (Schmid, 2003 p. 22)
688 Also: "§ 28. Über die beim Schuß in den Lauf übergehende Wärme. Arbeitsbilanz. Nut-zeffekt. Temperatur des Laufs." (§28. Temperature transfer to the barrel when a round is

ENDNOTES

fired. Work balance. Efficiency. Barrel temperature. (Cranz, et al., 1926 pp. 226–234) [translated by PD]

[689] "The energy [..] at the end of a process is equal to the energy [..] at the beginning of the process, increased by the work input [..] and reduced by the work output [..] during the process." (Böge, et al., 2019 p. 239) [translated by PD]

[690] *Bock* recommends $m_P m_S E_0 / (m_S+m_B)^2$ to calculate the slide racking work. E_0 stands for the muzzle energy of the projectile. (Bock, et al., 1989 p. 682) Calculating W_{fl} in this way results in higher values than the proposed 3.5 joules. The latter corresponds to the average of the values of the slide racking work, determined by planimetry in Appendix 17.

The cocking work of a possibly existing hammer is included in the 3.5 joules. *Brukner* assumes 0.3 joules would be required to cock the hammer; i.e. a value that may be neglected. (Brukner, 2003 p. 391)

[691] (Böge, et al., 2019 p. 221)

[692] (Maier, 1993a p. 1)

Related to what Maier calls the *semi-rigid roller lock* for rifles chambered in 8 mm × 33 Maier explains:

> "I had the idea to build the steering piece [Steuerstück; PD] preferably heavy, which would cause an increase in time for the rollers to slide along the slopes and in turn, this would delay the unlocking. Already, a rough calculation had shown that the breech's recoil velocity after unlocking ranged from 6 to 8 m/s, which used to be the preferred range." (Maier, 1993b p. 5) [translated by PD]

Dr. Karl Wilhelm Maier (1912 – ?) was the leading mathematician at MAUSER's Waffenforschungsanstalt [Weapons Research Center; PD], (Stevens, 2006 p. 10f) reporting to director Niemann, Leiter Versuch [head of testing; PD]. All various departments of the Waffenforschungsanstalt were under director Ott-Helmuth von Loßnitzer. The departments had been a part of MAUSER, and comprised of about 720 employees in the spring of 1945. (Maier, 1994 p. 1)

The source cited is a private letter from Karl W. Maier addressed to Hinderikus (Henk) Lucas Visser. During Visser's days at the NEDERLANDSCHE WAPEN MUNITIEFABRIEK Visser held the manufacturing license for the CETME rifle in a variety of countries. (Johnston, et al., 2010 pp. 390–416)

[693] (Rheinmetall, 1982 p. 288)

[694] (Dobrinski, et al., 2010 p. 376)

> "The task of mechanics is to describe and predetermine the movements of bodies and the forces associated with these movements. One can divide mechanics into *kinematics* and *dynamics*. Kinematics is the study of the geometric and temporal sequence of movements without considering forces as cause or effect of the movement. Dynamics, on the other hand, deals with the interaction of forces and movements: Forces are physical quantities that cause a change in the state of motion of bodies. Dynamics, in turn, is divided into statics and kinetics. Statics deals with the forces at resting bodies (equilibrium), while kinetics investigates actual movements under the effect of forces.
>
> The origin of statics lies in antiquity. Kinetics, on the other hand, is a much more recent science. The first systematic studies were carried out by Galileo Galilei (1564–1642). With the help of ingenious experiments he found the laws of falling and throwing and formulated the law of inertia in 1638. In order to appreciate Galileo's achievements, one should bear in mind that differential and integral calculus were still unknown at that time and that there was not yet a device for the precise measurement of time.
>
> Kinetics was scientifically founded by Isaac Newton (1643–1727), who gave the first formulation of the laws of motion in 1687. Newton's Basic Laws are a summary of all

experimental experience; all conclusions drawn from them are consistent with experience. We regard these laws – without being able to prove them – as correct: they have axiomatic character.

[...] The starting point of all [...] considerations of kinetics are Newton's fundamental equations. Here we limit ourselves to the treatment of the movements of mass points or of rigid bodies. With the help of these idealizations many technically important problems can be described and solved. [...]

In kinetics, many of the terms introduced in statics (e.g. space, mass, force, torque) and idealizations (e.g. mass point, rigid body, single force) are still used. [...] In order to describe movements, [...] time must be introduced as a new basic quantity [...]. This allows further terms (e.g. speed, acceleration, momentum, kinetic energy) to be defined and new laws (e.g. the principle of linear momentum, the law of energy) to be specified [...]." (Gross, et al., 2019b p. XIIIf) [translated by PD]

[695] (Gross, et al., 2019b pp. 211–263)

[696] (Gross, et al., 2019a p. XIII and 5f)

[697] The resting position of the system under consideration is defined as the pistol with slide fully forward. During firing, the slide leaves its rest position, is kicked back to the rear stop, bounces off from it and returns to battery position soon after. The instantaneous distance from the rest position is called displacement, deflection or elongation in case of mechanical oscillations. (Dobrinski, et al., 2010 p. 379)

[698] Should this effect nevertheless be of interest, please refer to Kneubuehl. (Kneubuehl, 2018 pp. 131–146)

[699] *Brukner* mentions an approximate value of 0.3 joules for hammer-fired pistols, which, if necessary, can be taken into account as energy needed to cock the hammer as the slide cycles. (Brukner, 2003 p. 391)

[700] (Meissner, et al., 2007 p. 345), (Assmann, et al., 2011 p. 345f)

[701] (Stroppe, et al., 2018 pp. 52, 356f)

[702] Also known as natural frequency or eigenfrequency. (Assmann, et al., 2011 p. 348)

[703] Equation A19.36 reveals a mathematical relationship which essentially describes the temporal course of the displacement of the mass as a function of the slide mass and the spring rate. The equation is thus suitable for a discussion of the effect of the recoil spring during the slide's recoil motion and makes it clear to the observer that the recoil spring does indeed have an influence on the slide's dynamics.

Lewisch discusses the influence of spring rate and the slide's mass and identifies three conclusions:

1) The recoil spring has a "restraining effect". Amplitude and oscillation period are both affected by the recoil spring. (Lewisch, 1971 p. 1054)
2) If, on the one hand, the spring rate is increased n-fold for a given slide mass and, on the other hand, the n-fold slide mass is used for the same recoil spring, the same amplitude results in both cases. (Lewisch, 1971 p. 1054)
3) The pretension of the recoil spring influences the slide's dynamics. (Lewisch, 1971 p. 1055)

[704] (Assmann, et al., 2011 p. 349)

[705] "The extreme value of the periodically changing quantity (i.e. vibration quantity) is called amplitude." (Dobrinski, et al., 2010 p. 379) [translated by PD]

[706] *Phase* describes the instantaneous state of an oscillation. Its initial shift along the time axis is defined by the initial phase angle. (Dobrinski, et al., 2010 p. 379) Values are given in radians, abbreviated "rad". In our application the phase angle is described by muzzle velocity, projectile weight and recoiling mass as well as spring rate and deflection s_1 of the recoil spring.

ENDNOTES

[707] In the present application, an exact calculation of the velocity after the collision at the back-stop is difficult, since the frame, the shooter's posture, his hand, forearm, stature, etc. have a decisive influence on the dynamics. Therefore, a significantly simplified relationship was chosen. This takes into account only one factor, i.e. a simplified sort of *coefficient of restitution*, and the slide velocity before the impact. Please note further explanations in Endnote 720.

[708] (Wittel, et al., 2019 p. 406)

[709] (Dobrinski, et al., 2010 p. 379)

[710] (Lewisch, 1971 p. 1054)

[711] (Chinn, 1955 p. 23) The formula is presented in a form adapted for our purposes. The source quoted uses imperial units and the resulting cyclic rate is given in shots per minute.
   A fire rate calculated with this equation represents a theoretical quantity. *Chinn* deals with rigidly mounted or heavy long guns. Pistols, on the other hand, are comparatively light, are fired off hand and therefore move significantly during recoil. This results in a relative movement between slide and frame. In reality a lower rate of fire will be found. This can be explained by mechanical processes in the gun and friction.

[712] Cf. Endnote 154 (Hatcher, 1966 p. 396), (Kneubuehl, 2013 p. 191) *Brukner* specifies 0.0038 to 0.0105 seconds for hammer-fired pistols. (Brukner, 2003 p. 384f)
   Calculation using the oscillation equation gave a lock time of approximately 0.003 seconds for striker-fired pistols (see Appendix 24).

[713] (Rinker, 2011 p. 404), (Brukner, 2003 p. 386) *Kneubuehl* mentions 0.001 to 0.003 seconds. (Kneubuehl, 2013 p. 191) *Rheinmetall* specifies "less than one millisecond" for small caliber guns and more than 10 milliseconds for larger calibers. (Rheinmetall, 1982 p. 80)

[714] (Dannecker, 2016 pp. 11, 21, 573) We differentiate between controlled travel (x_ℓ) and safe-guarded displacement (x_{u1}). The *controlled travel* is the displacement of the slide at the instant when the base of the projectile exits the muzzle. The controlled travel is the yardstick for assessing whether the case is safely supported when a gun is fired. In the case of inertia type simple blowback pistols, it tells us how far the unsupported portion of the rear of the fired case emerges from the chamber while chamber pressure is still at a high level.

[715] (Sweeney, 2008 p. 26)

[716] The TR "Pistolen" defines the *safeguarded displacement* of locked breech pistols as the displacement of slide and barrel until uncoupling is initiated. The safeguarded displacement "should be large enough that no unlocking of the system is initiated until the gas pressure in the system has dropped to a level that is not dangerous for the shooter, with respect to the maximum possible projectile mass." The TR emphasizes "that even a projectile of maximum weight must have exited the barrel before the uncoupling is initiated." For the calculation of the controlled travel, the TR takes into account the mass of barrel, slide and cartridge case as well as the 100 percent of the recoil spring's mass, plus the length of travel of the projectile's base to the muzzle, and twice the mass of the heaviest projectile of ammo qualified for police use ($2\, m_P = 12.8$ g). (TR "Pistolen", 2008 p. 26) and (ER "Pistolen", 2008 p. 8)

[717] Using twice the mass of the heaviest projectile is done to consider a bullet obstruction.
   The standard way of calculation uses just the mass of the heaviest projectile possible; see for example: (Bock, et al., 1989 p. 681)
   Peter Dannecker states that the standard formula does not take mechanical losses into account and therefore supplies the maximum value for the displacement x_ℓ. For hammer-fired blowback pistols, a factor of 0.8 can be used in the formula, i.e. $x_\ell = 0.8\, s_\ell\, m_P/m_r$. This factor takes into account the effect of the hammer cocking impact, powder weight and the masses of the hammer, the case, and springs as well as obturational friction between case and chamber wall, etc. The result of the formula would be very close to reality. High-speed recording can prove this factual. [face-to-face conversation, December 17, 2019]

ENDNOTES

[718] (Rinker, 2011 p. 63)

[719] *Heydenreich* provides a method for calculating gas pressure and velocity as a function of the projectile travel. Furthermore, projectile displacement, velocity and time at pressure peak. The method leads to good results with commonly available calculation parameters. (Heydenreich, 1908 pp. 56–64) *Oerlikon* names updated operands for the calculation process. (Oerlikon, 1981 p. 103f)

The Heydenreich approach in combination with the updated OERLIKON factors provides reasonably accurate and quickly obtainable results for our purposes.

If specific data concerning powder characteristics are available or accessible, a calculation according to Rheinmetall can be performed alternatively. (Rheinmetall, 1982 pp. 77–113) However, this approach is way more demanding.

[720] The calculation assumes a reversing action in an infinitely short time. Due to the elastic deformation of the frame, the shooter's hand, his arm, his stature, how well the gun is supported, etc., only a limited amount of the mechanical energy of the slide is returned back to it. From a physical point of view, the collision is neither a completely elastic nor an inelastic impact. The phenomenon occurring here is known as a *partially elastic* or *real impact*.

In contrast to the procedure in physics, where mass points are considered for reasons of simplification, real bodies behave neither completely elastic nor completely inelastic. The processes in real impact depend on the material pairing as well as the speeds and masses of the elements involved. *Böge* and others tell us how to calculate the velocities of the impact partners in real impact. A coefficient of restitution is used, which can be taken from various sources and for various material pairings. A coefficient of restitution $k = 1$ corresponds to elastic impact, $k = 0$ to the calculation of a plastic impact.

For the real impact, the coefficient of restitution k is determined from the ratio of the relative speeds of the impact partners after impact. This is usually determined by drop tests. Drop height and rebound height are set in relation to each other. For steel at 20 °C, the coefficient of restitution is 0.7. For the ram of a drop forging hammer *Böge* suggests $k = 0.5$ (Böge, et al., 2019 p. 247f)

An example of a real impact would be a ball bouncing off a wall. In comparison to a ball, the wall has a theoretically infinite mass and is at rest. In our application case, however, the boundary conditions are different. The slide bounces with a velocity v_{S2} against a frame of finite mass. Before the impact, the frame already started to move in the same direction as the slide. This frame movement is caused by mechanical processes and recoil transmission (e.g. recoil spring, uncoupling or hammer cocking impact), but at a lower speed. The effect of the shooter on the pistol makes it almost impossible to determine the rebound in theory. A more or less accurate calculation of the velocities of the impact partners after a real impact would have to take into account the mass of the frame, the loaded magazine, the shooter's hand and his arm and body. Furthermore, how tightly the shooter holds on to the handle of the pistol in order to support the gun.

In any case, the fact remains: After the collision the velocity of the slide is lower than before, because the deformation work done to the frame is not completely returned. A test arrangement for our application case could be to fire a shot from a hammer-fired pistol, after removing the recoil spring. Under some circumstances, the slide comes to a complete stop at the backstop, because certain components of the pistol deform elastically during the firing process and absorb a part of the energy by internal friction. In addition, the shooter consumes mechanical energy.

Chinn explains: "[…] the impact is accompanied by a loss of energy with the result that the bolt after impact will be at best approximately 70 per cent of its striking velocity." (Chinn, 1955 p. 116) Chinn avoids any complex calculation of the real impact. However, the proposed 70 % refer to heavy and rigidly mounted long guns. Consequently, ideal conditions

ENDNOTES

prevail. We adopt the formula structure from Chinn, but assign a lower value to the coefficient of restitution.

The figures for k_S mentioned in the corresponding notes of this book are approximate values. They have been roughly determined by evaluating high-speed recordings. In the case of hammer-fired pistols, k_S ranged from 0 to 0.1, in the case of striker-fired guns, they were approximately 0 to 0.2. These are by no means scientifically proven results. Interestingly, an influence of the material of the impact partners on k_S could not be proven. The values given are therefore valid irrespective of the material.

Conclusion: For the calculation of the process at the turning point, we suggest to use $k_S = 0.1$ for hammer-fired and $k_S = 0.2$ for striker-fired pistols. The difference in the calculation result is marginal. If you want to take the worst-case scenario as a basis, choose $k_S = 0$.

[721] Following *Chinn*: (Chinn, 1955 p. 116)

[722] Here the calculation assumes an elastic impact. An impact is called elastic if the two bodies are neither heated nor permanently deformed by the impact. In addition to the law of energy of mechanics, the principle of linear momentum applies. (Stroppe, et al., 2018 p. 101)

[723] Experience has shown that numbers computed with this equation for the cyclic rate are too high. Apart from the negligence of frictional losses, this is due to the influence of the firearm operator; and correspondingly to shooting the pistol off-hand, whereby a lower relative speed between the slide and the moving frame is noticeable.

[724] (Glock IM, 2013 p. 11)

[725] (Glock, 2013 p. 8), (PTOOMA Productions, 2004 p. 23)

[726] (Heydenreich, 1908 pp. 56–64), (Oerlikon, 1981 pp. 103–105) and (Kneubuehl, 2018 pp. 96–100); more information on Willy Heydenreich's approach: (Berz, 2001 pp. 537–550, 591, 628)

[727] (Heydenreich, 1908 p. 60)

[728] (Heydenreich, 1908 p. 57)

[729] (Heydenreich, 1908 p. 57)

[730] Instead of the factors originally communicated by *Heydenreich*, updated values are given here. In the middle of the 20th century OERLIKON was running tests using more recent powder types which were commonly used in those days. The evaluation of these tests provided new numerical values which replace Heydenreich's factors. (Oerlikon, 1981 p. 103) Factor values for $\eta = 0.55$ taken from *Kneubuehl* (Kneubuehl, 2018 p. 97).

[731] (Brukner, 2003 p. 386)

[732] (Kneubuehl, 2013 p. 72f cf. A.1.5.1) For the computation of the barrel time *Heydenreich* suggests to use $t_\ell = 2 \cdot s_\ell \cdot T(\eta) / v_0$. (Heydenreich, 1908 p. 57)

Heydenreich's computation method is based on an empirical approach. Various internal ballistic characteristics are obtained by multiplying known quantities with an individual factor, where each factor depends on the pressure ratio η. Heydenreich lists values for these factors in two tables. (Heydenreich, 1908 p. 58f)

According to Heydenreich, the pressure ratio η is calculated as the quotient of the mean gas pressure and the highest pressure when firing a shot. The amount of the peak pressure should be determined by pressure testing. The mean pressure is a pressure which would propel the projectile to the actual muzzle velocity observed if it would theoretically act constantly. Heydenreich suggests an equation to calculate this mean pressure $\bar{p}$. The formula takes into account the bore cross-sectional area A, mass of the powder charge m_C, projectile mass m_P and muzzle velocity v_0. It reads: $\bar{p} = (m_P + 0.5\,m_C) \cdot v_0^2 / (2\,g \cdot s_\ell \cdot 1.0333 \cdot A)$. (Heydenreich, 1908 p. 11) Heydenreich's studies took place at the end of the 19th century and thus before the introduction of the International System of Units. There is a factor $g = 9.808$, which presumably establishes the connection between kilopond and force (acceleration due to gravity g, defined since 1901: $g = 9.80665$ m/s²) as well as a factor 1.0333 for the conversion of pressure from technical to physical atmospheres.

ENDNOTES

After calculating the mean and determining the peak pressure, the pressure ratio η is available. Based on this, the relevant multiplier can be taken from the corresponding table at *Heydenreich*. Intermediate values of $T(\eta)$ can be determined by interpolation for any η given and thus t_ℓ can be calculated, as Heydenreich points out. (Heydenreich, 1908 p. 57)

OERLIKON carried out series of tests using powder types which were popular in more recent times. The evaluation of these tests led to new results which replace Heydenreich's multipliers. (Oerlikon, 1981 p. 103) *Oerlikon* lists factors for η which are in the range of 0.25 to 0.5. If a diagram is drawn up for the multiplier $T(\eta)$ based on *Oerlikon* as a function of the pressure ratio η, it can be seen that the factor values $T(\eta)$ form almost a straight line. The straight-line equation was determined and this was taken into account in Equation A23.12. The application of the straight-line equation enables an algorithm for calculating the barrel time which avoids reading the required factor value $T(\eta)$ from the chart.

Based on *Oerlikon*, the following data was used to derive the straight-line equation: Pressure ratio $\eta = 0.25$ and corresponding factor 0.725 as well as ratio 0.5 with factor $T(\eta) = 0.91$. Formula A23.12 also takes into account base units according to the International System of Units, e.g. MPa for pressure.

[733] (Rheinmetall, 1982 p. 98)

[734] The bore cross-sectional area is the area where the pressure exerts on the base of the projectile. (Oerlikon, 1981 p. 82)

The bore cross-sectional area is listed in the C.I.P. tables under the symbol Q. According to the C.I.P., barrel profiles whose shape deviates from the usual fields and grooves (e.g. polygonal rifling) may fall below the value Q given in the C.I.P. tables by a maximum of 0.7 percent.

Examples for the internal cross-sectional area of the barrel and the Maximum Average Pressure p_{Tmax} according to C.I.P. (CIP, 2011):

Cartridge type	A [mm²]	p_{Tmax} [MPa]
.22 Long Rifle	24.07	170
.25 ACP	31.14	120
.32 ACP	47.37	160
.380 ACP	63.26	135
9 mm Luger	62.61	235
.40 S&W	79.55	225
.45 ACP	101.33	130

[735] (Cranz, et al., 1926 p. 231f)

[736] (Cranz, et al., 1926 p. 231f)

[737] (Heydenreich, 1908 p. 58)

[738] OERLIKON conducted series of tests using more recent powder types. The evaluation of these tests led to new numerical values which replace Heydenreich's factors. (Oerlikon, 1981 p. 103) Multipliers taken from *Oerlikon* (Oerlikon, 1981 p. 104). Factor values for $\lambda = 35$ see Kneubuehl (Kneubuehl, 2018 p. 98).

There is no indication of how to proceed with λ for values $\lambda < 0.25$. We suggest an interpolation where the corresponding values for $\lambda = 0$ are selected to be zero.

NB: For F8(20) *Kneubuehl* specifies 3.816, Oerlikon states 3.861 (Oerlikon, 1981 p. 104).

[739] In two examples *Kneubuehl* compares values determined using the *Heydenreich/Oerlikon* method with graphs from actual testings. The shape of the respective pressure and velocity curves are almost alike, but have a slight offset to each other. (Kneubuehl, 2018 p. 99)

[740] (Rinker, 2011 p. 406)

[741] (TR "Patrone", 2009 p. 12 and Anlage 13)

[742] (TR "Patrone", 2009 p. Anlage 4)

ACKNOWLEDGMENTS

A fundamental idea of this book is to give the reader a deeper insight into the prevailing and technically sophisticated subject of weapon technology and, at the same time, build a bridge to the historical development. In some instances, questions may come up and I look forward to sharing my ideas. My particular interest in this book is to deliver practice-oriented descriptions and explanations on the subject of pistol making. In this context, my special thanks goes to the students of the classes that I teach, as well as to all interested readers. I welcome their feedback and suggestions.

Greg Cornett, Adam Dugdale, Jason Haddock, Dickson Ly, Peter Suciu, Bret Vorhees and Adam Walker deserve to be mentioned above all others; and I would like to thank all those who followed their example by investing their time and energy from the first idea through to the publication of this book: Dr. Joachim Boßlet, Peter Dannecker, S.P. Fjestad (†), Dr. Ilya Gutmann, Bas Martens, Dr. Vladimir Jakub Mrvik, Kristóf Nagy, Dr. Rolf Wirtgen, and many more dear friends whose names can't be mentioned individually. I'm also appreciative of companies including Carl Walther GmbH, and their excellent staff, who provided images and input.

At the Helmut-Schmidt-Universität, where this work first saw the light of day as a doctoral dissertation, I was encouraged by Univ.-Prof. Dr.-Ing. habil. Hendrik Rothe to get the project started. Dr. Arash Ramezani and other members of the faculty of mechanical engineering were always there to provide assistance and advice. Thank you.

Most certainly not least, I would like to recognize that this work would have never happened without the support of my beloved wife, Niki Dallhammer. She encouraged me all those years, beginning with the first vague idea and following through all the way to this book's successful completion. You own my heart.

Trust I seek and I find in you
Every day for us something new
Open mind for a different view
And nothing else matters
(METALLICA – "Nothing Else Matters")

Sincerely
Peter Dallhammer

WALTHER *PPQ 45*